FOUNDATIONS
OF MATHEMATICS

11

SECOND EDITION

THE McGRAW-HILL RYERSON MATHEMATICS PROGRAM

MATH 1 SOURCE BOOK
MATH 2 SOURCE BOOK
MATH 3
MATH 4
MATH 5
MATH 6

LIFE MATH 1
LIFE MATH 2
LIFE MATH 3

INTERMEDIATE MATHEMATICS 1
INTERMEDIATE MATHEMATICS 2
INTERMEDIATE MATHEMATICS 3

APPLIED MATHEMATICS 9
APPLIED MATHEMATICS 10
APPLIED MATHEMATICS 11
APPLIED MATHEMATICS 12

FOUNDATIONS OF MATHEMATICS 9
FOUNDATIONS OF MATHEMATICS 10
FOUNDATIONS OF MATHEMATICS 11
FOUNDATIONS OF MATHEMATICS 12

TEACHER'S EDITIONS FOR:
MATH 3
MATH 4
MATH 5
MATH 6
INTERMEDIATE MATHEMATICS 1
INTERMEDIATE MATHEMATICS 2
INTERMEDIATE MATHEMATICS 3
APPLIED MATHEMATICS 9

TEACHER'S GUIDES FOR:
AM 10
AM 11
AM 12
FM 10
FM 11
FM 12

FOUNDATIONS
OF MATHEMATICS

11

SECOND EDITION

Dino Dottori, B.Sc., M.S.Ed.
George Knill, B.Sc., M.S.Ed.
James Stewart, Ph.D.
Gerry Gadoury, B.Sc., M.Ed.

McGRAW-HILL RYERSON LIMITED

TORONTO MONTREAL NEW YORK AUCKLAND BOGOTÁ CAIRO HAMBURG LISBON
LONDON MADRID MEXICO MILAN NEW DELHI PANAMA PARIS SAN JUAN
SÃO PAULO SINGAPORE SYDNEY TOKYO

FOUNDATIONS OF MATHEMATICS 11
SECOND EDITION

ISBN 0-07-548730-6

34567890   JD   6543210987

Cover Art Direction/Display Headings by Dan Kewley
Cover Design by Marc Mireault
Cover Photography by Imtek Imagineering
Technical Illustrations by Frank Zsigo

A complete list of photo credits appears on page 479.

Printed and bound in Canada

Canadian Cataloguing in Publication Data

Dottori, Dino, date —
Foundations of mathematics 11

(The McGraw-Hill Ryerson mathematics program)
First ed. published under title: F M T : foundations of mathematics for tomorrow : intermediate.
ISBN 0-07-548730-6

1. Mathematics — 1961-   .   I. Knill, George, date –
II. Stewart, James.   III. Title.   IV. Title: F M T : foundations of mathematics for tomorrow : intermediate.   V. Series.

QA39.2.D68 1986   510   C86-093823-9

The Metric Office has granted use of the National Symbol for Metric Conversion.

TABLE OF CONTENTS

REAL NUMBERS

CHAPTER

I often say that when you can measure what you are speaking about, and express it in numbers, you know something about it, but when you cannot express it in numbers, your knowledge is of a meagre and unsatisfactory kind; it may be the beginning of knowledge, but you have scarcely, in your thoughts, advanced to the stage of Science, whatever the matter may be.

Lord Kelvin

REVIEW AND PREVIEW TO CHAPTER 1

ROUNDING AND APPROXIMATING

$$2572 \doteq \begin{cases} 2570 \text{ to the nearest ten} \\ 2600 \text{ to the nearest hundred} \\ 3000 \text{ to the nearest thousand} \end{cases}$$

Where the key digit is 5 followed by zeros, we round to the nearest even digit.

$0.365 \doteq 0.36$ to the nearest hundredth
$0.375 \doteq 0.38$ to the nearest hundredth
$3.05 \ \doteq 3.0$ to the nearest tenth

To find an estimate, we round and perform the calculation.

$$38 \times 23 \div 41 \doteq 40 \times 20 \div 40$$
$$\doteq 20$$

The actual answer is   21 R 13.

EXERCISE

1. Round off to the nearest hundred.
(a) 36 521   (b) 36 145   (c) 4052
(d) 255 642   (e) 4520   (f) 67 456

2. Round off to the nearest hundredth.
(a) 0.2564   (b) 0.056 42   (c) 2.625
(d) 1.205 46   (e) 32.652   (f) 0.375

3. Round to the nearest thousandth.
(a) 1.265 25   (b) 32.5658   (c) 0.5645
(d) 0.567 504 (e) 0.568 32   (f) 1.0055

4. Round to the nearest ten.
(a) 25 065   (b) 365.365   (c) 27.5
(d) 2.75   (e) 12 257   (f) 63.5

5. Estimate the following.
(a) $64 \times 24 \div 12$
(b) $58 \times 28 \div 14$
(c) $523 \div 46 \times 36$
(d) $3.8 \times 2.7 \times 5.5$
(e) $0.625 \times 5.375 \div 0.125$

6. Estimate the following.
(a) $\dfrac{36 \times 72 \times 85}{42}$
(b) $\dfrac{5.34 \times 2.75 \times 63.2}{63 \times 36}$
(c) $\dfrac{625 \times 526 \times 285}{22 \times 31}$
(d) $\dfrac{625 \times 256}{315 \times 1.414}$
(e) $\dfrac{6350 \times 256 \times 307}{1256}$

SET NOTATION

$$\{\, x \mid x < 5 \,,\, x \in I \,\}$$

the set   rule   domain of
of all x   the variable
such that

EXERCISE

1. List the members of the following sets.
(a) $\{\, x \mid x > -2 \,,\, x \in I \,\}$
(b) $\{\, x \mid x > 7 \,,\, x \in N \,\}$
(c) $\{\, x \mid x \leqslant 5 \,,\, x \in W \,\}$
(d) $\{\, x \mid x = 3 \,,\, x \in N \,\}$
(e) $\{\, x \mid x \geqslant -2 \,,\, x \in I \,\}$
(f) $\{\, x \mid x \leqslant 3 \,,\, x \in W \,\}$
(g) $\{\, x \mid -3 \leqslant x \leqslant 3 \,,\, x \in I \,\}$
(h) $\{\, x \mid -4 < x < 3 \,,\, x \in I \,\}$
(i) $\{\, x \mid x < -3 \text{ or } x > 2 \,,\, x \in I \,\}$
(j) $\{\, x \mid x \leqslant 3 \text{ and } x \geqslant -2 \,,\, x \in I \,\}$

INTERSECTION AND UNION

A = { 2, 3, 4, 6, 8 }
B = { 4, 5, 6, 7, 9, 11 }

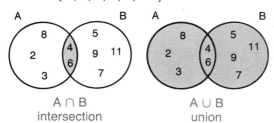

A ∩ B
intersection

A ∪ B
union

A ∩ B = { 4, 6 } both sets have # in
A ∪ B = { 2, 3, 4, 5, 6, 7, 8, 9, 11 } (common)
everything contained in set
A + B

EXERCISE

1. A = { 2, 4, 6, 8, 10 }
   B = { 6, 7, 8, 9 }
List the members of the following sets.
(a) A ∪ B   (b) A ∩ B

2. P = { 5, 10, 15, 20, 25 }
   Q = { 10, 20, 30, 40, 50, 60 }
List the members of the following sets.
(a) P ∩ Q   (b) P ∪ Q

3. A = { 1, 2, 3, 4, 5, 6 }
   B = { 2, 4, 6, 8, 10 }
   C = { 4, 6, 7, 9 }
List the members of the following sets.
(a) A ∩ B   (b) A ∪ B
(c) B ∩ A   (d) B ∪ A
(e) A ∩ B ∩ C   (f) A ∪ B ∪ C
(g) A ∩ (B ∪ C)   (h) (A ∩ B) ∪ C
(i) (A ∩ B) ∪ (A ∩ C)
(j) (A ∪ C) ∩ (B ∪ C)

4. List the members of the following
sets for x ∈ I.
(a) { x | x < 3 } ∩ { x | x > −2 }
(b) { x | x > 2 } ∪ { x | x < 7 }
(c) { x | x > 3 } ∪ { x | x < −3 }
(d) { x | x = 2 } ∪ { x | x ≤ 1 }
(e) { x | x ≥ −2 } ∩ { x | x < 5 }
(f) { x | x ≥ 0 } ∪ { x | x < −2 }

CALCULATOR MATH

Calculators have algebraic rules specifically programmed into them. Without rules, expressions such as

3 × 5 + 4 × 6

can have several meanings.

3 × (5 + 4) × 6 = 162
(3 × 5) + (4 × 6) = 39
((3 × 5) + 4) × 6 = 114
3 × (5 + (4 × 6)) = 87

The rules of algebra state that multiplication and division are performed from left to right before addition and subtraction. Therefore,

3 × 5 + 4 × 6 = 15 + 24
            = 39

Using a calculator, press

C 3 × 5 + 4 × 6 =   { 39
                        { 114

A calculator producing the answer 39 is programmed with a priority level function, and is said to have algebraic logic. If your calculator gave the answer 114, then the priority level must be determined by the person using the calculator. In order to perform this calculation on such a machine, we use the memory keys as follows.

This key puts the display into the memory.

C 3 × 5 = M C

4 × 6 = + MR =

This key recalls what is in the memory.

Check your user's manual for more information on memory keys.

EXERCISE

1. (2.4 + 3.7) × (5.8 − 2.6)
2. (6.7 + 7.4) ÷ (2.8 − 2.2)
3. 6.3 × 5.1 + 2.6 × 7.2
4. (7.2 + 5.3) − (4.6 + 2.7)
5. 15.12 ÷ 5.6 − 6.84 ÷ 3.8

1.1 RATIONAL NUMBERS

Integers, mixed numbers, and terminating decimals can be expressed as the quotient of an integer, a, and a nonzero integer, b. These numbers are written in the form $\frac{a}{b}$ and are called rational numbers.

$\frac{7}{8}, \frac{5}{3}, \frac{-3}{7}, \frac{4}{-5}$, and $\frac{-3}{-2}$ are examples of rational numbers.

> A rational number is a number that can be written in the form $\frac{a}{b}$ where a and b are integers, and b $\neq$ 0.
> Q is the set of rational numbers.
> $$Q = \left\{ \frac{a}{b} \mid a,b \in I, b \neq 0 \right\}$$

EXAMPLE 1. Write (a) $\frac{7}{4}$ and (b) $\frac{-2}{9}$ as decimals.

SOLUTION:

(a) $\frac{7}{4}$ means 7 ÷ 4
Using a calculator, we press

7 ÷ 4 =

and the display is

1.75

1.75 is called a terminating decimal because the division terminates due to a zero remainder.

(b) $\frac{-2}{9}$ means −2 ÷ 9
On a calculator, the result is

−2 ÷ 9 = **−0.2222222**

Since a calculator, whether it rounds off, or truncates, only shows a finite number of decimal places we understand that −0.2222222 means −0.222 ... since division could continue. The repeating decimal −0.2222222 ... can be written
$$-0.222 ... = -0.\overline{2}$$

Repeating decimals are also called periodic decimals. The sequence of digits that repeat is called the period of the decimal. The following chart illustrates how we can write a repeating decimal using its period.

Number	Period	Periodic Form
2.353 535 353 ...	35	$2.\overline{35}$
3.427 527 527 ...	275	$3.4\overline{275}$
6.302 530 253 ...	3025	$6.\overline{3025}$

EXAMPLE 2. Express (a) $0.\overline{54}$ and (b) $2.1\overline{453}$ in the form $\frac{a}{b}$.

SOLUTION:

(a) Let $x = 0.5454\ldots$ ⎫
$100x = 54.54\ldots$ ⎬ Step 1.
$x = 0.54\ldots$ ⎭

$99x = 54$ } Step 2.

$x = \dfrac{54}{99}$ ⎫ Step 3.

$= \dfrac{6}{11}$ ⎭

(b) Let $x = 2.145\,345\,3\ldots$
$10\,000x = 21\,453.453\ldots$
$10x = 21.453\ldots$

$9990x = 21\,432$

$x = \dfrac{21\,432}{9990}$

$= 2\dfrac{242}{1665}$

EXERCISE 1.1

B **1.** Express in decimal form.

(a) $\frac{1}{8}$ (b) $\frac{1}{4}$ (c) $\frac{3}{8}$

(d) $\frac{5}{8}$ (e) $-\frac{7}{8}$ (f) $\frac{21}{50}$

(g) $\frac{1}{9}$ (h) $\frac{5}{11}$ (i) $-\frac{7}{13}$

(j) $\frac{15}{11}$ (k) $\frac{5}{3}$ (l) $\frac{3}{5}$

2. Express in the form $\frac{a}{b}$.

(a) 0.375 (b) 0.215 (c) 0.618

(d) 0.035 (e) $0.\overline{36}$ (f) $0.\overline{5}$

(g) $0.\overline{235}$ (h) $4.7\overline{54}$ (i) $0.4\overline{9}$

(j) $0.\overline{142\,857}$ (k) $4.\overline{773}$ (l) $0.0\overline{18}$

C **3.** Express in the form $\frac{a}{b}$.

(a) $(0.5)(0.\overline{6})$ (b) $(0.25)(0.\overline{3})$

(c) $(0.\overline{18})(0.5)$ (d) $(0.\overline{3})^3$

(e) $(0.\overline{4})(0.5)$ (f) $(0.2)(0.\overline{36})$

4. Express in the form $\frac{a}{b}$.

(a) $0.4\overline{9}$ (b) $0.3\overline{9}$

(c) $0.6\overline{9}$ (d) $0.\overline{9}$

5. Express in the form $\frac{a}{b}$.

(a) $0.1 + 0.01 + 0.001 + \ldots$
(b) $0.3 + 0.03 + 0.003 + \ldots$
(c) $0.36 + 0.0036 + 0.000\,036 + \ldots$
(d) $0.156 + 0.000\,156 + 0.000\,000\,156 + \ldots$

CALCULATOR MATH

Some calculators round off, while others truncate, or "chop off" the extra digits without rounding. The following exercise demonstrates how your calculator handles the extra digits. The displays from two different kinds of calculators are shown.

Press	Calculator #1 Display	Calculator #2 Display
C	0	0.00
2	2.	2.
÷	2.	2.00
3	3.	3.
=	0.6666666	0.6666667
×	0.6666666	0.6666667
3	3.	3.
=	1.9999998	2

Notice that the two calculators give slightly different answers to the problem
$$2 \div 3 \times 3$$
The difference arises because Calculator #1 truncates and drops the digits after dividing. Calculator #2 does two things: (i) only the display is rounded off; (ii) digits are saved in the memory and not dropped off.

The solution is $2 \div 3 \times 3 = 2$ as we multiply and divide in order from left to right.

Experiment with the following questions to determine whether the calculator you are using truncates, or rounds.

1. $5 \div 3 \times 3$ 4. $2 \div 6 \times 6$
2. $6 \div 7 \times 7$ 5. $8 \div 3 \times 3$
3. $1 \div 6 \times 6$ 6. $3 \div 7 \times 7$

1.2 IRRATIONAL NUMBERS

Some numbers have interesting patterns that are easily recognized.
Numbers such as

0.121 122 111 222 ...
0.122 112 221 112 222 ...
0.123 124 125 126 127 ...

have patterns that permit you to continue writing the number.
However, they are not periodic since there is no definite pattern that
repeats, and hence do not represent rational numbers.

Decimal numbers that are non-terminating, or non-periodic cannot
be expressed in the form $\frac{a}{b}$ and are called irrational numbers.

> An irrational number is a number that cannot be written in the
> form $\frac{a}{b}$, where a and b are integers, and b $\neq$ 0.
>
> $\overline{Q}$ is the set of irrational numbers.

Other familiar examples of non-periodic decimals are $\sqrt{2}$, $\sqrt{3}$, and
π. In the following diagram, we take a circle with a diameter of 1 unit,
and roll it along the number line as shown. The circumference of the
circle is $\pi \times 1 = \pi$.

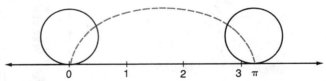

The number $\sqrt{2}$ has been located on the number line by placing a
right triangle with sides 1, 1, and $\sqrt{2}$ on the number line as shown
and drawing an arc to cut the line at $\sqrt{2}$.

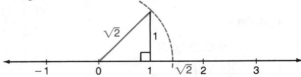

From these two examples, we see that the irrational numbers $\sqrt{2}$
and π have positions on the number line. We can conclude that every
irrational number has a position on the number line.

If n is an integer that is not a perfect square, then $\sqrt{n}$ is an
irrational number. For this reason, $\sqrt{3}$ and $\sqrt{5}$ are irrational numbers,
whereas $\sqrt{4}$ is not an irrational number, because $4 = 2^2$. Since
$\sqrt{4} = 2$, $\sqrt{4}$ is a rational number.

When we add, subtract, multiply, or divide any irrational number by
a nonzero rational number, the result is an irrational number.
For example,

$$3 + \sqrt{5}, \qquad \sqrt{5} - 2, \qquad 5\sqrt{7}, \qquad \frac{6 - \sqrt{5}}{10},$$

are all irrational numbers since none can be written in the form $\frac{a}{b}$.

EXERCISE 1.2

A 1. Which of the following are irrational numbers?

(a) -7 (b) 0.777...
(c) 0.202 202 22...   (d) $\sqrt{25}$
(e) $\sqrt{15}$ (f) 0.235 235...
(g) 0.65   (h) $-\sqrt{5}$

B 2. Make diagrams to show how you would locate the following positions on a number line.

(a) $\sqrt{3}$ (b) $\sqrt{5}$
(c) $\sqrt{10}$ (d) $\sqrt{13}$

1.3 RADICALS

Expressions of the form $\sqrt{3}$, $\sqrt[4]{7}$, and $\sqrt[5]{-16}$ are called radicals. 4 and 5 are called indices, where in $\sqrt{3}$, the index 2 is understood. The 3, 7, and -16 are called the radicands.

$x = \sqrt{a}$ means $x^2 = a$, and $a \neq 0$

$x = \sqrt[3]{a}$ means $x^3 = a$

In general,

$$x = \sqrt[n]{a} \text{ means } x^n = a$$
If n is even, the convention is that $a > 0$ and $x > 0$.

Following are two general rules for nth roots.

I.	$\sqrt[n]{a} \times \sqrt[n]{b} = \sqrt[n]{ab}$	If n is even then $a, b \geqslant 0$
II.	$\dfrac{\sqrt[n]{a}}{\sqrt[n]{b}} = \sqrt[n]{\dfrac{a}{b}}$	

Proof of Rules I. and II.

Let $\quad x = \sqrt[n]{a}$ and $y = \sqrt[n]{b}$

Then $\quad x^n = a$ and $y^n = b$

Rule I.

$(xy)^n = x^n y^n = ab$

$xy = \sqrt[n]{ab}$

$\sqrt[n]{a} \times \sqrt[n]{b} = \sqrt[n]{ab}$

Rule II.

$\left(\dfrac{x}{y}\right)^n = \dfrac{x^n}{y^n} = \dfrac{a}{b}$

$\dfrac{x}{y} = \sqrt[n]{\dfrac{a}{b}}$

$\dfrac{\sqrt[n]{a}}{\sqrt[n]{b}} = \sqrt[n]{\dfrac{a}{b}}$

EXAMPLE 1.   Express as entire radicals.

(a) $3\sqrt{2}$

(b) $-5\sqrt{7}$

(c) $2\sqrt[3]{3}$

SOLUTION:

(a) $3\sqrt{2} = \sqrt{9} \times \sqrt{2}$
$\quad\quad = \sqrt{18}$

(b) $-5\sqrt{7} = -\sqrt{25} \times \sqrt{7}$
$\quad\quad\quad = -\sqrt{175}$

(c) $2\sqrt[3]{3} = \sqrt[3]{8} \times \sqrt[3]{3}$
$\quad\quad\quad = \sqrt[3]{24}$

EXAMPLE 2.   Express as mixed radicals.

(a) $\sqrt{18}$

(b) $\sqrt[3]{-54}$

(c) $5\sqrt{12}$

SOLUTION:

(a) $\sqrt{18} = \sqrt{9} \times \sqrt{2}$
$\quad\quad = 3\sqrt{2}$

(b) $\sqrt[3]{-54} = \sqrt[3]{-27} \times \sqrt[3]{2}$
$\quad\quad\quad = -3\sqrt[3]{2}$

(c) $5\sqrt{12} = 5 \times \sqrt{4} \times \sqrt{3}$
$\quad\quad\quad = 5 \times 2 \times \sqrt{3}$
$\quad\quad\quad = 10\sqrt{3}$

EXAMPLE 3. Simplify.

(a) $3\sqrt{12} + \sqrt{18} - \sqrt{27} + 3\sqrt{8}$

(b) $3\sqrt{2}(2\sqrt{3} - 5)$

SOLUTION:

(a) $3\sqrt{12} + \sqrt{18} - \sqrt{27} + 3\sqrt{8}$
$= 3(2\sqrt{3}) + 3\sqrt{2} - 3\sqrt{3} + 3(2\sqrt{2})$
$= 6\sqrt{3} - 3\sqrt{3} + 3\sqrt{2} + 6\sqrt{2}$
$= 3\sqrt{3} + 9\sqrt{2}$

(b) $3\sqrt{2}(2\sqrt{3} - 5) = 3\sqrt{2} \times 2\sqrt{3} - 3\sqrt{2} \times 5$
$= 6\sqrt{6} - 15\sqrt{2}$

EXERCISE 1.3

B 1. Express as entire radicals.

(a) $5\sqrt{5}$ (b) $2\sqrt{7}$ (c) $-3\sqrt{11}$
(d) $4\sqrt{2}$ (e) $3\sqrt{7}$ (f) $3\sqrt[3]{2}$
(g) $-2\sqrt[3]{5}$ (h) $2\sqrt[4]{3}$ (i) $5\sqrt{6}$

2. Express as mixed radicals.

(a) $\sqrt{27}$ (b) $\sqrt{98}$ (c) $\sqrt{288}$
(d) $\sqrt{75}$ (e) $\sqrt{200}$ (f) $\sqrt[3]{-27}$
(g) $\sqrt[3]{-128}$ (h) $\sqrt[4]{48}$ (i) $\sqrt[3]{54}$

3. Collect like radicals.

(a) $5\sqrt{3} + 2\sqrt{3}$
(b) $3\sqrt{5} - 2\sqrt{5}$
(c) $6\sqrt{3} + 2\sqrt{3} - 4\sqrt{3}$
(d) $4\sqrt{6} - 3\sqrt{5} + 5\sqrt{6}$
(e) $4\sqrt{2} - 5\sqrt{2} + \frac{1}{2}\sqrt{2}$
(f) $5\sqrt{2} - 7\sqrt{3} + 3\sqrt{2}$

4. Expand and simplify.

(a) $\sqrt{2}(\sqrt{6} - \sqrt{3})$ (b) $2\sqrt{3}(\sqrt{5} - 2)$
(c) $3\sqrt{5}(2 - \sqrt{2})$ (d) $3\sqrt{2}(\sqrt{2} + 1)$
(e) $\sqrt{a}(\sqrt{b} - 2)$ (f) $(\sqrt{a} + \sqrt{b})\sqrt{a}$
(g) $(\sqrt{3} - 1)^2$ (h) $(\sqrt{2} - 3)^2$
(i) $(\sqrt{2} - \sqrt{5})2\sqrt{3}$

5. Express as mixed radicals.

(a) $\sqrt{12}$ (b) $\sqrt{125}$ (c) $\sqrt{50}$
(d) $\sqrt{20}$ (e) $\sqrt{288}$ (f) $\sqrt{98}$
(g) $\sqrt{8}$ (h) $\sqrt[3]{16}$ (i) $\sqrt[3]{128}$
(j) $\sqrt{27}$ (k) $\sqrt[3]{250}$ (l) $\sqrt[3]{24}$

6. Express as entire radicals.

(a) $3\sqrt{5}$ (b) $2\sqrt{7}$ (c) $4\sqrt{6}$
(d) $2\sqrt{5}$ (e) $2\sqrt[3]{2}$ (f) $3\sqrt[3]{3}$
(g) $-5\sqrt{3}$ (h) $-2\sqrt[3]{3}$ (i) $3\sqrt{6}$
(j) $2\sqrt[4]{3}$ (k) $3\sqrt[3]{5}$ (l) $-2\sqrt{11}$

7. Simplify and collect like radicals.

(a) $3\sqrt{2} + 5\sqrt{8} - 2\sqrt{18} + 2\sqrt{32}$
(b) $2\sqrt{20} - \sqrt{125} + 2\sqrt{45} + 6\sqrt{5}$
(c) $3\sqrt{3} + 2\sqrt{20} - \sqrt{12} + \sqrt{80}$
(d) $3\sqrt{72} - 5\sqrt{27} + 4\sqrt{8} + 2\sqrt{48}$
(e) $4\sqrt{99} - 7\sqrt{12} + 3\sqrt{44} - 2\sqrt{22}$
(f) $4\sqrt{50} + 7\sqrt{32} - 5\sqrt{7} - 2\sqrt{98}$

C 8. Expand and simplify.

(a) $3\sqrt{2}(2 - \sqrt{6})$ (b) $5\sqrt{3}(2 - \sqrt{6})$
(c) $2\sqrt{5}(\sqrt{10} - 2)$ (d) $(3\sqrt{2} - 1)^2$
(e) $(\sqrt{3} - \sqrt{2})^2$ (f) $(2\sqrt{3} - 5\sqrt{2})^2$
(g) $(3\sqrt{2} - 1)(3\sqrt{2} + 1)$
(h) $(4 - \sqrt{5})(4 + \sqrt{5})$
(i) $(3\sqrt{7} - 1)(3\sqrt{7} + 1)$
(j) $(3\sqrt{2} + 7)(3\sqrt{2} + 5)$

CALCULATOR MATH

USING THE $\sqrt{}$ KEY

To find $\sqrt{25}$ on a calculator press

C **2** **5** **$\sqrt{}$**

and the display is **5.00**

EXERCISE

1. Evaluate.

(a) $\sqrt{6}$ (b) $\sqrt{11}$ (c) $\sqrt{20}$

2. Show that the following are true.

(a) $\sqrt{2} + \sqrt{3} \neq \sqrt{5}$
(b) $\sqrt{1.2^2 + 1.6^2} = 2$

1.4 REAL NUMBERS

We have established the following number systems.

$N = \{1, 2, 3, 4, ...\}$

$W = \{0, 1, 2, 3, ...\}$

$I = \{..., -3, -2, -1, 0, 1, 2, 3, ...\}$

$Q = \left\{ \dfrac{a}{b} \,\middle|\, a, b, \in I, b \neq 0 \right\}$

$\overline{Q}$ is the set of irrational numbers.

EXAMPLE 1. Graph $\{x \mid x \leq 3\}$ on an appropriate number line.
(a) $x \in N$ (b) $x \in W$ (c) $x \in I$ (d) $x \in Q$

SOLUTION:

(a) $\{x \in N \mid x \leq 3\}$

(b) $\{x \in W \mid x \leq 3\}$

(c) $\{x \in I \mid x \leq 3\}$

(d) $\{x \in Q \mid x \leq 3\}$

Note that we use a dotted line when graphing on the Q-line, since there are gaps where the irrational numbers are located. If we take the union of the rational numbers and the irrational numbers, we have a complete set where there is a one-to-one correspondence between this new set and the points on the number line.

The set of real numbers, R, is the set of all periodic and non-periodic decimals. We can also define R as the union of the set of rational numbers, Q, and the set of irrational numbers, $\overline{Q}$.

$$R = Q \cup \overline{Q}$$

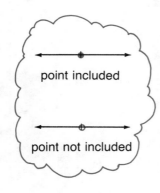

point included

point not included

EXAMPLE 2. Graph each set on a Real number line.
(a) $\{x \in R \mid x > -3\}$
(b) $\{x \in R \mid x \leq 3\}$
(c) $\{x \in R \mid -2 \leq x < 3\}$

SOLUTION:

(a) $\{x \in R \mid x > -3\}$

(b) $\{x \in R \mid x \leq 3\}$

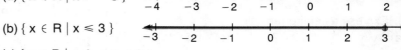

(c) $\{x \in R \mid -2 \leq x < 3\}$

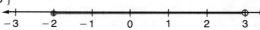

EXAMPLE 3. Given $A = \{\, x \in R \mid x > -2 \,\}$
$B = \{\, x \in R \mid x \leq 1 \,\}$
$C = \{\, x \in R \mid x \geq 2 \,\}$

Graph each of the following on a Real number line.
(a) $A \cap B$
(b) $B \cup C$

SOLUTION:

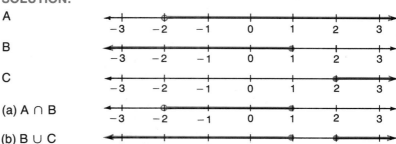

In set-builder notation, $A \cap B$ can be written $\{\, x \in R \mid -2 < x \leq 1 \,\}$
$B \cup C$ can be written $\{\, x \in R \mid x \leq 1 \text{ or } x \geq 2 \,\}$

intersection

union

EXERCISE 1.4

B 1. Describe each set using set-builder notation.

(a)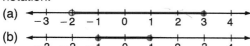

(b)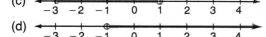

(c)
(d)

2. Describe the following using set-builder notation.

(a)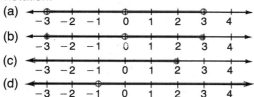

(b)
(c)
(d)

3. Graph the following sets on an appropriate number line.

(a) $\{\, x \in N \mid x \leq 4 \,\}$ (b) $\{\, x \in W \mid x < 5 \,\}$
(c) $\{\, x \in W \mid x > 3 \,\}$ (d) $\{\, x \in I \mid x \leq 2 \,\}$
(e) $\{\, x \in I \mid x \geq -3 \,\}$ (f) $\{\, x \in Q \mid x < 5 \,\}$
(g) $\{\, x \in R \mid x \geq -2 \,\}$ (h) $\{\, x \in R \mid x > -3 \,\}$

4. Graph the following.

(a) $\{\, x \in W \mid 2 \leq x \leq 5 \,\}$
(b) $\{\, x \in I \mid -2 \leq x < 3 \,\}$

(c) $\{\, x \in Q \mid -3 < x < 3 \,\}$
(d) $\{\, x \in Q \mid 0 \leq x \leq 4 \,\}$
(e) $\{\, x \in R \mid -1 < x \leq 4 \,\}$
(f) $\{\, x \in R \mid -3 \leq x \leq 3 \,\}$
(g) $\{\, x \in I \mid -3 \leq x \leq 5 \,\}$
(h) $\{\, x \in R \mid -2 \leq x < 3 \,\}$

5. Graph the following for $x \in R$.

(a) $\{\, x \mid x > -2 \,\} \cap \{\, x \mid x \leq 3 \,\}$
(b) $\{\, x \mid x \leq 0 \,\} \cup \{\, x \mid x \geq 4 \,\}$
(c) $\{\, x \mid x \leq -3 \,\} \cup \{\, x \mid x > 0 \,\}$
(d) $\{\, x \mid x \leq 3 \,\} \cap \{\, x \mid x \geq 3 \,\}$

6. Graph the following for $x \in R$.

(a) $\{\, x \mid -1 \leq x \leq 2 \,\} \cap \{\, x \mid 1 < x \leq 4 \,\}$
(b) $\{\, x \mid -2 \leq x < 1 \,\} \cup \{\, x \mid 1 \leq x \leq 3 \,\}$
(c) $\{\, x \mid -3 \leq x \leq 2 \,\} \cap \{\, x \mid x \neq 2 \,\}$
(d) $\{\, x \mid -2 \leq x < 3 \,\} \cup \{\, x \mid x - 3 = 0 \,\}$

If you switch the S and V in the word RESERVE, you get REVERSE. Switch two letters in LIMESTONE to make a new word.

1.5 PROPERTIES OF REAL NUMBERS

The operations $+$, $-$, $\times$, $\div$, and $\sqrt{}$ are called algebraic operations. Because $-$, $\div$, and $\sqrt{}$ can be defined in terms of $+$ and $\times$ we can summarize the properties of the real numbers in terms of addition and multiplication.

For a, b, c $\in$ R	$+$	$\times$
Closure	$(a + b) \in R$	$ab \in R$
Associative	$(a + b) + c = a + (b + c)$	$(ab)c = a(bc)$
Neutral Element	$a + 0 = a$	$a \times 1 = a$
Inverse Element	$a + (-a) = 0$	$a \times \frac{1}{a} = 1, a \neq 0$
Commutative	$a + b = b + a$	$ab = ba$
Distributive	$a(b + c) = ab + ac$	

There is a one-to-one correspondence between the real numbers, R, and the points on the number line. We determine whether one number is greater than, equal to, or less than another number according to their positions on the number line. Given any two real numbers a and b, one and only one of (i) $a < b$ (ii) $a = b$ (iii) $a > b$ is true. This is called the trichotomy property of order.

$$\text{If } a - b > 0, \text{ then } a > b.$$
$$\text{If } a - b = 0, \text{ then } a = b.$$
$$\text{If } a - b < 0, \text{ then } a < b.$$

We can now state some properties of inequality in R.

Transitive Property of Order a, b, c $\in$ R
If $a < b$, and $b < c$, then $a < c$.   If $a > b$, and $b > c$, then $a > c$.

If we add the same positive or negative quantity to both sides of an inequality, the sense of the inequality is not changed.

For a, b, c $\in$ R
If $a < b$, then $a + c < b + c$.
If $a > b$, then $a + c > b + c$.

If we multiply both sides of an inequality by the same quantity, then the sense of the inequality is
  (i) unchanged if the quantity is positive.
  (ii) reversed if the quantity is negative.

For a, b, c $\in$ R
If $a < b$ and $c > 0$, then $ac < bc$. If $a > b$, and $c > 0$, then $ac > bc$.
If $a < b$ and $c < 0$, then $ac > bc$. If $a > b$, and $c < 0$, then $ac < bc$.

An important property of real numbers is the axiom of equality which we state in three parts.

Axiom of Equality	For a, b, c $\in$ R
I. Reflexive Property	$a = a$
II. Symmetric Property	If $a = b$, then $b = a$.
III. Transitive Property	If $a = b$, and $b = c$, then $a = c$.

R has the property that between any two real numbers we can find a third real number. If we assume that $a < b$, then we can prove

$$a < \frac{a + b}{2} < b$$

as follows.

(i) $b - \dfrac{a + b}{2} = \dfrac{2b - a - b}{2}$

$= \dfrac{b - a}{2}$

> 0

because $\qquad a < b$

$\therefore \dfrac{a + b}{2} < b$

(ii) $\dfrac{a + b}{2} - a = \dfrac{a + b - 2a}{2}$

$= \dfrac{b - a}{2}$

> 0

because $\qquad a < b$

$\therefore a < \dfrac{a + b}{2}$

and $\qquad a < \dfrac{a + b}{2} < b$

This shows that between any two real numbers we can find another real number.

EXERCISE 1.5

A 1. State the property that justifies each statement.

(a) $5 + x \in$ R
(b) $t \times 1 = t$
(c) $5(x + 3) = (x + 3)5$
(d) $(2 + x) + 4 = 2 + (x + 4)$
(e) $4 + (2 - a) = (2 - a) + 4$
(f) $x + (-x) = 0$
(g) $3a \in$ R
(h) $3x - 6 = 3(x - 2)$

B 2. State the property that justifies each statement.

(a) $1 < \sqrt{3}$ or $1 = \sqrt{3}$ or $1 > \sqrt{3}$
(b) $x = x$
(c) Every point on the number line corresponds to a real number.
(d) If $3 < 5$, and $5 < 7$, then $3 < 7$.
(e) If $3 < x$, and $x < 8$, then $3 < 8$.
(f) If $x = 8$, and $8 = y$, then $x = y$.
(g) If $a = 5 + b$, then $5 + b = a$.
(h) If $(a + b)x = ax + bx$, then $ax + bx = (a + b)x$.

3. Make a conclusion from each of the following using the axiom of equality and the trichotomy property of order.

(a) $a < 5$ and $5 < b$ $\qquad$ (b) $a = ?$
(c) $a \neq b$, $a \not< b$
(d) $3(a + b) = 3a + 3b$, hence $3a + 3b = ?$
(e) $x + 3 = y$ and $2x + 1 = y$
(f) $a + b > c$ and $c > a - b$
(g) $-1 < a$ and $a < 3c$
(h) $0.25 \not< \frac{1}{4}$ and $0.25 \not> \frac{1}{4}$

4. State the condition ($y < 0$, $y = 0$, $y > 0$) for each of the following to be a true statement.

(a) $x = x + y$ (b) $2y < y$ $\qquad$ (c) $3y > y$
(d) $y = -y$ $\qquad$ (e) $-y < 0$ $\qquad$ (f) $y^2 = y$

5. Find a real number between each of the following pairs of real numbers.

(a) 2, 3 $\qquad$ (b) -1, 2 $\qquad$ (c) -3, -4
(d) 33, 34 $\qquad$ (e) 1.4, 1.5 $\qquad$ (f) -2.5, -2.6
(g) $\frac{1}{2}, \frac{1}{3}$ $\qquad$ (h) $-\frac{2}{3}, -\frac{3}{4}$ $\qquad$ (i) $\frac{1}{4}, \frac{1}{4}$

1.6 EXPONENTS AND POWERS

If a $\in$ R and n $\in$ N, then the nth power of a is written

$$a^n = \underbrace{a \times a \times a \times ... \times a}_{n \text{ factors}}$$

n is called the exponent, and a is called the base. For example,

$$8^2 = 8 \times 8 \qquad \text{and} \qquad 2^8 = 2 \times 2 \times 2 \times 2 \times 2 \times 2 \times 2 \times 2$$
$$= 64 \qquad\qquad\qquad\qquad = 256$$

EXAMPLE 1.   Express 64 as a power of 2, 4, and 8.

SOLUTION:

$$64 = 2 \times 2 \times 2 \times 2 \times 2 \times 2 = 2^6$$
$$64 = 4 \times 4 \times 4 = 4^3$$
$$64 = 8 \times 8 = 8^2$$

EXAMPLE 2.   Evaluate for x = 3, and y = 5.
(a) y^3 (b) $(x + y)^2$ (c) x^2y^4

SOLUTION:

(a) $y^3 = 5^3$ (b) $(x + y)^2 = (3 + 5)^2$ (c) $x^2y^4 = 3^2 \times 5^4$
   $\quad\ = 125$ $= 8^2$ $= 9 \times 625$
                                 $= 64$ $= 5625$

EXERCISE 1.6

A 1. State the following numbers as powers of 2.

(a) 16   (b) 2   (c) 32
(d) $2^8 \times 2^7$ (e) $2^5 \times 2^4$ (f) 2×2^9
(g) 4×32 (h) 4×8 (i) 64×64

2. State the following numbers as powers of 3.

(a) 27   (b) 243   (c) 729
(d) $3^8 \times 3$ (e) 3×3^3 (f) $3^4 \times 3^5$
(g) $3^8 \div 3$ (h) 27×81 (i) 81^3

3. State the following numbers as powers of 5.

(a) 25   (b) 125   (c) 625
(d) 25×25 (e) 5×125 (f) 25×625
(g) $5^3 \times 5^2$ (h) $5^6 \div 5^4$ (i) 625^3

B 4. If x = 2, and y = 3, evaluate the following.

(a) x^3y^2 (b) xy^3 (c) x^4y
(d) x^2y^2 (e) x^3y^3 (f) $(xy)^2$
(g) $5x$ (h) $2y$ (i) y^3

5. If x = −3, evaluate.

(a) x^2 (b) x^3 (c) x^4
(d) $3x^3$ (e) $(3x)^3$ (f) $x^2 + 1$
(g) $(x + 1)^2$ (h) $x^2 + 3^2$ (i) $-x^2$

6. If x = 3, y = −2, and z = 4, evaluate.

(a) xyz (b) $(xyz)^2$ (c) $x^2 + y^2$
(d) $y^2 + z^2$ (e) x^2y^2 (f) $(x + 1)^2$
(g) $x^2 + yz$ (h) $x^3 - yz$ (i) $x^3 + x^2$

C 7. Evaluate.

(a) 1.05^2 (b) 1.1^4 (c) 1.7^3
(d) 1.25^2 (e) 1.001^3 (f) 1.4^4
(g) 1.2^5 (h) 1.015^4 (i) 1.25^5

1.7 APPLICATION: COMPOUND AMOUNT

When money is deposited in a bank, or other financial institution, interest is paid for the use of the money. If interest is payable at regular intervals for the duration of the deposit, and it is not actually paid but added to the principal for the next interest period, it is called compound interest.

EXAMPLE 1. $1000 is deposited in an account that pays interest at the rate of 0.75% per month on the money on deposit. Find the amount of money on deposit after 12 months if there have been no withdrawals.

SOLUTION:
After 1 month, $1000 accumulates to
$$1000 + 1000(0.0075) = 1000(1 + 0.0075)$$
$$= 1000(1.0075)$$
This amount, $1000(1.0075) now becomes the principal for the next month's investment. After 2 months, we have,
$$1000(1.0075) + 1000(1.0075)(0.0075) = 1000(1.0075)(1 + 0.0075)$$
$$= 1000(1.0075)(1.0075)$$
$$= 1000(1.0075)^2$$
We continue this process until we have $1000(1.0075)^{12}$ for 12 months. The results of Example 1 are generalized in the following formula.

> $$A = P(1 + i)^n$$
>
> A is the amount.   P is the principal.
> i is the rate of interest per interest period.
> n is the total number of interest periods.

> a is the SI symbol for year.

EXAMPLE 2. Find the amount after $2500 is invested for 3 a at an annual rate of 11% compounded semi-annually.

SOLUTION:
$P = 2500$, $i = 0.055$ since it is a semi-annual compounding, and $n = 6$ since there are 6 interest periods in 3 a.
$$A = P(1 + i)^n$$
$$A = 2500(1 + 0.055)^6$$
$$= 2500 \times (1.378\ 84)$$
$$= 3447.1$$
The compound amount is $3447.10

EXERCISE 1.7

B 1. Find the amount of each of the following investments.

(a) $2000 at 1% per month for 6 months compounded monthly.
(b) $3000 at 12% per annum for 6 months compounded monthly.
(c) $5000 at 10% per annum for 3 a compounded semi-annually.

2. Find the amount of each of the following investments. When only the rate of interest is stated, it is assumed that this is an annual, or per annum rate.

(a) $5000 at 9% for 1 a compounded annually.
(b) $1500 at 9% for 1 a compounded semi-annually.
(c) $5000 at 15% for 5 a compounded semi-annually.

1.8 EXPONENT LAWS

The following examples illustrate the five exponent laws.

I. Exponent Law for Multiplication

$$5^4 \times 5^2 = (5 \times 5 \times 5 \times 5)(5 \times 5)$$
$$= 5 \times 5 \times 5 \times 5 \times 5 \times 5$$
$$= 5^6$$

$$a^m \times a^n = \underbrace{(a \times a \times \dots \times a)}_{m \text{ factors}}\underbrace{(a \times a \times \dots \times a)}_{n \text{ factors}}$$
$$= \underbrace{a \times a \times a \times \dots \times a}_{m + n \text{ factors}}$$
$$= a^{m + n}$$

II. Exponent Law for Division

$$\frac{3^5}{3^3} = \frac{3 \times 3 \times 3 \times 3 \times 3}{3 \times 3 \times 3}$$
$$= 3 \times 3$$
$$= 3^2$$

$$\frac{a^m}{a^n} = \frac{\overbrace{a \times a \times a \times \dots \times a}^{m \text{ factors}}}{\underbrace{a \times a \times \dots \times a}_{n \text{ factors}}} \quad (m > n, a \neq 0)$$
$$= \underbrace{a \times a \times \dots \times a}_{m - n \text{ factors}}$$
$$= a^{m - n}$$

III. Power Law

$$(7^3)^2 = (7 \times 7 \times 7)^2$$
$$= (7 \times 7 \times 7) \times (7 \times 7 \times 7)$$
$$= 7 \times 7 \times 7 \times 7 \times 7 \times 7$$
$$= 7^6$$

$$(a^m)^n = \underbrace{(a \times a \times \dots \times a)}_{m \text{ factors}}{}^{n}$$
$$= \underbrace{\underbrace{(a \times a \times \dots \times a)}_{m \text{ factors}} \times \underbrace{(a \times a \times \dots \times a)}_{m \text{ factors}} \times \dots \times \underbrace{(a \times a \times \dots \times a)}_{m \text{ factors}}}_{n \text{ times}}$$
$$= \underbrace{a \times a \times a \times \dots \times a}_{mn \text{ factors}}$$
$$= a^{mn}$$

IV. Power of a Product

$$(4 \times 5)^3 = (4 \times 5) \times (4 \times 5) \times (4 \times 5)$$
$$= 4 \times 4 \times 4 \times 5 \times 5 \times 5$$
$$= 4^3 \times 5^3$$

$$(ab)^n = \underbrace{(ab) \times (ab) \times \dots \times (ab)}_{n \text{ factors}}$$
$$= \underbrace{(a \times a \times \dots \times a)}_{n \text{ factors}} \times \underbrace{(b \times b \times \dots \times b)}_{n \text{ factors}}$$
$$= a^n b^n$$

V. Power of a Quotient

$$\left(\frac{2}{3}\right)^4 = \frac{2}{3} \times \frac{2}{3} \times \frac{2}{3} \times \frac{2}{3}$$
$$= \frac{2 \times 2 \times 2 \times 2}{3 \times 3 \times 3 \times 3}$$
$$= \frac{2^4}{3^4}$$

$$\left(\frac{a}{b}\right)^n = \underbrace{\frac{a}{b} \times \frac{a}{b} \times \dots \times \frac{a}{b}}_{n \text{ factors}}$$
$$= \frac{\overbrace{a \times a \times \dots \times a}^{n \text{ factors}}}{\underbrace{b \times b \times \dots \times b}_{n \text{ factors}}}$$
$$= \frac{a^n}{b^n}$$

EXERCISE 1.8

A 1. State the following as powers of 2.
   (a) $2^5 \times 2^4$ (b) $2^7 \div 2^4$
   (c) $2^3 \times 2^a$ (d) $2^b \div 2^4$
   (e) $(2^5)^3$ (f) $(2^a)^3$

2. State the following as powers of 3.
   (a) $3^4 \times 3^2$ (b) $3^5 \div 3^2$
   (c) $3^a \times 3^2$ (d) $3^m \div 3^5$
   (e) $(3^3)^3$ (f) $(3^2)^m$
   (g) $(4^m)^n$ (h) $3^a \div 3$

B 3. Use the Laws of Exponents to express the following in another form, if possible.
   (a) $(9 \times 13)^4$ (b) $(9x)^4$
   (c) $\left(\frac{5}{6}\right)^{18}$ (d) $\left(\frac{x}{6}\right)^5$
   (e) $7^8 \times 9^8$ (f) $(6 \times 72)^5$
   (g) $2^6 \times 5^4$ (h) $(xy)^3$
   (i) $\pi^2 \times \pi^3$ (j) $(\pi^2)^3$
   (k) $\left(\frac{2}{3}\right)^{10}$ (l) $\left(\frac{x}{y}\right)^6$
   (m) $\frac{(2.78)^{12}}{(2.78)^4}$ (n) $\frac{(-2)^{100}}{(-2)^{93}}$
   (o) $\left(\frac{3}{a}\right)^2$ (p) $\frac{a^8}{b^8}$
   (q) $4^9 \times 5^6$ (r) $(2x^2)^3$

4. Simplify the following expressions.
   (a) $(-3x^8)^3$ (b) $(-3x^6)(4x^3)$
   (c) $3^8 \times 3^6 \times 3^4$ (d) $a^m \times a^n \times a^p$
   (e) $((2^3)^2)^4$ (f) $((a^m)^n)^p$
   (g) $\frac{4^7 \times 4^3}{4^6}$ (h) $\frac{x^{m+n} \times x^{2m}}{x^n}$

5. Simplify the following.
   (a) $\frac{7^{10} \times 7^{12} \times 7^6}{(7^3)^9}$ (b) $\frac{3^6 \times 3^5 \times 3^4}{3^8 \times 3^9}$
   (c) $\frac{56^4}{14^4}$ (d) $\frac{2^{n+2} \times 4^{n+1}}{8^n}$
   (e) $\frac{125^{16}}{5^{47}}$ (f) $\frac{a^{27} \times a^{18} \times a^4}{a^7 \times (a^3)^{14}}$

6. Simplify the following.
   (a) $a^{2+p} + a^{2p} \times (-a)^7$
   (b) $x^n \times (x^2)^{n+1} \times (x^3)^{n+2}$
   (c) $\frac{(3^6)^n \times (81)^{2n}}{(3^n)^4}$ (d) $\frac{2^n \times 4^{n-1} \times 8^{3n-2}}{16^{2n-1}}$
   (e) $\frac{32^n \times 16^{1-n} \times 8^{2n}}{(4^2)^{n+1}}$

7. Simplify if possible.
   (a) $(x^5y^3)(x^2y^6)$ (b) x^8y^9
   (c) $\left(\frac{1}{2}x^3\right)^2 (x^8y^2)^4$ (d) $(-xy)^3(x^4y)^9$
   (e) $\frac{2xy^4}{3x^3y^2} \times \frac{15x^2y^3}{12x^4y^2}$ (f) $\frac{a^2b^4}{a^3b^2} \times \left(\frac{a^4}{b^2}\right)^3$
   (g) $\frac{(2x^2y^2)^5}{8x^4y^3(x^2y)^3}$ (h) $\left(\frac{x}{y}\right)^5\left(\frac{2}{x}\right)^4\left(\frac{y}{4}\right)^3$

CALCULATOR MATH

USING THE EXPONENTIAL KEY y^x

Some calculators have an exponential key y^x .

In order to calculate a^b, we press

C **a** **y^x** **b** **=**

Check the use of this key with the following examples.

3^2 **C** **3** **y^x** **2** **=**

5^3 **C** **5** **y^x** **3** **=**

2^5 **C** **2** **y^x** **5** **=**

EXERCISE

1. Simplify.
   (a) 1.015^7 (b) 1.15^3
   (c) 1.025^4 (d) 1.0375^6
   (e) 1.01^{10} (f) 1.1^{10}

1.9 NEGATIVE EXPONENTS

We have defined a^n if $a \in R$ and $n \in N$, but how can we give a meaning to a^n if n is 0 or a negative integer? For example, $5^4 = 5 \times 5 \times 5 \times 5$, but is it impossible to give a reasonable definition of 5^0 or 5^{-3}? If so, we would still like all the Laws of Exponents to be true. If the second law of exponents $\dfrac{a^m}{a^n} = a^{m-n}$ is to be true, then we would

have
$$\frac{2^3}{2^3} = 2^{3-3} = 2^0$$

But
$$\frac{2^3}{2^3} = 1 \qquad \text{So } 2^0 \text{ must be 1.}$$

Similarly, a^0 must be 1 if $a \neq 0$. Note that if we define $a^0 = 1$, then
$$a^m \times a^0 = a^m \times 1 = a^m = a^{m+0}$$
So the first law of exponents is also true.

$$\boxed{\text{If } a \neq 0, \text{ we define } a^0 = 1}$$

If the first law is to hold for negative exponents, then

$$
\begin{aligned}
5^3 \times 5^{-3} &= 5^{3+(-3)} \\
&= 5^0 \\
&= 1
\end{aligned}
\qquad \therefore 5^{-3} = \frac{1}{5^3} \text{ Divide both sides by } 5^3.
$$

Similarly, under the same assumption,

$$
\begin{aligned}
a^n \times a^{-n} &= a^{n+(-n)} \\
&= a^0 \\
&= 1 \qquad \text{If } a \neq 0.
\end{aligned}
\qquad \therefore a^{-n} = \frac{1}{a^n} \text{ Divide both sides by } a^n.
$$

$$\boxed{\begin{array}{c}\text{If } a \neq 0 \text{ and } n \in N, \text{ we define} \\ a^{-n} = \dfrac{1}{a^n}\end{array}}$$

0⁰ is not defined because
$$
\begin{aligned}
0^0 &= 0^{1-1} \\
&= \frac{0^1}{0^1} \\
&= \frac{0}{0} \quad \text{which is} \\
&\qquad \text{undefined.}
\end{aligned}
$$

With these definitions we can show that all the previous laws of exponents still hold for any integer exponents. For example,

$$
\begin{aligned}
(3 \times 5)^{-4} &= \frac{1}{(3 \times 5)^4} \\
&= \frac{1}{3^4} \times \frac{1}{5^4} \\
&= \frac{1}{3^4 \times 5^4} \\
&= 3^{-4} \times 5^{-4} \qquad \therefore (3 \times 5)^{-4} = 3^{-4} \times 5^{-4}
\end{aligned}
$$

Laws of Exponents for Integral Exponents
If $a \in R$, $a \neq 0$, and $m, n \in I$, then

I. $a^m \times a^n = a^{m+n}$	IV. $(ab)^n = a^n b^n$
II. $\dfrac{a^m}{a^n} = a^{m-n}$	V. $\left(\dfrac{a}{b}\right)^n = \dfrac{a^n}{b^n}$
III. $(a^m)^n = a^{mn}$	

EXAMPLE. Evaluate.

(a) $\left(\dfrac{2}{3}\right)^{-2}$

(b) $\dfrac{(-12)^0}{2^{-4}}$

(c) $\dfrac{2^{-8} + 2^{-10}}{2^{-9} + 2^{-7}}$

SOLUTION:

(a) $\left(\dfrac{2}{3}\right)^{-2} = \dfrac{1}{\left(\dfrac{2}{3}\right)^2}$

$= \dfrac{1}{\dfrac{4}{9}}$

$= \dfrac{9}{4}$

(b) $\dfrac{(-12)^0}{2^{-4}} = \dfrac{1}{2^{-4}}$

$= \dfrac{1}{\dfrac{1}{2^4}}$

$= 2^4$

(c) $\dfrac{2^{-8} + 2^{-10}}{2^{-9} + 2^{-7}} = \dfrac{2^{10}}{2^{10}} \times \dfrac{2^{-8} + 2^{-10}}{2^{-9} + 2^{-7}}$

$= \dfrac{2^{10} \times 2^{-8} + 2^{10} \times 2^{-10}}{2^{10} \times 2^{-9} + 2^{10} \times 2^{-7}}$

$= \dfrac{2^2 + 2^0}{2^1 + 2^3}$

$= \dfrac{4 + 1}{2 + 8}$

$= \dfrac{5}{10}$

$= \dfrac{1}{2}$

EXERCISE 1.9

A 1. State the value of each of the following.

(a) 6^0 (b) 9^{-1} (c) 2^{-3} (d) 3^{-2}

(e) $(-1)^{-1}$ (f) $(-\pi)^0$ (g) 10^{-4} (h) 837^{-1}

2. State the following using only positive exponents.

(a) x^{-8} (b) $x^2 y^{-2}$ (c) $a^{-3} b^{-4}$

(d) $a^3 \times a^{-5}$ (e) $\dfrac{1}{a^{-10}}$ (f) $\left(\dfrac{x}{y}\right)^3$

3. State the following with the variables in the numerator.

(a) $\dfrac{1}{x^3}$ (b) $2\dfrac{a}{b^4}$ (c) $\pi\dfrac{x^2}{y^{-1}}$

B 4. Evaluate.

(a) $2^{-1} \times 3^0 \times 4^2 \times 5^{-2}$ (b) $(2^3 \times 4^{-4})^{-2}$

(c) $(3^{-3} \times 5^2)^{-1}$ (d) $4^{10} \times 2^{-18}$

(e) $\dfrac{(27)^{-2}}{3^{-8}}$ (f) $\dfrac{(5^3 + 3^5)^0}{2^{-1}}$

(g) $\dfrac{3^{-12} + 3^{-14}}{3^{-12} - 3^{-14}}$ (h) $\dfrac{(7 \times 5^{-1})^{-1}}{(2 \times 3^{-1})^{-1}}$

(i) $\left(\dfrac{3^{-1} - 4^{-1}}{2^{-1} - 3^{-1}}\right)^{-1}$ (j) $[5^{-4} \times (25)^3]^2$

(k) $1 + \dfrac{1}{2^{-1} + \dfrac{1}{3^{-1} + \dfrac{1}{4^{-1}}}}$

(l) $\{1 - [(2 + 3^{-1})^{-1}]\}^{-1}$

5. Rewrite the following using only positive exponents. Simplify where possible.

(a) $(7x^2 y^{-3})^3$ (b) $(3a^{-1} b^{-2})^{-5}$

(c) $(4ab^{-2} c^3 d^{-4})^3$ (d) $\dfrac{a^{-3} b^2}{a^{-5} b^5}$

(e) $\dfrac{a^2 x^3 y^{-2}}{b^{-2} xy^{-6}}$ (f) $(a^2 b^{-1} - 1)^2$

(g) $a^4(a^2 + a - 5a^{-2})$ (h) $(x^2 - 1)(x^{-2} + 2)$

(i) $(b^{-2})^{n-2} \div b^4$

(j) $(x^{-n} + y^{-m})(x^{-n} - y^{-m})$

CALCULATOR MATH

Some calculators carry digits beyond the display.

1. Perform the calculation
$$2 \div 3 \times 3 - 2$$
on a calculator.
Can you explain your result?

2. Press [+/−]

and explain your result.

3. Press

[×] [1] [0] [yˣ] [1] [0] [=]

4. Summarize the results of 1, 2, 3.

1.10 APPLICATION: PRESENT VALUE

In some situations we need to know how much money to invest today to produce a desired amount at a later date. This is called the present value (PV) of an amount.

EXAMPLE 1. What principal invested now at 10% per annum compounded semi-annually will produce an amount of $5000 in 3 a?

SOLUTION:
Use the formula learned in Section 1.7.

$$A = P(1 + i)^n$$

$A = 5000, \quad i = 0.05, \quad P = PV, \quad n = 6$

$$5000 = PV (1 + 0.05)^6$$
$$\frac{5000}{(1.05)^6} = PV$$
$$\frac{5000}{1.3400956} = PV$$
$$3731.08 = PV$$

∴ $3731.08 invested today will produce $5000 in 3 a.

The results of Example 1 can be generalized in the following formula.

$$PV = \frac{A}{(1 + i)^n} \text{ or } PV = A(1 + i)^{-n}$$

PV is the present value.
A is the amount to be achieved.
i is the rate of interest per interest period.
n is the total number of interest periods.

EXAMPLE 2. How much money must be invested now in a fund that pays 11% per annum compounded semi-annually in order to have $50 000 in 4 a?

SOLUTION:
$A = 50\,000, \quad i = 0.055, \quad n = 8$

$$PV = A(1 + i)^{-n}$$
$$PV = 50\,000(1.055)^{-8}$$
$$= 50\,000(0.651\,60)$$
$$= 32\,580$$

∴ $32 580 should be invested now to produce $50 000 in 4 a.

EXERCISE 1.10

B 1. Find the present value of each of the following amounts.
(a) $15 000 in 5 a at 9% compounded semi-annually.
(b) $25 000 in 3 a at 10% compounded semi-annually.
(c) $10 000 in 2 a at 12% compounded monthly.
(d) $5000 in 3 a at 11% compounded annually.
(e) $6000 in 4.5 a at 10% compounded semi-annually.
(f) $500 in 9 months at 12% compounded monthly.
(g) $8000 in 3 a at 9% compounded semi-annually.

2. How much money should Bill invest now in order to have $8000 in 3.5 a if the funds are invested at 11% compounded semi-annually?

3. Josephine Bouton has an investment certificate that will pay her $8000 at age 65. What is the value of this certificate when Josephine is 58 if the funds are invested at 9% compounded semi-annually?

4. Adam Jones has a savings bond that will pay $10 000 in 7 a. What is the present value of the bond if the current interest rate is 10% compounded annually?

MICRO MATH

In the BASIC computer language, we use the ↑ operator for exponentiation. For small integral values of the exponent, we can use multiplication as in

$$x^2 = x*x$$
$$x^3 = x*x*x$$

However, when evaluating powers with larger integral components, it is more convenient to use the ↑ operator of BASIC, and we write

$$x ↑ 2 \text{ for } x^2, \text{ and } x ↑ 3 \text{ for } x^3$$

The following program computes the value of a^x.

NEW

```
10 PRINT "COMPUTING THE VALUE OF A↑X."
20 INPUT "    BASE = ";A
30 INPUT "EXPONENT = ";X
40 PRINT "THE ANSWER IS ";A↑X
50 END
```

RUN

EXERCISE

Run the program to evaluate the following expressions.

1. 1.025^{12}
2. 1.0075^6
3. 1^{125}
4. 1^{444}
5. $(-1)^2$
6. $(-1)^5$
7. $(-1)^{45}$
8. $(-1)^{54}$

MIND BENDER

Three men, Mike, Allen, and Ed and their wives Jane, Carol, and Sue, were playing tennis. In a doubles match, Mike partnered Allen's wife, and Carol's husband partnered Jane. Ed sat out with his sister. Who is married to whom?

1.11 RATIONAL EXPONENTS

So far we have defined a^n where $a \neq 0$ and $n \in I$. We shall see in this section that it is also possible to define powers where the exponent is a rational number, such as $5^{\frac{1}{2}}$ or $8^{\frac{3}{5}}$, in such a way that the laws of exponents are still true.

If the third law of exponents

$$(a^m)^n = a^{mn}$$

is to be true, then putting $m = \dfrac{1}{n}$ we would have

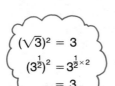

$$(\sqrt{3})^2 = 3$$
$$(3^{\frac{1}{2}})^2 = 3^{\frac{1}{2} \times 2}$$
$$= 3$$

$$(a^{\frac{1}{n}})^n = a^{\frac{1}{n} \times n} = a^1 = a$$

If $a \geqslant 0$, we can take nth roots of both sides of the equation

$$(a^{\frac{1}{n}})^n = a$$

we get

$$a^{\frac{1}{n}} = \sqrt[n]{a}$$

This suggests the following definition.

$$\boxed{a^{\frac{1}{n}} = \sqrt[n]{a}, \ n \in N}$$

Note that if n is even, then we must have $a \geqslant 0$, but if n is odd, then a can be any real number.

EXAMPLE 1. Find.

(a) $16^{\frac{1}{4}}$ (b) $(-27)^{\frac{1}{3}}$

SOLUTION:

(a) $16^{\frac{1}{4}} = \sqrt[4]{16}$ | (b) $(-27)^{\frac{1}{3}} = \sqrt[3]{-27}$

$\qquad\quad = 2$ | $= -3$

To see how we should define $a^{\frac{m}{n}}$ when $m, n \in N$ we again use the third law of exponents.

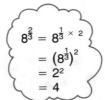

$$8^{\frac{2}{3}} = 8^{\frac{1}{3} \times 2}$$
$$= (8^{\frac{1}{3}})^2$$
$$= 2^2$$
$$= 4$$

$$a^{\frac{m}{n}} = a^{m \times \frac{1}{n}} \qquad\qquad a^{\frac{m}{n}} = a^{\frac{1}{n} \times m}$$

$$= (a^m)^{\frac{1}{n}} \qquad\qquad\quad = (a^{\frac{1}{n}})^m$$

$$= \sqrt[n]{a^m} \qquad\qquad\quad = (\sqrt[n]{a})^m$$

Therefore, we make the following definition for rational exponents.

$$\boxed{a^{\frac{m}{n}} = \sqrt[n]{a^m} = (\sqrt[n]{a})^m, \ m, n \in N}$$

Again, if n is even then we must have $a \geqslant 0$.

Notice that in calculating $a^{\frac{m}{n}}$ you can either raise a to the mth power and then take the nth root, or first take the nth root and then raise the result to the mth power. With these definitions we can show that all of the previous laws of exponents are still true for rational exponents.

$$\boxed{\begin{array}{l}
\qquad\qquad\text{Laws of Exponents for Rational Exponents} \\[4pt]
\text{If } a \in R,\ a \geqslant 0,\ \text{and } r, s \in Q,\ \text{then} \\[10pt]
\begin{array}{llll}
\text{I.} & a^r \times a^s = a^{r+s} & \text{V.} & \left(\dfrac{a}{b}\right)^r = \dfrac{a^r}{b^r} \ \ (b \neq 0) \\[14pt]
\text{II.} & \dfrac{a^r}{a^s} = a^{r-s} & \text{VI.} & a^0 = 1 \\[14pt]
\text{III.} & (a^r)^s = a^{rs} & \text{VII.} & a^{-r} = \dfrac{1}{a^r} \\[14pt]
\text{IV.} & (ab)^r = a^r b^r & \text{VIII.} & a^{\frac{r}{s}} = \sqrt[s]{a^r} = (\sqrt[s]{a})^r
\end{array}
\end{array}}$$

EXAMPLE 2. Find $4^{\frac{3}{2}}$.

SOLUTION:
There are several methods.

(i) $4^{\frac{3}{2}} = (\sqrt{4})^3$
$= 2^3$
$= 8$

(ii) $4^{\frac{3}{2}} = \sqrt{4^3}$
$= \sqrt{64}$
$= 8$

(iii) $4^{\frac{3}{2}} = (4^{\frac{1}{2}})^3$
$= 2^3$
$= 8$

(iv) $4^{\frac{3}{2}} = (2^2)^{\frac{3}{2}}$
$= 2^{2 \times \frac{3}{2}}$
$= 2^3$
$= 8$

EXAMPLE 3. Simplify.

(a) $125^{-\frac{2}{3}}$

(b) $(2x^2y^4)^{\frac{3}{2}}$

SOLUTION:

(a) $125^{-\frac{2}{3}} = \dfrac{1}{125^{\frac{2}{3}}}$

$= \dfrac{1}{(\sqrt[3]{125})^2}$

$= \dfrac{1}{5^2}$

$= \dfrac{1}{25}$

(b) $(2x^2y^4)^{\frac{3}{2}} = 2^{\frac{3}{2}}(x^2)^{\frac{3}{2}}(y^4)^{\frac{3}{2}}$

$= (\sqrt{2})^3 x^{2 \times \frac{3}{2}} y^{4 \times \frac{3}{2}}$

$= 2\sqrt{2}x^3y^6$

EXAMPLE 4. Find an approximation of $5^{-0.3}$ using a calculator.

SOLUTION:

Press ▮5▮ ▮yˣ▮ ▮.▮ ▮3▮ ▮+/−▮ ▮=▮

The display is ▮0.617033862▮

The number 0.617 033 862 is a rational approximation of the irrational number $5^{-0.3}$.

EXERCISE 1.11

A 1. State the following using radicals.

(a) $2^{\frac{1}{9}}$ (b) $37^{\frac{1}{2}}$ (c) $x^{\frac{1}{3}}$ (d) $2^{\frac{2}{3}}$

(e) $2^{\frac{3}{2}}$ (f) $3^{\frac{3}{4}}$ (g) $a^{\frac{2}{5}}$ (h) $x^{\frac{4}{7}}$

(i) $2^{-\frac{1}{2}}$ (j) $7^{-\frac{1}{5}}$ (k) $a^{-\frac{3}{2}}$ (l) $9^{\frac{2}{11}}$

2. State the following using exponents.

(a) $\sqrt{3}$ (b) $\sqrt{19}$ (c) $\sqrt[7]{23}$ (d) $\sqrt[4]{x}$

(e) $(\sqrt[3]{7})^2$ (f) $\sqrt[3]{7^2}$ (g) $(\sqrt[5]{6})^4$ (h) $(\sqrt[3]{13})^5$

(i) $\sqrt[5]{a^2}$ (j) $(\sqrt[6]{a})^5$ (k) $\dfrac{1}{\sqrt{5}}$ (l) $\dfrac{1}{(\sqrt[4]{7})^3}$

3. Evaluate.

(a) $25^{\frac{1}{2}}$ (b) $64^{\frac{1}{3}}$ (c) $9^{\frac{3}{2}}$ (d) $1^{\frac{9}{7}}$

(e) $36^{-\frac{1}{2}}$ (f) $98^{\frac{2}{3}}$ (g) $9^{0.5}$ (h) $8^{-\frac{1}{3}}$

(i) $(-8)^{\frac{2}{3}}$ (j) $4^{-\frac{3}{2}}$ (k) $(-8)^{\frac{1}{3}}$ (l) $(-32)^{\frac{2}{5}}$

B 4. Evaluate.

(a) $32^{\frac{4}{5}}$ (b) $8^{2\frac{1}{3}}$

(c) $100\,000^{\frac{2}{5}}$ (d) $64^{-\frac{1}{3}}$

(e) $81^{\frac{3}{4}}$ (f) $625^{-\frac{3}{4}}$

(g) $128^{\frac{8}{7}}$ (h) $3^{\frac{2}{7}} \times 3^{\frac{5}{7}}$

(i) $(6^{0.4})^5$ (j) $(49^6)^{\frac{1}{4}}$

(k) $2^{\frac{1}{5}} \times 4^{\frac{2}{5}}$ (l) $9^{\frac{3}{2}} \div 36^{-\frac{1}{2}}$

(m) $\left(\dfrac{8}{27}\right)^{\frac{1}{3}}$ (n) $\left(\dfrac{49}{144}\right)^{-\frac{1}{2}}$

(o) $\left(\dfrac{25}{64}\right)^{\frac{3}{2}}$ (p) $\dfrac{64^{\frac{2}{3}}}{216^{-\frac{1}{3}}}$

(q) $(0.16)^{\frac{1}{2}}(0.008)^{\frac{1}{3}}$ (r) $3^{\frac{1}{2}} \times 9^{\frac{1}{4}}$

(s) $256^{0.375}$ (t) $0^{1.356}$

(u) $[(\sqrt{343})^{\frac{1}{9}}]^6$ (v) $\dfrac{(0.09)^{-\frac{1}{2}}}{(0.125)^{-\frac{2}{3}}}$

(w) $(81^{-1})^{-\frac{1}{4}}$ (x) $\dfrac{(0.81)^{\frac{1}{2}} \times 6^{-3}}{(0.027)^{\frac{2}{3}}}$

5. Simplify.

(a) $2^{\frac{1}{2}} \times 2^{\frac{1}{3}}$ (b) $3^{\frac{2}{9}} \times 9^{\frac{1}{3}}$

(c) $(x^{\frac{2}{3}}y^{\frac{1}{6}})^3$ (d) $(a^{\frac{1}{4}}b^{\frac{1}{3}})^{12}$

(e) $(a^3b^6c^9)^{\frac{1}{3}}$ (f) $(x^{\frac{2}{3}} + 3x^{\frac{1}{3}})x^{\frac{1}{3}}$

(g) $(16x^8y^2)^{\frac{1}{4}}$ (h) $(64x^9y^{-3})^{\frac{2}{3}}$

(i) $(20x^2y^3z^{-1})^{\frac{3}{2}}$ (j) $\left(\dfrac{a^3b^{-4}}{x^{-1}y^2}\right)^2 \times \dfrac{x^{-1}b^{-1}}{a^{\frac{3}{2}}y^{\frac{4}{5}}}$

(k) $\sqrt[4]{\dfrac{y^{\frac{1}{2}}\sqrt{xy}}{x^{\frac{2}{3}}}}$ (l) $\sqrt[4]{a^{2n+1}} \times \sqrt[4]{a^{-1}}$

C 6. Find a rational approximation of each of the following to the nearest thousandth.

(a) $2^{0.4}$ (b) $3^{-1.6}$

(c) $5^{2.8}$ (d) $3^{\frac{2}{3}}$

(e) $6^{-\frac{3}{5}}$ (f) $10^{\frac{5}{7}}$

7. Determine which is the larger of each of the following pairs of numbers to the nearest thousandth.

(a) $5^{\frac{1}{3}}, 3^{\frac{1}{2}}$ (b) $7^{\frac{1}{4}}, 4^{\frac{1}{3}}$

(c) $3^{\frac{3}{2}}, 9^{\frac{2}{3}}$ (d) $6^{-\frac{1}{2}}, 14^{-\frac{1}{3}}$

Each letter represents a number. Solve the following cryptograms.

1.
```
  FORTY
    TEN
+   TEN
-------
  SIXTY
```

2.
```
   SLED
+ SNOW
------
  RIDE
```

3.
```
   SLED
-  SNOW
------
   BOB
```

1.12 EQUATIONS INVOLVING EXPONENTS

Equations such as $2^{2x-1} = 2^9$ are solved by setting the exponents equal to each other and solving the equation.

$$2x - 1 = 9$$
$$2x = 10$$
$$x = 5$$

> The bases must be the same.

This method of solving an exponential equation is based on the property that if $a^x = a^y$, then $x = y$, for $x \neq -1, 0, 1$.

EXAMPLE 1. Solve. $3^x = 81$

SOLUTION:
$$3^x = 81$$
$$3^x = 3^4$$
$$x = 4$$

∴ the solution is $x = 4$.

EXAMPLE 2. Solve. $4^{8x} = \dfrac{1}{16}$

SOLUTION:
$$4^{8x} = \frac{1}{16}$$
$$(2^2)^{8x} = 2^{-4}$$
$$2^{16x} = 2^{-4}$$
$$16x = -4$$
$$x = \frac{-4}{16}$$
$$x = -\frac{1}{4}$$

∴ the solution is $x = -\dfrac{1}{4}$.

> Check these solutions.

EXAMPLE 3. Solve. $5^{x-2} = 625$

SOLUTION:
$$5^{x-2} = 625$$
$$5^{x-2} = 5^4$$
$$x - 2 = 4$$
$$x = 6$$

∴ the solution is 6.

EXERCISE 1.12

1. Solve the following equations for $x \in N$.
(a) $2^x = 32$ (b) $3^x = 27$
(c) $2^x = 64$ (d) $5^x = 25$
(e) $3^x = 81$ (f) $7^x = 49$
(g) $(-3)^x = -27$ (h) $(-2)^x = -8$
(i) $(-2)^x = 16$ (j) $(-5)^x = 25$

2. Solve the following equations for $x \in N$.
(a) $4^x = 256$ (b) $6^{x+3} = 6^{2x}$
(c) $9^x = 729$ (d) $2^x = 16^4$
(e) $2^x = 4^{x-1}$ (f) $2(5^x) = 1250$
(g) $9^{2x-6} = 3^{x+6}$ (h) $4^{2x-1} = 64$
(i) $1^x = 1$ (j) $(-1)^x = 1$

3. Solve the following equations for $x \in I$.
(a) $6^{3x-6} = 1$ (b) $2^{-x} = 128$

(c) $5^{4-x} = \frac{1}{5}$ (d) $(-1)^x = 1$
(e) $3^{2-x} = 1$ (f) $4^{3x} = 64$
(g) $4^{x-1} = 1$ (h) $(-1)^{2x} = 1$
(i) $7^{x-2} = 49$ (j) $2^{-2x} = 32$

4. Solve the following for $x \in R$.
(a) $4^x = 8$ (b) $2^{9x} = \frac{1}{8}$
(c) $64^x = 16$ (d) $9^{6x} = \frac{1}{27}$
(e) $9^{2x+1} = 27$ (f) $5^{2x+1} = \frac{1}{125}$
(g) $32^{3x-2} = 64$ (h) $3^{3x-1} = \frac{1}{81}$
(i) $10^x = 10\,000$ (j) $10^{x-2} = \frac{1}{10\,000}$
(k) $3(5^{x+1}) = 15$ (l) $2(3^{x-2}) = 18$
(m) $5(4^x) = 10$ (n) $3^{2x-1} + 1 = 2$

1.13 PRINCIPLES OF PROBLEM SOLVING

Problem solving is a creative activity that requires more than routines, recipes, and formulas. There are no hard and fast rules that will ensure success in solving problems. However, it is possible to outline some general steps in the problem solving process and to give some principles which may be useful in the solution of certain problems. These steps and principles are just common sense made explicit. In this section we begin with the READ – PLAN – SOLVE – ANSWER model for problem solving.

EXAMPLE 1. The eight square sheets of paper overlap as shown. Number the square sheets from the top layer to the bottom.

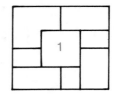

READ

Read the problem carefully and devote sufficient time to understanding the problem before you try to solve it. Identify the relevant and irrelevant information.

You may find it helpful to paraphrase statements, summarize information, or make lists.

PLAN

Think of a plan. Find a connection between the given information and the unknown, which will enable you to calculate the unknown. If the connection is not immediately seen, the following strategies may be helpful in devising the plan.

1. Classify information. Study the information carefully to determine what is needed to solve the problem. Is there insufficient information or conflicting information? Some information may be extraneous or redundant.

2. Search for a pattern. Try to recognize patterns. Some problems are solved by recognizing that some kind of pattern is occurring. The pattern could be geometric, numerical, or algebraic. If you can see that there is some sort of regularity or repetition in a problem, then you might be able to guess what the continuing pattern is, and then prove it.

3. Draw a diagram or flow chart. For many problems it is useful to draw a diagram and identify the given and required quantities on the diagram. A flow chart can be used to organize a series of steps that must be performed in a definite order.

4. Estimate, Guess, and Check. This is a valid method to solve a problem where a direct method is not apparent. You may find it necessary to improve your guess and "zero in" on the correct answer.

5. Sequence operations. To solve some problems, several operations performed in a definite order are needed.

6. Work backwards. Sometimes it is useful to imagine that your problem is solved and work backwards step by step until you arrive at

the given data. Then you may be able to reverse your steps and thereby, construct a solution to the original problem.

7. Use a formula or an equation.   In some problems, after analyzing the data, the problem can be written as an equation, or the data can be substituted into a formula.

8. Solve a simpler problem.   Try to think of a similar or a related problem, one that is easier than the original problem. If you can solve the similar, simpler problem, then it might give you the clues you need to solve the original, more difficult problem.

9. Account for all possibilities.   Take cases. You may sometimes have to split a problem into several cases and give a different argument for each of the cases.

10. Make a table.   In some problems, it is helpful to organize the data into a table, chart, or grid.

11. Check for hidden assumptions.   In some problems, the information concerning what is given is presented in a subtle manner that may not attract your attention. Re-read the problem carefully and look for the implied information.

12. Conclude from assumptions.   In some problems, it will be necessary to make assumptions. The conclusions that you draw from these assumptions should be those that you have made in the past, from the same types of information.

13. Introduce something extra.   Sometimes it may be necessary to introduce something new, an auxiliary aid, to help make the connection between the given and the unknown. For instance in geometry, the auxiliary could be a new line drawn in the diagram. In algebra, it could be a new unknown which is related to the original unknown.

SOLVE

Before solving the problem, look at the reasons for selecting your strategy. If you have more than one strategy available, you should consider familiarity, efficiency, and ease of implementation in making your final choice. In carrying out your strategy, work with care and check each step as you proceed. Remember to present your ideas clearly.

ANSWER

State the answer in a clear and concise manner. Check your answer in the original problem and use estimation to decide if your answer is reasonable. In checking your answer, you may discover an easier way to solve the problem. You may wish to generalize your method of solution so that it can be applied to similar problems.

SOLUTION:

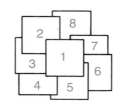

EXAMPLE 2. Two classrooms are next to each other on the same side of the hall. The product of the room numbers is 1224. Find the room numbers.

PLAN

SOLUTION:
The product of the room numbers is 1224.
We are not told whether the numbers differ by 1, or 2.
Since the product, 1224, is even, one or both of the room numbers is even.
Assume that the numbers are consecutive integers and use Guess and Check.

n	n + 1	n(n + 1)
32	33	1056
33	34	1122
34	35	1190
35	36	1260
36	37	

SOLVE

From the table, we see that $1190 < 1224 < 1260$. This suggests that the room numbers are not consecutive integers. Let us now assume that the room numbers are consecutive even integers.

n	n + 2	n(n + 2)
32	34	1088
34	36	1224
36	38	

ANSWER

Since $34 \times 36 = 1224$, the rooms are numbered using the consecutive even numbers 34 and 36.

We can make up some interesting generalizations in answering these follow-up questions.
(i) How are the rooms numbered if the product was given as 992?
(ii) How could your knowledge of square roots have been useful in solving this problem?
(iii) How are the rooms numbered if the product is odd?

EXAMPLE 3. Many whole numbers can be written as the difference of two squares. For example,

$$0 = 0^2 - 0^2$$
$$5 = 3^2 - 2^2$$
$$13 = 7^2 - 6^2$$

(a) Which of the whole numbers 0, 1, 2, 3, ..., 20 cannot be expressed as the difference of two squares?
(b) Generalize your results.

PLAN

SOLUTION:

(a)

Number	Difference of Squares
0	$0^2 - 0^2$
1	$1^2 - 0^2$
2	?
3	$2^2 - 1^2$
4	$2^2 - 0^2$
5	$3^2 - 2^2$
6	?
7	$4^2 - 3^2$
8	$3^2 - 1^2$
9	$3^2 - 0^2$
10	?

Number	Difference of Squares
11	$6^2 - 5^2$
12	$4^2 - 2^2$
13	$7^2 - 6^2$
14	?
15	$4^2 - 1^2$
16	$4^2 - 0^2$
17	$9^2 - 8^2$
18	?
19	$10^2 - 9^2$
20	$6^2 - 4^2$

SOLVE

The numbers 2, 6, 10, 14, and 18 cannot be expressed as the difference of two squares.

(b) Starting with 2, every fourth number in our list was not expressed as a difference of two squares. There are different patterns for the odd numbers and for the even numbers.

CASE 1	*CASE 2*
ODD NUMBERS The bases of the squares are consecutive numbers and they add to the number.	**EVEN NUMBERS** The bases of the squares are two apart. The even numbers are also multiples of 4.
The odd numbers are	The even numbers are

CASE 1 — ODD NUMBERS:
$$(n + 1)^2 - n^2$$
$$= n^2 + 2n + 1 - n^2$$
$$= 2n + 1$$

CASE 2 — EVEN NUMBERS:
$$(n + 1)^2 - (n - 1)^2$$
$$= (n^2 + 2n + 1) - (n^2 - 2n + 1)$$
$$= 4n$$

ANSWER

The numbers which can be expressed as the difference of two squares are all numbers of the form $2n + 1$, and $4n$.

READ

EXAMPLE 4. Each edge of a cubical box has length 1 m. The box contains 9 spherical balls with the same radius r. The centre of one ball is at the centre of the cube and it touches the other 8 balls. Each of the other 8 balls touches 3 sides of the box, that is, the balls are tightly packed in the box. Find r.

PLAN

SOLUTION:
Since 3-dimensional situations are often difficult to visualize and work with, let us first try to find an analogous problem in 2 dimensions.

The analogue of a cube is a square and the analogue of a sphere is a circle. Thus, a similar problem in 2 dimensions is the following.

If 5 circles with the same radius r are contained in a square of side 1 m so that the circles touch each other and 4 of the circles touch 2 sides of the square, find r.

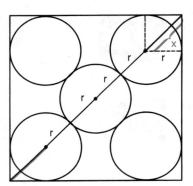

The diagonal of the square is $\sqrt{2}$.
The diagonal is also $4r + 2x$.
But x is the diagonal of a smaller square of side r.

$$\therefore x = \sqrt{2}r$$
$$\therefore \sqrt{2} = 4r + 2x$$
$$= 4r + 2\sqrt{2}r$$
$$= (4 + 2\sqrt{2})r$$
$$\therefore r = \frac{\sqrt{2}}{4 + 2\sqrt{2}}$$

SOLVE

Let us use these ideas to solve the original 3-dimensional problem.

The diagonal of the cube is $\sqrt{1^2 + 1^2 + 1^2} = \sqrt{3}$.
The diagonal of the cube is also $4r + 2x$, where x is the diagonal of a smaller cube with edge r.

$$\therefore x = \sqrt{r^2 + r^2 + r^2}$$
$$= \sqrt{3}r$$
$$\therefore \sqrt{3} = 4r + 2x$$
$$= 4r + 2\sqrt{3}r$$
$$= (4 + 2\sqrt{3})r$$
$$\therefore r = \frac{3}{4 + 2\sqrt{3}}$$
$$= \frac{2\sqrt{3} - 3}{2}$$

ANSWER

The radius of each ball is $\sqrt{3} - \frac{3}{2}$ m.

EXERCISE 1.13

B 1. The Diamond Cottages are connected to each other and to the central lodge through an intercom system. How many cottage to cottage, and cottage to central lodge connections are there if there are 15 cottages?

2. How many ways can you spell mathematics in the following pattern?

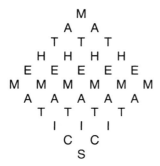

```
                M
             A     A
           T    T    T
         H    H    H    H
       E    E    E    E    E
     M    M    M    M    M    M
       A    A    A    A    A
         T    T    T    T
           I    I    I
             C     C
                S
```

3. The units digit of 2^5 when written in standard form is 2 ($2^5 = 32$). Find the units digit of 3^{99} when it is written in standard form.

4. What is the units digit of 7^{99} when it is written in standard form?

5. 3000 cubes each having a volume of 1 cm³ are arranged to form a rectangular block measuring 10 cm by 15 cm by 20 cm. The 15 cm by 20 cm base of the rectangular block sits on a table. The block is spray painted blue on all sides other than the base. How many of the 1 cm³ have exactly one face painted blue?

6. At the Calgary Stampede, Peter counted cowboys and horses. There were 35 heads and 102 legs. How many cowboys and how many horses were there?

7. Sally mailed out 25 photographs at a total cost of $14.14. It costs $0.42 to mail a plain photograph and $0.68 to mail a photograph with a frame. How many of each did Sally mail?

8. Mary buys a golf ball for $4.33 including tax. Her change from a five dollar bill is 10 coins consisting of pennies, nickels, and quarters. Did Mary get the correct change? Explain.

9. If 3093 digits are used to number the pages in a book, how many pages are in the book?

10. What is the units digit when 4567^{535} is written in standard form?

11. A clock has just struck 04:00. At what time in the next hour will the hands coincide?

12. The population of Elk Horn in 1976 was a perfect square. By 1978, the population had increased by 100 and was 1 more than a perfect square. With a further increase of 100 the population was again a perfect square in 1980. Find the population of Elk Horn in 1976.

13. The square formed by joining the centres has a side of 4. Find the area of the shaded region.

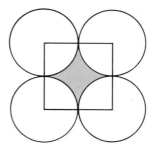

14. Today a pair of jeans cost $50. Due to inflation, the price increases at an annual rate of 6%? What will the cost be in 100 a?

(a) Make a guess.
(b) Find an estimate.
(c) Calculate the answer using fast exponentiation.

1.14 REVIEW EXERCISE

1. Express in decimal form.

(a) $\frac{3}{5}$ (b) $\frac{3}{4}$ (c) $\frac{3}{8}$

(d) $\frac{2}{7}$ (e) $\frac{2}{11}$ (f) $\frac{2}{15}$

(g) $\frac{7}{50}$ ✓   (h) $\frac{6}{7}$ (i) $\frac{8}{11}$ ✓

2. Express in the form $\frac{a}{b}$.

(a) 0.265   (b) 3.25   (c) 0.065
(d) 0.35   (e) $0.2\overline{56}$ ✓   (f) $0.2\overline{56}$
(g) $0.\overline{7}$ (h) $0.7\overline{75}$ (i) $0.\overline{27}$
(j) $0.2\overline{9}$ (k) $(0.\overline{36})(0.5)$
(l) $0.2 + 0.02 + 0.002 + \ldots$
(m) $0.45 + 0.0045 + 0.000\,045 + \ldots$ ✓

3. Express as entire radicals.

(a) $3\sqrt{5}$ (b) $x\sqrt{y}$ (c) $3x^2\sqrt{yz}$

(d) $-x^2y\sqrt[3]{z}$ (e) $2x\sqrt[4]{x}$ (f) $2x^2y\sqrt[4]{z}$ ✓

4. Express as mixed radicals.

(a) $\sqrt{40}$ (b) $\sqrt{x^2y^3}$ (c) $\sqrt{16x^4y^3}$

(d) $\sqrt{125xy^2z^3}$ (e) $\sqrt[3]{-27x^3y^2}$ ✓(f) $\sqrt[5]{32x^6}$

5. Simplify.

(a) $\dfrac{\sqrt{8}}{\sqrt{2}}$ (b) $\dfrac{\sqrt{24}}{\sqrt{3}}$ (c) $\dfrac{\sqrt{12x}}{\sqrt{3}}$

(d) $\dfrac{\sqrt[3]{81}}{\sqrt[3]{3}}$ ✓   (e) $\dfrac{\sqrt[3]{-54}}{\sqrt[3]{-2}}$ (f) $\dfrac{\sqrt{3}}{\sqrt{243}}$ ✓

6. Simplify.

(a) $5\sqrt{2} + 3\sqrt{2} - 2\sqrt{3} - 4\sqrt{3}$
(b) $3\sqrt{x} - 4\sqrt{y} - 5\sqrt{x} + 2\sqrt{y}$
(c) $3\sqrt{8} - 2\sqrt{32} + \sqrt{50}$ ✓
(d) $2\sqrt{3} - 4\sqrt{12} + 5\sqrt{27}$
(e) $3\sqrt{5} + 2\sqrt{50} - 5\sqrt{8} + \sqrt{45}$
(f) $\sqrt{24} - \sqrt{54} + \sqrt{12} - \sqrt{27}$ ✓

7. Simplify.

(a) $3\sqrt{5}(2 + \sqrt{10})$ (b) $2\sqrt{3}(\sqrt{6} - \sqrt{2})$
(c) $2\sqrt{3}(\sqrt{3} - \sqrt{5})$ ✓   (d) $2\sqrt{3}(\sqrt{3} - \sqrt{6})$
(e) $\sqrt{xy}(\sqrt{x} + \sqrt{y})$ (f) $3\sqrt{x}(2\sqrt{x} - \sqrt{y})$
(g) $3\sqrt{x}(\sqrt{x} + 2\sqrt{x})$ (h) $2\sqrt{6x}(\sqrt{6x} + 1)$ ✓

8. Graph the following for $x \in R$.

(a) $\{\,x \mid x \leqslant 5\,\}$
(b) $\{\,x \mid -2 \leqslant x \leqslant 4\,\}$
(c) $\{\,x \mid -2 \leqslant x < 5\,\}$ ✓
(d) $\{\,x \mid x > -3\,\} \cap \{\,x \mid x \leqslant 5\,\}$
(e) $\{\,x \mid x < 0\,\} \cup \{\,x \mid x \geqslant 3\,\}$
(f) $\{\,x \mid -1 < x \leqslant 2\,\} \cup \{\,x \mid 2 < x \leqslant 4\,\}$ ✓

9. Name the property that justifies each statement.

(a) $x + y \in R$ (b) $ab = ba$
(c) $xy \in R$ (d) $(a + b)x = x(a + b)$ ✓
(e) $-x + x = 0$ (f) $x + 1 = 1 + x$
(g) $y \times 1 = y$ (h) $4x + 2 = 2(2x + 1)$ ✓

10. State the condition on x, $x < 0$, $x = 0$, or $x > 0$ for each of the following to be true statements.

(a) $y = y - x$ (b) $2x < x$ ✓   (c) $5x = 0$ ✓
(d) $2 + x = 2$ (e) $-x > 0$ (f) $-3x < x$

11. State the following numbers using powers.

(a) 32   (b) 25   (c) 49
(d) 125   (e) 243 ✓   (f) 128 ✓
(g) 9×27 (h) 5×625 (i) 27×3

12. Evaluate for $x = 2$, and $y = -3$.

(a) x^2y^2 (b) $3x^3$ (c) $2y^3$
(d) $-2x^3$ (e) x^3y^3 ✓   (f) $2y^3$
(g) $(x + y)^3$ (h) $(x + y)^4$ (i) y^x ✓

13. Simplify the following expressions.

(a) $(2x^5)^2$ (b) $(2x^5)(-x^4)$
(c) $((x^2)^2)^2$ ✓   (d) $(3x^2)^3$
(e) $a^xa^ya^z$ (f) $((a^x)^y)^z$
(g) $(-2x^2)^3$ (h) $(3x^4)(-4x^3)$ ✓
(i) $\dfrac{27x^4}{9x^2}$ (j) $\dfrac{(3x^2)^3}{9x^3}$

14. Simplify the following.

(a) $\dfrac{5^5 \times 5^2}{5^4}$ ✓   (b) $\dfrac{x^{a+b} \times x^{2a}}{x^b}$

(c) $\dfrac{3^{n+1} \times 9^{2n}}{27^n}$ ✓   (d) $\dfrac{81^7}{243^4}$

(e) $x^n(x^{2n+3}) \div x^{2n}$ (f) $(x^{n+1})^2$

(g) $\dfrac{(2^5)^n \times (16^n)^3}{8^{n-2}}$ (h) $\dfrac{3^{2n} \times 9^{1-n} \times 27^{2n}}{81^{2n+3}}$

15. Evaluate.

(a) 5^0　　　　(b) -3^0　　　　(c) 5^{-1}

(d) -1^{-1}　　(e) $(-1)^{-1}$　(f) 10^{-2}

16. Evaluate.

(a) $\dfrac{5^{-1} \times 5^0}{5^{-2}}$

(b) $\dfrac{2^{-1} + 2^2}{2^2 - 2^{-1}}$

(c) $\left(\dfrac{2^{-1} + 3^{-1}}{5}\right)^{-1}$

(d) $\left(\dfrac{7^{-1} + 7^{-2}}{7^0 + 7^{-1}}\right)^0$

(e) $\dfrac{2^{-1} + 3^{-1}}{2 + 3}$

(f) $\dfrac{a^{-1} + b^{-1}}{a + b}$

17. Find the amount of each of the following investments.

(a) $3000 at 1% per month for 6 months.
(b) $2500 at 8% per annum for 2 a, compounded semi-annually.
(c) $5000 at 12% per annum for 1 a, compounded monthly.
(d) $10 000 at 10% per annum for 2 a, compounded semi-annually.
(e) $1000 at 10.5% per annum for 3 a, compounded annually.
(f) $5000 at 12% per annum for 3 a, compounded semi-annually.

18. Find the amount of $1000 invested at 12% per annum for 3 a if the interest is compounded as follows.

(a) annually
(b) quarterly ✓
(c) semi-annually
(d) monthly

19. Find the present value of each of the following amounts.

(a) $1000 in 3 a at 8%, compounded annually.
(b) $1500 in 5 a at 10% per annum, compounded semi-annually.
(c) $3500 in 4 a at 12% per annum, compounded quarterly.
(d) $5000 in 6 a at 9% per annum, compounded semi-annually.
(e) $6500 in 3 a at 12% per annum, compounded semi-annually.

20. How much should David invest today at 10% per annum compounded semi-annually ✓ in order to have $15 000 for the purchase of a new car in 3 a?

21. How much should have been invested on the day you were born if you are to have $16 000 on your eighteenth birthday and the funds were invested at 14% per annum compounded semi-annually?

22. State the following using radicals.

(a) $5^{\frac{1}{2}}$　　(b) $34^{\frac{1}{3}}$　　(c) $6^{\frac{1}{5}}$ ✓

23. Write the following using exponents.

(a) $\sqrt{7}$　　(b) $\sqrt{a^3}$　　(c) $\sqrt[5]{a^2}$

24. Evaluate.

(a) $\sqrt[3]{125}$　　(b) $243^{\frac{2}{5}}$　　(c) $25^{0.5}$

(d) $625^{\frac{3}{4}}$　　(e) $\left(\dfrac{125}{8}\right)^{-\frac{1}{3}}$　　(f) $(-32)^{-\frac{1}{5}}$

25. Solve the following equations for $x \in R$.

(a) $5^{2+x} = 1$　　　　(b) $4^{2x+1} = 8$ ✓
(c) $2(3^{x+2}) = 18$　　(d) $4^{x-2} + 1 = 5$ ✓

26. The sum of the squares of two consecutive integers is 113. Find the integers.

27. The sum of the squares of two consecutive odd numbers is 74. Find the numbers.

28. Find two different whole numbers x and y so that

$$x^y = y^x$$ ✓

29. Find the sum of the following series of numbers.

(a) $1 + 2 + 3 + \ldots + 97 + 98 + 99$
(b) $2 + 4 + 6 + \ldots + 96 + 98 + 100$
(c) $1 + 3 + 5 + \ldots + 95 + 97 + 99$
(d) $1 + 2 + 3 + \ldots + 99 + 100 + 99$
$\quad + 98 + \ldots + 3 + 2 + 1$

30. An automobile tire is guaranteed for a total of 60 000 km of normal driving. During this driving period, it was found that the tire tread had worn down 1.15 cm. The diameter of the tire is 63 cm.

(a) Find the average amount of tread wear for each revolution of the tire.
(b) Find the amount of tread wear for each kilometre driven. ✓

1. (a) Express $\frac{3}{11}$ as a decimal fraction.

(b) Express $0.\overline{63}$ as a common fraction in simplest form.

2. (a) Express $\sqrt{288}$ as a mixed radical.

(b) Express $7\sqrt{5}$ as an entire radical.

(c) Simplify. $2\sqrt{3} + 3\sqrt{12} - 5\sqrt{20} + \sqrt{45}$

3. Graph the following sets on an appropriate number line.

(a) $\{\, x \in N \mid x \leqslant 5 \,\}$

(b) $\{\, x \in I \mid -3 < x \leqslant 3 \,\}$

(c) $\{\, x \in R \mid -2 < x < 3 \,\}$

(d) $\{\, x \in R \mid -3 \leqslant x < 1 \,\} \cup \{\, x \in R \mid 1 \leqslant x < 4 \,\}$

4. (a) Evaluate $3x^2$ for $x = 5$.

(b) Evaluate $-2x^2y^3$ for $x = -2$, $y = 1$.

(c) Simplify. $-2x^4 + (-2x)^4$

(d) Simplify. $\dfrac{16^n \times 8^{1-n} \times 2^{3n+2}}{4^{3-2n} \times 32^n}$

(e) Simplify. $(-3x^3y^4)(2x^2y^2)^2$

(f) Simplify. $\dfrac{3^{-2} + 2^{-3}}{3^{-1} + 2^{-1}}$

(g) Simplify. $2^{\frac{1}{2}} \times 2^{\frac{1}{3}}$

5. Solve. $16^{2x-2} = 32$

6. Sam needs to buy some wire. He needs between 15 and 20 pieces that are 7 m long, and one piece that is 80 m long. The wire is purchased in multiples of 12 m. How much wire should he buy to eliminate as much waste as possible?

7. How much money should Lori invest today at 12% per annum compounded semi-annually in order to have $10 000 in 3 a?

8. How much money will Marlene have in 3 a if she invests $5000 at 10% per annum compounded annually?

ALGEBRA

CHAPTER

Mathematics is the gate and key of the sciences.
Roger Bacon

THE DISTRIBUTIVE PROPERTY

$a(b + c) = ab + ac$

EXERCISE

1. Expand.
(a) $2(x + 6)$ (b) $3(x + 4)$
(c) $4(x - 5)$ (d) $7(5 - m)$
(e) $(4 - 3x)4$ (f) $(x - y)5$
(g) $-2(2x - 7)$ (h) $-7(3m - 4n)$
(i) $-(3m - 7)$ (j) $3(2x^2 - 4x + 5)$
(k) $-3(2t^2 - 5t + 4)$
(l) $-(3x^2 - 4xy + y^2)$
(m) $2x(x - 7)$ (n) $5x(x^2 - 3x - 4)$
(o) $-m(1 - 3m - m^2)$
(p) $3m(m - n)$

2. Expand.
(a) $2xy(3x - 4y)$
(b) $-2xy(x^2 - xy + y^2)$
(c) $3t^3(t^2 - 2t - 4)$ (d) $2\pi r(w - x)$
(e) $-\pi(3m - 2n)$ (f) $0.4(3x - 5y)$
(g) $-0.2(3x^2 - 4x)$ (h) $-1.4(10x - y)$
(i) $0.1x(3m - 7n)$ (j) $10t(0.3t - 0.7)$
(k) $-10x^2(4.8x - 7.3y)$
(l) $0.01(0.4m - 0.5)$
(m) $-0.3m(0.1m^2 - 0.2m)$
(n) $1.5x^2y(20x^2y - 5xy^2)$
(o) $4mnt(3m^2 - 4n^2 + 5t^2)$
(p) $-0.5x(8x - 7y)$

3. Expand and simplify.
(a) $2(x + 4) + 3(x + 8)$
(b) $3(2x + 7) + 4(5x + 8)$
(c) $7(3w - 4) + 5(5w - 1)$
(d) $6(m + 4) + 2(m - 5) + 4(3m - 9)$
(e) $2(4x - 3) - 3(2x - 1)$
(f) $5(3t - 4) - 2(t + 7) - (t - 6)$
(g) $4(3x^2 - 2x + 1) - 5(x^2 - x - 1)$
(h) $4(x^2 - 3x + 4) - 3(x^2 - 7x + 6)$
(i) $3(x - y) - 2(2x + 5y) - (x - 3y)$
(j) $2(x^2 - 4xy + y^2) + 6(x^2 - 2xy - y^2)$

(k) $5(x^2 - 8x - 4) - 3(2x - 7)$
(l) $(1 - 3t + 4t^2) - (8 + 5t - 6t^2)$
(m) $4(3m - n) - (4m - 7n + 6)$

4. If $w = 2$, $x = -1$, and $y = -2$, evaluate.
(a) $3wx + 2wy + 4xy$
(b) $w^2 + x^2 + y^2$
(c) $2wx^2 - w^2y^2 - x$
(d) $3w^2x^3 - w^3x - 2wxy$
(e) $2(w - x) - (x - y) - 2(x - w)$
(f) $3(1 - 2x) - 2(2w - 1) - 3(2y - 3)$
(g) $2w(w - x) - 3x(w + x) + 6wx$
(h) $0.1(w - 2x) + 0.2(2x - y)$
     $- 0.1(2y - w)$
(i) $2x(2x - 1) - x(1 - 3x)$
     $+ 4x(x - 1) + x^2$

FUNCTION NOTATION

$f(x) = 2x^2 + 3x - 5$
$f(3) = 2(3^2) + 3(3) - 5$
$\quad\ \ = 18 + 9 - 5$
$\quad\ \ = 22$

EXERCISE

1. If $f(x) = x + 4$, evaluate.
(a) $f(1)$ (b) $f(2)$ (c) $f(8)$
(d) $f(20)$ (e) $f(0)$ (f) $f(-4)$

2. If $g(x) = 3x - 2$, evaluate.
(a) $g(2)$ (b) $g(7)$ (c) $g(0)$
(d) $g(-5)$ (e) $g(14)$ (f) $g(-15)$

3. If $h(x) = x^2 + 6x - 24$, evaluate.
(a) $h(2)$ (b) $h(-3)$ (c) $h(4)$
(d) $h(0)$ (e) $h(-6)$ (f) $h(10)$

4. If $f(x) = x^3 - 2x^2 - 5x + 6$, evaluate.
(a) $f(1)$ (b) $f(2)$ (c) $f(0)$
(d) $f(-2)$ (e) $f(3)$ (f) $f(-3)$

MONOMIALS

$$a^m \times a^n = a^{m+n} \qquad a^m \div a^n = a^{m-n}$$

EXERCISE

1. If $x = 3$, evaluate.
(a) $3x^3$ (b) $5x$ (c) x^2
(d) $-2x^2$ (e) $-6x$ (f) $-3x^3$
(g) $(-x)^2$ (h) $-x^2$ (i) 2^x

2. If $x = -2$, evaluate.
(a) $3x^2$ (b) $4x^3$ (c) $-2x$
(d) $-x^2$ (e) $(-x)^2$ (f) $-2x^2$
(g) $5x^1$ (h) $2x^0$ (i) 3^x

3. Multiply.
(a) $x(4x)$ (b) $(3x)(5x)$
(c) $(-2x)(-4x)$ (d) $(-5x)(6x)$
(e) $(2x^2)(4x)$ (f) $(3x^4)(2x^3)$
(g) $(-2x^2)(4x^2)$ (h) $(5x^4)(2x^3)$
(i) $(3x^5)(2x^2)$ (j) $(x^3)(-5x^2)$
(k) $(-3x^3)(-2x^2)$ (l) $(0.5x^2)(-4x^5)$

4. Divide, $x \neq 0$.
(a) $6x \div 3$ (b) $5x \div x$
(c) $x^5 \div x^2$ (d) $x^4 \div x^4$
(e) $8x^3 \div 2x^2$ (f) $(12x^6) \div (-3x^3)$
(g) $-15x^5 \div (5x^3)$ (h) $(21x^7) \div (-3x^3)$
(i) $-18x^6 \div (-3x^3)$ (j) $(30x^7) \div (-5x^5)$
(k) $36x^9 \div 9x^4$ (l) $(36x^9) \div 4x^9$

5. Simplify.
(a) $(3x^4)(5x^2)$ (b) $8x^4 \div 2x^3$
(c) $24x^5 \div 6x^3$ (d) $(-8x^6)(-4x^3)$
(e) $(3x^4)(-2x^7)$ (f) $(-27x^3)(-3x^3)$
(g) $-18x^4 \div 6x$ (h) $28x^5 \div (-4x^4)$
(i) $(16x^6)(4x^4)$ (j) $-16x^6 \div 4x^4$
(k) $32x^4 \div 2x^4$ (l) $(-5x^3)(-5x^3)$

6. Evaluate for the given values of the variables, $x = 2$, $y = -3$, and $z = 5$.
(a) xyz (b) $x^2 + y^2 + z^2$
(c) $(x + y + z)^2$ (d) $(5xy)(-3yz)$
(e) $-(3x^2y)(4x^2yz)$
(f) $48x^4y^3z^2 \div 12x^2y^2z$
(g) $(-10xy^4z^5) \div (-5x^2y^3z)$

Calculator keys are divided into three groups.

Number Keys	Operation Keys	Special Function Keys
7 8 9 4 5 6 1 2 3 0	× ÷ + −	M+ M− MR MC % C CE + − ·

In the following exercise, we check your calculator for the constant functions using the ▇ key.

EXERCISE

Press the keys as indicated and note the patterns.
1. Addition.
(a) $2 + 3 = = = =$
(b) $4 + 3 = = = =$
(c) $1 + 5 = = = =$

2. Subtraction.
(a) $30 - 5 = = = =$
(b) $23 - 2 = = = =$
(c) $45 - 5 = = = =$

3. Multiplication.
(a) $2 \times 3 = = = =$
(b) $5 \times 3 = = = =$
(c) $3 \times 2 = = = =$

4. Division.
(a) $81 \div 3 = = = =$
(b) $125 \div 5 = = = = =$
(c) $1024 \div 2 = = = = =$

5. Start with 2 and make your calculator count by 3.

6. Start with 81 and make your calculator count backwards by 10.

7. Start with 5 and multiply by 3 four times.

8. Divide 16 384 by 2 five times.

2.1 POLYNOMIALS

A term is a mathematical expression using numbers and variables to indicate a product. Examples of terms are

$$3x, 5xy, -7xyz, x^2, 5x^3.$$

The numerical part of the term is called the numerical coefficient. A variable is a symbol which may represent any member of a particular set. The degree of a term is the sum of the exponents of its variables.

$5x^3$ is a term of degree 3

$-2x^4y^2$ is a term of degree 6

A polynomial is a monomial, or a sum of monomials. A polynomial such as $3x^2 + (-2x) + (-5)$ is written $3x^2 - 2x - 5$. We classify polynomials by the number of terms.

$6x^2$:	one term :	monomial
$2x + 3y$:	two terms :	binomial
$4x^2 + 3x - 7$:	three terms :	trinomial

The degree of a polynomial is the greatest degree of its terms after it has been simplified.

$3x + 7y$ is a first degree polynomial, or a linear polynomial.

$4y^2 - 2y + 3$ is a second degree polynomial, or a quadratic polynomial.

$3x^3 - 12$ is a third degree polynomial, or cubic polynomial.

> A polynomial is an expression of the form
> $$a_nx^n + a_{n-1}x^{n-1} + \ldots + a_1x^1 + a_0$$
> where n is a non-negative integer, x is a variable, and a_i (i = n, n - 1, ... , 1, 0) is a real number.

Monomials that differ only in their numerical coefficients are called similar, or like. For example, $5x^2y^5$ and $-3x^2y^5$ are similar and are called like terms.

In order to facilitate our work, we can write P(x) in place of a polynomial in x. When we write P(2), we mean the value of the polynomial when x = 2.

EXAMPLE 1. State the degree of the polynomial $3x^2y^4 - 2x^5 + x^4z^3$.

SOLUTION:
The degrees of the terms in order are

$$\overset{6}{3x^2y^4} - \overset{5}{2x^5} + \overset{7}{x^4z^3}$$

The degree of the polynomial is 7.

EXAMPLE 2. $P(x) = 3x^4 - 2x^2 + 7x - 8$, find
(a) P(2)
(b) P(-3)
(c) P(a)

SOLUTION:

$$P(x) = 3x^4 - 2x^2 + 7x - 8$$

(a) $\begin{aligned} P(2) &= 3(2^4) - 2(2^2) + 7(2) - 8 \\ &= 3(16) - 2(4) + 14 - 8 \\ &= 48 - 8 + 14 - 8 \\ &= 46 \end{aligned}$

(b) $\begin{aligned} P(-3) &= 3(-3)^4 - 2(-3)^2 + 7(-3) - 8 \\ &= 3(81) - 2(9) - 21 - 8 \\ &= 243 - 18 - 21 - 8 \\ &= 196 \end{aligned}$

(c) $P(a) = 3a^4 - 2a^2 + 7a - 8$

EXERCISE 2.1

A 1. Classify each of the following as either monomial, binomial, or trinomial, and state the degree.

(a) $5x^3$ (b) $-8x^2y^4$
(c) $7xy$ (d) $4x^2 + 3$
(e) $x + y + z$ (f) $3x^2 - 5x + 3$
(g) $3x^7 - 2x^2y + 5$ (h) $-16m^2n^2t$
(i) $7xy^3 - 3x^4$ (j) $3x^2 + 5x - 7$

2. Arrange the terms from lowest to highest degree.

(a) $16x^3 - 5x^5 + 4x^2 - 3x$
(b) $17x^2 + 15x^4 - 3x^3 + 5x$
(c) $5x^2y^3 - 4x^3y^3 + 5x^3y$
(d) $8x^3y - 4x^2y^2 + 2x^2y - 3x^5y^2$
(e) $4x^3y - 5x^6 + 11x^5y^2 - x^5$

3. Each of the following expressions represents the perimeter of a geometric figure. Name the figure and state the degree of the expression.

(a) $2\ell + 2w$ (b) $a + b + c$
(c) $4x$ (d) $2\pi r$
(e) $a + b + c + d$ (f) πd

4. Each of the following expressions represents the area of a geometric figure. Name the figure and state the degree of the expression.

(a) ℓw (b) πr^2 (c) $\frac{1}{2}bh$

(d) $\frac{1}{2}(a + b)h$ (e) $\pi R^2 - \pi r^2$

5. Each of the following expressions represents the volume of a geometric figure. Name the figure and state the degree of the polynomial.

(a) ℓwh (b) $\pi r^2 h$ (c) $\frac{4}{3}\pi r^3$

B 6. If $w = 2$, $x = 3$, and $y = -1$, evaluate.
(a) $3x + 4y + 2w$ (b) $6wxy$
(c) $2xy + 3wx$ (d) $2x - 3y - 2w$
(e) $w - x - y$ (f) $x^2 + y^2 + w^2$
(g) $2w^2 - 3x^2 - 4y^2$ (h) $x^2 - 2xy + y^2$
(i) $2x^2y - y^2 - w^2$ (j) $w^2 - x^2 - y^2$
(k) $w^3 - y^3 - x^2$ (l) $3y^5 - 13$

7. $P(x) = 3x^2 + 2x + 5$, find
(a) $P(2)$ (b) $P(4)$ (c) $P(8)$

8. $P(x) = -2x^3 + 3x - 4$, find
(a) $P(2)$ (b) $P(5)$ (c) $P(4)$

9. $P(x) = 3(x + 2)$, find
(a) $P(2)$ (b) $P(-2)$ (c) $P(0)$

10. $P(x) = 2x^3 + 3x^2 - 5x + 2$, find
(a) $P(3)$ (b) $P(-2)$ (c) $P(1)$

11. $P(x) = -3x^4 - 4$, find
(a) $P(3)$ (b) $P(-2)$ (c) $P(2)$

12. $P(x) = x^3 - 2x^2 + 1$, find
(a) $P(1)$ (b) $P(0)$ (c) $P(5)$
(d) $P(-1)$ (e) $P(3)$ (f) $P(6)$

13. $P(x) = 4x^3 - x + 3$, find
(a) $P(0)$ (b) $P(1)$ (c) $P(2)$
(d) $P(-1)$ (e) $P(-2)$ (f) $P(3)$

14. $P(x) = x^3 - 2x^2 + 3x - 5$
(a) Find $P(2)$.   (b) Find $P(3)$.
(c) Find $P(5)$.   (d) Find $P(6)$.
(e) Does $P(2) + P(3)$ equal $P(5)$?
(f) Does $P(2) \times P(3)$ equal $P(6)$?

15. $P(x) = 5x^2 - 3x + 4$
(a) Find $P(0)$.   (b) Find $P(1)$.
(c) Does $P(0)$ equal $P(1)$?
(d) Does $0 \times P(0)$ equal $P(0)$?
(e) Does $1 \times P(0)$ equal 1?
(f) Replace ■ by $<$, $=$, or $>$.
     $P(0) + P(1)$ ■ $P(1)$

16. $P(x) = 3x^2 - 5x + 2$
(a) Find $P(1)$.   (b) Find $P(\frac{2}{3})$.
(c) Does $P(1)$ equal $P(\frac{2}{3})$?
(d) Replace ■ by $<$, $=$, or $>$.
     $P(1) + P(\frac{2}{3})$ ■ $P(\frac{5}{3})$

17. $P(x) = x^3 + 3x^2 + 3x + 1$
     $Q(x) = (x + 1)^2$
Show that $P(x) = (x + 1)Q(x)$ for the following values of x.
(a) $x = 0$ (b) $x = 1$ (c) $x = -2$
(d) $x = -1$ (e) $x = 2$ (f) $x = 5$

18. $P(x) = x^3 - 8$
Show also that $P(x) = (x - 2)(x^2 + 2x + 4)$ for the following values of x.
(a) $x = 3$ (b) $x = 2$ (c) $x = 4$ (d) $x = 6$

2.2 ADDING AND SUBTRACTING POLYNOMIALS

Terms such as 3xy and 7xy that have the same variable factors are similar, and are called like terms. We use the distributive property to simplify polynomials containing like terms.

$$5a^2 + 3a^2 = (5 + 3)a^2$$
$$= 8a^2$$

EXAMPLE 1. Simplify.

(a) $5x + 7x - 4x$ (b) $3x^2 + 5x - x^2 + 4x + 3$ (c) $2x - 5xy + 3x + 4xy - 6y$

SOLUTION:

(a) $5x + 7x - 4x$
$= (5 + 7 - 4)x$
$= 8x$

(b) $3x^2 + 5x - x^2 + 4x + 3$
$= (3 - 1)x^2 + (5 + 4)x + 3$
$= 2x^2 + 9x + 3$

(c) $2x - 5xy + 3x + 4xy - 6y$
$= (2 + 3)x + (-5 + 4)xy - 6y$
$= 5x - xy - 6y$

EXAMPLE 2. Evaluate for $x = -2$. $3x^3 - 4x^2 + 2x - x^2 + 6x^2 - 5x + 7$

SOLUTION:

First simplify the polynomial.

$3x^3 - 4x^2 + 2x - x^2 + 6x^2 - 5x + 7$
$= 3x^3 + (-4 - 1 + 6)x^2 + (2 - 5)x + 7$
$= 3x^3 + x^2 - 3x + 7$

Then, solve for $x = -2$.

$3x^3 + x^2 - 3x + 7$
$= 3(-2)^3 + (-2)^2 - 3(-2) + 7$
$= 3(-8) + (4) - 3(-2) + 7$
$= -24 + 4 + 6 + 7$
$= -7$

EXAMPLE 3. Add.

(a) $2x^3 - 7x^2 - 6$ to $3x^3 + 4x^2 - 5x + 3$
(b) $5xy - 3yz - 4xz$ to $3xy - 2yz + 6xz$

SOLUTION:

We can add these polynomials using either a vertical or a horizontal arrangement as shown in (a).

(a) Vertical Arrangement

We write the polynomials with the terms of the same degree in the same columns. Note the space that is left for the x-term.

$$\begin{array}{r} 3x^3 + 4x^2 - 5x + 3 \\ 2x^3 - 7x^2 \qquad - 6 \\ \hline - 3 \end{array}$$

$$\begin{array}{r} 3x^3 + 4x^2 - 5x + 3 \\ 2x^3 - 7x^2 \qquad + 3 \\ \hline 5x^3 - 3x^2 - 5x - 3 \end{array}$$

Horizontal Arrangement

Using brackets.

$(3x^3 + 4x^2 - 5x + 3) + (2x^3 - 7x^2 - 6)$
$= 3x^3 + 4x^2 - 5x + 3 + 2x^3 - 7x^2 - 6$
$= (3 + 2)x^3 + (4 - 7)x^2 + (-5 + 0)x + (3 - 6)$
$= 5x^3 - 3x^2 - 5x - 3$

(b) Vertical Arrangement

$$\begin{array}{r} 3xy - 2yz + 6xz \\ 5xy - 3yz - 4xz \\ \hline + 2xz \end{array}$$

$$\begin{array}{r} 3xy - 2yz + 6xz \\ 5xy - 3yz - 4xz \\ \hline 8xy - 5yz + 2xz \end{array}$$

EXAMPLE 4. Subtract.
(a) $5x^3 - 2x^2 + 4x - 3$ from $7x^3 + 3x^2 - 4x + 3$
(b) $5xy + 4yz - 3xz$ from $2xy + yz - 4xz$

SOLUTION:
These polynomials can be subtracted by writing them in either a vertical or horizontal arrangement as shown in (a).

(a) Vertical Arrangement

$$7x^3 + 3x^2 - 4x + 3$$
$$\underline{5x^3 - 2x^2 + 4x - 3}$$
$$+ 6$$

$$7x^3 + 3x^2 - 4x + 3$$
$$\underline{5x^3 - 2x^2 + 4x - 3}$$
$$2x^3 + 5x^2 - 8x + 6$$

Horizontal Arrangement
Using brackets.

$$(7x^3 + 3x^2 - 4x + 3) - (5x^3 - 2x^2 + 4x - 3)$$
$$= 7x^3 + 3x^2 - 4x + 3 - 5x^3 + 2x^2 - 4x + 3$$
$$= (7 - 5)x^3 + (3 + 2)x^2 + (-4 - 4)x + (3 + 3)$$
$$= 2x^3 + 5x^2 - 8x + 6$$

(b) Vertical Arrangement

$$2xy + yz - 4xz$$
$$\underline{5xy + 4yz - 3xz}$$
$$- xz$$

$$2xy + yz - 4xz$$
$$\underline{5xy + 4yz - 3xz}$$
$$-3xy - 3yz - xz$$

EXAMPLE 5. Given $P(x) = (3x^2 - 5x) + (3x + 7) - (x^2 - 2)$, find.
(a) $P(2)$ (b) $P(-3)$

SOLUTION:
In this example, it is more efficient to simplify the polynomial before substituting for the variable.

$$P(x) = (3x^2 - 5x) + (3x + 7) - (x^2 - 2)$$
$$= 3x^2 - 5x + 3x + 7 - x^2 + 2$$
$$= 2x^2 - 2x + 9$$

(a) $P(2) = 2(2)^2 - 2(2) + 9$
$\quad\quad = 8 - 4 + 9$
$\quad\quad = 13$

(b) $P(-3) = 2(-3)^2 - 2(-3) + 9$
$\quad\quad\quad = 18 + 6 + 9$
$\quad\quad\quad = 33$

> To subtract, mentally change the sign and add.

EXERCISE 2.2

B 1. Add.
(a) $2x + 3y - 6z$
$\quad 4x - 5y + 2z$

(b) $\quad 3x^2 - 4x + 7$
$\quad -5x^2 + 6x - 8$

(c) $-5xy + 3xz$
$\quad\quad 3xy - 4xz - yz$
$\quad\quad\quad\quad\quad 3xz - yz$

(d) $-2x^3 - 3x^2 + 7x$
$\quad -6x^3 + 4x^2 - 7x$
$\quad\quad 4x^3 - 2x^2 + 2x$

(e) $\quad 2x + 3y - z + 2$
$\quad -5x + 4y - 3z$
$\quad\quad\quad\quad - 6y + z - 5$
$\quad\quad 4x - 5y \quad\quad - 5$

(f) $\quad 6x^2 - 5x$
$\quad\quad 3x^2 + 2x + 7$
$\quad -4x^2 - 3x - 6$
$\quad\quad 5x^2 + x - 2$

2. Add the sum of $3x^2 - 5x + 3$ and $-2x^2 + 6$ to $-x^2 - x - 1$.

3. Add $5x + 3y - 7$ to the sum of $-2x + 3y - 2$ and $x + y + 4$.

4. Subtract.

(a) $8x - 5y + 2$
$6x + 3y - 3$

(b) $3x^2 - 4x + 5$
$x^2 + 3x - 2$

(c) $5x + 2y - 3z$
$-2x - 3y + 2z$

(d) $2x^2 + 3x - 7$
$-2x^2 - 3x + 7$

(e) $6x - 2y + 3$
$-4x + 3y - 4$

(f) $3x^2 + 2x$
$x^2 - 5x + 3$

5. Subtract $3x^2 + 2x - 5$ from the sum of $2x^2 - 3x + 2$ and $4x^2 + 6x + 3$.

6. Subtract the sum of $x^2 + 2x - 4$ and $4x^2 - 5x + 3$ from $3x^2 + 4x - 1$.

7. Simplify.

(a) $3w - 4x + 7y + 4w - 3x + 8y$
(b) $2xy - 3yz + 4xz + 6xy + 12yz - 6xz$
(c) $5r - 4s + 11t + 6r - 7s - 14t$
(d) $4x^2 - 3x + 7 + 2x - 3x^2 + 6 + 4x$
(e) $m - 3m^2 - 2 + 4m^2 - 5 + 3m + 6m^2$
(f) $11xy - 4yz + 6x + 5x - 5yz + 11xy$
(g) $3x^2 - 4x + 7 - 2x^2 + 5x - 3$
(h) $1.4u - 3.2v + 6.1w + 5.2u - 8v + w$
(i) $5.3x^2 - 4.2x + 3.1 + 0.1x^2 + 3x$

8. Add.

(a) $(3x^2 + 5x - 7) + (2x^2 - 4x + 6)$
(b) $(-4x^2 + 2x - 6) + (3x^2 - 6x + 7)$
(c) $(3xy - 2x + 7) + (6xy + 5x - 3)$
(d) $(5x + 3y - 8xy) + (2x - 5y + 6xy)$
(e) $(3x^2 - 4x) + (2x^2 - 5x + 3) + 6$

9. Subtract.

(a) $(3x + 4y + 5z) - (x + y + z)$
(b) $(x - 3x^2 + 6) - (2x^2 - x - 4)$
(c) $(5x + 4y + 6z) - (3x - 4y + 2z)$
(d) $(6u - 5v + w) - (4u - 3v + 2w)$
(e) $(5m - 3n) - (2m - 7n + 4)$
(f) $8x - (7x - 3y + 4z)$
(g) $(4x - 3y) - (4x - 6xy + y)$
(h) $(4.7x - 2.7y) - (5.2x - 3.1y)$
(i) $(2.3x - 6.2y - 0.4z) - (7.1x - 4y)$
(j) $(6.1x^2 - 4.3x + 5) - (4.3x^2 + 5.9x)$

10. Evaluate P(3) for each of the following.

(a) $P(x) = (2x + 3) + (5x - 7) - (4x + 6)$
(b) $P(x) = 6x^2 + 5x - 7 + 2x^2 + 6x - 3$
(c) $P(x) = (4x - 3x^2 + 2) - (2x^2 + 5x + 3)$
(d) $P(x) = (3x^2 + 6x - 2) + (-5x^2 + 6x - 7)$
(e) $P(x) = (5 - 3x + 2x^2) + (4x - 6 + 5x^2)$

11. Evaluate for $x = 2$, $y = -3$, and $z = 4$.

(a) $2x - 3y + 4z - 6x + 2y - z$
(b) $x^2 + y^2 - z^2 + 2xyz$
(c) $(3x - 2y) + (z - 5x) - (2y - 3z)$
(d) $(4xy + 2yz - zx) - (-2yz + 4yz)$
(e) $(3x^2 - 2xy) - (y^2 + 4yz) + (z^2 + 1)$

12. Find $P(-2)$ for each of the following.

(a) $P(x) = (2x^2 - 5x + 3) + (x^2 + 3x - 5)$
(b) $P(x) = (x^2 - 3x - 4) - (3x^2 + x - 1)$
(c) $P(x) = (2x^2 - 3x) + (4x - 2) - (3 - x^2)$
(d) $P(x) = (1 - x + x^2) - (3x - 2) + 3x^2$

13. Simplify.

(a) $3x^2 - (5x^2 - 2x + 3) + 7x - 5$
(b) $(3x - 2y + 3) - (-4x + 3y - 5)$
(c) $(3x - 5) + (2x + 8) - (5x - 3)$
(d) $(4x^2 + 3x - 2) - (2x^2 - 5x + 4)$
(e) $(w - x) + (y - 3w) - (2x + 3y)$

14. Subtract $4u - 3v + 2w$ from the sum of $4u - 3w + 2v$ and $6w - 3u + v$.

15. By how much does $4x^2 - 3x + 6$ exceed $3x - 2 + 3x^2$?

16. What must be added to $3x^2 - 2xy + y^2$ to give $7xy - 3y^2 + 4x^2$?

17. What polynomial decreased by $4xy - 3xz + 2yz$ equals $5yz - 3xy + 2xz$?

18. Subtract the sum of $3x^2 - 2x + 7$ and $4x^2 - x - 12$ from the sum of $5x + 2x^2 - 6$ and $3x^2 - 5$.

19. By how much does $3m - 4mn + 5n$ exceed $6m$?

20. Find the perimeter of each figure in terms of x.

(a)

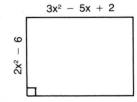

$3x^2 - 5x + 2$

$2x^2 - 6$

(b)

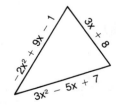

$3x - 1$

$-2x^2 + 9x$

$3x + 8$

$3x^2 - 5x + 7$

2.3 PROGRAMMING IN BASIC

A program is a set of instructions that a computer carries out in a definite order. To program a computer, we use a language that the computer understands called BASIC. BASIC is an acronym for Beginner's All-purpose Symbolic Instructional Code. The following chart gives the BASIC notation for the operations we use in mathematics.

Operation	Mathematics	BASIC
Addition	$+$	$+$
Subtraction	$-$	$-$
Multiplication	$\times$	$*$
Division	$\div$	$/$
Exponentiation		$\uparrow$ or $\wedge$
$3x^2 - 5x + 3$ is written $3*x\uparrow 2 - 5*x + 3$		

Mathematical statements are combined with system commands that have a special meaning to the computer. The following are some of the more useful system commands.

NEW : The NEW command clears the memory to accept a new program.
RUN : The RUN command tells the computer to run the program.
LIST : The LIST command tells the computer to print in order a list of all the statements in the program.
PRINT : The PRINT statement causes the computer to print what follows.
END : The END statement tells the computer that the program is finished.
INPUT : The INPUT statement is used to enter data.
GO TO : A GO TO command (also GOTO) tells the computer to go to another statement.
IF...THEN : If the given condition is true, the computer will continue to the given statement. If the condition is false, the next statement is executed.
$: The dollar sign, $, indicates to the computer that the data are alphabetic.

EXAMPLES.

1.
```
NEW
10 PRINT "HELLO STUDENT"
20 END
RUN
```
2.
```
NEW
10 C=A+B
20 INPUT A
30 INPUT B
40 PRINT C
50 END
RUN
```
3.
```
NEW
10 PRINT "ONE MORE TIME"
20 GOTO 10
30 END
RUN
```

Statement 20 causes the computer to go back to statement 10 in an endless loop. Press the STOP key to terminate.

4. To evaluate $P(x) = 3x^2 - 5x + 7$ for $x = 5$, we use the following program.
```
NEW
10 PRINT "EVALUATING"
20 INPUT "X = ";X
30 P=3*X↑2-5*X+7
40 PRINT "THE VALUE IS"
50 PRINT P
60 END
RUN
```
Run the program and input $x = 5$.

2.4 MULTIPLYING POLYNOMIALS

In Section 2.2 we were able to add and subtract like terms using the distributive property.

$$3x + 7x = (3 + 7)x$$
$$= 10x$$

In this section we shall use the distributive property to multiply polynomials.

$$(3x + 2)(2x - 5) = (3x + 2)(2x) + (3x + 2)(-5)$$
$$= 6x^2 + 4x - 15x - 10$$
$$= 6x^2 - 11x - 10$$

To find the product of two polynomials, multiply each term of one of the polynomials by each term of the other and then add the products.

EXAMPLE 1.   Expand and simplify. $(2x^2 - x - 1)(x^2 - 2x + 3)$

SOLUTION:

$$(2x^2 - x - 1)(x^2 - 2x + 3) = 2x^4 - 4x^3 + 6x^2 - x^3 + 2x^2 - 3x - x^2 + 2x - 3$$
$$= 2x^4 - 5x^3 + 7x^2 - x - 3$$

Squaring a polynomial is simply finding the product of two polynomials.

$(x^2 - 3x + 1)^2$
$= (x^2 - 3x + 1)(x^2 - 3x + 1)$
$= x^4 - 3x^3 + x^2 - 3x^3 + 9x^2 - 3x + x^2 - 3x + 1$
$= x^4 - 6x^3 + 11x^2 - 6x + 1$

Three special cases of binomial products are now illustrated.

$(x + 3y)^2 = (x + 3y)(x + 3y)$
$\qquad = x^2 + 3xy + 3xy + 9y^2$ $\boxed{(a + b)^2 = a^2 + 2ab + b^2}$
$\qquad = x^2 + 6xy + 9y^2$

$(3x - 4)^2 = (3x - 4)(3x - 4)$
$\qquad = 9x^2 - 12x - 12x + 16$ $\boxed{(a - b)^2 = a^2 - 2ab + b^2}$
$\qquad = 9x^2 - 24x + 16$

$(3x - 2y)(3x + 2y) = 9x^2 - 6xy + 6xy - 4y^2$ $\boxed{(a - b)(a + b) = a^2 - b^2}$
$\qquad = 9x^2 - 4y^2$

Squaring a trinomial produces a pattern similar to the one resulting when a binomial is squared.

$(2r - 3s + t)^2$
$= (2r - 3s + t)(2r - 3s + t)$
$= 4r^2 - 6rs + 2rt - 6rs + 9s^2 - 3st + 2rt - 3st + t^2$
$= 4r^2 + 9s^2 + t^2 - 12rs + 4rt - 6st$

$$\boxed{(a + b + c)^2 = a^2 + b^2 + c^2 + 2ab + 2ac + 2bc}$$

EXAMPLE 2.   Expand and simplify. $2(2x - 3)(x + 1) - 3(2x - 1)^2$

SOLUTION:

$$2(2x - 3)(x + 1) - 3(2x - 1)^2 = 2(2x^2 + 2x - 3x - 3) - 3(4x^2 - 4x + 1)$$
$$= 2(2x^2 - x - 3) - 3(4x^2 - 4x + 1)$$
$$= 4x^2 - 2x - 6 - 12x^2 + 12x - 3$$
$$= -8x^2 + 10x - 9$$

EXERCISE 2.4

A 1. Expand.

(a) $3(x + y)$ (b) $x(x + y)$ (c) $3(x - 2)$
(d) $x(x + 7)$ (e) $4x(x - 5)$ (f) $2x(3 - 2x)$

B 2. Expand.

(a) $(x + 3)^2$ (b) $(x - 2)(x + 2)$
(c) $(m - x)^2$ (d) $(r + 7)(r - 7)$
(e) $(x - y)(x + y)$ (f) $(2m + 1)^2$
(g) $(2x - 3y)(2x + 3y)$ (h) $(1 - x)^2$
(i) $(3x - 4y)^2$ (j) $(5s + 3t)^2$
(k) $(2 - 3st)^2$
(l) $(3x^2 - 2y)(3x^2 + 2y)$

3. Expand.

(a) $(x + 3)(x + 2)$ (b) $(y + 4)(y + 5)$
(c) $(y - 3)(y - 7)$ (d) $(t + 3)(t - 4)$
(e) $(x - 7)(x + 3)$ (f) $(m - 5)(m - 7)$
(g) $(t + 5)(t + 11)$ (h) $(x + 10)(x + 11)$
(i) $(x^2 - 3)(x^2 + 6)$ (j) $(x^2 + 1)(x^2 - 2)$
(k) $(x^3 - 8)(x^3 + 6)$ (l) $(1 - x)(5 + x)$
(m) $(10 - x)(8 - x)$ (n) $(7 - t)(8 + t)$

4. Expand and simplify.

(a) $2(x - 4) - 3(x + 2)$
(b) $2(x^2 - 7x + 5) - 3(x - 4)$
(c) $5(3x - 4y) - (2x - 5y) + 7$
(d) $3(r - 2s - t) - 3(4r + 2s - 6t)$
(e) $3(2x - 4) - 3 - (2x + 1) + 5$
(f) $5(3x - 1) - 4(5y + 2) - 6$
(g) $2(2x^2 - 3x + 1) - 4(3x + 5)$
(h) $2x(3x - 5) - 4(2x + 7) + x^2$
(i) $2(1 - 3x + 2x^2) - (1 - 4x + 5x^2)$
(j) $2m(1 - 3m) - m(2m - 3) + m$
(k) $3(x_1 - 2x_2 + 3x_3) - 2(x_2 - x_3)$
(l) $4(2x^2 - 3xy + 4y^2) - 2(x^2 - 3y^2)$

5. Expand and simplify.

(a) $(3x + 4)(x + 5)$ (b) $(2t + 1)(3t + 7)$
(c) $(3x - 4)(2x - 1)$ (d) $(3m - 8)(2m - 3)$
(e) $(4x + 3)(5x - 4)$ (f) $(2r + 7)(3r - 1)$
(g) $(3 - 5y)(1 - 6y)$ (h) $(1 - 3m)(2m + 5)$
(i) $(3x + y)(2x - 3y)$
(j) $(4x - 5y)(3x - 10y)$
(k) $(6w - 11x)(w + 3x)$
(l) $(7x + 2y)(8x - 7y)$
(m) $(5x^2 - 4x)(3x^2 + 2x)$
(n) $(2m - 3m^2)(m^2 + 2m)$

6. Expand and simplify.

(a) $(x + y + z)^2$ (b) $(w - x - y)^2$
(c) $(2x + y + z)^2$ (d) $(2w - 3x + y)^2$

(e) $(1 - 3x - 4x^2)^2$ (f) $(5m - 3n + 4)^2$

7. Find the following products.

(a) $(2x + 3)(x^2 + 2x + 1)$
(b) $(3w^2 - 4w - 3)(2w - 1)$
(c) $(2m^2 + 3m - 1)(4m^2 - 2m + 3)$
(d) $(2w - 3x + 2y)(4w - x + 4y)$
(e) $(1 - 3x - x^2)(2 + 4x - 5x^2)$
(f) $(3x - 4y + 2z)(x + 3y - z)$
(g) $(x^3 - x^2 + x - 1)(x^2 - x - 3)$
(h) $(x^3 - x^2 - 2x - 3)(x^3 + 2x^2 + 3x + 1)$
(i) $(m^3 - 2m^2 - 3m - 1)(2m - 5)$
(j) $(3x - 4)(x^3 - 2x^2 + 5x - 4)$

8. Expand and simplify.

(a) $2(x - 4)(x + 3) + 5(2x - 1)(x + 6)$
(b) $3(2t - 5)(t - 4) - 3(5t - 3)(t + 4)$
(c) $2(m - 3)(m - 4) - 3(m + 5)^2$
    $- 2(2m - 1)(2m + 1)$
(d) $3(2m + 3)^2 - (m - 5)^2 - (2m - 4)$
    $(m - 5)$
(e) $5(2x - 5)(2x + 5) - 4(x - 2)(x + 3)$
    $- (2x + 1)^2$
(f) $(1 - 3x)(2 + 5x) - (x - 4)(2x - 5)$
    $- (2x + 3)^2$
(g) $5(2x - 3) - 2(x - 4)(x - 5) + 3x^2$
    $- (x - 6)$
(h) $5x^2 - (x - 3)^2 - 2(x^2 - 5x)$
    $+ 2(2x - 3)^2$
(i) $1 - (1 - 3x) - (x + 5)^2 - (3 - 4x)^2$
    $+ 6x^2$
(j) $(x - y)(x + 2y) - 3(2x - 3y)(x - 4y)$
    $+ 3(x + y)^2$
(k) $(2w + 3x)(w - x) - 4(w - 2x)^2$
    $+ 5(w^2 - x^2)$
(l) $4(x^2 - 3xy) - (x + y)^2$
    $- 2(x - y)(x + y) + 5$
(m) $2(x - 1)(x^2 - 3x + 2)$
    $- (2x^2 - 3x - 4)(2x + 3)$
(n) $5(r - s + t)(r - 2s - 3t)$
    $- (r + s + t)^2 - (r - s - 3t)$

C 9. Expand and simplify.

(a) $(2x - 1)(x + 4)(3x - 5)$
(b) $(x - 2y)(x + 3y)(2x - 5y)$
(c) $(w + x + y + z)^2$

(d) $\left(x + \dfrac{1}{x}\right)\left(x - \dfrac{1}{x}\right)$ (e) $\left(m - \dfrac{2}{m}\right)\left(m + \dfrac{3}{m}\right)$

(f) $\left(1 - x + \dfrac{1}{x}\right)\left(2 + x - \dfrac{1}{x}\right)$

2.5 COMMON FACTOR

Since $4 \times 6 = 24$ we say that 4 and 6 are factors of 24. If we limit the factors to prime numbers we get: $2 \times 2 \times 2 \times 3 = 24$

A prime number is an integer greater than 1 whose only integer factors are 1 and itself.

A positive integer is said to be completely factored when it is written as the product of prime numbers. A negative integer is factored completely when it is written as a product of -1 and prime numbers.

By the distributive property,

$$2x(x + 7) = 2x^2 + 14x$$

Reversing the procedure, we get:

$$2x^2 + 14x = 2x(x + 7)$$

This is called factoring the polynomial, where $2x$ is called a common factor since it is common to both terms of the polynomial.

$$\text{——EXPANDING→}$$
$$2x(x + 7) = 2x^2 + 14x$$
$$\text{←FACTORING——}$$

To factor a polynomial is to express the polynomial as the product of polynomials. We shall agree that a polynomial is completely factored when no more variable factors can be removed and no more integer factors, other than 1 or -1, can be removed.

This agreement is necessary since it is always possible to factor out something. For example,

$$2x + 3 = 2\left(\frac{2x + 3}{2}\right) \qquad\qquad 2x + 3 = x\left(\frac{2x + 3}{x}\right)$$
$$= 2\left(x + \frac{3}{2}\right) \qquad\qquad\qquad = x\left(2 + \frac{3}{x}\right)$$

$$2x + 3 = 3\left(\frac{2x + 3}{3}\right)$$
$$= 3\left(\frac{2}{3}x + 1\right)$$

In factoring polynomials with integer coefficients, unless otherwise stated, we will find factors with integer coefficients. This is called factoring over the integers.

EXAMPLE 1. Factor completely.

(a) $4xy + 6y$ (b) $12x^3 - 9x^2 + 3x$

SOLUTION:

(a) $4xy + 6y$
$\quad = 2y(2x) + 2y(3)$
$\quad = 2y(2x + 3)$

(b) $12x^3 - 9x^2 + 3x$
$\quad = 3x(4x^2) + 3x(-3x) + 3x(1)$
$\quad = 3x(4x^2 - 3x + 1)$

EXAMPLE 2. Factor completely. $x^2(x - 1) + 3x(x - 1) - 5(x - 1)$

SOLUTION:

Think of $(x - 1)$ as one number.
$$x^2(x - 1) + 3x(x - 1) - 5(x - 1)$$
$$= (x - 1)(x^2) + (x - 1)(3x) + (x - 1)(-5)$$
$$= (x - 1)(x^2 + 3x - 5)$$

Prime Numbers
2, 3, 5, 7, 11, 13, 17, ...

EXAMPLE 3. Factor completely. $10x^2 - 5xy - 6x + 3y$

SOLUTION:

The terms of the polynomial may be grouped to produce common factors.

$$10x^2 - 5xy - 6x + 3y$$
$$= (10x^2 - 5xy) - (6x - 3y)$$
$$= 5x(2x - y) - 3(2x - y)$$
$$= (2x - y)(5x - 3)$$

EXERCISE 2.5

A 1. Factor completely, where possible.

(a) $5x + 10$
(b) $2x - 6$
(c) $2x - 4y$
(d) $5x + 7$
(e) $3xy - 4x$
(f) $10m - 5$
(g) $3xy + 4xw$
(h) $2t^2 + 4t - 6$
(i) $5mn - 6mnt$
(j) $8x^2 - 4x$
(k) $7r - 14r^2$
(l) $2sw + 4sx + 6sy$
(m) $r(x - 1) + t(x - 1)$
(n) $7(w - z) - 5x(w - z)$

B 2. Factor completely, where possible.

(a) $5x^2 + 25x^3 - 30x^4$
(b) $x^5 - x^4 + x^3 - x^2$
(c) $28x^3 + 24x^2$
(d) $15x^5 - 7y^3 + 4z^2$
(e) $6m^3 - 10m^2t + 6mt^2$
(f) $14rst - 7r^2st + 21rs^2t - 28rst^2$
(g) $x^3y^3 - x^2y + xy^2$
(h) $13m^2n - 26m^2n^2$
(i) $16p^4q^3 - 8p^3q^4 - 4p^3q^3$
(j) $18x^2y^2 + 9xy - 3x^3y^4$
(k) $11pqr - 15x + 23$
(l) $x^6 - x^5 + x^4 - x^3 - x^2$
(m) $36p^2x^4 - 16p^3x^5 + 24p^5x^4$
(n) $m^{15}n^{11} - 3m^{10}n^{12} - 4m^9n^{15}$
(o) $24x^9y^7 - 48x^{11}y^{13} + 36x^{10}y^{10}$
(p) $2x^8y^4 - 6x^{17}y^{18} + 4x^8y^9 - 10x^{14}y^{21}$

3. Factor.

(a) $3(p + q) + 2x(p + q)$
(b) $5(x - 1) + 2m(x - 1)$
(c) $9m(x - y) - 14n(x - y)$
(d) $8(m + 4) - 9x(m + 4)$
(e) $5ty(m + n) - 6x(m + n)$
(f) $(x - 3) + y(x - 3)$
(g) $x(m + 7) - (m + 7)$
(h) $2(x^2 - 2x - 1) - y(x^2 - 2x - 1)$
(i) $14m^2(x - y) - 7m(x - y)$
(j) $(x - y)(x + y) + 3(x + y)$
(k) $2x(x - 4) - 3(x - 4)$
(l) $5m(2m + 1) - 4(2m + 1)$
(m) $(x - 5)^2 - m(x - 5)$
(n) $(x - y)^2 - (x - y)$
(o) $(y + 2)x - (y + 2)7$
(p) $(m - n)^3 - 2(m - n)^2$

4. Factor.

(a) $mx + my + nx + ny$
(b) $tx + 3t + wx + 3w$
(c) $mx + my - 3x - 3y$
(d) $x^3 + x^2 + x + 1$
(e) $x^3 - x^2y + xy^2 - y^3$
(f) $am + 3m + an + 3n$
(g) $1 - x + x^2 - x^3$
(h) $5x - 5y - x^2 + xy$
(i) $3x^2 + y + 3xy + x$
(j) $4mx + ny - 4nx - my$
(k) $3ny + 2mx - 2my - 3nx$
(l) $2mx - 3ny - 2my + 3nx$

C 5. Express the total area of the right circular cylinder in terms of r, h, and π. Factor the expression.

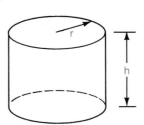

6. Factor.

(a) $2m(x + y) - 3(y + x)$
(b) $4x(m + n) - 2(n + m)$
(c) $6n(x - y) + 5(y - x)$
(d) $7m(x - 4) - 3(4 - x)$
(e) $5m(3x - y) - 2(y - 3x)$
(f) $3x(m - n) - (n - m)$

A perfect number is an integer which is equal to the sum of all its factors except itself.
Example. $6 = 1 + 2 + 3$
Find another perfect number.

2.6 POLYNOMIALS IN NESTED FORM

When using a calculator, a polynomial such as $P(x) = 3x^3 - 5x^2 + 7$ can be evaluated more simply and more quickly if it is factored first. This is illustrated in the following example, where we evaluate $P(x)$ for $x = 4$.

Normal Method

$$\begin{aligned} P(x) &= 3x^3 - 5x^2 + 7 \\ &= 3(4)^3 - 5(4)^2 + 7 \\ &= 3(64) - 5(16) + 7 \\ &= 192 - 80 + 7 \\ &= 119 \end{aligned}$$

Factored Method

$$\begin{aligned} P(x) &= 3x^3 - 5x^2 + 7 \\ &= [3x^2 - 5x]x + 7 \\ &= [(3x - 5)x]x + 7 \\ &= [(\langle 3 \rangle x - 5)x]x + 7 \\ P(4) &= [(\langle 3 \rangle 4 - 5)4]4 + 7 \end{aligned}$$

To evaluate this expression, we say

3×4	$12 - 5$	7×4	28×4	$112 + 7$
$= 12$	$= 7$	$= 28$	$= 112$	$= 119$

When using this factored method, x^3 and x^2 are not calculated directly. Instead, the factoring of the polynomial has included some operations in another operation. The x is cubed indirectly when we multiply $(3x - 5)$ by x, and again by x. A polynomial factored as above is said to be a polynomial in nested form.

$$P(x) = [(3)x - 5x]x + 7$$

is usually written in nested form using only right hand brackets as follows.

$$P(x) = 3]x - 5]x]x + 7$$

Calculations are done from left to right. Each operation is completed in turn, and its result then included in the next operation.

EXAMPLE. Write the given polynomials in nested form and find $P(2)$.
(a) $3x^3 - 5x^2 + 2x - 3$
(b) $4x^3 - 2x + 1$

SOLUTION:

(a)
$$\begin{aligned} & 3x^3 - 5x^2 + 2x - 3 \\ &= [3x^2 - 5x + 2]x - 3 \\ &= [(3x - 5)x + 2]x - 3 \\ &= [(\langle 3 \rangle x - 5)x + 2]x - 3 \end{aligned}$$
We write this polynomial in nested form by dropping the left brackets and writing only the right brackets as follows.
$$P(x) = 3]x - 5]x + 2]x - 3$$
In order to find $P(2)$ using a calculator, we enter

[3] [×] [2] [=] [−] [5] [=] [×] [2] [=]

[+] [2] [=] [×] [2] [=] [−] [3] [=]

and the display is `5.00`

(b)
$$\begin{aligned} & 4x^3 - 2x + 1 \\ &= 4x^3 + 0x^2 - 2x + 1 \\ &= [4x^2 + 0x - 2]x + 1 \\ &= [(4x + 0)x - 2]x + 1 \\ &= [(\langle 4 \rangle x + 0)x - 2]x + 1 \end{aligned}$$
We write this polynomial in nested form using only the right brackets.
$$\begin{aligned} P(x) &= [(\langle 4 \rangle x + 0)x - 2]x + 1 \\ &= 4]x + 0]x - 2]x + 1 \\ P(2) &= 4]2 + 0]2 - 2]2 + 1 \\ &= 8 + 0]2 - 2]2 + 1 \\ &= 8]2 - 2]2 + 1 \\ &= 16 - 2]2 + 1 \\ &= 14]2 + 1 \\ &= 28 + 1 \\ &= 29 \end{aligned}$$

A general polynomial of degree 4 in nested form is:
$$P(x) = Ax^4 + Bx^3 + Cx^2 + Dx + E$$
$$= A]x + B]x + C]x + D]x + E$$

EXERCISE 2.6

A 1. Evaluate for x = 3.
(a) 3]x + 5]x + 2   (b) 4]x + 3]x − 5
(c) 2]x − 3]x − 4   (d) − 5]x + 3]x − 2
(e) 7]x − 6]x − 5]x
(f) − 5]x + 6]x + 0]x + 4

2. Evaluate for x = − 2.
(a) 2]x + 3]x − 1]x + 3
(b) 3]x − 5]x + 2]x + 4
(c) − 2]x + 1]x − 3]x + 0
(d) 5]x + 0]x + 3]x − 5
(e) − 3]x − 2]x + 4
(f) 5]x + 0]x + 0]x + 4

B 3. Express in nested form.
(a) $5x^2 − 7x + 3$
(b) $3x^3 + 4x^2 − 5x + 2$
(c) $3x^3 − 2x + 5$
(d) $4x^3 + 3x^2 − 7$
(e) $3x^3 − 7x^2 + 5x − 3$
(f) $− 2x^3 + 7x − 11$

4. Express in nested form.
(a) $3x^3 + 2x^2 − 5x + 3$
(b) $− 2x^3 + 3x^2 − 4x + 7$
(c) $5x^3 − 2x^2 + 3x − (−6)$
(d) $x^3 − 3x^2 + 2x − 1$
(e) $6x^3 − 3x + 5$
(f) $− 2x^3 − 3x^2 + 6$

5. Express in nested form, and evaluate for x = 2.
(a) $3x^2 + 5x + 3$ (b) $2x^2 − 3x + 1$
(c) $− x^2 + 5x − 1$ (d) $3x^2 − 5x + 1$
(e) $6x^2 + x − 3$ (f) $5x^2 − 3x + 4$

6. Express in nested form, and evaluate for x = −3.
(a) $5x^3 − 2x^2 + 3x − 1$
(b) $3x^3 − 5x^2 + 2x + 7$
(c) $− x^3 + 5x^2 − 3x + 2$
(d) $2x^3 − 6x^2 − 9$
(e) $4x^3 − 3x^2 + 5x$
(f) $3x^3 − 2x − 6$

7. Express in nested form, and evaluate for the given value.
(a) $2x^3 − 5x + 7$, for x = 2
(b) $3x^3 + 7x^2 − 5x + 3$, for x = −2
(c) $2x^3 − 5x^2 + 6$, for x = −1
(d) $6x^3 − 3x^2 + 5$, for x = −2
(e) $− 2x^3 + 5x^2 − 2x + 3$, for x = 3
(f) $3x^4 − 2x^2 + 3x − 1$, for x = 3

8. Evaluate as indicated.
(a) $P(x) = 3x^2 − 5x + 7$, find P(4)
(b) $P(x) = 2x^2 + 4x − 5$, find P(−2)
(c) $P(x) = − x^3 + 3x^2 − 1$, find P(3)
(d) $P(x) = 2x^3 − x + 6$, find P(−2)
(e) $P(x) = − x^3 − x^2 − x + 6$, find P(−3)

9. A ball is thrown vertically upward with an initial velocity of 24.4 m/s. The formula for the height, d, of the ball from the point of release in time, t, seconds is
$$d = −4.9t^2 + 24.4t + 1.4$$
(a) Express the polynomial in nested form.
(b) Find d for t = 1, t = 2, t = 3, t = 4, and t = 5.
(c) What is the greatest height the ball reaches?

CALCULATOR MATH

$$P(x) = 3.7x^4 − 2x^3 + 7.5x − 2.6$$

EXERCISE

1. Find.
(a) P(1.375)
(b) P(3.075)

2. Replace ■ by <, >, or =
P(4.6) ■ P(5.2)

2.7 FACTORING $x^2 + bx + c$

Many polynomials appear in the form $ax^2 + bx + c$, such as $2x^2 + 7x + 3$, where $a = 2$, $b = 7$, and $c = 3$. Many of these quadratic polynomials can be written as the product of two linear polynomials.

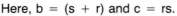

$$\xleftarrow{\qquad\text{EXPANDING}\qquad\longrightarrow}$$
$$(2x + 1)(x + 3) = 2x^2 + 7x + 3$$
$$\xleftarrow{\qquad\text{FACTORING}\qquad}$$

First we will consider quadratics of the form $ax^2 + bx + c$, where $a = 1$. An analysis of a general expansion of two linear polynomials will simplify factoring.

$$\begin{aligned}(x + r)(x + s) &= x^2 + sx + rx + rs \\ &= x^2 + (s + r)x + rs \\ &= x^2 + bx + c\end{aligned}$$

Here, $b = (s + r)$ and $c = rs$.

If we can write $x^2 + bx + c$ as $(x + r)(x + s)$ where r and s are integers, we can say we have factored $x^2 + bx + c$ over the integers.

To write $x^2 + 7x + 12$ in the form $(x + r)(x + s)$,

$b = r + s = 7$
$c = rs = 12$

We are looking for two integers whose sum is 7 and whose product is 12.

> Why are negative factors not considered here?

Trial Factors:	(1)(12)	(2)(6)	(3)(4)
Corresponding Sums:	13	8	7

The two integers that add to give 7 and multiply to give 12 are 4 and 3. Hence $r = 3$ and $s = 4$ and the factors are $(x + 4)$ and $(x + 3)$. Therefore, $x^2 + 7x + 12 = (x + 4)(x + 3)$.

EXAMPLE 1. Factor over the integers. $x^2 - x - 6$

SOLUTION:
For $x^2 - x - 6$,
$b = r + s = -1$
$c = rs = -6$

We are looking for two integers whose sum is -1 and whose product is -6.

Trial Factors:	$(-1)(6)$	$(1)(-6)$	$(-3)(2)$	$(3)(-2)$
Corresponding Sums:	5	-5	-1	1

The two integers that add to give -1, and multiply to give -6 are -3 and 2. Therefore, $x^2 - x - 6 = (x - 3)(x + 2)$.

EXAMPLE 2. Factor. $3x^2 - 6x - 15$

SOLUTION:
$3x^2 - 6x - 45 = 3(x^2 - 2x - 15)$
$b = r + s = -2$
$c = rs = -15$

> Always check for a common factor first.

Trial Factors:	$(-1)(15)$	$(-15)(1)$	$(-3)(5)$	$(3)(-5)$
Corresponding Sums:	-14	-14	2	-2

$3(x^2 - 2x - 15)$
$= 3(x + 3)(x - 5)$

EXERCISE 2.7

A 1. Complete the factoring.

(a) $x^2 + 6x + 8 = (x + 4)(\blacksquare)$
(b) $x^2 - 7x + 12 = (x - 4)(\blacksquare)$
(c) $t^2 - 13t + 22 = (t - 2)(\blacksquare)$
(d) $y^2 + 17y + 72 = (\blacksquare)(y + 9)$
(e) $m^4 - 12m^2 + 20 = (m^2 - 2)(\blacksquare)$
(f) $x^2 - 6x - 16 = (x + 2)(\blacksquare)$
(g) $2 + 3x + x^2 = (2 + x)(\blacksquare)$
(h) $12 + x - x^2 = (\blacksquare)(4 - x)$
(i) $28 + 3m - m^2 = (4 + m)(\blacksquare)$

2. If possible, determine integer values for m and n.

(a) $m + n = 7, mn = 12$
(b) $m + n = 20, mn = 36$
(c) $m + n = -3, mn = -18$
(d) $m + n = 6, mn = 7$
(e) $m + n = 3, mn = -40$
(f) $m + n = -8, mn = 16$
(g) $m + n = 20, mn = 75$
(h) $m + n = -6, mn = -27$
(i) $m + n = -5, mn = 4$

B 3. Factor over the integers, if possible.

(a) $x^2 + 7x + 10$
(b) $y^2 - 8y + 15$
(c) $m^2 + 10m + 21$
(d) $w^2 - w - 56$
(e) $x^2 + 5x + 7$
(f) $s^2 - 4s - 96$
(g) $t^2 + 36t + 320$
(h) $x^2 + x + 1$
(i) $x^2 + 26x + 165$
(j) $w^2 - 13w + 42$
(k) $z^2 + 3z - 40$
(l) $m^2 - 7m - 18$
(m) $x^2 - 8x + 16$
(n) $9 + 8t + t^2$
(o) $n^2 - 13n + 36$
(p) $20 - 9x + x^2$
(q) $42 - x - x^2$
(r) $88 + 3y - y^2$

4. Factor over the integers, if possible.

(a) $2x^2 + 4x + 2$
(b) $3q^2 - 6q - 45$
(c) $m^2 + 11m + 28$
(d) $5p^2 - 45p + 70$
(e) $2x^2 - 4x - 20$

(f) $x^2 - 22x + 117$
(g) $60 + 7m - m^2$
(h) $10y^2 - 30y - 400$
(i) $x^2 + 12x + 30$
(j) $x^2 + 14x + 48$
(k) $120 + 14x - 2x^2$
(l) $s^2 - 6s + 9$
(m) $3x^2 - 36x + 84$
(n) $36 - 12m + m^2$
(o) $w^2 + 4w - 77$
(p) $x^2 - 18x + 32$

5. Factor over the integers, if possible.

(a) $x^2 + 6xy - 160y^2$
(b) $s^2 + 9st + 14t^2$
(c) $x^2 + xy - 12y^2$
(d) $56 - 15t + t^2$
(e) $x^2 - 6xy - 55y^2$
(f) $m^2 - 41m + 408$
(g) $x^2 + 20x + 22$
(h) $x^2 + 6xy - 91y^2$
(i) $42 - 13x + x^2$
(j) $x^4 - x^2 - 90$
(k) $a^6 + 2a^3 - 35$
(l) $s^2 - 9st + 20t^2$
(m) $2y^2 - 20y - 48$
(n) $112 + 22z + z^2$
(o) $m^2 - 7m - 744$
(p) $m^2 - 8mn - 240n^2$
(q) $x^2 - 8xy - 660y^2$
(r) $x^4 - 8x^2 - 48$

MIND BENDER

Place 8 queens on a chessboard so that no queen attacks another queen.

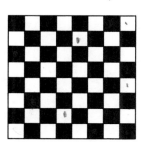

2.8 FACTORING ax² + bx + c

Factoring trinomials in the form $ax^2 + bx + c$, where $a \neq 1$, is simplified if we factor by grouping. To do this we must break up the middle term into two parts.

$$12x^2 + 17x + 6 = 12x^2 + 9x + 8x + 6$$
$$= (12x^2 + 9x) + (8x + 6)$$
$$= 3x(4x + 3) + 2(4x + 3)$$
$$= (4x + 3)(3x + 2)$$

A problem arises in trying to determine how to break up the middle term. For 17x in the above example, we might have used $10x + 7x$, or $2x + 15x$, or $-3x + 20x$, etc. The decision as to which two terms must be used will be clarified if we analyse a general expansion.

$$(px + r)(qx + s) = pqx^2 + psx + qrx + rs$$
$$= pqx^2 + (ps + qr)x + rs$$
$$= ax^2 + bx + c$$

If we break up the middle term bx into two terms, say mx and nx, then it is clear that

$$m + n = ps + qr = b$$
and
$$m \times n = pqrs = ac$$

EXAMPLE 1. Factor $6x^2 + 5x - 6$ over the integers.

SOLUTION:
For $6x^2 + 5x - 6$, $a = 6$, $b = 5$, and $c = -6$. To factor by grouping, we will replace 5x by mx + nx where
$$m + n = b = 5$$
$$m \times n = ac = -36$$

Trial Factors: (18)(−2)   (12)(−3)   (9)(−4)   (6)(−6) ...
Corresponding Sums: 16   9   5   0 ...

Therefore, m and n are 9 and −4.

$$6x^2 + 5x - 6 = 6x^2 + 9x - 4x - 6$$
$$= 3x(2x + 3) - 2(2x + 3)$$
$$= (2x + 3)(3x - 2)$$

EXAMPLE 2. Factor $24x^2 - 14xy - 20y^2$ over the integers.

SOLUTION:
Always check for common factors.
$$24x^2 - 14xy - 20y^2 = 2[12x^2 - 7xy - 10y^2]$$
We now factor $12x^2 - 7xy - 10y^2$ by breaking $-7xy$ into two parts.
$m + n = -7$ and $mn = -120$. Therefore, m and n are 8 and −15.

$$2[12x^2 - 7xy - 10y^2] = 2[12x^2 + 8xy - 15xy - 10y^2]$$
$$= 2[(12x^2 + 8xy) - (15xy + 10y^2)]$$
$$= 2[4x(3x + 2y) - 5y(3x + 2y)]$$
$$= 2[(3x + 2y)(4x - 5y)]$$
$$= 2(3x + 2y)(4x - 5y)$$

EXERCISE 2.8

A 1. Complete the factoring.

(a) $x^2 + 10x + 16 = (x + 8)(\blacksquare)$ $x + 2$

(b) $m^2 - 8m + 12 = (m - 2)(\blacksquare)$ $m - 6$

(c) $n^2 - n - 2 = (\blacksquare)(n + 1)$

(d) $x^2 + 3x - 18 = (\blacksquare)(x + 6)$

(e) $x^6 - 20x^3 + 36 = (\blacksquare)(x^3 - 2)$

(f) $x^2 - x - 20 = (x - 5)(\blacksquare)$

(g) $5 + 6s + s^2 = (5 + s)(\blacksquare)$

(h) $56 - 15t + t^2 = (7 - t)(\blacksquare)$

(i) $27 - 12x + x^2 = (\blacksquare)(3 - x)$

2. If possible, determine integer values for m and n.

(a) $m + n = -8$, $mn = 15$

(b) $m + n = 5$, $mn = 4$

(c) $m + n = 7$, $mn = -18$ $9 - 2$

(d) $m + n = -1$, $mn = -12$ $-4 + 3$

(e) $m + n = -2$, $mn = -10$

(f) $m + n = 3$, $mn = -28$ $7 - 4$

(g) $m + n = -7$, $mn = -18$ $-9 + 2$

(h) $m + n = 6$, $mn = -27$ $+9 - 3$

(i) $m + n = 7$, $mn = 15$ $+ \quad +$

B 3. Factor over the integers, where possible.

(a) $6x^2 - 7x - 20$ (b) $12x^2 + 23x + 5$

(c) $22t^2 + 13t + 1$ (d) $6r^2 + 11r + 6$

(e) $6m^2 - 7m - 3$ (f) $2y^2 - y - 28$

(g) $12q^2 + 29q + 15$ (h) $20x^2 - 3x - 9$

(i) $12x^2 - 40x - 7$ (j) $3m^2 - 19m - 20$

(k) $10t^2 - 31t + 15$ (l) $10x^2 - 16x + 3$

(m) $6q^2 - 23q + 7$ (n) $4m^2 + 23m - 35$

(o) $3y^2 + 22y - 16$ (p) $3x^2 + 25x + 42$

(q) $4s^2 + 31s - 45$ (r) $20x^2 + 11x - 4$

(s) $20x^2 - 64x + 35$ (t) $36m^2 - 7m - 15$

(u) $16t^2 - 18t - 9$ (v) $15x^2 + 27x + 8$

(w) $4s^2 + 21s + 27$ (x) $8x^2 - 2x - 3$

4. Factor over the integers, where possible.

(a) $24x^2 - 2x - 2$ (b) $10x^2 + 29x - 21$

(c) $6x^2 - 27x - 15$ (d) $6x^2 + 17xy + 5y^2$

(e) $12x^2 + 13xy - 35y^2$

(f) $60m^2 + 370mn + 60n^2$

(g) $24x^2 - 47xy + 20y^2$

(h) $15t^2 + 22st + 8s^2$

(i) $2 + 2x - 84x^2$

(j) $5yx^2 + 18yx - 8y$

(k) $4x^4 + 35x^2 + 49$

(l) $2x^2 + 2xy + 2y^2$

(m) $10 + 17x + 17x^2$

(n) $15 - 44x - 20x^2$

5. Factor over the integers.

(a) $8x^2 + 38x + 45$ (b) $20y^2 + 44y - 15$

(c) $40t^2 - 47t + 12$ (d) $48x^2 + 74x + 21$

(e) $42m^2 - 51m + 15$ (f) $40x^2 + 38x + 7$

(g) $8m^2 + 46m + 63$ (h) $48y^2 - 26y + 3$

(i) $20s^2 - 29s - 33$ (j) $15 - 53x + 42x^2$

(k) $42 + t - 56t^2$

(l) $28m^2 + 107m + 99$

MICRO MATH

The following program checks your factoring by multiplication. Use the program to check your answers.

NEW

```
10 PRINT "QUADRATIC FACTOR CHECKER"
20 PRINT "(PX + M)(QX + N)"
30 PRINT "ENTER P,Q,M,N IN ORDER."
40 INPUT "P = ";P
50 INPUT "Q = ";Q
60 INPUT "M = ";M
70 INPUT "N = ";N
80 PRINT "THE POLYNOMIAL IS"
90 PRINT (P*Q);"X↑2 + ("P*N+Q*M;")X
   + (";M*N")"
95 END
```

RUN

Statement 90 in the program above should be entered on one line.

MIND BENDER

A clock is stopped 1 min every 5 min. How long will it take the minute hand to complete one revolution?

2.9 FACTORING SPECIAL QUADRATICS

In this section we shall consider two special cases of factoring.

The Complete Square	The Difference of Squares
$x^2 + 2xy + y^2 = (x + y)^2$	$x^2 - y^2 = (x + y)(x - y)$
$x^2 - 2xy + y^2 = (x - y)^2$	
$a^2x^2 + 2abxy + b^2y^2 = (ax + by)^2$	

EXAMPLE. Factor.
(a) $9x^2 - 25$ (b) $2x^2 + 12x + 18$
(c) $4x^2 - 12xy + 9y^2$

SOLUTION:

(a) $9x^2 - 25 = (3x)^2 - (5)^2$
$\qquad\qquad = (3x - 5)(3x + 5)$

(c) $4x^2 - 12xy + 9y^2 = (2x)^2 - 2(2x)(3y) + (3y)^2$
$\qquad\qquad\qquad\quad = (2x - 3y)(2x - 3y)$
$\qquad\qquad\qquad\quad = (2x - 3y)^2$

(b) $2x^2 + 12x + 18 = 2(x^2 + 6x + 9)$
$\qquad\qquad\qquad = 2(x + 3)(x + 3)$
$\qquad\qquad\qquad = 2(x + 3)^2$

EXERCISE 2.9

A 1. Factor.
(a) $x^2 - 9$ (b) $x^2 - 6x + 9$
(c) $x^2 + 8x + 16$ (d) $x^2 - 36$
(e) $x^2 - 12x + 36$ (f) $x^2 - 64$

2. Factor.
(a) $4x^2 - 9$ (b) $4x^2 - 12x + 9$
(c) $9x^2 - 25$ (d) $16x^2 - 49$
(e) $25x^2 - 36$ (f) $25x^2 - 1$

B 3. Factor.
(a) $x^2 - 121$ (b) $x^2 - 20x + 100$
(c) $25x^2 + 20xy + 4y^2$ (d) $81x^2 - 18x + 1$
(e) $100x^2 + 60xy + 9y^2$ (f) $25x^2 - 144$

4. Factor completely.
(a) $x^4 - 1$ (b) $81x^4 - 1$
(c) $625x^2 - 81$ (d) $x^4 - 2x^2 + 1$
(e) $16x^4 - 625$ (f) $x^4 - 6x^2y^2 + 9y^4$

5. Use factoring to simplify these expressions. An example is shown for you.
$$99 \times 101 = (100 - 1)(100 + 1)$$
$$= 10\ 000 - 1$$
$$= 9999$$
(a) 98×102 (b) 105×95
(c) 93×107 (d) 85×115
(e) 999×1001 (f) 995×1005

6. Factor.
(a) $5x^2 - 20$ (b) $3x^2 - 12x + 12$
(c) $7x^2 - 343$ (d) $18x^2 + 12x + 8$
(e) $45x^2 - 60x + 20$ (f) $8x^2 - 50$

MINDBENDER

WORD LADDER

Start with the word "flour" and change one letter at a time to form a new word until you reach "bread." The best solution has the fewest steps.

f l o u r
_ _ _ _ _
_ _ _ _ _
_ _ _ _ _
_ _ _ _ _
b r e a d

2.10 FACTORING BY GROUPING

In some polynomials, we need to group the factors first in order to factor. For example, in the expression below, we see that a is a common factor of (ax + ay), and b is a common factor of (bx + by).

$$ax + by + bx + ay = (ax + ay) + (bx + by).$$
$$= a(x + y) + b(x + y)$$
$$= (x + y)(a + b)$$

We apply this grouping principle in the following examples.

EXAMPLE 1. Factor $a^2 + b^2 + 2ab - 1$ by grouping.

SOLUTION:

$$a^2 + b^2 + 2ab - 1$$
$$= [a^2 + 2ab + b^2] - 1$$
$$= [a + b]^2 - 1^2$$
$$= ([a + b] + 1)([a + b] - 1)$$
$$= (a + b + 1)(a + b - 1)$$

EXAMPLE 2. Factor. $-a^2 - b^2 - 2ab + 1$

SOLUTION:

$$-a^2 - b^2 - 2ab + 1$$
$$= 1 - a^2 - 2ab - b^2$$
$$= 1 - [a^2 + 2ab + b^2]$$
$$= 1 - [a + b]^2$$
$$= (1 + [a + b])(1 - [a + b])$$
$$= (1 + a + b)(1 - a - b)$$

EXERCISE 2.10

B 1. Factor.
(a) $ax - by + bx - ay$
(b) $ax + bx + ay + by$
(c) $2ax - 3bx + 2ay - 3by$
(d) $2xy + 3y - 10x - 15$
(e) $6xy + 3x - 4y - 2$
(f) $12xy - 6x + 2y - 1$

2. Factor.
(a) $9 - (a + b)^2$
(b) $(a - b)^2 - 1$
(c) $25 - (x + y)^2$
(d) $(2x - 1)^2 - y^2$
(e) $x^2 - (y - z)^2$
(f) $(a + b)^2 - c^2$

3. Factor.
(a) $a^2 - 2ab + b^2 - 25$
(b) $x^2 - 6x + 9 - y^2$
(c) $x^2 - y^2 + 10x + 25$
(d) $4x^2 - 9y^2 - 4x + 1$
(e) $16 - 4x^2 + 4xy - y^2$
(f) $100 - a^2 - 6a - 9$

4. Factor.
(a) $ax^2 + ay^2 + 2axy - ab^2$
(b) $b^2 - x^2 - y^2 + 2xy$
(c) $3a^2 + 12a + 12b - 3b^2$
(d) $36 - x^2 - y^2 - 2xy$

EXTRA

AMICABLE NUMBERS

Two numbers are amicable if each is equal to the sum of the proper divisors of the other.

Ancient scholars knew that 220 and 284 were amicable numbers. They were able to show this as follows.

The proper divisors of 220 are
    1, 2, 4, 5, 10, 11, 20, 22, 44, 55, 110
The sum of the proper divisors is
    1 + 2 + 4 + 5 + 10 + 11 + 20
    + 22 + 44 + 55 + 110 = 284
The proper divisors of 284 are
    1, 2, 4, 71, 142
The sum of the proper divisors is
    1 + 2 + 4 + 71 + 142 = 220

EXERCISE

1. Prove that the following are pairs of amicable numbers.
(a) 17 296 and 18 416
(b) 9 363 584 and 9 437 056
(c) 1184 and 1210

2. If 12 285 is one of a pair of amicable numbers, find the other number.

2.11 DIVIDING A POLYNOMIAL BY A MONOMIAL

The rule for exponents in division can be illustrated as follows:

$$\frac{x^m}{x^n} = x^{m-n}$$

thus,

$$\frac{x^5}{x^3} = x^{5-3} = x^2, \qquad x \neq 0$$

Since division by zero is not defined, replacements for variables which make the denominator zero will not be allowed.

EXAMPLE 1. Simplify. $\dfrac{24x^5y^3}{6x^2y^2}$

SOLUTION:

$$\frac{24x^5y^3}{6x^2y^2} = \frac{24}{6} \times \frac{x^5}{x^2} \times \frac{y^3}{y^2}$$
$$= 4x^3y, \qquad x, y \neq 0$$

These are the restrictions on the variables. Why?

The distributive property applies to division as well as multiplication.

$$\frac{a + b}{c} = \frac{1}{c}(a + b) = \frac{a}{c} + \frac{b}{c}$$

To divide a polynomial by a monomial, each term of the polynomial is divided by the monomial.

EXAMPLE 2. Simplify. $\dfrac{25x^5 - 15x^4 - 10x^3}{5x^2}$

SOLUTION:

$$\frac{25x^5 - 15x^4 - 10x^3}{5x^2} = \frac{25x^5}{5x^2} - \frac{15x^4}{5x^2} - \frac{10x^3}{5x^2}$$
$$= 5x^3 - 3x^2 - 2x, \qquad x \neq 0$$

EXAMPLE 3. Perform the indicated division.
$$\frac{36x^5 + 24x^5 - 4x^2 + 8}{2x^2}$$

SOLUTION:

$$\frac{36x^5 + 24x^3 - 4x^2 + 8}{2x^2} = \frac{36x^5}{2x^2} + \frac{24x^3}{2x^2} - \frac{4x^2}{2x^2} + \frac{8}{2x^2}$$
$$= 18x^3 + 12x - 2 + \frac{4}{x^2}, \qquad x \neq 0$$

EXERCISE 2.11

A 1. Divide and state restrictions on the variables.

(a) $\dfrac{16x^2}{2x}$

(b) $\dfrac{24t^5}{12t^2}$

(c) $\dfrac{15xy}{3xy}$

(d) $\dfrac{-48r^2s^2}{4rs}$

(e) $\dfrac{54m^2n^4}{-6mn^2}$

(f) $\dfrac{-100m^9n^{10}}{-20m^8n^9}$

(g) $\dfrac{56x^2y^4z^6}{-7xy^3z}$

(h) $\dfrac{-90m^{10}n^8}{9m^7n}$

(i) $\dfrac{-75mn}{-25mn}$

(j) $\dfrac{-30r^7s^9t}{-6r^2s^3}$

(k) $\dfrac{-12x^6y^8m^3}{2xym^3}$

(l) $\dfrac{-20xyz^5}{4xyz^3}$

2. Reduce the following fractions to lowest terms and state restrictions on the variables.

(a) $\dfrac{2}{2x}$ (b) $\dfrac{x}{xy}$ (c) $\dfrac{2t}{4t^3}$

(d) $\dfrac{30y}{6y^6}$ (e) $\dfrac{5xy}{20x^2y^2}$ (f) $\dfrac{12mn^2}{24mn}$

B 3. Perform the indicated divisions and state restrictions on the variables.

(a) $\dfrac{2m + 2n}{2}$ (b) $\dfrac{4x - 10y}{2}$

(c) $\dfrac{15x^2 - 10x}{5x}$ (d) $\dfrac{40s + 8}{8}$

(e) $\dfrac{-7x^4 + 14x^3}{7x^3}$ (f) $\dfrac{2m^4 - 4m^3 - 8m^2}{2m^2}$

4. Divide and state restrictions on the variables.

(a) $\dfrac{4x^2y + 8xy^2}{4xy}$

(b) $\dfrac{20m^4n^2 - 15m^2n^3 - 5mn}{5mn}$

(c) $\dfrac{3x^4m^6 - 9x^3m^4 - 18x^2m^3}{3x^2m^2}$

(d) $\dfrac{21x^4y^2 - 7x^3y^4 + 14x^5y^2}{7x^3}$

5. Divide and state restrictions on the variables.

(a) $\dfrac{2x - 4y}{-2}$

(b) $\dfrac{7t^2 - 14t^3}{-7t}$

(c) $\dfrac{21rs^2t - 15r^2st - 6rst^2}{-3rst}$

(d) $\dfrac{16m^4n^5 - 4m^3n^2 - 12m^3n^6}{-4m^3n^2}$

(e) $\dfrac{45x^7y^6z^5 - 9x^8y^5z^3}{-3x^5y^5z^3}$

(f) $\dfrac{12x^3y^4z^3 - 8x^2y^2z^4 - 4x^2y^2z^3}{-4x^2y^2z^3}$

C 6. Perform the indicated division and state restrictions on the variables.

(a) $\dfrac{8x^3 + 4x^2 + 2x}{2x^2}$

(b) $\dfrac{12x^4 - 8x^3 + 4x^2 + 2}{2x^2}$

(c) $\dfrac{m^5 - 6m^4 + 9m^3}{3m^3}$

(d) $\dfrac{21t^5 - 14t^4 - 7t^3 - t^2}{7t^3}$

(e) $\dfrac{10x^3y^4 - 5x^2y^5 + 15xy}{10x^2y^2}$

(f) $\dfrac{51w^4 - 34w^3 - 17w + 3}{17w^2}$

(g) $\dfrac{20m^4n^3 - 8m^3n^4 - 4mn + 5}{-2m^2n^2}$

(h) $\dfrac{63r^3s^4 - 81r^4s^5 + 18r^3s^2 - 9r^2}{-3r^2s}$

(i) $\dfrac{x^5 + 24x^4y^5 - 8x^3y^6 - y^3}{-8x^3y^3}$

A sheet of paper is 0.07 mm thick. We fold the paper repeatedly so that we have folded the paper 50 times. How thick is the stack of paper?

one fold

(a) Make a guess.
(b) Find the answer to the nearest kilometre using fast exponentiation.

2.12 DIVIDING A POLYNOMIAL BY A POLYNOMIAL

$$\overset{\text{quotient}}{\text{divisor}\,)\overline{\text{dividend}}}$$

remainder

To illustrate the division statement two integers are divided.

$$\begin{array}{r} 2 \\ 23\overline{)49} \\ 46 \\ \hline 3 \end{array}$$

To check the division we verify that

$$(2 \times 23) + 3 = 49$$

Division statement: (quotient) × (divisor) + remainder = dividend

The method used to divide a polynomial by a polynomial is similar to the method used for long division in arithmetic.

EXAMPLE 1. Divide $(6x^2 + 16x - 1)$ by $(3x - 1)$ and write the division statement.

SOLUTION:

Divide the first term of the dividend by the first term of the divisor.

$$3x - 1\overline{)6x^2 + 16x - 1}^{\displaystyle 2x}$$

Multiply the first term of the quotient by the divisor.

$$\begin{array}{r} 2x \\ 3x - 1\overline{)6x^2 + 16x - 1} \\ 6x^2 - 2x \end{array}$$

Subtract the resulting product from the dividend.

$$\begin{array}{r} 2x \\ 3x - 1\overline{)6x^2 + 16x - 1} \\ 6x^2 - 2x \\ \hline 18x - 1 \end{array}$$

Divide the first term of the remainder by the first term of the divisor.

$$\begin{array}{r} 2x + 6 \\ 3x - 1\overline{)6x^2 + 16x - 1} \\ 6x^2 - 2x \\ \hline 18x - 1 \end{array}$$

Multiply the second term of the quotient by the divisor.

$$\begin{array}{r} 2x + 6 \\ 3x - 1\overline{)6x^2 + 16x - 1} \\ 6x^2 - 2x \\ \hline 18x - 1 \\ 18x - 6 \end{array}$$

Subtract the resulting product from the previous remainder.

$$\begin{array}{r} 2x + 6 \\ 3x - 1\overline{)6x^2 + 16x - 1} \\ 6x^2 - 2x \\ \hline 18x - 1 \\ 18x - 6 \\ \hline 5 \end{array}$$

The restriction is $x \neq \frac{1}{3}$.

Why?

Stop when the remainder is zero or the degree of the remainder is less than the degree of the divisor.

Division statement: (quotient) × (divisor) + remainder = dividend
$$(2x + 6)(3x - 1) + 5 = 6x^2 + 16x - 1$$

EXAMPLE 2. Divide
$(6x^2 - 6 + x^3 + 7x)$ by $(x + 3)$.

SOLUTION:
Arrange the terms of the dividend and divisor in descending powers of the variable.

$$
\begin{array}{r}
x^2 + 3x - 2 \\
x + 3\overline{)x^3 + 6x^2 + 7x - 6} \\
\underline{x^3 + 3x^2} \\
3x^2 + 7x \\
\underline{3x^2 + 9x} \\
- 2x - 6 \\
\underline{- 2x - 6} \\
0
\end{array}
$$

What is the restriction on x?

EXAMPLE 3. Divide
$(x^2y^2 + y^4 - xy^3 + x^4)$ by $(x^2 + y^2 - xy)$ and write the division statement.

SOLUTION:
Arrange the terms in ascending or descending powers of one of the variables. Where there are missing terms in the dividend represent them by using zero as the coefficient.

$$
\begin{array}{r}
x^2 + xy + y^2 \\
x^2 - xy + y^2\overline{)x^4 + 0x^3y + x^2y^2 - xy^3 + y^4} \\
\underline{x^4 - x^3y + x^2y^2} \\
x^3y + 0x^2y^2 - xy^3 \\
\underline{x^3y - x^2y^2 + xy^3} \\
x^2y^2 - 2xy^3 + y^4 \\
\underline{x^2y^2 - xy^3 + y^4} \\
- xy^3
\end{array}
$$

Division statement:

$$(x^2 + xy + y^2)(x^2 - xy + y^2) - xy^3$$
$$= x^4 + x^2y^2 - xy^3 + y^4$$

EXERCISE 2.12

A 1. Arrange the following in descending or ascending powers of a variable.

(a) $x^3 + 3x^5 - 4x^2 + 7$
(b) $1 - x^4 + 2x^2 - 5x$
(c) $3m^2 + 2m^5 - 4m^3 + 5m$
(d) $r^4 + rs^3 - s^4 + 2r^2s^2$
(e) $3x - 5x^4 + 4x^2 + 2x^5$
(f) $xy^3 + 2x^2y^2 + 4x^4 - 3x^3y$
(g) $1 - t^5 + 4t^3 - 5t^2 + t^4$
(h) $m - 3m^3 + 2m^2 + 5m^5$

2. Write the division statement for the following.

(a) $19 \div 3$ (b) $7 \div 2$ (c) $11 \div 7$
(d) $46 \div 20$ (e) $21 \div 9$ (f) $72 \div 7$
(g) $65 \div 12$ (h) $98 \div 9$ (i) $30 \div 7$

B 3. Divide, then write the division statement and state any restrictions on the variables.

(a) $(2x^2 + 7x + 5) \div (x + 1)$
(b) $(x^3 - 2x^2 - 11x + 13) \div (x - 4)$
(c) $(2y^3 - 14y^2 + 31y - 30) \div (y - 4)$
(d) $(6w^2 - 29w + 18) \div (2w - 7)$
(e) $(9m^2 - 3m - 20) \div (3m + 4)$
(f) $(2t^3 - t^2 - 5t + 20) \div (2t + 5)$
(g) $(4m^3 - 4m^2 - 11m + 5) \div (2m + 1)$
(h) $(12x^3 - 23x^2 + 38x - 35) \div (4x - 5)$
(i) $(8y^3 + 12y^2 - 10y + 1) \div (2y + 3)$

4. Divide.

(a) $(16x^3 + 40x^2 - 2x - 11) \div (2x + 5)$
(b) $(x^2 + x^3 + 8 - 10x) \div (x^2 - 3x + 2)$
(c) $(m^2 - 5m + m^3 + 7) \div (m + 3)$
(d) $(4x^2 - 53) \div (2x - 7)$
(e) $(7t - t^3 + t^4 - 5t^2 - 8) \div (2 + t^2 - 3t)$
(f) $(8r^3 + 7) \div (2r - 1)$
(g) $(2 + 15s^4 - 6s - 5s^3) \div (3s - 1)$
(h) $(x^4 - 19) \div (x - 2)$
(i) $(x^5 - 2) \div (x - 1)$

5. Divide.

(a) $(x^2 + 2xy + 2y^2) \div (x + y)$
(b) $(x^2 - 2xy + 4y^2) \div (x - y)$
(c) $(x^2 - 7xy + 10y^2) \div (x - 4y)$
(d) $(x^3 - y^3) \div (x - y)$
(e) $(3x^2y + y^3 + 3xy^2 + 2x^3) \div (2x + y)$
(f) $(2x^2y^2 + 2y^4 + xy^3 + x^4) \div (xy + x^2 + y^2)$
(g) $(x^3y + x^4 - xy^3 - 8x^2y^2 + 3y^4)$
     $\div (x^2 + 3xy - y^2)$
(h) $(10m^2n^2 - n^4 + 6m^4 - 15m^3n)$
     $\div (3m^2 - 3mn - n^2)$

6. The area of a rectangle is given by $6x^2 - 13x - 28$. If the width is $2x - 7$, find an expression for the length.

7. Divide $(x^7 - 128)$ by $(x - 2)$.

2.13 SYNTHETIC DIVISION

We can divide $3x^3 - 11x^2 + 12x - 6$ by $x - 2$, using a short-cut called synthetic division. With this shorter method, we work only with the coefficients as indicated in the following example.

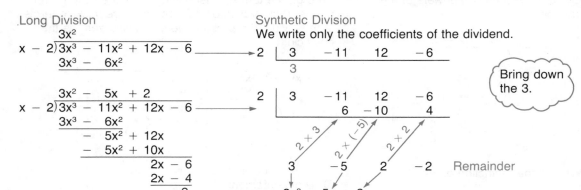

Long Division

$$
\begin{array}{r}
3x^2 \\
x - 2{\overline{\smash{\big)}\,3x^3 - 11x^2 + 12x - 6}} \\
\underline{3x^3 - 6x^2}
\end{array}
$$

Synthetic Division
We write only the coefficients of the dividend.

Bring down the 3.

$$
\begin{array}{r}
3x^2 - 5x + 2 \\
x - 2{\overline{\smash{\big)}\,3x^3 - 11x^2 + 12x - 6}} \\
\underline{3x^3 - 6x^2 } \\
- 5x^2 + 12x \\
\underline{- 5x^2 + 10x } \\
2x - 6 \\
\underline{2x - 4} \\
2
\end{array}
$$

EXAMPLE 1. Divide $2x^3 + x^2 - 7x + 30$ by $x + 3$.

SOLUTION:

$$
\begin{array}{r}
2x^2 - 5x + 8 \\
x + 3{\overline{\smash{\big)}\,2x^3 + x^2 - 7x + 30}} \\
\underline{2x^3 + 6x^2 } \\
- 5x^2 - 7x \\
\underline{- 5x^2 - 15x } \\
8x + 30 \\
\underline{8x + 24} \\
6
\end{array}
$$

In order to divide $2x^3 + x^2 - 7x + 30$ by $x + 3$, we use -3 since $x + 3 = x - (-3)$.

$$
\begin{array}{r|rrrr}
-3 & 2 & 1 & -7 & 30 \\
   &   & -6 & 15 & -24 \\
\hline
   & 2 & -5 & 8 & 6
\end{array}
$$

The quotient is $2x^2 - 5x + 8$, with a remainder of 6.

The following example illustrates how we can use synthetic division to divide a polynomial of the form $ax + b$. We write $a(x + \frac{b}{a})$ and divide by $x + \frac{b}{a}$ as before.

EXAMPLE 2. Divide $15x^3 + 8x^2 + 9$ by $3x - 2$.

SOLUTION:

$$3x - 2 = 3(x - \tfrac{2}{3})$$

$$
\begin{array}{r|rrrr}
\frac{2}{3} & 15 & 8 & 0 & 9 \\
   &   & 10 & 12 & 8 \\
\hline
   & 15 & 18 & 12 & 17
\end{array}
$$

Division statement: $= 15x^3 + 8x^2 + 9$
$= \frac{1}{3}(3x - 2)(15x^2 + 18x + 12) + 17$
$= (3x - 2)(5x^2 + 6x + 4) + 17$

Why must we multiply by $\frac{1}{3}$?

EXERCISE 2.13

B 1. Complete the following synthetic divisions.

(a) $3 \underline{\rvert\ 2 \quad 5 \quad 4 \quad 7}$

(b) $-3 \underline{\rvert\ 1 \quad 5 \quad 0 \quad -3}$

(c) $\frac{3}{4} \underline{\rvert\ 8 \quad 6 \quad -5 \quad 1}$

2. Divide the following polynomials.

(a) $(x^3 + 4x^2 + 3x - 5) \div (x - 3)$
(b) $(x^3 - 2x^2 - x - 2) \div (x + 2)$
(c) $(2x^3 - 5x + 7) \div (x + 3)$
(d) $(5x^3 - 7x^2 + 3x - 2) \div (x - 1)$
(e) $(-2x^3 + 3x^2 - 8) \div (x + 3)$

3. Divide the following.

(a) $\dfrac{x^3 - 3x^2 + 7x + 5}{x + 3}$

(b) $\dfrac{x^3 + 5x^2 + x + 3}{x - 2}$

(c) $\dfrac{x^3 + 5x^2 + 6}{x - 1}$

(d) $\dfrac{x^3 + 30}{x + 2}$

(e) $\dfrac{8x^3 - 27}{2x - 3}$

(f) $\dfrac{x^3 + 3x - 5}{x + 1}$

4. Divide the following.

(a) $(x^4 + 3x^3 - 2x^2 + x + 3) \div (x + 2)$
(b) $(2x^4 + 7x^3 - 3x^2 + 5x + 2) \div (x - 4)$
(c) $(3x^4 - 2x^2 + 7x + 5) \div (x + 5)$
(d) $(x^4 + x^3 - 7x + 4) \div (x - 3)$
(e) $(x^3 - 120) \div (x - 5)$

(f) $\dfrac{x^4 - 3x^3 + 5x^2 + 3x - 2}{x + 1}$

(g) $\dfrac{x^4 + 2x^2 + 7x + 5}{x - 3}$

(h) $\dfrac{x^5 + 3x^3 - 2x + 7}{x + 1}$

(i) $\dfrac{x^5 + 2x^4 - 7x^2 + 3}{x - 5}$

5. Divide the following.

(a) $(3x^3 + 10x^2 - 12x - 15) \div (3x + 1)$
(b) $(-8x^3 + 2x^2 + 19x - 6) \div (2x + 3)$
(c) $(2x^3 - 3x^2 + 6x - 8) \div (2x - 3)$
(d) $(2x^4 - 3x^3 - 10x^2 + 29x - 24) \div (2x - 3)$
(e) $(4x^4 + 9x^3 - 9x^2 - 20x - 15) \div (4x - 3)$
(f) $(4x^4 + 16x^3 + x^2 - 15x + 12) \div (2x + 3)$
(g) $(9x^4 - 13x^3 - 11x^2 + 8x - 12) \div (3x + 2)$

MICRO MATH

We can write a BASIC program to factor third degree polynomials using synthetic division. Let us take a general polynomial.

$$(Ax^3 + Bx^2 + Cx + D) \div (x - x_1)$$

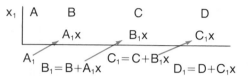

Since D_1 is the remainder, the polynomial is divisible by $(x - x_1)$ when $D_1 = 0$. These ideas are applied in the program listed below.
In the BASIC computer language, subscripted variables such as x_1 are used by writing x1. This program uses synthetic division to factor a polynomial of degree 3.

```
NEW
100 PRINT "FACTORING OF A"
110 PRINT "POLYNOMIAL OF DEGREE 3."
120 PRINT "AX↑3 + BX↑2 + CX + D"
130 PRINT "ENTER THE COEFFICIENTS AND"
140 PRINT "CONSTANT TERM SEPARATED BY"
150 PRINT "COMMAS."
160 INPUT A,B,C,D
170 PRINT
180 PRINT "ENTER A TRIAL VALUE FOR X>"
190 INPUT X
200 A1=A
210 B1=B+A1*X
220 C1=C+B1*X
230 D1=D+C1*X
240 IF D1<>0 PRINT "TRY AGAIN":GOTO 170
250 PRINT
260 X1=-X
270 PRINT "THE FACTORED POLYNOMIAL:"
280 PRINT
290 PRINT "(X + "X1")
    ("A1"X↑2 + "B1"X + "C1")"
300 END
RUN
```

Statement 290 in the program above should be entered on one line.

EXERCISE

1. Is it helpful in selecting the next trial value for x if you print the remainder D1?

2. Modify the program to factor a polynomial of degree 4.

2.14 SIMPLIFYING RATIONAL EXPRESSIONS

If x and y are two integers, then the quotient $\frac{x}{y}$, where $y \neq 0$, is a rational number. Examples of rational numbers are

$$\frac{3}{7}, \qquad -8 = \frac{-8}{1}, \qquad 0.23 = \frac{23}{100}$$

We define rational expressions in a similar manner. A rational expression is an expression that can be written as the quotient of two polynomials.

> If P and Q are polynomials and $Q \neq 0$, then $\dfrac{P}{Q}$ is a rational expression.

Examples of rational expressions are

$$\frac{3x}{5 + x}, \qquad \frac{x^2 + 2x}{x^2}, \qquad \frac{2x}{x - y}, \qquad \frac{x^2 + y^2}{x^2 - y^2}$$

A rational number is said to be reduced or simplified if the numerator and denominator have no common factors other than 1.

To simplify rational numbers

I. Factor the numerator and denominator into prime factors.
II. Divide both numerator and denominator by common factors.

$$\frac{30}{42} = \frac{2 \times 3 \times 5}{2 \times 3 \times 7} = \frac{5}{7}$$

Rational expressions are in simplest form, or lowest terms, when expressed as the quotient of two polynomials whose greatest common factor is 1.

To simplify rational expressions

I. Factor the numerator and denominator.
II. Divide both numerator and denominator by every factor common to both.

Since division by zero is not defined, restrictions must be placed on the variables in rational expressions in order to prevent division by zero.

EXAMPLE 1. Reduce $\dfrac{x^2 - 9}{x^2 - x - 6}$ to lowest terms and state any restrictions on the variable.

SOLUTION:

3 and -2 are called the "zeros" of the denominator.

$$\frac{x^2 - 9}{x^2 - x - 6} = \frac{(x - 3)(x + 3)}{(x - 3)(x + 2)}$$
$$= \frac{x + 3}{x + 2} \qquad \text{Restrictions: } x \neq 3, -2$$

EXAMPLE 2. Simplify $\dfrac{3x^2 - 3x^3}{x^3 - x}$ and state restrictions on the variable.

SOLUTION:

$$\dfrac{3x^2 - 3x^3}{x^3 - x} = \dfrac{3x^2(1 - x)}{x(x^2 - 1)}$$

$$1 - x = (-1)(x - 1)$$

$$= \dfrac{3x \times x(-1)(x - 1)}{x(x + 1)(x - 1)}$$

$$= \dfrac{-3x}{x + 1} \qquad \text{Restrictions: } x \neq 0, 1, -1$$

EXAMPLE 3. Simplify.

(a) $\dfrac{x^2 - 9}{3 - x}$

(b) $\dfrac{x^2 - x - 20}{15 + 2x - x^2}$

SOLUTION:

(a) $\dfrac{x^2 - 9}{3 - x} = \dfrac{(x - 3)(x + 3)}{(3 - x)}$

$$= \dfrac{(x - 3)(x + 3)}{(-1)(x - 3)}$$

$$= (-1)(x + 3)$$

$$= -x - 3$$

Restriction: $x \neq 3$

(b) $\dfrac{x^2 - x - 20}{15 + 2x - x^2} = \dfrac{(x - 5)(x + 4)}{(5 - x)(3 + x)}$

$$= \dfrac{(x - 5)(x + 4)}{(-1)(x - 5)(3 + x)}$$

$$= \dfrac{(-1)(x + 4)}{(x + 3)}$$

$$= \dfrac{-x - 4}{x + 3}$$

Restrictions: $x \neq 5, -3$

EXERCISE 2.14

A 1. State the missing numerator or denominator to obtain an equivalent rational expression.

(a) $\dfrac{mx}{x^2} = \dfrac{\blacksquare}{x}$

(b) $\dfrac{x}{y} = \dfrac{wx}{\blacksquare}$

(c) $\dfrac{mx}{m} = \dfrac{x}{\blacksquare}$

(d) $\dfrac{rst}{rstu} = \dfrac{\blacksquare}{u}$

(e) $\dfrac{x + y}{2} = \dfrac{\blacksquare}{6}$

(f) $\dfrac{1}{m - n} = \dfrac{7}{\blacksquare}$

(g) $\dfrac{(x + 2)(x + 3)}{(x + 3)(x - 1)} = \dfrac{x + 2}{\blacksquare}$

(h) $\dfrac{3}{x - y} = \dfrac{-3}{\blacksquare}$

(i) $\dfrac{-2}{m - n} = \dfrac{\blacksquare}{n - m}$

(j) $\dfrac{2x(x + 4)}{3x^2(x + 4)} = \dfrac{\blacksquare}{3x}$

2. State the restrictions on the variables.

(a) $\dfrac{5}{x}$

(b) $\dfrac{6}{x - 4}$

(c) $\dfrac{8}{t + 3}$

(d) $\dfrac{1}{(x - 1)(x + 3)}$

(e) $\dfrac{r - 4}{r + 5}$

(f) $\dfrac{m + 2}{m(m + 8)}$

(g) $\dfrac{16}{x - m}$

(h) $\dfrac{x - 4}{(2x + 1)(3x - 2)}$

B 3. Simplify, where possible.

(a) $\dfrac{12}{30}$

(b) $\dfrac{24x}{10}$

(c) $\dfrac{-6rst}{3rs}$

(d) $\dfrac{4m}{3n}$

(e) $\dfrac{mx}{ny}$

(f) $\dfrac{-15xy}{-3w}$

(g) $\dfrac{-21x^2y^4z}{3x^4y^2}$

(h) $\dfrac{x^5}{x^{10}}$

(i) $\dfrac{2xy}{8x^4y^3}$

(j) $\dfrac{-2m^3n}{6m^2n^7}$

(k) $\dfrac{(x + 2)(x + 5)}{(x + 5)(x - 2)}$

(l) $\dfrac{(m - 3)^2}{(m - 3)(m + 3)}$

4. Simplify, where possible, and state any restrictions on the variables.

(a) $\dfrac{x^2 - 3x - 28}{x^2 - 4x - 21}$

(b) $\dfrac{x^2 + 7x}{x^2 + 14x + 49}$

(c) $\dfrac{2t - 4}{t^2 - 3t + 2}$

(d) $\dfrac{2m + 6}{m^2 - 4}$

(e) $\dfrac{6x + 6y}{3x - 3y}$

(f) $\dfrac{2x - 10}{x^2 + 2x - 35}$

(g) $\dfrac{x^2 - 49}{x^2 + 6x - 7}$

(h) $\dfrac{x^2 - 5x + 6}{x^2 - 5x - 6}$

(i) $\dfrac{x^2 - 4}{2 - x}$

(j) $\dfrac{x - 4}{4 - x}$

(g) $\dfrac{x^2 - y^2}{x^3 + y^3}$

(h) $\dfrac{4 - x^2}{2x^3 - x^2 - 8x + 4}$

(i) $\dfrac{2(x - 3) - 3(x - 4)}{2x^2 - 6x - 36}$

(j) $\dfrac{15x^2 + 14xy - 8y^2}{6x^2 + 5xy - 4y^2}$

5. Simplify, where possible, and state restrictions.

(a) $\dfrac{2x^2 + 3x - 2}{x^2 - 4}$

(b) $\dfrac{6x^2 + 11x + 3}{6x^2 - 7x - 3}$

(c) $\dfrac{x^2 - y^2}{y - x}$

(d) $\dfrac{x^2 - 5x}{x^2 - x - 20}$

(e) $\dfrac{2m^2 + 7m + 3}{2m^2 + 5m + 3}$

(f) $\dfrac{12 - 5m - 2m^2}{2m^2 - 7m + 6}$

(g) $\dfrac{6m^2 - 54}{2m^2 + 7m + 3}$

(h) $\dfrac{x^3 - 8}{3x^2 - x - 10}$

(i) $\dfrac{-xm - ym}{x^2 + 2xy + y^2}$

(j) $\dfrac{3x + 6}{x^4 - 16}$

(k) $\dfrac{4(x^2 - y^2)}{(x - y)^2}$

(l) $\dfrac{6x^2 + 21x - 12}{9x^2 - 3x - 30}$

C 6. Simplify the following.

(a) $\dfrac{x^2 - 4}{x^3 + 4x^2 - 4x - 16}$

(b) $\dfrac{x^3 - 7x - 6}{x^2 - 2x - 3}$

(c) $\dfrac{3x + 4(x - 2)}{4x + 5(x - 2)}$

(d) $\dfrac{(x - 2)(x + 3) + (x - 2)}{(x - 2)(x + 3)(x - 4)}$

(e) $\dfrac{1 - (10 - x^2)}{2x^2 + 9x + 9}$

(f) $\dfrac{x^3 + 8}{3x^2 - 6x + 12}$

EXTRA

Multiplication does not distribute over multiplication.
$$3(x \times y) = 3 \times x \times y$$
Division distributes over addition.
$$\frac{x + y}{3} = \frac{x}{3} + \frac{y}{3}$$
Division does not distribute over multiplication.
$$\frac{x \times y}{3} = \frac{x}{3} \times y = x \times \frac{y}{3}$$
Knowing the above rules may eliminate errors such as

ERROR

$$\frac{x + \cancel{3}}{\cancel{3}} = x + 1$$

Since division distributes over addition, then
$$\frac{x + 3}{3} = \frac{x}{3} + \frac{3}{3} = \frac{x}{3} + 1$$
Similarly,
$$\frac{x + 2}{x + 3} \neq \frac{\cancel{x} + 2}{\cancel{x} + 3} \neq \frac{1 + 2}{1 + 3}$$
but
$$\frac{x + 2}{x + 3} = \frac{\dfrac{x}{x} + \dfrac{2}{x}}{\dfrac{x}{x} + \dfrac{3}{x}}$$
$$= \frac{1 + \dfrac{2}{x}}{1 + \dfrac{3}{x}}$$

2.15 MULTIPLICATION AND DIVISION OF RATIONAL EXPRESSIONS

Rational expressions are multiplied in the same way that we multiply rational numbers. Recall that

$$\frac{3}{5} \times \frac{2}{7} = \frac{3 \times 2}{5 \times 7} = \frac{6}{35}$$

> For rational expressions $\dfrac{P}{Q}$ and $\dfrac{R}{S}$,
>
> $$\frac{P}{Q} \times \frac{R}{S} = \frac{PR}{QS}, \qquad Q, S \neq 0$$

When multiplying rational expressions, first factor the numerators and denominators and divide by any common factors. Then express the product as a rational expression.

EXAMPLE 1. Multiply $\dfrac{x + 1}{x^2 - x - 6}$ by $\dfrac{x^2 - 7x + 12}{x^2 - 16}$.

SOLUTION:

$$\frac{x + 1}{x^2 - x - 6} \times \frac{x^2 - 7x + 12}{x^2 - 16}$$

$$= \frac{(x + 1)}{(x - 3)(x + 2)} \times \frac{(x - 4)(x - 3)}{(x - 4)(x + 4)}$$

$$= \frac{(x + 1)}{(x + 2)(x + 4)}, \qquad x \neq 3, -2, 4, -4$$

Rational expressions are divided in the same way that we divide rational numbers. Recall that

$$\frac{3}{4} \div \frac{11}{5} = \frac{3}{4} \times \frac{5}{11} = \frac{15}{44}$$

> For rational expressions $\dfrac{P}{Q}$ and $\dfrac{R}{S}$,
>
> $$\frac{P}{Q} \div \frac{R}{S} = \frac{P}{Q} \times \frac{S}{R} = \frac{PS}{QR}, \qquad Q, R, S \neq 0$$

To divide one fraction by another, invert the divisor and multiply by the dividend.

EXAMPLE 2. Divide $\dfrac{2x - 4}{x^2 + 9x + 20}$ by $\dfrac{x^2 + x - 6}{x^2 + 7x + 12}$.

SOLUTION:

$$\frac{2x - 4}{x^2 + 9x + 20} \div \frac{x^2 + x - 6}{x^2 + 7x + 12}$$

$$= \frac{2x - 4}{x^2 + 9x + 20} \times \frac{x^2 + 7x + 12}{x^2 + x - 6}$$

$$= \frac{2(x - 2)}{(x + 5)(x + 4)} \times \frac{(x + 4)(x + 3)}{(x + 3)(x - 2)}$$

$$= \frac{2}{x + 5}, \qquad x \neq -5, -4, -3, 2$$

EXAMPLE 3. Simplify.

$$\frac{x^2 - 9y^2}{x^2 + 4xy} \times \frac{x^2 + 9xy + 20y^2}{x^2 + 10xy + 21y^2} \div x^2 + 2xy - 15y^2$$

SOLUTION:

It is helpful to introduce 1 as a denominator.

$$\frac{x^2 - 9y^2}{x^2 + 4xy} \times \frac{x^2 + 9xy + 20y^2}{x^2 + 10xy + 21y^2} \div x^2 + 2xy - 15y^2$$

$$= \frac{x^2 - 9y^2}{x^2 + 4xy} \times \frac{x^2 + 9xy + 20y^2}{x^2 + 10xy + 21y^2} \div \frac{x^2 + 2xy - 15y^2}{1}$$

$$= \frac{x^2 - 9y^2}{x^2 + 4xy} \times \frac{x^2 + 9xy + 20y^2}{x^2 + 10xy + 21y^2} \times \frac{1}{x^2 + 2xy - 15y^2}$$

$$= \frac{(x - 3y)(x + 3y)}{x(x + 4y)} \times \frac{(x + 4y)(x + 5y)}{(x + 3y)(x + 7y)} \times \frac{1}{(x + 5y)(x - 3y)}$$

$$= \frac{1}{x(x + 7y)}, \qquad x \ne 0, \, -4y, \, -3y, \, -7y, \, -5y, \, 3y$$

EXERCISE 2.15

Express each of the following as rational expressions in lowest terms. Assume that all variables are restricted so that no denominator or divisor is equal to zero.

A 1. (a) $\dfrac{x}{y} \times \dfrac{y}{w}$ (b) $\dfrac{2}{x} \times \dfrac{y}{2}$ (c) $\dfrac{1}{2} \times \dfrac{5}{6}$

(d) $2 \times \dfrac{x}{y}$ (e) $\dfrac{rs}{3} \times \dfrac{7}{rst}$ (f) $\dfrac{3x^2}{y} \times \dfrac{2y^2}{6x}$

(g) $\dfrac{3}{(x + 2)} \times \dfrac{(x + 2)}{5}$ (h) $\dfrac{1}{2} \div \dfrac{x}{y}$

(i) $x \div \dfrac{1}{x}$ (j) $\dfrac{(m - 3)}{x} \div (m - 3)$

B 2. (a) $\dfrac{14x^2}{21y^2} \times \dfrac{15y^4}{7x^3}$ (b) $\dfrac{6x^5y^3}{5xy} \times \dfrac{10x^2y}{4x^3y^4}$

(c) $\dfrac{21x^3y}{5xy} \div \dfrac{7x^2y^6}{10xy^4}$ (d) $\dfrac{56x^4y^6z}{7x^3yz} \div 2xyz$

(e) $\dfrac{20r^3s^4}{8x^2y} \div \dfrac{5rs^2}{2xy}$ (f) $\dfrac{6m^2n^2}{5x^2y^4} \times \dfrac{3xy}{12mn}$

3. (a) $\dfrac{x + 3}{5x} \times \dfrac{10x}{x + 3}$ (b) $\dfrac{m - 2}{x} \times \dfrac{9x^2}{3m - 6}$

(c) $\dfrac{7t - 7}{4t} \div \dfrac{2t - 2}{16t^4}$ (d) $\dfrac{x^2 - 9}{3x + 9} \times \dfrac{2}{3 - x}$

(e) $\dfrac{x^2 + 3x + 2}{4 - x^2} \times \dfrac{10x}{5x + 5}$

4. (a) $\dfrac{x^2 + 7x + 12}{x^2 + 3x + 2} \times \dfrac{x^2 + 5x + 6}{x^2 + 6x + 9}$

(b) $\dfrac{x^2 + 2x - 15}{x^2 - 6x + 8} \times \dfrac{x^2 + 2x - 8}{x^2 - 6x + 9}$ $\times \dfrac{x^2 - 7x + 12}{x^2 - x - 30}$

(c) $\dfrac{x + 7}{x^2 - 5x - 36} \times \dfrac{x^2 - 15x + 54}{x^2 - 36}$ $\div \dfrac{x^2 - 2x - 63}{x + 4}$

5. (a) $\dfrac{2x^2 + 7x - 4}{6x^2 + x - 2} \times \dfrac{15x^2 + 7x - 2}{5x^2 + 19x - 4}$

(b) $\dfrac{16 - 9x^2}{6x^2 + 13x - 5} \times \dfrac{2x^2 + 7x + 5}{6x^2 - 29x + 28}$ $\div \dfrac{3x^2 + 7x + 4}{6x^2 - 19x - 7}$

(c) $\dfrac{2x^2 + 5x - 3}{4x^2 - 12x + 5} \times \dfrac{6x^2 - 7x - 20}{3x^2 + 13x + 12}$

C 6. (a) $\dfrac{x^3 + 8}{2x^2 + x - 6} \times \dfrac{12x - 18}{3x^2 - 6x + 12}$

(b) $\dfrac{x^3 + x^2 - 14x - 24}{3x^2 + 17x - 6} \times \dfrac{3x^2 + 14x - 5}{x^2 - x - 12}$

7. (a) $\dfrac{\dfrac{x^2 - 4}{4x^2}}{\dfrac{x + 2}{8x}}$ (b) $\dfrac{\dfrac{m^2 + 2m + 1}{m + 1}}{m}$

2.16 LEAST COMMON MULTIPLE

In order to add or subtract rational expressions it is necessary to know how to determine the least common multiple (LCM) of two or more polynomials.

The LCM for 12 and 30 is the smallest number that has 12 and 30 as factors. In order to determine the LCM for 12 and 30 we first factor 12 and 30 into primes.

$$12 = 2 \times 2 \times 3$$
$$30 = 2 \times 3 \times 5$$

The LCM for 12 and 30 must include all the separate factors that make up 12 and 30.

The LCM is $\overbrace{2 \times 2}^{12} \times 3 \times 5$ or 60.

$$\underbrace{}_{30}$$

EXAMPLE. Find the LCM for $x^2 + x - 6$, $x^2 + 7x + 12$, and $x^2 - 2x$.

SOLUTION:
Factor each polynomial.
$$x^2 + x - 6 = (x + 3)(x - 2)$$
$$x^2 + 7x + 12 = (x + 3)(x + 4)$$
$$x^2 - 2x = x(x - 2)$$

The LCM is $x\underbrace{(x - 2)\overbrace{(x + 3)}^{x^2 + x - 6}(x + 4)}_{x^2 - 2x \quad x^2 + 7x + 12}$.

EXERCISE 2.16

B Find the LCM for each of the following. Answer may be left in factored form.

1. (a) 6, 14
(b) 8, 6, 3
(c) 15, 24, 45
(d) 20, 25, 30
(e) 7, 16, 18
(f) 4, 6, 9, 28

2. (a) x^2, xy, y^2
(b) $2x^2$, $4xy$, y^2
(c) $6m^2$, $2m^2n$, $3mn^2$
(d) $8m^3n$, $3m^2n^2$, $2mn^3$, $6n^4$
(e) $20xy$, $15y$, $30x^3$
(f) $18x^5y$, $27txy$, $15xy^4$

3. (a) $x^2 - x - 12$, $x^2 - 9$
(b) $mx - my$, $nx - ny$
(c) $2x + 6$, $x^2 + 6x + 9$
(d) $x^2 - xy$, $x^2 - y^2$, $x^2 + xy$
(e) $5x - 10$, $x^2 - 4$, $6x + 12$
(f) $5x + 15$, $x^2 + 8x + 15$, $x^2 + 4x - 5$
(g) $t^2 - 1$, $t^2 + 7t + 6$, $t^2 + 6t - 7$
(h) $m^2 + m - 20$, $m^2 - 6m + 8$, $m^2 + 3m - 10$

4. (a) $x^2 - y^2$, $x^2 - 2xy + y^2$
(b) $m^2 + 2mn + n^2$, $m^2 - 2mn - 3n^2$
(c) $2x^2 + 5x - 3$, $6x^2 + x - 2$
(d) $10m^2 + 9m - 9$, $6m^2 + m - 12$
(e) $15r^2 - 17r - 4$, $9r^2 - 16$, $15r - 20$
(f) $6x^2 - 5x - 6$, $6x^2 - 11x - 10$, $4x^2 - 16x + 15$

2.17 ADDITION AND SUBTRACTION OF RATIONAL EXPRESSIONS

To write the sum of fractions with equal denominators as a single fraction we use the distributive property. Division, like multiplication, distributes over addition.

$$\frac{a}{b} + \frac{c}{b} = \frac{1}{b} \times a + \frac{1}{b} \times c = \frac{1}{b}(a + c) = \frac{a + c}{b}, \qquad b \neq 0$$

For rational expressions $\frac{P}{Q}$ and $\frac{R}{Q}$ with the same denominator

$$\frac{P}{Q} + \frac{R}{Q} = \frac{P + R}{Q} \quad \text{and} \quad \frac{P}{Q} - \frac{R}{Q} = \frac{P - R}{Q}, \qquad Q \neq 0$$

EXAMPLE 1. Simplify.

(a) $\dfrac{5}{13} + \dfrac{8}{13} - \dfrac{2}{13}$

(b) $\dfrac{3x}{x + 2} + \dfrac{x + 5}{x + 2} - \dfrac{x}{x + 2}$

SOLUTION:

(a) $\dfrac{5}{13} + \dfrac{8}{13} - \dfrac{2}{13} = \dfrac{5 + 8 - 2}{13}$

$= \dfrac{11}{13}$

(b) $\dfrac{3x}{x + 2} + \dfrac{x + 5}{x + 2} - \dfrac{x}{x + 2} = \dfrac{3x + x + 5 - x}{x + 2}$

$= \dfrac{3x + 5}{x + 2}, \qquad x \neq -2$

To add or subtract rational expressions with different denominators

I.   Find the least common denominator (LCD) of the fractions. (The LCD is the LCM of the denominators.)

II. Express each fraction as an equivalent fraction having the LCD as the denominator.

III. Perform the indicated operations and simplify.

EXAMPLE 2. Simplify. $\dfrac{2w - 3}{4} + \dfrac{3w - 1}{5} - \dfrac{w - 5}{2}$

SOLUTION:
The LCD is 20.

$$\dfrac{2w - 3}{4} + \dfrac{3w - 1}{5} - \dfrac{w - 5}{2} = \dfrac{5(2w - 3)}{5 \times 4} + \dfrac{4(3w - 1)}{4 \times 5} - \dfrac{10(w - 5)}{10 \times 2}$$

$$= \dfrac{5(2w - 3)}{20} + \dfrac{4(3w - 1)}{20} - \dfrac{10(w - 5)}{20}$$

$$= \dfrac{5(2w - 3) + 4(3w - 1) - 10(w - 5)}{20}$$

$$= \dfrac{10w - 15 + 12w - 4 - 10w + 50}{20}$$

$$= \dfrac{12w + 31}{20}$$

EXAMPLE 3. Simplify. $\dfrac{5}{3x^2} - \dfrac{1}{2x} + \dfrac{3}{5x^3}$

SOLUTION:

The LCD is $30x^3$.

$$\dfrac{5}{3x^2} - \dfrac{1}{2x} + \dfrac{3}{5x^3} = \dfrac{10x \times 5}{10x \times 3x^2} - \dfrac{15x^2 \times 1}{15x^2 \times 2x} + \dfrac{6 \times 3}{6 \times 5x^3}$$

$$= \dfrac{50x}{30x^3} - \dfrac{15x^2}{30x^3} + \dfrac{18}{30x^3}$$

$$= \dfrac{50x - 15x^2 + 18}{30x^3}, \qquad x \neq 0$$

> Simplify expressions where possible.

EXAMPLE 4. Simplify. $\dfrac{3x - 12}{x^2 - x - 12} - \dfrac{2}{x^2 + 6x + 9} - \dfrac{1}{x^2 - 4x - 21}$

SOLUTION:

$$\dfrac{3x - 12}{x^2 - x - 12} - \dfrac{2}{x^2 + 6x + 9} - \dfrac{1}{x^2 - 4x - 21}$$

$$= \dfrac{3(x - 4)}{(x - 4)(x + 3)} - \dfrac{2}{(x + 3)(x + 3)} - \dfrac{1}{(x - 7)(x + 3)}$$

> LCD is $(x + 3)(x + 3)(x - 7)$

$$= \dfrac{3}{(x + 3)} - \dfrac{2}{(x + 3)(x + 3)} - \dfrac{1}{(x - 7)(x + 3)}$$

$$= \dfrac{3(x + 3)(x - 7)}{(x + 3)(x + 3)(x - 7)} - \dfrac{2(x - 7)}{(x + 3)(x + 3)(x - 7)} - \dfrac{1(x + 3)}{(x - 7)(x + 3)(x + 3)}$$

$$= \dfrac{3(x + 3)(x - 7) - 2(x - 7) - (x + 3)}{(x + 3)(x + 3)(x - 7)}$$

$$= \dfrac{3(x^2 - 4x - 21) - 2(x - 7) - (x + 3)}{(x + 3)(x + 3)(x - 7)}$$

$$= \dfrac{3x^2 - 12x - 63 - 2x + 14 - x - 3}{(x + 3)(x + 3)(x - 7)}$$

$$= \dfrac{3x^2 - 15x - 52}{(x + 3)(x + 3)(x - 7)}, \qquad x \neq 4, -3, 7$$

EXERCISE 2.17

A 1. State the LCD for each of the following.

(a) $\dfrac{3}{7} - \dfrac{5}{6}$

(b) $\dfrac{3}{4} - \dfrac{5}{6} + \dfrac{1}{3}$

(c) $\dfrac{5}{8} - \dfrac{1}{3} + \dfrac{3}{4} - \dfrac{5}{6}$

(d) $\dfrac{3}{x} - \dfrac{4}{y} + \dfrac{5}{xy}$

(e) $\dfrac{2}{3x^2} - \dfrac{3}{2x^3} + \dfrac{4}{x}$

(f) $\dfrac{a}{x^2y} - \dfrac{b}{xy^2} + \dfrac{c}{3x^3y^2}$

(g) $\dfrac{1}{2} + \dfrac{3}{m + 2}$

(h) $\dfrac{3}{x - 1} + \dfrac{4}{x + 2} + \dfrac{3}{4}$

(i) $\dfrac{5}{x} - \dfrac{4}{x - 3}$

(j) $\dfrac{2}{m - 1} + \dfrac{3}{m + 2} - \dfrac{4}{m + 3}$

(k) $\dfrac{5}{(x + 1)(x + 2)} - \dfrac{7}{(x + 2)(x - 4)}$

(l) $\dfrac{1}{t(t + 5)} - \dfrac{2}{(t + 5)(t + 7)} - \dfrac{3}{t^2(t + 7)}$

2. Simplify. Assume that all variables are restricted so there are no zero denominators.

(a) $\dfrac{5}{x} + \dfrac{3}{x}$

(b) $\dfrac{7}{y} - \dfrac{3}{y}$

(c) $\dfrac{7}{m} - \dfrac{3}{m} + \dfrac{4}{m}$

(d) $\dfrac{3}{x + 1} + \dfrac{2}{x + 1}$

(e) $\dfrac{3x}{x - 4} - \dfrac{5}{x - 4}$

(f) $\dfrac{4x + 3y}{8} + \dfrac{x + 2y}{8}$

(g) $\dfrac{5x^2}{(x + 5)(x + 3)} - \dfrac{3x^2}{(x + 5)(x + 3)}$

(h) $\dfrac{2m^2}{m - 5} - \dfrac{3m}{m - 5} + \dfrac{4}{m - 5}$

B Perform the indicated operations and simplify. Assume that all variables are restricted so that no denominator is equal to zero.

3. (a) $\dfrac{3}{4} - \dfrac{1}{3} + \dfrac{5}{6}$

(b) $\dfrac{7}{8} + \dfrac{1}{12} - \dfrac{2}{3}$

(c) $\dfrac{3x}{2} - \dfrac{x}{3} + \dfrac{5x}{6}$

(d) $\dfrac{2m + 3n}{2} + \dfrac{m - n}{3}$

(e) $\dfrac{x - 1}{4} - \dfrac{x + 1}{5} + \dfrac{x + 3}{10}$

(f) $\dfrac{4x^2}{5} - \dfrac{2x^2 + 1}{6} - \dfrac{3x^2 - 5}{15}$

(g) $5 + \dfrac{m}{7} + \dfrac{n}{2}$

(h) $\dfrac{2t - 3}{8} - 2 + \dfrac{1 - 3t}{3}$

4. (a) $\dfrac{5}{2x} + \dfrac{3}{3x} - \dfrac{4}{6x}$

(b) $\dfrac{7}{5x} - \dfrac{2}{3x} + \dfrac{9}{10x}$

(c) $\dfrac{2}{x} - \dfrac{3}{x^2} + \dfrac{4}{x^3}$

(d) $\dfrac{2}{3y^2} - \dfrac{1}{2y} + \dfrac{5}{9y^3}$

(e) $\dfrac{2}{s^2} - \dfrac{s - 3}{st} + \dfrac{t - 2}{t^2}$

(f) $\dfrac{2x - 1}{xy} - \dfrac{3}{y} + \dfrac{4}{x}$

(g) $m - \dfrac{x}{y} + y$

(h) $w - \dfrac{x}{w} + x$

(i) $3x + \dfrac{1}{x}$

(j) $4 - \dfrac{1}{x} + x$

5. (a) $\dfrac{3}{x + 1} + \dfrac{4}{x - 3}$ ✓

(b) $\dfrac{6}{w + 4} + \dfrac{5}{w}$

(c) $\dfrac{6}{m - 4} - \dfrac{3}{m - 2}$ ✓

(d) $\dfrac{3}{t - 2} - \dfrac{4}{2 - t}$

(e) $\dfrac{5x}{3} - \dfrac{2}{x + 5} + \dfrac{x}{4}$ ✓

(f) $\dfrac{2}{x - y} - \dfrac{1}{x + y}$

(g) $\dfrac{7}{2r - 1} - \dfrac{3}{1 - 2r}$ ✓

(h) $\dfrac{2t}{t - 4} - \dfrac{3t}{t + 5}$

(i) $\dfrac{x + 3}{x + 2} + \dfrac{x + 2}{x + 3}$

(j) $\dfrac{m - 5}{m - 1} - \dfrac{m - 1}{m - 3}$

(k) $7 + \dfrac{3x}{x - 4} - \dfrac{x}{x + 2}$ ✓

(l) $\dfrac{2x}{x - y} - \dfrac{3y}{x + y} + 1$

6. (a) $\dfrac{2}{(x - 4)(x + 3)} + \dfrac{5}{(x + 3)(x + 5)}$

(b) $\dfrac{4}{x(x + 1)} - \dfrac{2}{(x + 1)(x + 2)}$

(c) $\dfrac{2w}{(w - 1)(w - 5)} - \dfrac{3w}{(w - 5)(w - 6)}$

(d) $\dfrac{1}{x + 4} + \dfrac{3}{x^2 + 7x + 12}$

(e) $\dfrac{1}{t^2 + 5t + 4} - \dfrac{2}{t^2 + 9t + 20}$

(f) $\dfrac{5}{x^2 - 5x} - \dfrac{2}{x^2 - 4x - 5}$

(g) $\dfrac{3x - 3}{x^2 - 5x + 4} + \dfrac{4}{x^2 - 16}$

(h) $\dfrac{2}{m - 3} + \dfrac{4m - 12}{9 - m^2}$

(i) $\dfrac{3}{x^2 - xy} - \dfrac{4}{y^2 - xy}$

(j) $\dfrac{w}{w - x} - \dfrac{x}{w + x} + \dfrac{wx}{x^2 - w^2}$

7. (a) $\dfrac{x + 6}{x^2 + 9x + 18} + \dfrac{x - 3}{x^2 - 2x - 3}$

(b) $\dfrac{x^2 + x - 6}{x^2 - 3x + 2} - \dfrac{x^2 + 3x + 2}{x^2 - x - 2}$

(c) $\dfrac{x + 2}{x^2 - x - 6} - \dfrac{x - 4}{x^2 - x - 12} + \dfrac{x + 3}{x^2 + 2x - 3}$

(d) $\dfrac{1}{(w - x)(x - y)} - \dfrac{2}{(x - y)(y - w)}$
$+ \dfrac{3}{(w - x)(w - y)}$

(e) $\dfrac{x - 3}{x + 3} - \dfrac{2}{x - 3} + \dfrac{3}{x^2 - 9}$

(f) $\dfrac{x}{x + 1} + \dfrac{x + 1}{2} - \dfrac{x}{x^2 - 1} + \dfrac{2}{1 - x}$

8. (a) $\dfrac{2}{2x^2 + 3x - 2} - \dfrac{3}{3x^2 + 4x - 4}$

(b) $\dfrac{3t + 1}{6t^2 - 7t - 3} - \dfrac{2t - 1}{2t^2 + 9t - 5}$
$- \dfrac{5}{2t^2 + 7t - 15}$

(c) $\dfrac{2}{x^2 - 3x - 4} \times \dfrac{x - 4}{x - 2} + \dfrac{3}{x + 1}$

(d) $\dfrac{x + 1}{x + 2} \div \dfrac{x^2 - 2x - 15}{x^2 + 5x + 6} - \dfrac{4}{x^2 - 25}$

(e) $\dfrac{1 - \dfrac{2}{x} + \dfrac{3}{x^2}}{2 - \dfrac{1}{x} - \dfrac{1}{x^2}}$

(f) $\dfrac{\dfrac{r}{st} + \dfrac{s}{rt} + \dfrac{t}{rs}}{\dfrac{1}{r^2 s^2} - \dfrac{1}{s^2 t^2} - \dfrac{1}{r^2 t^2}}$

(g) $x - \dfrac{x}{x + \dfrac{1}{3}}$

(h) $\dfrac{\dfrac{x - 1}{x + 1} + \dfrac{x + 1}{x - 1}}{\dfrac{x - 1}{x + 1} - \dfrac{x + 1}{x - 1}}$

MICRO MATH

Find a four-digit number of the form BBAA which is a perfect square.

Use the following program to solve this problem.

NEW
```
10 PRINT ''BBAA''
20 FOR N=1 TO 100
30 PRINT N, N*N
40 NEXT N
50 END
RUN
```

2.18 REVIEW EXERCISE

1. Expand.

(a) $(x + 7)(x + 11)$
(b) $(2x + 5)^2$
(c) $(3x - 1)(3x + 1)$
(d) $(t - 3)(t + 7)$
(e) $(m - 8)(m + 9)$
(f) $(m^2 - 8)(m^2 + 2)$

2. Factor.

(a) $5x + 30$
(b) $8m^2 - 4m$
(c) $3m(x - y) + 2n(x - y)$
(d) $x^2 + 9x + 18$
(e) $m^2 - 14m + 49$
(f) $n^2 - 2n - 24$
(g) $x^2 - 16$
(h) $x^2 - 11x + 28$
(i) $y^2 - y - 20$
(j) $4x^2 - 25$
(k) $t^2 + 8t + 16$
(l) $1 - 25w^2$

3. State the restrictions on the variables in each of the following.

(a) $\dfrac{7}{m}$

(b) $\dfrac{5}{x - 5}$

(c) $\dfrac{8}{n + 7}$

(d) $\dfrac{x}{(x - 5)(x + 2)}$

4. Simplify and state restrictions.

(a) $\dfrac{5}{x} \times \dfrac{3}{x}$

(b) $\dfrac{5}{x} + \dfrac{3}{x}$

(c) $\dfrac{5}{x} \div \dfrac{3}{x}$

(d) $\dfrac{5}{x} - \dfrac{3}{x}$

(e) $\dfrac{3}{x + 2} - \dfrac{4}{x + 2}$

(f) $\dfrac{3}{x + 2} \div \dfrac{4}{x + 2}$

(g) $\dfrac{3}{x + 2} \times \dfrac{4}{x + 2}$

(h) $x \div \dfrac{1}{x}$

(i) $\dfrac{1}{x} \div x$

(j) $\dfrac{(x + 2)(x + 1)}{(x + 1)(x - 3)} \times \dfrac{(x - 3)}{(x + 2)}$

5. Expand and simplify.

(a) $2(x - 7) + 3(x + 5)$
(b) $4(3m - 2n) - 5(m + 3n)$
(c) $2x(3x - 5) - x(x - 11)$
(d) $(4x - 3)(3x + 8)$
(e) $(1 - 7t)(3t - 4)$
(f) $(3x^2 - 5x)(4x^2 - 7x)$
(g) $(x^2 - 3x + 7)^2$
(h) $(3m + 2)(2m^2 - m - 4)$
(i) $(3s^2 - 4s + 5)(1 - 2s)$
(j) $(3w^2 - 4w + 1)(w^2 - w - 5)$
(k) $2(x - 3)(x + 5) - 4(3x - 1)(2x + 5)$
(l) $3(x - 3y)(x - 4y) - (x - 2y)^2 + 16x^2$
(m) $2(x - 3)(x^2 - 6x + 5)$
    $- 3(2x - 1)(3x + 7)$

6. Factor over the integers.

(a) $m^2 + 8m + 15$
(b) $2x^2 - 2x - 24$
(c) $2x^2 - 17x + 21$
(d) $6x^2 + 41x + 63$
(e) $15x^2 - 17x + 4$ ～
(f) $12m^2 - 7m - 10$
(g) $12t^2 - 43t + 35$
(h) $8n^2 - 38n - 60$ ―
(i) $12r^2 + 28r - 5$
(j) $mx + nx + my + ny$ ―
(k) $4y + mx - 4x - my$ ―
(l) $sw + 5s + tw + 5t$
(m) $8m^2 + 14mn - 15n^2$ ⌁
(n) $20p^2 + 11pq - 3q^2$
(o) $6x^2 + 23xy + 20y^2$
(p) $18s^2 + 15st - 75t^2$ ―

7. Factor.

(a) $25x^2 - 36y^2$
(b) $1 - 100m^2$
(c) $t^4 - 36x^2y^2$

(d) $9x^2 - 30x + 25$
(e) $16m^2 - 24mn + 9n^2$
(f) $4r^2 + 20rs + 25s^2$
(g) $(x + y)^2 - 36$
(h) $25m^2 - (x - y)^2$
(i) $(m + 2n)^2 - 4(r + s)^2$
(j) $x^2 + 4xy + 4y^2 - m^2 - 6mn - 9n^2$
(k) $x^2 - 36m^2 - 8xy + 16y^2$
(l) $9t^2 - 9y^2 - x^2 + 6xy$

8. Divide and state restrictions.

(a) $\dfrac{30x + 5}{5}$

(b) $\dfrac{12x^4 - 8x^3 + 4x^2}{4x^2}$

(c) $\dfrac{25x^4y^5 - 30x^3y^7 + 35x^4y^6}{-5x^2y}$

(d) $\dfrac{3m^2n^4 - 9m^3n^5 + 12m^5n^6}{3m^2n^4}$

(e) $\dfrac{40r^2t^4 - 30r^3t^5 + 50r^2t^7}{-10r^2t^4}$

(f) $\dfrac{-7m^3n^4 - 14m^5n^2 + 21m^6n^7}{-7m^3n^2}$

9. Divide.
(a) $(x^3 + 2x^2 - 4x - 3) \div (x + 3)$
(b) $(2x^3 - 7x^2 - 7x + 5) \div (2x - 1)$
(c) $(15m - 4m^3 - 9m^2 + 3m^4 - 4)$
$\div (3m - 4)$
(d) $(14t^2 - 7t^3 + t^4 - 7t + 1) \div (t^2 - 4t + 1)$
(e) $(6x^3 - 13x^2y + 8xy^2 - 3y^3) \div (2x - 3y)$

10. Simplify the following and state restrictions.

(a) $\left(\dfrac{25x^4}{7y^3}\right)\left(\dfrac{21y^4}{5x^2}\right)$

(b) $\dfrac{35x^4y^7}{7x^3y^4} \div \dfrac{5x^6y}{2xy^2}$

(c) $\dfrac{x^2 + 11x + 28}{x^2 + 4x - 21} \times \dfrac{x^2 - 2x - 3}{x^2 + 6x + 5}$

(d) $\dfrac{t^2 + 9t + 20}{t^2 + 7t + 12} \div \dfrac{t^2 + 3t - 10}{t^2 + t - 6}$

(e) $\dfrac{x^2 - 4x}{x^2 + 4x + 4} \times \dfrac{x^2 - 6x - 16}{x^2 - 12x + 32}$

(f) $\dfrac{x^2 + 5x + 4}{x^2 - 5x - 14} \times \dfrac{x^2 - 4x - 21}{x^2 + 7x + 12}$
$\times \dfrac{x^2 - 4x - 12}{x^2 - 6x - 7}$

(g) $\dfrac{x^2 - 25}{x^2 - 6x} \times \dfrac{x^2 - 12x + 36}{x^2 + 2x - 15}$
$\div \dfrac{x^2 - 11x + 30}{x^2 + 4x - 21}$

11. Simplify and state restrictions.

(a) $\dfrac{1}{x + 3} + \dfrac{1}{x + 4}$

(b) $\dfrac{2}{t - 3} - \dfrac{3}{t - 4}$

(c) $\dfrac{5}{2x + 3} - \dfrac{6}{2x - 3}$

(d) $\dfrac{2}{x^2 + 3x + 2} + \dfrac{4}{x^2 + 5x + 6}$

(e) $\dfrac{2m}{m^2 + 2m - 3} - \dfrac{3m}{m^2 + m - 6}$

(f) $\dfrac{3}{x^2 - x - 6} - \dfrac{4}{x^2 - 2x - 8}$

(g) $\dfrac{6}{x^2 - 4} + \dfrac{4}{x^2 + 4x + 4}$

(h) $\dfrac{5}{2x^2 - 7x - 4} - \dfrac{2}{3x^2 - 10x - 8}$

(i) $\dfrac{3x - 3}{x^2 + x - 2} + \dfrac{4x + 20}{x^2 + 6x + 5}$

(j) $\dfrac{5m - 10}{m^2 - 4} - \dfrac{2m - 6}{m^2 - 7m + 12}$

(k) $\dfrac{x^2 + 3x + 2}{x^2 - 3x - 4} + \dfrac{x^2 + 2x - 15}{x^2 + 3x - 10}$

12. Simplify and state restrictions.

(a) $\dfrac{6x^2 - 13x - 5}{8x^2 - 22x + 5} \times \dfrac{4x^2 - x - 14}{3x^2 - 5x - 2}$

(b) $\dfrac{49x^2 - 1}{21x^2 + 11x - 2} \div \dfrac{14x^2 + 9x + 1}{6x^2 + 7x + 2}$

(c) $\dfrac{15x^2 - 14x - 8}{9x^2 + 26x - 3} \times \dfrac{x^2 + 9x + 18}{3x^2 + 14x - 24}$
$\div \dfrac{5x^2 + 32x + 12}{9x^2 - 46x + 5}$

2.19 CHAPTER 2 TEST

1. $P(x) = 3x^3 - 4x^2 + 7$
(a) Find $P(3)$.
(b) Find $P(-2)$.

2. (a) Add.
$$4x^2 + 5x - 3$$
$$\underline{3x^2 - 7x + 4}$$
(b) Subtract.
$$4x^3 + 5x - 6$$
$$\underline{5x^3 - 3x + 4}$$

3. Simplify. $(3x^3 - 5x^2 + 8) - (-2x^3 + 7x - 5)$

4. Expand and simplify.
(a) $(3x - 2)(3x + 2)$
(b) $(5x - 2)(3x + 7)$
(c) $(x + 3)(2x - 5) + (x - 4)(3x + 1)$

5. Factor.
(a) $3x^3 - 12x^2 + 15x$
(b) $x^2 - 16y^2$
(c) $x^2 - 2x - 24$
(d) $2x^2 - 9x - 18$

6. Divide. $(6x^3 - x^2 - 11x + 10) \div (2x + 3)$

7. Simplify and state restrictions.
(a) $\dfrac{12x^3 - 15x^2 + 21x}{3x}$
(b) $\dfrac{x^2 - x - 20}{x + 4}$  $\quad x - 5$

8. Simplify and state restrictions.
(a) $\dfrac{x + 2}{x^2 - x - 6} \times \dfrac{x^2 - 9}{x + 3} \div \dfrac{x^2 - 4}{x + 3}$
(b) $\dfrac{x + 3}{x^2 - 16} + \dfrac{x - 2}{x^2 - x - 12} - \dfrac{x + 1}{x^2 + 7x + 12}$

THE STRAIGHT LINE
AND
LINEAR SYSTEMS

CHAPTER

If a man's wit be wandering, let him study the mathematics.
Francis Bacon

SLOPE

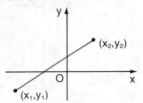

$$\text{Slope} = m = \frac{y_2 - y_1}{x_2 - x_1}$$

$$= \frac{\Delta y}{\Delta x}$$

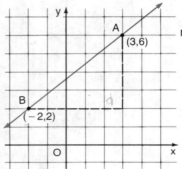

$$m_{AB} = \frac{6 - 2}{3 - (-2)}$$

$$= \frac{4}{5}$$

Horizontal lines have a slope of 0.

Vertical lines have no slope.

Lines that rise from left to right have a positive slope.

Lines that fall from left to right have a negative slope.

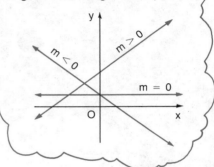

EXERCISE

1. Determine the slope of the line containing the given points.

(a) (3,2), (10,14)   (b) (1,3), (4,7)
(c) (3,3), (−5,2)   (d) (−1,0), (−3,4)
(e) (−6,−7), (−1,−4)
(f) (−4,6), (8,−3)
(g) (−6,10), (−11,7)
(h) (5,−9), (−4,−11)
(i) (5,−12), (0,−6)
(j) (7,−3), (7,6)
(k) (−3,−2), (9,−2)
(l) (5,−7), (16,3)
(m) (−4,−3), (−4,7)

2. Graph the line containing the given point and having the given slope.

(a) (3,2); $m = \frac{3}{4}$ (b) (1,2); $m = \frac{2}{3}$

(c) (−1,3); $m = \frac{2}{5}$ (d) (3,0); $m = -\frac{1}{2}$

(e) (0,−4); $m = -\frac{2}{3}$ (f) (−1,−2); $m = 3$

(g) (−5,4); $m = -2$ (h) (−2,5); $m = 0$
(i) (2,−4); $m = 1$ (j) (3,6); no slope

3. Determine the value of k so that the line containing the given points has the given slope.

(a) (3,2), (5,k); m = 2
(b) (−1,k), (3,15); m = 3
(c) (−2,4), (k,−2); $m = \frac{1}{2}$
(d) (k,5), (6,−1); m = −3
(e) (−3,4), (2,k); m = 0
(f) (k,7), (−3,−2); no slope

4. Determine the missing coordinate so that the given point lies on the given line.

(a) x + y = 7; (3,■), (■,6), (−3,■)
(b) x + y = −2; (5,■), (■,3), (−3,■)
(c) x − y = 3; (4,■), (■,6), (−4,■)
(d) x − y = −4; (6,■), (■,5), (−3,■)
(e) y − x = 4; (3,■), (■,3), (−2,◦)
(f) y = 2x + 3; (4,■), (■,5), (−2,■)
(g) y = −x − 2; (5,■), (■,10), (−8,■)
(h) 2x + 3y = 12; (3,■), (■,4), (−3,■)
(i) 4x − 5y = 20; (0,■), (■,4), (−5,■)

GEOMETRY

ANGLE THEOREMS

(OAT) Opposite Angle Theorem. If two lines intersect, then the opposite angles are equal.

(CAT) Complementary Angle Theorem. If two angles are equal, then their complements are equal.

(SAT) Supplementary Angle Theorem. If two angles are equal, then their supplements are equal.

(SATT) Sum of the Angles of a Triangle Theorem. The sum of the interior angles of a triangle is 180°.
When two angles of one triangle are respectively equal to two angles of another triangle, the third angles are equal.

(ITT) Isosceles Triangle Theorem. The angles opposite the equal sides are equal.

(EAT) Exterior Angle Theorem. An exterior angle of a triangle is equal to the sum of the interior and non-adjacent angles. .

POSTULATES

(SAS) Side-Angle-Side Congruence. If two sides and the contained angle of one triangle are respectively equal to two sides and the contained angle of another triangle, then the triangles are congruent.

(SSS) Side-Side-Side Congruence. If three sides of one triangle are respectively equal to three sides of another triangle, then the triangles are congruent.

(ASA) Angle-Side-Angle Congruence. If two angles and the contained side of one triangle are respectively equal to two angles and the contained side of another triangle, then the triangles are congruent.

EXERCISE

1. Find the value of x in each of the
following.

(a)

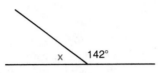

56° x

(b)

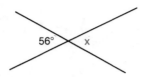

x   142°

(c)

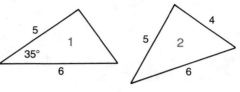

64°

x

(d)

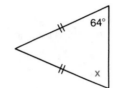

76°

41°   x

(e)

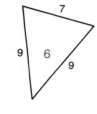

3x

2x

x

(f)

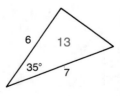

50°

3x   40°

2. Name the pairs of congruent
triangles below and give the reason.

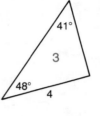

5
35°   1   6

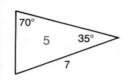

5   2   4
6

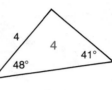

41°
3
48°   4
4

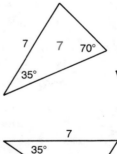

4
4   41°
48°

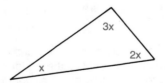

70°
5   35°
7

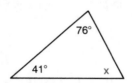

7
9   6   9

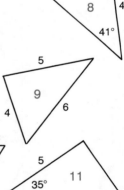

7   7   70°
35°

48°
8   4
41°

5
9   6
4

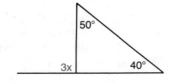

7
35°
10
6

5
35°   11
6

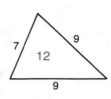

7   9
12
9

6   13
35°   7

PARALLEL LINES, TRANSVERSALS, AND ANGLES

When a transversal intersects two parallel lines,

(i) the alternate angles are equal;

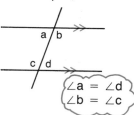

$$\angle a = \angle d$$
$$\angle b = \angle c$$

(ii) the corresponding angles are equal;

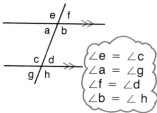

$$\angle e = \angle c$$
$$\angle a = \angle g$$
$$\angle f = \angle d$$
$$\angle b = \angle h$$

(iii) the interior angles on the same side of the transversal are supplementary.

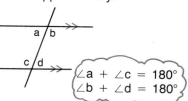

$$\angle a + \angle c = 180°$$
$$\angle b + \angle d = 180°$$

The three cases are remembered in the Z, F, and C patterns given below.

alternate angles

corresponding angles

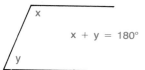

$$x + y = 180°$$

interior angles on the same side

EXERCISE

1. Find the values of a, b, and c.

(a) 70° a, b, c

(b) 65° c, b, a

(c) a, 80°, c, 30°, b

(d) a, 30°, c, 70°, b

(e) 55°, 50°, b, a, c

(f) 80°, 60°, c, b, a

(g) 60°, 40°, b, a, c

(h) 2a, b, a, c

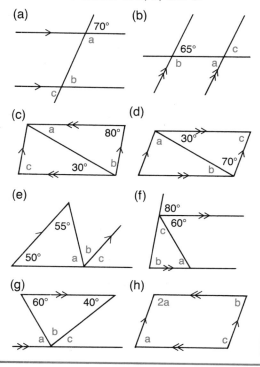

2. Find the values of a, b, c, and d.

(a) b, a, c, 80°, 50°, d

(b) a, d, 60°, 85°, c, b

(c) b, a, c, 50°, 40°, d

(d) b, a, d, 70°, 45°

(e) 85°, 75°, a, b, c, d

(f) b, a, d, c, 30°

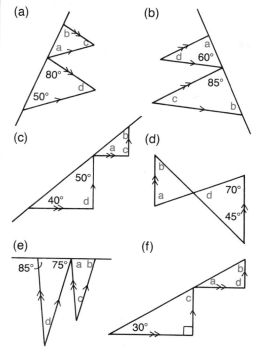

3.1 DEFINING AND GRAPHING THE STRAIGHT LINE

A straight line is a set of all points such that for any two points on the line $\frac{\Delta y}{\Delta x}$ is the same. We have called the constant value $\frac{\Delta y}{\Delta x}$ the slope. In the next section we will prove the following.

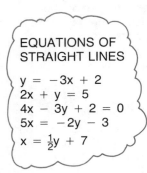

EQUATIONS OF STRAIGHT LINES

$y = -3x + 2$
$2x + y = 5$
$4x - 3y + 2 = 0$
$5x = -2y - 3$
$x = \frac{1}{2}y + 7$

The graph of every equation in the standard form $Ax + By + C = 0$, where A, B, and C are constants and A and B are not both zero, is always a straight line.

To graph a straight line we must find ordered pairs that satisfy the equation of the line. There are two common methods of doing this.

METHOD I. Solving for y

The equation is solved for y and ordered pairs are determined by substituting values for x.

EXAMPLE. Draw the graph of
$3x + 2y - 4 = 0$.

SOLUTION:
Solve the equation for y.
$$3x + 2y - 4 = 0$$
$$2y = 4 - 3x$$
$$y = \frac{4 - 3x}{2}$$

Unless otherwise stated, $x, y \in R$

Two points are sufficient to determine a straight line.

x	y
-2	5
4	-4

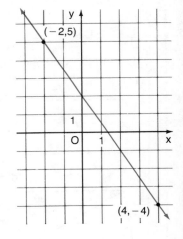

METHOD II. Determining the Intercepts

A particular graph may cross either one or both of the axes. The x-intercept, a, is the x-coordinate of the point where the line intersects the x-axis. At this point, y = 0. The y-intercept, b, is the y-coordinate where the line intersects the y-axis. At this point, x = 0. When the equation of a straight line is expressed in the form $Ax + By + C = 0$, then the x-intercept is $-\frac{C}{A}$ and the y-intercept is $-\frac{C}{B}$, where $A, B \neq 0$.

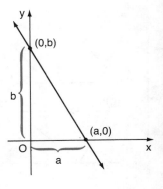

EXAMPLE. Draw the graph of $3x - 2y - 6 = 0$ using the intercept method.

SOLUTION:

To find the x-intercept, let $y = 0$.
$$3x - 2y - 6 = 0$$
$$3x - 2(0) - 6 = 0$$
$$3x = 6$$
$$x = 2$$

To find the y-intercept, let $x = 0$.
$$3x - 2y - 6 = 0$$
$$3(0) - 2y - 6 = 0$$
$$-2y = 6$$
$$y = -3$$

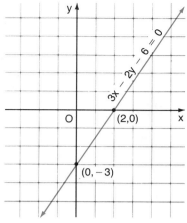

EXERCISE 3.1

A 1. State 3 ordered pairs that satisfy each of the following.

(a) $y = 3x + 2$ (b) $y = 2x - 5$
(c) $y = -4x + 5$ (d) $y = -7x$
(e) $y = \dfrac{x + 3}{2}$ (f) $y = \dfrac{1 - 2x}{3}$
(g) $y = 5$ (h) $x = -4$
(i) $y = -\dfrac{x - 1}{4}$ (j) $y = -\dfrac{x}{3} + 5$

2. State the x- and y-intercepts, if they exist, for each of the following.

(a) $5x + 3y = 15$ (b) $x + 3y = 6$
(c) $2x - y = 8$ (d) $4x - 7y = 28$
(e) $4x - 3y - 12 = 0$ (f) $y + 6 = 0$
(g) $x = -2$ (h) $2x = 5y + 10$
(i) $x - 2y = 2$ (j) $2x - 3y - 1 = 0$

3. Describe the graph of the line that passes through $(5,7)$ and $(5,-2)$.

4. Describe the graph of the line that passes through $(6,-2)$ and $(-4,-2)$.

5. State the equation of the horizontal line that passes through $(-5,6)$.

6. State the equation of the vertical line that passes through $(4,7)$.

B 7. Write each of the following in standard form, $Ax + By + C = 0$, where A, B, and C are integers.

(a) $y = 3x - 4$ (b) $3x = 2y + 7$

(c) $y - 5 = 2(x - 3)$
(d) $y + 3 = -4(x + 1)$

(e) $y - \dfrac{1}{2} = -\dfrac{1}{3}(x + 5)$

(f) $\dfrac{x}{3} = \dfrac{y}{4} - 6$

8. Solve for y in each of the following.

(a) $2x + y = 7$ (b) $x + 2y = 3$
(c) $3x + y - 4 = 0$ (d) $2x - y = 6$
(e) $5x - y - 1 = 0$ (f) $3x - 2y = 0$
(g) $5x + 2y - 9 = 0$ (h) $x - 5y = 6$
(i) $2x = 3y + 4$ (j) $3y + 7 = 0$

9. Graph each of the following.

(a) $y = 3x + 2$ (b) $2x - 3y = 6$
(c) $4x + 5y - 20 = 0$ (d) $x + 3y = 7$
(e) $x - \frac{1}{2}y - 4$ (f) $y - 2 = 3(x - 1)$
(g) $2x = 3$ (h) $5y + 10 = 0$
(i) $\dfrac{x}{2} + \dfrac{y}{3} = 1$ (j) $\dfrac{x}{4} - \dfrac{y}{2} = 1$

10. State the value of k so that the given point lies on the graph of the given equation.

(a) $2x + ky = 21$; $(3,5)$
(b) $kx - 3y = 11$; $(-1,-5)$
(c) $2kx + 5y - 6 = 0$; $(-1,2)$
(d) $x + 4y = 2k - 3$; $(1,1)$
(e) $(1 + k)x + ky = k + 2$; $(-2,4)$
(f) $3x + 7y = k^2 + 6$; $(0,1)$

THE STRAIGHT LINE AND LINEAR SYSTEMS   **81**

3.2 DETERMINING EQUATIONS OF LINES

The line graphed at the right passes through the point (5,2) and has a slope $m = -\frac{1}{2}$. To determine the equation of the line let (x,y) be any point on the line other than (5,2). By the definition of slope,

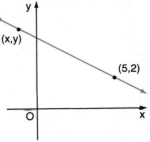

$$\frac{y - 2}{x - 5} = -\frac{1}{2}$$

$$y - 2 = -\frac{1}{2}(x - 5)$$

$$2y - 4 = -x + 5$$

$$x + 2y - 9 = 0$$

In general, if we let (x_1, y_1) be the given point, m the given slope, and (x,y) any other point on the line, then $\dfrac{y - y_1}{x - x_1} = m$

$$\boxed{y - y_1 = m(x - x_1)}$$

This is called the point-slope form for the equation of a straight line.

EXAMPLE 1. Determine the equation of the line passing through the points (4,3) and (−2,5). Express the equation in the form Ax + By + C = 0.

SOLUTION:

$$\text{Slope} = m = \frac{\Delta y}{\Delta x} = \frac{5 - 3}{-2 - 4} = \frac{2}{-6}$$
$$= -\frac{1}{3}$$

Select either point, say (4,3), to be (x_1, y_1) and use the point-slope form.

$$y - y_1 = m(x - x_1)$$

$$y - 3 = -\frac{1}{3}(x - 4)$$

$$y - 3 = -\frac{1}{3}x + \frac{4}{3}$$

$$3y - 9 = -x + 4$$

$$x + 3y - 13 = 0$$

When the given point lies on the y-axis, the result is an important case of the point-slope form.

The line graphed at the right passes through the point (0, −3) and has a slope m = 2. To determine the equation of the line we use the point-slope form for the equation of a line.

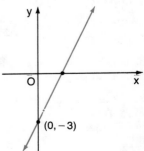

$$y - y_1 = m(x - x_1)$$

$$y + 3 = 2(x - 0)$$

$$y + 3 = 2x$$

$$y = 2x - 3$$

In general, if we let (0,b) be the point on the y-axis and m the given slope, then

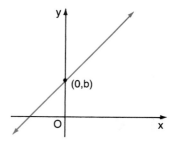

$$y - y_1 = m(x - x_1)$$
$$y - b = m(x - 0)$$
$$y - b = mx$$
$$y = mx + b$$

In each case the constant term (-3 or b) is the y-intercept of the line and the coefficient of x, (2 or m), is the slope of the line.

The general equation

$$y = mx + b$$

is called the slope y-intercept form of the equation of the line.

What we have shown above is that if (x,y) is a point on the line with slope m and y-intercept b, then it satisfies the equation $y = mx + b$.

Conversely, we can show that any equation of the form $y = mx + b$ represents a straight line.

Let $P_1(x_1,y_1)$ and $P_2(x_2,y_2)$ be points which lie on $y = mx + b$.

Then
$$y_1 = mx_1 + b$$
and
$$y_2 = mx_2 + b$$

$$\therefore \text{ the slope of } P_1P_2 = \frac{y_2 - y_1}{x_2 - x_1}$$
$$= \frac{(mx_2 + b) - (mx_1 + b)}{x_2 - x_1}$$
$$= \frac{m(x_2 - x_1)}{x_2 - x_1}$$
$$= m$$

Therefore, for any two points P_1 and P_2 on the line, $\frac{\Delta y}{\Delta x}$ is the same; so $y = mx + b$ represents a straight line.

$Ax + By + C = 0$ is a straight line since we can write it in the form $y = mx + b$.

There are two cases, namely when $B = 0$ and $B \neq 0$.

If $B = 0$, $Ax + By + C = 0$ reduces to $x = -\frac{C}{A}$ whose graph is a straight line parallel to the y-axis.

If $B \neq 0$, then $Ax + By + C = 0$ can be written in the form $y = mx + b$, namely $y = -\frac{A}{B}x - \frac{C}{B}$.

There are three other forms for the equation of a straight line.

If a line passes through the points (x_1,y_1) and (x_2,y_2), then its equation can be written as

$$y - y_1 = \frac{y_2 - y_1}{x_2 - x_1}(x - x_1)$$

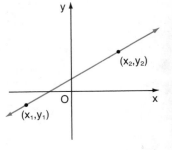

This is called the two-point form for the equation of a line.

If a line passes through a point on the x-axis, say (a,0), with slope m, then
$$y - y_1 = m(x - x_1)$$
$$y - 0 = m(x - a)$$

$$y = m(x - a)$$

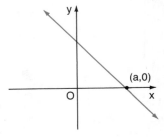

This is called the slope x-intercept form for the equation of a line.

If a line has x-intercept a and y-intercept b, then the slope of the line $m = -\frac{b}{a}$ and
$$y - y_1 = m(x - x_1)$$
$$y - 0 = -\frac{b}{a}(x - a)$$
$$y = -\frac{b}{a}x + b$$
$$\frac{b}{a}x + y = b$$

$$\frac{x}{a} + \frac{y}{b} = 1$$

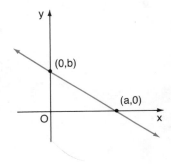

This is called the intercept form for the equation of a line.

EXAMPLE 2. Determine the slope and y-intercept of the line $2x + 3y = 10$.

SOLUTION:
Solve the equation for y.
$$2x + 3y = 10$$
$$3y = -2x + 10$$
$$y = -\frac{2}{3}x + \frac{10}{3}$$

The equation is now in the form $y = mx + b$ where $m = -\frac{2}{3}$ and $b = \frac{10}{3}$. Therefore, the slope is $-\frac{2}{3}$ and the y-intercept is $\frac{10}{3}$.

EXERCISE 3.2

A 1. State the point-slope form of the equation of the line through the given point and having the given slope.

(a) (3,2); m = 4   (b) (1,4); m = 6
(c) (−2,5); m = 1   (d) (−7,−2); m = 5
(e) (−3,−4); m = −2 (f) (6,0); m = −4
(g) (0,5); m = 2   (h) (−1,−2); m = 0

2. State the slope y-intercept form of the equation of the line with the given slope m and y-intercept b.

(a) m = 2; b = 3   (b) m = −4; b = 5
(c) m = 5; b = −3   (d) m = $\frac{1}{2}$; b = −7

3. State the slope and y-intercept of each of the following.

(a) $y = 3x - 7$ (b) $y = -\frac{1}{2}x + 5$
(c) $y = -5x - 4$ (d) $y = 7x$
(e) $y = \dfrac{2x - 1}{3}$ (f) $y + 7 = 2x$
(g) $y + 3x = 4$ (h) $2x - 3 = y$
(i) $5x + y - 2 = 0$ (j) $7x = 3 + y$
(k) $4x - y = 5$ (l) $3x + 2y = 5$
(m) $x + 3y = -2$ (n) $6x - y - 4 = 0$

B 4. Determine an equation of the line through the given point and having the given slope. Express the equation in the form Ax + By + C = 0.

(a) (2,1); m = 3   (b) (−3,4); m = 5
(c) (4,−5); m = −1   (d) (−1,−2); m = 7
(e) (5,0); m = −6   (f) (0,−5); m = $\frac{1}{2}$
(g) (−1,6); m = 0   (h) (4,−3); no slope
(i) (−6,4); m = $-\frac{1}{2}$ (j) (−7,−1); m = $\frac{1}{3}$

5. Write the equation of the line containing the point (−4,5) having the given slope.

(a) 2   (b) −4
(c) 1.5   (d) −0.5

6. Write the equation of the line having slope $\frac{4}{3}$ and containing the given point.

(a) (0,0)   (b) (−3,2)
(c) (5,−3)   (d) (−1,−3)

7. Determine an equation of the line through the given point and having the given slope. Express the equation in the form y = mx + b and determine two other points on the line.

(a) (1,2); m = 3   (b) (−1,−1); m = −2
(c) (3,0); m = 2   (d) (4,−2); m = −0.5
(e) (0,−3); m = $\frac{1}{2}$ (f) (−5,−6); m = 0
(g) (−3,4); m = $\frac{3}{4}$ (h) (10,20); m = 0.2

8. Determine an equation of the line through the given points. Express the equation in the form Ax + By + C = 0.

(a) (1,2), (2,4)   (b) (5,6), (7,8)
(c) (2,3), (−2,−1)   (d) (−1,4), (5,1)
(e) (−2,−1), (4,−2)   (f) (−6,4), (−1,−2)
(g) (10,8), (−3,1)   (h) (2,5), (7,−4)
(i) (3,−1), (−6,−5)   (j) (−6,7), (1,−3)
(k) $(2\frac{1}{2}, \frac{1}{2})$, $(−3, −2\frac{1}{2})$ (l) $(−\frac{3}{4}, −1\frac{1}{2})$, $(2,2\frac{1}{4})$
(m) (−3.8,1.2), (1.2,2.7)
(n) (−4.5,6), (−1.5,−4)
(o) (−4,0), (0,−8)   (p) (4,6), (4,−2)

9. Determine the slope and y-intercept for each of the following. Determine two other points on each line.

(a) $4x + y = 7$ (b) $3x + 2y = 5$
(c) $2x - 3y - 1 = 0$ (d) $x - 2y = 4$
(e) $4y - x = 2$ (f) $2x + 3y = 0$

10. Determine an equation of the line with the given slope and x-intercept.

(a) m = 3; a = 4   (b) m = −2; a = 5
(c) m = −1; a = −3 (d) m = 5; a = −2
(e) m = $\frac{1}{4}$, a = 0   (f) m = 0.2; a = −1
(g) m = −0.5; a = 1.4
(h) m = −2.6; a = −5.2

11. Determine an equation of the line whose x- and y-intercepts are the following.

(a) 3 and 6   (b) 2 and −5   (c) −8 and 4
(d) −2 and 4   (e) −1 and −6 (f) 3 and −7

12. An elevator has a maximum load of 1000 kg. The average mass of a child is 36 kg. The average mass of an adult is 77 kg. If the elevator is filled to capacity, write an equation in two variables to represent the number of adults and children in it.

3.3 PARALLEL AND PERPENDICULAR LINES

When two different lines in a plane have the same slope or no slope (vertical lines), the lines are parallel. This can be proved as follows.

> The length of a line segment AB may be written as AB, $\overline{AB}$, or |AB|. In this text we use the convention AB.

Given: slope ℓ_1 = slope ℓ_2 = m
Required: To prove $\ell_1 \parallel \ell_2$
Proof: If ℓ_1 and ℓ_2 have no slope, both are parallel to the y-axis. Hence $\ell_1 \parallel \ell_2$.
Likewise if m = 0, ℓ_1 and ℓ_2 are parallel to the x-axis. Hence $\ell_1 \parallel \ell_2$.
If m ≠ 0, then ℓ_1 and ℓ_2 meet the x-axis at A and D respectively.
Locate B and E to the right of A and D respectively, so that
AB = DE = 1

Draw perpendiculars at B and E to intersect ℓ_1 and ℓ_2 at C and F respectively, so that
∠ABC = ∠DEF = 90°
Since AB = DE = 1
∴ CB = m = FE
∴ △ABC ≅ △DEF (SAS)
∴ ∠CAB = ∠FDE

∴ $\ell_1 \parallel \ell_2$ (corresponding angles are equal)

Conversely, by reversing the above argument, it can be proved that two parallel lines in a plane either have the same slope or have no slope (vertical lines).

> If two non-vertical lines are parallel, they have the same slope; conversely, if two distinct non-vertical lines have the same slope, they are parallel.

EXAMPLE 1.   Write an equation, in standard form, of the line containing (-4,6) and parallel to the graph of 3x + 2y = 7.

SOLUTION:
Determine the slope of 3x + 2y = 7 by expressing the equation in the form y = mx + b.
$$3x + 2y = 7$$
$$2y = -3x + 7$$
$$y = -\tfrac{3}{2}x + \tfrac{7}{2}$$

The slope of the required line containing (-4,6) is $-\tfrac{3}{2}$. Finding the equation.
$$y - y_1 = m(x - x_1)$$
$$y - 6 = -\tfrac{3}{2}(x + 4)$$
$$2y - 12 = -3x - 12$$
$$3x + 2y = 0$$

Two lines which intersect at right angles are called perpendicular lines. The slopes of perpendicular lines are negative reciprocals of each other. This can be proved as follows.

Given: $\ell_1 \perp \ell_2$

Required: To prove $m_1 = -\dfrac{1}{m_2}$

Proof: If ℓ_1 and ℓ_2 are the original lines, consider the parallel lines that pass through the origin. Since the lines are parallel the angles don't change.
Locate A and C as shown so that

$$OA = OC = 1$$

$\angle BOA = \angle COD$ (each is the complement of $\angle COB$)

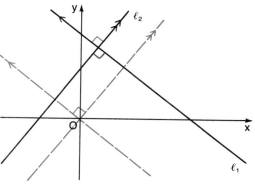

$$\angle BAO = \angle DCO = 90°$$
$$\therefore \triangle ABO \cong \triangle CDO \text{ (ASA)}$$
$$\therefore AB = DC$$

The coordinates of B are $(1, m_2)$ since the slope of ℓ_2 is m_2.

Since AB = CD, the coordinates of D are $(-m_2, 1)$. The slope of the line joining $D(-m_2, 1)$ to $O(0,0)$ is

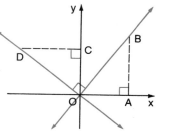

$$m_1 = \frac{\Delta y}{\Delta x} = \frac{1 - 0}{-m_2 - 0}$$
$$= -\frac{1}{m_2}$$

Conversely, it can be proved that if the slopes of two lines are negative reciprocals of each other, the lines are perpendicular. This theorem may also be stated as follows.

> If two lines ℓ_1 and ℓ_2 with slopes $m_1 \neq 0$ and $m_2 \neq 0$ respectively are perpendicular, then $m_1 m_2 = -1$; conversely, if the slopes of two lines are such that $m_1 m_2 = -1$, then the lines are perpendicular.

EXAMPLE 2. Write an equation, in standard form, of the line containing $(4, -2)$ and perpendicular to the graph of $3x - 2y + 4 = 0$.

SOLUTION:
Determine the slope of $3x - 2y + 4 = 0$.
$$3x - 2y + 4 = 0$$
$$-2y = -3x - 4$$
$$y = \tfrac{3}{2}x + 2$$

Slope of given line is $\tfrac{3}{2}$.

Slope of required line is $-\dfrac{1}{\frac{3}{2}} = -\tfrac{2}{3}$.

The slope of the line containing $(4, -2)$ is $-\tfrac{2}{3}$. Determining the equation:
$$y - y_1 = m(x - x_1)$$
$$y + 2 = -\tfrac{2}{3}(x - 4)$$
$$3y + 6 = -2x + 8$$
$$2x + 3y - 2 = 0$$

EXERCISE 3.3

A 1. Given the slope m of a line ℓ, state the slope of a line

(a) parallel to ℓ.

(b) perpendicular to ℓ.

(i) $m = 2$ (ii) $m = \frac{1}{4}$ (iii) $m = -\frac{3}{4}$

(iv) $m = -4$ (v) $m = -\frac{3}{2}$ (vi) $m = \frac{5}{6}$

2. Given the slopes of two lines, classify the lines as parallel, perpendicular, or neither parallel nor perpendicular.

(a) $m_1 = 3$, $m_2 = \frac{1}{3}$ (b) $m_1 = \frac{3}{4}$, $m_2 = \frac{6}{8}$

(c) $m_1 = -2$, $m_2 = \frac{1}{2}$ (d) $m_1 = -\frac{1}{2}$, $m_2 = \frac{1}{2}$

(e) $m_1 = \frac{4}{3}$, $m_2 = -\frac{3}{4}$ (f) $m_1 = 0$, $m_2 = 1$

3. Classify the lines whose equations are given as parallel, perpendicular, or neither parallel nor perpendicular.

(a) $y = 2x + 5$, $y = 2x - 1$

(b) $y = \frac{1}{2}x + 3$, $y = -2x - 3$

(c) $y = 3x - 1$, $y = -3x + 1$

(d) $y = x + 5$, $y = -x - 4$

(e) $2y = 6x + 7$, $y = 3x - 5$

B 4. Classify the lines determined by two pairs of points as parallel, perpendicular, or neither parallel nor perpendicular.

(a) $(-2,8)$, $(3,7)$ and $(4,3)$, $(9,2)$

(b) $(0,1)$, $(-5,4)$ and $(5,3)$, $(0,5)$

(c) $(2,5)$, $(8,7)$ and $(-3,1)$, $(-2,-2)$

(d) $(4,6)$, $(-3,-1)$ and $(6,-3)$, $(4,5)$

5. Write an equation, in standard form, of the line satisfying the following conditions:

(a) through $(4,6)$ and parallel to the graph of $y = 3x + 4$.

(b) through $(-2,-3)$ and parallel to the graph of $y + 2x = 6$.

(c) through $(-1,5)$ and perpendicular to the graph of $y = -3x + 7$.

(d) through $(-3,-2)$ and perpendicular to $y - 2x + 6 = 0$.

(e) through $(5,2)$ and parallel to $3x - 5y = 6$.

(f) through $(-5,6)$ and perpendicular to $2x + 3y - 7 = 0$.

(g) having the same x-intercept as $3x + 5y - 15 = 0$ and parallel to $5x + 2y = 7$.

(h) having the same y-intercept as $2x - 3y = -6$ and perpendicular to $4x - y = 6$.

6. Find an equation for the line satisfying the following conditions:

(a) with slope $-\frac{3}{2}$ and x-intercept 4.

(b) through $(-1,4)$ and perpendicular to the x-axis.

(c) with slope $\frac{1}{5}$ and y-intercept -2.

(d) through $(5,-2)$ and parallel to the x-axis.

7. Show that the triangle with vertices $(3,3)$, $(8,17)$, and $(11,5)$ is a right triangle.

8. Prove using slopes that $(-8,4)$, $(3,2)$, $(9,10)$, and $(-2,12)$ are the vertices of a parallelogram.

9. Prove using slopes that $(-1,1)$, $(1,-1)$, $(6,4)$, and $(4,6)$ are the vertices of a rectangle.

10. For each pair of equations, find a value of k so that the graph of the first equation is parallel to that of the second.

(a) $y = kx + 3$; $3y = 2x + 4$

(b) $3x - ky = 7$; $y = -2x + 5$

(c) $4x + ky - 2 = 0$; $3x - 2y = 5$

11. For each pair of equations, find a value of k so that the graph of the first equation is perpendicular to that of the second.

(a) $y = kx - 4$; $3y = 2x + 7$

(b) $5x - ky = 3$; $y = -3x + 2$

(c) $4x + ky = 6$; $5x - 2y + 5 = 0$

12. Determine an equation of the line containing $(5,4)$ and having its x-intercept equal to its y-intercept.

C 13. Find an equation of the line containing (s,t) and parallel to the graph of $Ax + By + C = 0$, where $A \neq 0$ and $B \neq 0$.

14. Given $R(a,b)$, $T(2a,2b)$, $a \neq 0$, and $b \neq 0$, write an equation, in standard form, of the line through R that is perpendicular to RT.

15. Given $R(a,b)$, $S(3a,3b)$, $T(a,2b)$, $a \neq 0$, and $b \neq 0$, write an equation, in standard form, of the line through T that is parallel to RS.

3.4 APPLICATIONS: VARIATION

In this section we will study some applications of linear functions. Some liberties will be taken in using equations and graphs to solve problems since mathematical methods may not fit real world situations exactly. However, the models are close enough for most practical purposes.

In the problems that follow, assume that a straight line is the best mathematical model that fits the situation.

APPLICATIONS OF THE FORM $y = mx$: Direct Variation

An equation of the form $y = mx$ is called a direct variation. We say that y varies directly as x or y is directly proportional to x. The constant m is called the constant of proportionality or constant of variation. In this case, since $y = mx$, then $\frac{y}{x} = m$, or the ratio of $\frac{y}{x}$ is constant.

EXAMPLE 1. Since light travels much faster than sound, during a thunderstorm you see the lightning before you hear the thunderclap. The time interval between the flash and the thunderclap is directly proportional to the distance between you and the storm.

(a) Write an equation for this direct variation.
(b) If the thunderclap from lightning 1980 m away takes 6 s to reach you, determine the constant of proportionality.
(c) If the time interval is 10 s, how far away is the storm centre?
(d) Plot a graph of time versus distance.
(e) What quantity does the slope (constant) represent?
(f) Use the graph to determine the time interval if the storm is 2.5 km away.

SOLUTION:
(a) Distance is directly proportional to time, and we write:
$$d \propto t$$
$$d = mt$$

The general direct variation is given by $d = mt$, $m \neq 0$, where m is the constant of proportionality.

(b)
$$d = mt$$
$$1980 = 6\,m$$
$$m = 330$$

$E=mc^2$

$\propto$
is read
"varies as"

(c)

$$d = 330t \qquad \text{or} \qquad \frac{d}{t} = 330$$

$$d = 330(10) \qquad \frac{d}{10} = 330$$

$$= 3300 \qquad d = 3300$$

The storm is 3300 m away.

(d)

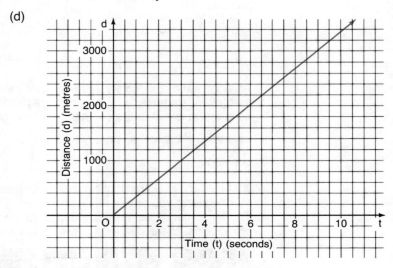

(e) The approximate speed of sound, 330 m/s.

(f) Approximately 7.5 s.

APPLICATIONS OF THE FORM y = mx + b: Partial Variation

An equation of the form y = mx + b is called a partial variation where b is constant and y varies directly as x.

EXAMPLE 2. The Mainway Restaurant has banquet facilities for up to 200 people. When the owner quotes a price for a banquet he is including the room rent plus the cost of the meal. A banquet for 60 people will cost $1300. For 90 people the price is $1900.

(a) Plot a graph of cost versus number of people.
(b) From the graph, determine the cost of a banquet for 80 people.
(c) Determine an equation in the form c = mn + b, where c represents the cost and n the number of people.

(d) Use the equation to determine the cost of a banquet for 70 people.

(e) What quantity does the slope of the line represent?

(f) What meaning does the y-intercept have?

SOLUTION:

(a)

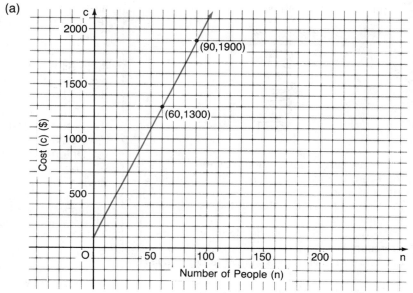

(b) The cost of a banquet for 80 people is $1700.

(c) $m = \dfrac{\Delta y}{\Delta x} = \dfrac{1900 - 1300}{90 - 60}$

$ = \dfrac{600}{30}$

$ = 20$

$c - c_1 = m(n - n_1)$

$c - 1300 = 20(n - 60)$

$c - 1300 = 20n - 1200$

$ c = 20n + 100$

(d) $c = 20n + 100$

$ = 20(70) + 100$

$ = 1400 + 100$

$ = 1500$

The cost of a banquet for 70 people is $1500.

(e) The slope represents the price per meal.

(f) The y-intercept represents the room rent.

EXERCISE 3.4

B 1. In order to determine the "law of stretch" for a particular coil spring the following test results were obtained. When a mass of 40 g was added, the length was 60 cm. For an 80 g mass, the length was 90 cm.

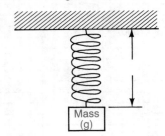

(a) Plot a graph of length versus mass.
(b) From the graph determine the length for a mass of 70 g.
(c) Determine an equation in the form
y = mx + b.
(d) What quantity does the slope of the line represent?
(e) What meaning does the y-intercept have?

2. In the 18th century a French scientist, Jacques Charles, discovered that when he plotted the volume of a fixed amount of gas versus the temperature, a straight line resulted. At 27°C, the volume was 400 cm³ and at 81°C the volume was 472 cm³.

(a) Plot a graph of volume versus temperature.
(b) From the graph determine the volume at 50°C.
(c) From the graph determine the temperature if the volume is 450 cm³.
(d) Determine an equation in the form
v = mt + b, where v represents the volume and t the temperature.
(e) Use the equation to determine the volume when the temperature is 63°C.
(f) What quantity does the slope of the line represent?
(g) What meaning does the y-intercept have?

3. Normally, property tax is directly proportional to the assessed value of the property. In Satellite City, the tax is $2400 on a property with an assessed value of $96 000.

(a) Write an equation for this direct variation.

(b) Determine the constant of proportionality.
(c) If the assessed value of a property is $80 000, what is the property tax?
(d) If the property tax is $2800, what is the assessed value of the property?
(e) If 1 mill = $0.001, what is the mill rate for Satellite City?

4. In Silver City, a driver must pay a fine for exceeding the speed limit. He is fined $60 for driving 55 km/h and $135 for driving 80 km/h.

(a) Plot a graph of fine versus speed.
(b) From the graph determine the fine for driving 70 km/h.
(c) Determine an equation.
(d) Use the equation to determine the fine for driving 60 km/h.
(e) What quantity does the slope of the line represent?
(f) What is the legal speed limit.

5. If there are 1.8 g of haemoglobin in 15 mL of normal human blood, how many grams of haemoglobin would 28 mL of the same blood contain?

6. The dividend on 500 shares of stock is $875. What is the dividend on 160 shares of the same stock?

7. The Pollution Control Department determined that a factory increased the pollution in a lake by 8 g/m³ in fourteen months. At the same rate, how long would it take to increase the pollution level by 26 g/m³?

8. A pharmacist read the following instructions on a bottle of medicine: "Mix with water in the ratio of 5 mL of water per 3 mL of medicine." How much water should be mixed with 37 mL of medicine?

9. An automobile company used the following procedure to determine the gasoline consumption of a new car. The tank was filled with gas and the test driver drove around the test track at a constant highway speed. After 90 km, 85 L of gas remained in the tank. After 360 km, 40 L of gas remained.

(a) Plot a graph of litres of gas remaining versus distance driven.
(b) From the graph, determine the number of litres remaining after 180 km.
(c) Determine an equation.
(d) Use the equation to determine the number of litres remaining after 240 km.
(e) What quantity does the slope of the line represent?
(f) What is the capacity of the gas tank?
(g) At the test speed, how many kilometres would you expect to drive on a full tank of gas?
(h) Express the gasoline consumption as L/100 km.

C INVERSE VARIATION

36 tiles can be arranged to form rectangles. Here are 2 examples.

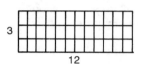

3 — 12

4 — 9

| L | 36 | 18 | 12 | 9 | 6 | 4 | 3 | 2 | 1 |
| W | 1 | 2 | 3 | 4 | 6 | 9 | 12 | 18 | 36 |

From the chart we see that as the width increases, the length decreases. This is called an inverse variation, and we write

$$L \propto \frac{1}{W}$$

and say "L is inversely proportional to W." We can show the relationship on a graph.

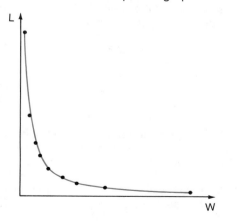

The general inverse variation is given by

$$xy = k \qquad \text{or} \qquad y = \frac{k}{x}$$

where $k \neq 0$ is the constant of variation.

EXAMPLE. The amount of force required on a wrench handle varies inversely as the length of the wrench handle. It takes 450 N of force to loosen a bolt with a wrench handle 20 cm long.
(a) Write an equation that expresses force in terms of length.
(b) Find the force needed for a wrench handle 30 cm long.

SOLUTION:
(a) Let the force be n newtons (N) and the length of the handle be d cm.
An equation is

$$n \propto \frac{1}{d}$$

$$n = \frac{k}{d} \qquad \text{and} \qquad nd = k$$

For n = 450 and d = 20
$$450 \times 20 = k$$
$$k = 9000$$

$$n = \frac{9000}{d} \qquad \text{or} \qquad nd = 9000$$

(b) For d = 30
$$n = \frac{9000}{30}$$
$$= 300$$

∴ a force of 300 N is required.

10. Plot the graph for the inverse variation
$$nd = 9000$$
for $10 \leqslant d \leqslant 90$.

11. Calculate the constant of variation for each of the following.
(a) x varies inversely as y.
$$x = 30 \text{ when } y = 1.5$$
(b) m is inversely proportional to n.
$$m = 125 \text{ when } n = 25$$
(c) V varies inversely as p.
$$V = 400 \text{ when } p = 40$$
(d) (4,100) is an ordered pair that belongs to a variation where b varies inversely as i.

12. Boyle's law states that the volume of a fixed amount of gas varies inversely as the pressure of the gas.

(a) Write an equation to express volume, V, in terms of pressure, p, if
$$V = 280 \text{ mL when } p = 80 \text{ kPa.}$$

(b) Draw the graph of the variation for
$$75 \leqslant p \leqslant 225.$$

(c) Prove that if (p_1, V_1) and (p_2, V_2) are ordered pairs that belong to the variation, then
$$p_1 V_1 = p_2 V_2$$

JOINT VARIATION

EXAMPLE. The cost of patrolling a national park during the summer months varies directly as the number of people working and directly as the number of days that they work.

(a) Write an equation for this joint variation.
(b) Determine the constant of proportionality if it costs $70 000 for twenty people to work for seventy days.
(c) What quantity does the constant of proportionality represent?
(d) Determine the cost if thirty people work for seventy-five days.

SOLUTION:

(a) Cost, C, varies directly as the number of people, p, and directly as the number of days worked, d.
$$C = kpd$$
(b)
$$70\,000 = k(20)(70)$$
$$70\,000 = 1400\,k$$
$$k = 50$$

(c) The constant of proportionality represents the daily wage.
(d)
$$C = 50\,pd$$
$$= 50(30)(75)$$
$$= 112\,500$$

It costs $112 500 for thirty people to work for seventy-five days.

13. P varies directly with Q and directly with R. When $Q = 30$ and $R = 25$, $P = 5250$.

(a) Write an equation for this joint variation and determine the value of k.
(b) Find the value of P when $Q = 65$ and $R = 15$.

14. D varies directly with E and inversely with F. When $E = 40$ and $F = 28$, $D = 30$.

(a) Write an equation for this joint variation and determine the value of k.
(b) Find the value of D when $E = 35$ and $F = 7$.

15. X varies directly with the square of Y and inversely with Z. When $Y = 5$ and $Z = 3$, $X = 150$.

(a) Write an equation for this joint variation and determine the value of k.
(b) Find the value of X when $Y = 11$ and $Z = 9$.

16. The cost of publishing a magazine varies directly as the number of pages in the magazine, and directly as the number of magazines printed.

(a) It costs $600 000 to publish 2000 copies of a 50 page magazine. Write an equation for this joint variation and determine the value of k.
(b) How much will it cost to publish 3500 copies of a 75 page magazine?

17. The resistance of a wire varies directly as its length, and inversely as the square of its diameter.

(a) A wire 0.6 m long and 0.005 m in diameter has a resistance of 7200 Ω. Write an equation for this joint variation and determine the value of k.
(b) Determine the resistance of a wire that is 1.8 m long having a diameter of 0.006 m.

18. The safe load for a horizontal beam varies directly as the width and the square of the depth, and inversely as the length.

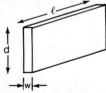

(a) A beam 3 m long and 5 cm wide with a depth of 20 cm, can support a load of 1000 kg. Write an equation for this joint variation and determine the value of k.
(b) What load will a beam 3 m long and 4 cm wide with a depth of 15 cm support?

MICRO MATH

3.5 EQUATION OF A LINE

We have seen that if a line passes through two points (x_1, y_1) and (x_2, y_2), then its equation can be written as

$$y - y_1 = \frac{y_2 - y_1}{x_2 - x_1}(x - x_1)$$

We can write this equation in the form $y = mx + b$ as follows.

$$y - y_1 = \frac{x(y_2 - y_1)}{x_2 - x_1} - \frac{x_1(y_2 - y_1)}{x_2 - x_1}$$

$$y = \frac{x(y_2 - y_1)}{x_2 - x_1} + y_1 - \frac{x_1(y_2 - y_1)}{x_2 - x_1}$$

$$y = \frac{x(y_2 - y_1)}{x_2 - x_1}$$
$$+ \frac{y_1(x_2 - x_1) - x_1(y_2 - y_1)}{x_2 - x_1}$$

$$y = \frac{x(y_2 - y_1)}{x_2 - x_1}$$
$$+ \frac{x_2y_1 - x_1y_1 - x_1y_2 + x_1y_1}{x_2 - x_1}$$

$$y = \frac{x(y_2 - y_1)}{x_2 - x_1} + \frac{x_2y_1 - x_1y_2}{x_2 - x_1}$$

$$y = \frac{y_2 - y_1}{x_2 - x_1}x + \frac{x_2y_1 - x_1y_2}{x_2 - x_1}$$

$$y = mx + b$$

The following BASIC program can be used to find the equation of a line in the form $y = mx + b$ when you are given two points on the line.

NEW
```
10 PRINT "GIVEN TWO POINTS"
20 PRINT "INPUT X1, Y1"
30 INPUT "X1=";X1
40 INPUT "Y1=";Y1
50 PRINT "INPUT X2, Y2"
60 INPUT "X2=";X2
70 INPUT "Y2=";Y2
80 IF X1 = X2 THEN 140
90 IF Y1 = Y2 THEN 160
100 M=(Y2-Y1)/(X2-X1)
110 B=(X2*Y1-X1*Y2)/(X2-X1)
120 PRINT "Y = ";M;"X + (";B;")"
130 END
140 PRINT "X = ";X1
150 GOTO 130
160 PRINT "Y = ";Y1
170 GOTO 130
```
RUN

EXERCISE 3.5

B 1. Use the program to find the equation of the line through the given points.

(a) (4,2) and (3,5)
(b) (6, −5) and (8, −3)
(c) (6,9) and (4, −1)
(d) (8, −4) and (8,5)
(e) (7, −11) and (−4, −11)

2. The following program can be used to find the equation of a line in the form $Ax + By + C = 0$ when you are given two points on the line and the line is neither horizontal nor vertical.

NEW
```
10 PRINT "AX + BY + C = 0"
20 PRINT "INPUT X1, Y1"
30 INPUT "X1=";X1
40 INPUT "Y1=";Y1
50 PRINT "INPUT X2, Y2"
60 INPUT "X2=";X2
70 INPUT "Y2=";Y2
80 A=Y2-Y1
90 B=X1-X2
100 C=X2*Y1-X1*Y2
110 PRINT A;"X+(" B ")Y+(" C ")=0"
120 END
```
RUN

3. Use the program to find the equation of the line in the form $Ax + By + C = 0$ through the given points.

(a) (1,4) and (2, −5)
(b) (−6,8) and (9, −11)
(c) (4,5) and (−3, −4)
(d) (−11, −13) and (−19, −24)
(e) (56, −18) and (−27, −43)
(f) (123, −56) and (−81, −66)
(g) (204, −333) and (−311, −367)

4. Modify the program in Question 2 to print the slope and the intercepts given two points on a line.

3.6 SOLVING LINEAR SYSTEMS GRAPHICALLY

A pair of linear equations in two variables is called a system of equations.

A solution of a system of equations is an ordered pair which satisfies both equations in the system. You can solve a system of equations by graphing each equation. The point of intersection of the two graphs, if any, is a solution of the system. Unless otherwise stated all variables represent real numbers.

EXAMPLE. Solve the following system graphically.

$$x - 2y = 2 \qquad ①$$
$$3x + 4y = 16 \qquad ②$$

SOLUTION:

Write each equation in the form $y = mx + b$.

① $x - 2y = 2$	② $3x + 4y = 16$
$2y = x - 2$	$4y = -3x + 16$
$y = \dfrac{x}{2} - 1$	$y = -\dfrac{3x}{4} + 4$

Draw the graph of each equation.

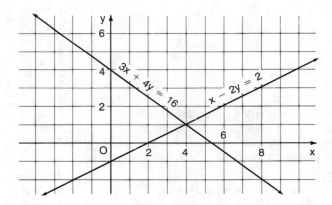

From the graph, the solution is the ordered pair (4,1). We can check the solution by substitution.

Check in ①.	Check in ②.
L.S. $= x - 2y$	L.S. $= 3x + 4y$
$= 4 - 2(1)$	$= 3(4) + 4(1)$
$= 4 - 2$	$= 12 + 4$
$= 2$	$= 16$
R.S. $= 2$	R.S. $= 16$

When you graph two linear equations in two variables, the resulting lines may (a) intersect at one point, (b) coincide, or (c) be parallel.

(a)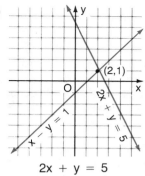

$$2x + y = 5$$
$$x - y = 1$$

(b)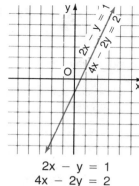

$$2x - y = 1$$
$$4x - 2y = 2$$

(c)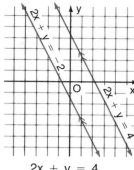

$$2x + y = 4$$
$$2x + y = -2$$

Systems that have solutions, such as (a) and (b), are called consistent. A system, such as (a), that has one solution is called independent. A system, such as (b), that has infinitely many solutions is called dependent. A system, such as (c), that has no solution is called inconsistent.

The following chart gives a summary of the possibilities for the graphs of two linear equations in two variables.

Graphs of Equations	Number of Solutions	Slopes of Lines	Name of System
Lines intersect	one	different slopes	consistent and independent
Lines coincide	infinitely many	same slopes, same intercepts	consistent and dependent
Lines are parallel	none	same slopes, different intercepts	inconsistent

EXERCISE 3.6

B 1. Solve each system of equations graphically. State whether the system is consistent and independent, consistent and dependent, or inconsistent.

(a) $y = x + 1$
    $y = 2x - 3$
(b) $y = -x - 8$
    $y = -3x + 4$
(c) $x + y = 5$
    $2x + 2y = 7$
(d) $x - y - 5 = 0$
    $3x = 15 + 3y$
(e) $2x - y - 1 = 0$
    $5x + 2y = -11$
(f) $2x + 3y - 6 = 0$
    $3x + 2 = y$

2. Write a system of two linear equations in two variables that has the given number of solutions.

(a) infinitely many solutions
(b) one solution
(c) no solution

C 3. Write an equation, in standard form, of the line passing through (2,3) and inconsistent with $2x - y = 5$.

4. Find the values of s and t so that $3x + sy = t$ and $9x + 12y = 15$ are a dependent and consistent system of equations.

3.7 SOLVING LINEAR SYSTEMS ALGEBRAICALLY

Graphing is not always the best way to find a solution, especially when the solutions are not integers. We will now investigate two algebraic methods of solving linear systems of equations. However, there are two important facts that should be reviewed before this is done.

MULTIPLES OF AN EQUATION

There are many ordered pairs that satisfy the equation $2x - y - 4 = 0$. The same ordered pairs satisfy $4x - 2y - 8 = 0$ or $2(2x - y - 4) = 0$. The equation $4x - 2y - 8 = 0$ is called a multiple of $2x - y - 4 = 0$.

In general, if (x_1, y_1) satisfies the equation $Ax + By + C = 0$, then it will also satisfy $k(Ax + By + C) = 0$, where k is any constant.

EQUIVALENT SYSTEMS

Two systems are said to be equivalent if they have the same solution set. Consider the following system of equations.

$$2x + y - 7 = 0 \qquad ①$$
$$x - 2y - 1 = 0 \qquad ②$$

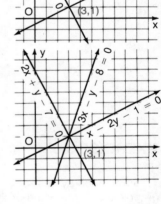

The ordered pair $(3,1)$ is the solution of this system.
If we add equations ① and ② we get
$$(2x + y - 7) + (x - 2y - 1) = 0$$

This equation is also satisfied by $(3,1)$ and may be simplified to
$$3x - y - 8 = 0$$

In other words,

$$\begin{cases} 2x + y - 7 = 0 \\ x - 2y - 1 = 0 \end{cases} \quad \begin{cases} 2x + y - 7 = 0 \\ 3x - y - 8 = 0 \end{cases} \quad \begin{cases} x - 2y - 1 = 0 \\ 3x - y - 8 = 0 \end{cases}$$

are equivalent systems since they have the same solution.

If we extend this idea to include the multiples of an equation, then we obtain the following.

> If $A_1x + B_1y + C_1 = 0$ and $A_2x + B_2y + C_2 = 0$ are the equations of two intersecting lines l_1 and l_2, and h and k are arbitrary constants, then the equation
> $$h(A_1x + B_1y + C_1) \pm k(A_2x + B_2y + C_2) = 0$$
> represents a system of lines passing through the point of intersection of l_1 and l_2.

Stated another way,

> If you add or subtract any multiples of two equations whose graphs intersect, you will get another equation whose graph contains their point(s) of intersection.

The set of all such sums or differences of multiples of two linear equations has as its graph a family of lines with a common point. The lines shown on the graph all belong to the family that contains (3,1).

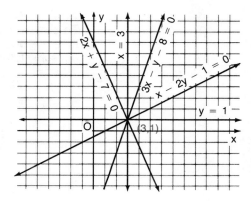

To solve a pair of linear equations by addition or subtraction, we choose multiples that eliminate one of the variables when the equations are added or subtracted. By doing this, we can reduce the original system to an equivalent one that can be solved by inspection.

ALGEBRAIC METHOD I
SOLVING BY ADDITION OR SUBTRACTION

EXAMPLE. Solve the following system.

$$2x - 3y = 13 \qquad ①$$
$$3x + 4y = -6 \qquad ②$$

SOLUTION:

$$2x - 3y = 13 \qquad ①$$
$$3x + 4y = -6 \qquad ②$$

① × 4
② × 3
Adding.

$$8x - 12y = 52$$
$$9x + 12y = -18$$
$$17x = 34$$
$$x = 2$$

Substituting x = 2 in ①.

$$2(2) - 3y = 13$$
$$-3y = 9$$
$$y = -3$$

We have now reduced the original system to an equivalent one, namely

$$x = 2$$
$$y = -3$$

From this system the solution set {(2, -3)} can be easily read.

ALGEBRAIC METHOD II
SOLVING BY SUBSTITUTION

EXAMPLE. Solve the following system of equations.

$$3x + y = 1 \qquad ①$$
$$2x + 5y = 18 \qquad ②$$

SOLUTION:

To eliminate a variable by substitution, you
(i) solve one of the equations for one of the variables and
(ii) substitute the result in the other equation.

$$3x + y = 1 \qquad ①$$
$$2x + 5y = 18 \qquad ②$$

Solving for y in ①.

$$3x + y = 1$$
$$y = 1 - 3x$$

Substituting for y in ②.

$$2x + 5y = 18$$
$$2x + 5(1 - 3x) = 18$$
$$2x + 5 - 15x = 18$$
$$-13x = 13$$
$$x = -1$$

Since

$$y = 1 - 3x$$
$$y = 1 - 3(-1)$$
$$y = 1 + 3$$
$$y = 4$$

Check in ①.	Check in ②.
L.S. $= 3x + y$	L.S. $= 2x + 5y$
$= 3(-1) + (4)$	$= 2(-1) + 5(4)$
$= -3 + 4$	$= -2 + 20$
$= 1$	$= 18$
R.S. $= 1$	R.S. $= 18$

$\therefore x = -1$ and $y = 4$.

EXERCISE 3.7

A 1. Solve the following for x and y.

(a) $x + y = 12$
$\quad x - y = 2$
(b) $x + y = 9$
$\quad x - y = 1$
(c) $x + y = 10$
$\quad x - y = 6$
(d) $x - y = 2$
$\quad x + y = 6$
(e) $3x = 6$
$\quad x + y = 7$
(f) $x - y = 7$
$\quad 2y = 4$
(g) $x + 2y = 5$
$\quad x - 2y = 1$
(h) $x - y = 0$
$\quad x + y = 10$
(i) $2x + y = 7$
$\quad x + y = 4$
(j) $x + 2y = 9$
$\quad x + y = 5$
(k) $x + 2y = 4$
$\quad x - y = 1$
(l) $x - 2y = 0$
$\quad 3x + 2y = 16$

B 2. Solve and check the following systems of equations.

(a) $3x + 2y = 14$
$\quad 5x - y = 6$
(b) $2x + 3y = 41$
$\quad 4x + 5y = 71$
(c) $7x - 2y = -21$
$\quad 8x - 3y = -19$
(d) $3x - 2y = 3$
$\quad 4x + 3y = -30$
(e) $5x + 2y = 24$
$\quad 2x + 3y = 25$
(f) $2x - 3y = 3$
$\quad 3x - 5y = 3$
(g) $4x + 7y = 61$
$\quad 3x - 2y = -5$
(h) $3x - 5y = 33$
$\quad 2x + 7y = -40$
(i) $5x - 2y = 3$
$\quad 2x - 7y = 26$
(j) $3x + 5y = -37$
$\quad 4x - 7y = 19$

3. Find the point of intersection for each of the following systems of equations.

(a) $5x = 3y - 29$
   $2x + 7y = 13$

(b) $3x - 2y = 6$
   $6x = 11 + 4y$

(c) $3x = 5y - 21$
   $2y = -4x - 28$

(d) $5x - 7y + 71 = 0$
   $8x - 9y + 96 = 0$

(e) $13 = 4x - 3y$
   $5x = 59 + 6y$

(f) $y = 4x + 7$
   $6x + 2y = 7$

(g) $5x + 3y = -14$
   $3x - 6y = -11$

(h) $2x = 9 - 6y$
   $6x + 18y - 27 = 0$

(i) $4x - 5 = 15y$
   $6x = 4 + 5y$

(j) $3x - 7y = 41$
   $5x - 3y = 51$

4. Solve each system by substitution.

(a) $2x + y = 5$
   $3x - 4y = 2$

(b) $x + 6y = 17$
   $3x - 7y = 1$

(c) $5x - 2y = -4$
   $4x + y = -11$

(d) $5x - y = 11$
   $3x + 4y = 2$

(e) $3y - x + 10 = 0$
   $3x + 4y = -22$

(f) $2x + 5y = 31$
   $x + 2 = 0$

5. Solve the following systems of equations.

(a) $\dfrac{x}{2} + \dfrac{y}{4} = 4$
   $\dfrac{x}{4} + \dfrac{y}{8} = 2$

(b) $\dfrac{x}{2} - \dfrac{4y}{3} = 5$
   $\dfrac{5x}{2} + y = 2$

(c) $0.3x + 0.5y = 6$
   $0.4x - y = -2$

(d) $8x + 7y = 3.7$
   $5x - 3y = 0.1$

(e) $5x - 3y + 3.6 = 0$
   $2x + 5y = -0.2$

(f) $8x = 3y - 1.9$
   $3x - 2 = 5y$

(g) $0.6x - 0.5y = 3.8$
   $1.4x + 0.1y = 3.8$

(h) $\dfrac{x}{2} - \dfrac{y}{3} = -1$
   $\dfrac{x}{4} + \dfrac{y}{2} = 5\frac{1}{2}$

(i) $3(x - 1) - 4(y + 2) = -5$
   $4(x + 5) - (y - 1) = 16$

(j) $2(1 - 2x) - 3(1 - y) = -16$
   $4(2y + 1) - 3(x + 3) = -22$

(k) $\dfrac{x + 3}{3} - \dfrac{y - 1}{2} = 0$
   $\dfrac{x - 6}{3} + \dfrac{y + 1}{6} = 0$

(l) $\dfrac{x - 1}{4} - \dfrac{y - 3}{5} = 2$
   $\dfrac{x + 1}{3} + \dfrac{y + 6}{2} = 4$

6. In each of the following determine whether the system is consistent and independent, consistent and dependent, or inconsistent.

(a) $2x + 3y = 12$
   $4x + 6y = 24$

(b) $3x - 2y = 8$
   $6x - 4y = 21$

(c) $x + 3y = 4$
   $5x - y = 4$

(d) $2x + 5y = -1$
   $2x - 3y = 7$

(e) $5x = 3y - 8$
   $15x - 9y + 24 = 0$

(f) $2x = y - 14$
   $6y - 11 = 12x$

7. Determine the vertices of the triangle whose sides lie on the lines $x + 3y = 13$, $x - y = 1$, and $5x - y = -15$.

C 8. Solve the following systems of equations.

(a) $\dfrac{1}{x} + \dfrac{1}{y} = 5$
   $\dfrac{2}{x} - \dfrac{3}{y} = -5$

(b) $\dfrac{2}{x} - y = 5$
   $\dfrac{1}{x} + 2y = 10$

(c) $\dfrac{3}{y} - \dfrac{2}{x} = 8$
   $\dfrac{4}{x} - \dfrac{2}{y} = -8$

(d) $\dfrac{3}{\frac{2}{x}} - \dfrac{4}{\frac{1}{y}} = 17$
   $\dfrac{1}{\frac{2}{y}} - \dfrac{2}{\frac{3}{x}} = -5$

Solve.
$6751x + 3249y = 26\ 751$
$3249x + 6751y = 23\ 249$

3.8 APPLICATIONS OF LINEAR SYSTEMS IN TWO VARIABLES

Many practical problems can be expressed as a system of linear equations in two variables.

The following terms are used in problems involving aircraft.

Air Speed: the speed of the plane in still air.

Ground Speed: the speed of the plane in relation to the ground.

Wind Speed: the speed of the wind in relation to the ground.

Tail Wind: a wind blowing in the same direction as the plane is flying.

Head Wind: a wind blowing in the opposite direction to that in which the plane is flying.

READ

EXAMPLE 1. With a tail wind a plane took 4 h to fly 1920 km. Flying back against the same wind and with the same air speed the plane took 1 h longer. Find the wind speed and the plane's air speed.

PLAN

SOLUTION:
Let x be the plane's air speed (in km/h).
Let y be the wind speed (in km/h).
Then x, y ⩾ 0.

We use the relationship Distance = Speed × Time, D = ST, and arrange the given facts in a chart.

	Distance (km)	Ground Speed (km/h)	Time (h)
with tail wind	1920	x + y	4
with head wind	1920	x − y	5

The distance with a tail wind is 1920 km.
$$4(x + y) = 1920 \qquad ①$$
The distance with a head wind is 1920 km.
$$5(x - y) = 1920 \qquad ②$$

① ÷ 4 $\qquad x + y = 480 \qquad ①$
② ÷ 5 $\qquad x - y = 384 \qquad ②$

Adding.
$$2x = 864$$
$$x = 432$$

Substitute in ①.
$$432 + y = 480$$
$$y = 48$$

ANSWER

The air speed is 432 km/h and the wind speed is 48 km/h.

Check.
When flying with the wind, the plane's ground speed is
432 km/h + 48 km/h = 480 km/h and the time to travel 1920 km is
$$\frac{1920}{480} = 4 \text{ h.}$$
When flying against the wind, the plane's ground speed is
432 km/h − 48 km/h = 384 km/h and the time to travel 1920 km is
$$\frac{1920}{384} = 5 \text{ h.}$$

READ

EXAMPLE 2. A certain alloy contains 9% silver. Another alloy is 14% silver. How much of each type should be combined to make 10 kg of an alloy that contains 12% silver?

PLAN

SOLUTION:
Let x represent the amount of 9% alloy used.
Let y represent the amount of 14% alloy used.
Then $x, y \geqslant 0$.

$$x + y = 10 \qquad ① \text{ alloy equation}$$
$$0.09x + 0.14y = (0.12)10 \qquad ② \text{ silver equation}$$

① × 9 $\qquad 9x + 9y = 90 \qquad ①$
② × 100 $\qquad 9x + 14y = 120 \qquad ②$

SOLVE

Subtracting.
$$-5y = -30$$
$$y = 6$$

Substituting in ①.
$$x + 6 = 10$$
$$x = 4$$

ANSWER

4 kg of 9% alloy and 6 kg of 14% alloy should be combined to give 10 kg of 12% alloy.

Check.
There must be 10 kg of alloy.
$$4 + 6 = 10$$
There must be 1.2 kg of silver in the alloy.
$$0.09(4) + 0.14(6) = 0.36 + 0.84$$
$$= 1.2$$

EXERCISE 3.8

B 1. Find two numbers whose sum is 763 and whose difference is 179.

2. Find two numbers whose sum is 947 and whose difference is 177.

3. The sum of two numbers is 181. Three times the larger plus twice the smaller is 459. Find the numbers.

4. Five times the larger of two numbers, plus four times the smaller is 271. Three times the larger less twice the smaller is 57. Find the numbers.

5. Six times the larger of two numbers less three times the smaller is 270. Five times the larger less twice the smaller is 261. Find the numbers.

6. Jill had $12 000 to invest. She invested part of it in bonds paying 8% per annum and the remainder in a second mortgage paying 9% per annum. After one year the total interest from these investments was $1043. How much did Jill invest at each rate?

7. Terry invested his inheritance of $335 000, part at 7% per annum and the remainder at 10% per annum. After one year the total interest from these investments was $24 500. How much did Terry invest at each rate?

8. A parking lot contained 102 vehicles (cars and buses). Each car was charged $3 and each bus $10. The total revenue was $418. How many buses were on the lot?

9. With a certain tail wind, a jet aircraft takes 3 h to travel 1890 km. Flying against the same wind, the jet makes the return trip in 3.5 h. Find the wind speed and the jet's air speed.

10. A private plane took 2 h to make a 600 km trip when flying with the wind. The return trip, flying against the same wind, took 2.5 h. Find the wind speed and the plane's air speed.

11. During March, stock X gained or lost x dollars in value per share and stock Y, y dollars in value per share. On stocks X and Y, two shareholders realized the following loss and gain during March.

Share-holder	Shares of X	Shares of Y	Monthly Gain or Loss ($)
Lee	50	100	− 50
Adams	100	120	+ 60

Find x and y.

12. During May, stock M gained or lost m dollars in value per share and stock N, n dollars in value per share. On stocks M and N, two shareholders realized the following gain and loss during May.

Share-holder	Shares of M	Shares of N	Monthly Gain or Loss ($)
Brown	210	130	+ 450
Reid	60	90	− 30

Find m and n.

13. Two bottles contain hydrochloric acid, one 40% strength and the other 30% strength. How much should be taken from each bottle to make 20 L of 34% strength?

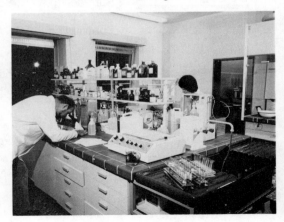

14. A chemist wants to get 100 L of 47% alcohol solution by volume by mixing 40% alcohol solution with 50% alcohol solution. How many litres of each should she use?

15. A chemical supply company received an order for 1500 L of 38% salt solution. To fill the order it was necessary to mix 40% salt solution with 30% salt solution. How many litres of each should be mixed?

16. It took 3 h for a Coast Guard patrol boat to travel 48 km up a river and 2 h for the return trip. Find the speed of the current and the speed of the boat in stillwater.

17. In still water, a boat can travel four times as fast as the current in the river. A trip up the river and back which totalled 150 km took 8 h. Find the speed of the current.

18. Raisins worth $4.20/kg are mixed with dried fruit worth $6.90/kg to make a mixture worth $5.28/kg. How many kilograms of raisins and how many kilograms of dried fruit must be used to make 10 kg of the mixture?

19. An apartment building contains sixty units. One-bedroom apartments rent for $725 per month. Two-bedroom apartments rent for $925 per month. When the building is entirely rented the total monthly rental income is $48 700. How many apartments of each type are there?

20. For delivering packages, the Cross Town Courier Service charges a set amount for the first 50 g and a fixed rate for each additional 10 g. The charge for a 210 g package is $5.60. The charge for a 330 g package is $7.40. What is the cost for each additional 10 g?

21. The tens digit of a two-digit number exceeds the units digit by two. The difference between the original number and the number formed when the digits of the original number are reversed is 18. Find the original number.

22. The units digit of a two-digit number is two more than three times the tens digit. When the digits are reversed, the new number is two less than three times the original number. Find the original number.

23. Find A and B so that the graph of $Ax + By = 20$ contains the points $(2,4)$ and $(-1,8)$.

24. Find A and B so that the graph of $Ax + By = -10$ contains the points $(-2,1)$ and $(2,9)$.

25. Find m and b so that the graph of $y = mx + b$ contains the points $(2,2)$ and $(-4, -16)$.

C 26. Find two numbers whose sum is m and whose difference is n.

27. Two kilometres upstream from his starting point, a rower passed a raft floating with the current. He rowed upstream for one more hour, and then rowed back and reached his starting point at the same time as the raft. Find the speed of the current.

1. (a) Choose two nonzero numbers s and t. Graph the three equations on the same set of axes.
$$3x + y = 7$$
$$x - y = 1$$
$$s(3x + y) + t(x - y) = 7s + t$$
(b) What is the point of intersection of the lines?
(c) Why must the point of intersection of the first two lines lie on the third line?

3.9 BREAK-EVEN POINT

Linear systems are used in a business context to determine the break-even point for a manufacturer.

Let x represent the number of units manufactured and sold. Then C represents the cost of manufacturing the units and R represents the revenue from selling the units.

The graph on the right shows the cost and revenue lines. Because of fixed overhead costs, the cost line is higher than the revenue line at the beginning. When only a few units are made and sold produced, the manufacturer suffers a loss.

When many units are made and sold, the revenue line is higher than the cost line and the manufacturer realizes a profit.

The point where the two lines cross is called the break-even point. Here, the cost and revenue are equal. The manufacturer breaks even and does not experience a profit or a loss.

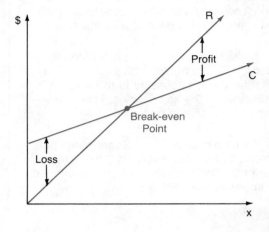

EXAMPLE. The Porter Company manufactures windsurfers. It costs $100 to make one windsurfer. The fixed overhead costs are $4500. Each windsurfer can be sold for $200.
(a) How many windsurfers must be made and sold for the Porter Company to break even?
(b) How many windsurfers must be made and sold to realize a profit of $5000?

SOLUTION:

Let x represent the number of windsurfers made and sold. The cost of making them is

$$C = 100x + 4500 \qquad x \geq 0$$

The revenue is given by

$$R = 200x \qquad x \geq 0$$

(a) To find the break-even point we let the revenue equal the cost and solve.

$$R = C$$

or

$$200x = 100x + 4500$$
$$100x = 4500$$
$$x = 45$$

Forty-five windsurfers must be made and sold to break even.

(b) The profit is the revenue minus the cost.

$$P = R - C$$

Substituting $P = 5000$.

$$5000 = 200x - (100x + 4500)$$
$$5000 = 200x - 100x - 4500$$
$$9500 = 100x$$
$$x = 95$$

Ninety-five windsurfers must be made and sold to realize a profit of $5000.

EXERCISE 3.9

A 1. What are some fixed overhead costs that manufacturers would have?

B 2. The Smith Company makes picnic tables. It costs $50 to make one table and the fixed overhead costs are $3000. Each table can be sold for $70.

(a) How many picnic tables must be made and sold for the Smith Company to break even?

(b) How many picnic tables must be made and sold to realize a profit of $10 000?

3. The Longhorn Company makes jeans. It costs $15 to make one pair of jeans. The fixed overhead costs are $100 000. Each pair of jeans can be sold for $40.

(a) How many pairs of jeans must be made and sold for the Longhorn Company to break even?

(b) How many pairs of jeans must be made and sold to realize a profit of $500 000?

4. The Body Shop makes and sells replicas of old luxury cars. It costs $40 000 to make one car. The fixed overhead costs are $400 000. Each car can be sold for $60 000.

(a) What is the profit or loss if 15 cars are made and sold?

(b) How many cars must be made and sold for The Body Shop to break even?

(c) How many cars must be made and sold to realize a profit of $1 000 000?

5. The Bird Place makes and sells bird feeders. It costs $6 to make one feeder. The fixed overhead costs are $450. Each feeder can be sold for $15.

(a) What is the profit or loss if 80 feeders are made and sold?

(b) How many feeders must be made and sold for the Bird Place to break even?

(c) How many feeders must be made and sold to realize a profit of $1800?

C 6. SUPPLY AND DEMAND

Linear systems are used to interpret the law of supply and demand. We will illustrate the law using corn as an example.

The demand curve is shown below. As the price of corn increases, the amount of corn bought decreases.

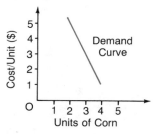

Now the supply curve is shown. As the cost of corn increases, the amount of corn grown increases.

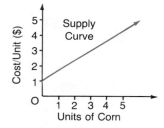

The point of intersection of the supply and demand curves is called the equilibrium point. This is the price and quantity where the supply and demand match.

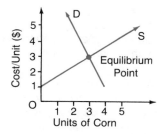

The law of supply and demand states that in a situation of competition, a commodity will be sold at its equilibrium price.

(a) What would result if corn was sold at $4 per unit?

(b) What would result if corn was sold at $2 per unit?

3.10 SYSTEMS WITH LITERAL COEFFICIENTS

In some systems of equations the coefficients of the variables for which you are to solve may not be numerical. These systems are solved in the same way as you solve "numerical systems," but the solutions are found in terms of the other variables, called parameters, in the equations.

EXAMPLE. Solve the following system for x and y in terms of c and d.

$$cx + dy = 2cd \qquad ①$$
$$dx + cy = c^2 + d^2 \qquad ②$$

SOLUTION:

$$cx + dy = 2cd \qquad ①$$
$$dx + cy = c^2 + d^2 \qquad ②$$

$① \times c$ $c^2x + cdy = 2c^2d$
$② \times d$ $d^2x + cdy = c^2d + d^3$

Subtracting. $c^2x - d^2x = c^2d - d^3$

Factoring. $x(c^2 - d^2) = d(c^2 - d^2)$

$x = d$ (when $c \neq \pm d$)

Substituting. $x = d$ in $①$

$$cx + dy = 2cd$$
$$c(d) + dy = 2cd$$
$$cd + dy = 2cd$$
$$dy = cd$$
$$y = c \text{ (when } d \neq 0)$$

Check in $①$.   Check in $②$.
L.S. $= cx + dy$ L.S. $= dx + cy$
$= c(d) + d(c)$ $= d(d) + c(c)$
$= cd + cd$ $= d^2 + c^2$
$= 2cd$ $= c^2 + d^2$
R.S. $= 2cd$ R.S. $= c^2 + d^2$
$\therefore x = d$ and $y = c$.

EXERCISE 3.10

A 1. If we wish to eliminate y by addition or subtraction, state what each equation should be multiplied by in each of the following systems.

(a) $x - 2y = a$ $①$
  $x + 3y = b$ $②$

(b) $3x + 4y = m$ $①$
  $5x + 6y = n$ $②$

(c) $ax + by = c$ $①$
  $dx + ey = f$ $②$

(d) $mx - ny = q$ $①$
  $nx + my = r$ $②$

(e) $ax + by = a^2 - b^2$
  $bx - ay = 2ab$

(f) $dx + cy = -2cd$
  $cx + dy = -c^2 - d^2$

(g) $mx + ny = m^2 + n^2$
  $my - nx = m^2 + n^2$

(h) $ax - by = a^2 + b^2$
  $x - y = 2a$

(i) $mx + ny = t$
  $rx + sy = w$

(j) $a_1x + b_1y = c_1$
  $a_2x + b_2y = c_2$

B 2. Solve the following systems for x and y in terms of the other variables.

(a) $x - y = a + b$
  $x + y = a - b$

(b) $x + y = a$
  $x - y = b$

(c) $2x + 3y = 13a$
  $3x - 4y = -6a$

(d) $ax + by = 2$
  $ax - by = 0$

3. (a) Solve for v and x.
  $av + bx = 1$
  $cv + dx = 0$

(b) Solve for w and y.
  $aw + by = 0$
  $cw + dy = 1$

3.11 LINEAR SYSTEMS IN TWO VARIABLES

We can write the general case for a linear system of equations in two variables as

$$ax + by = c \qquad ①$$
$$dx + ey = f \qquad ②$$

Solving for x.

① × e	$aex + bey = ce$
② × b	$bdx + bey = bf$
Subtracting.	$aex - bdx = ce - bf$
Factoring.	$x(ae - bd) = ce - bf$
	$x = \dfrac{ce - bf}{ae - bd}$

Solving for y.

① × d	$adx + bdy = cd$
② × a	$adx + aey = af$
Subtracting.	$bdy - aey = cd - af$
Factoring.	$y(bd - ae) = cd - af$
	$y = \dfrac{cd - af}{bd - ae}$

The following BASIC program can be used to solve a linear system in two variables in the form

$$Ax + By = C$$
$$Dx + Ey = F$$

where the solution is consistent and independent.

```
NEW
10 INPUT "A=";A
20 INPUT "B=";B
30 INPUT "C=";C
40 INPUT "D=";D
50 INPUT "E=";E
60 INPUT "F=";F
61 IF A*E <> B*D THEN 70
62 IF A*F = C*D THEN PRINT
   "DEPENDENT SYSTEM": GOTO 110
63 PRINT "INCONSISTENT SYSTEM":
   GOTO 110
70 X=(C*E-B*F)/(A*E-B*D)
80 Y=(C*D-A*F)/(B*D-A*E)
90 PRINT "X =";X;"Y =";Y
100 PRINT "CONSISTENT AND
    INDEPENDENT SYSTEM"
110 END
RUN
```

Statements 62, 63, and 100 in this program should be entered on one line.

EXERCISE

1. Use the program to solve the following linear systems.

(a) $x + y = 9$
    $x - y = 1$

(b) $2x + 3y = 35$
    $5x - 2y = 2$

(c) $7x - 8y = -10$
    $11x + 5y = -86$

(d) $-13x + 24y = 285$
    $21x - 17y = 308$

(e) $34x - 25y = -3$
    $-66x + 45y = 3$

(f) $30x - 25y - 4 = 0$
    $75x + 20y + 23 = 0$

(g) $8x + 12y = -17.6$
    $13x - 5y = 35.1$

3.12 SYSTEMS OF LINEAR EQUATIONS IN THREE VARIABLES

The ordered triple $(4, 2, -1)$ is a solution of the equation
$2x + 3y - 4z = 18$ since
$$2(4) + 3(2) - 4(-1) = 18$$
is a true statement. Some other solutions of this equation are $(9, 0, 0)$, $(5, 0, -2)$, and $(0, 2, -3)$. There are many others.

In solving problems it is sometimes necessary to solve systems of three equations in three variables. The method is similar to the method of elimination by addition or subtraction that was used in solving two equations in two variables.

EXAMPLE. Solve.

$$3x - 4y + 5z = 2 \qquad ①$$
$$4x + 5y - 3z = -5 \qquad ②$$
$$5x - 3y + 2z = -11 \qquad ③$$

SOLUTION:

We first reduce the system of three equations in three variables to a system of two equations in two variables by taking two different pairs of equations and eliminating the same variable from each. It doesn't matter which variable you decide to eliminate first.

Eliminating y from ① and ②.

$① \times 5 \qquad 15x - 20y + 25z = 10$
$② \times 4 \qquad \underline{16x + 20y - 12z = -20}$

Adding to get ④. $\qquad 31x + 13z = -10 \qquad ④$

Eliminating y from ② and ③.

$② \times 3 \qquad 12x + 15y - 9z = -15$
$③ \times 5 \qquad \underline{25x - 15y + 10z = -55}$

Adding to get ⑤. $\qquad 37x + z = -70 \qquad ⑤$

We have now reduced the system to one of two equations in two variables, which we can solve by the methods of Section 3.7.

$$31x + 13z = -10 \qquad ④$$
$$37x + z = -70 \qquad ⑤$$

$\qquad\qquad\qquad\qquad 31x + 13z = -10 \qquad ④$
$⑤ \times 13 \qquad \underline{481x + 13z = -910} \qquad ⑤$

Subtracting. $\qquad\qquad -450x = 900$
$\qquad\qquad\qquad\qquad x = -2$

Substituting $x = -2$ in ⑤.

$$37x + z = -70$$
$$37(-2) + z = -70$$
$$-74 + z = -70$$
$$z = 4$$

Substituting $x = -2$ and $z = 4$ in ①.

$$3x - 4y + 5z = 2$$
$$3(-2) - 4y + 5(4) = 2$$
$$-6 - 4y + 20 = 2$$
$$-4y = -12$$
$$y = 3$$

Check in ①.
L.S. $= 3x - 4y + 5z$
$= 3(-2) - 4(3) + 5(4)$
$= -6 - 12 + 20$
$= 2$
R.S. $= 2$

Check in ②.
L.S. $= 4x + 5y - 3z$
$= 4(-2) + 5(3) - 3(4)$
$= -8 + 15 - 12$
$= -5$
R.S. $= -5$

Check in ③.
L.S. $= 5x - 3y + 2z$
$= 5(-2) - 3(3) + 2(4)$
$= -10 - 9 + 8$
$= -11$
R.S. $= -11$

$x = -2$, $y = 3$, and $z = 4$.

EXERCISE 3.12

A 1. Is the given ordered triple a solution of the given equation?

(a) $3x + y + 2z = 6$; $(1,1,1)$
(b) $2x - 3y + z = 7$; $(3,0,1)$
(c) $2x + 3y - z = 10$; $(3,2,1)$
(d) $4x + 5y + 2z = 5$; $(2,0,-2)$
(e) $4x - 2y + 3z = -9$; $(0,3,-1)$
(f) $5x + 3y - 2z = 6$; $(2,-1,0)$
(g) $2x - 3y - 4z = 8$; $(1,-1,-1)$
(h) $3x - 3y - z = -5$; $(-2,1,-2)$

B 2. Solve the following systems of equations.

(a) $x + y + 3z = 12$
$x - y + 4z = 11$
$2x + y + 3z = 13$

(b) $3x + 2y + z = 14$
$4x + 3y + 2z = 20$
$5x + 4y + 3z = 26$

(c) $4x + 3y - z = -7$
$3x - 2y + 3z = -10$
$x + y - z = -2$

(d) $3x + 5y - z = 47$
$2x - y + 3z = -2$
$4x + y - 2z = 30$

(e) $3x - 4y + 5z = 26$
$6x - 2y - 3z = -39$
$x + 3y - 4z = -31$

(f) $4x - 3y - 2z = 31$
$5x - y - 4z = 28$
$7x - 3y + 5z = 26$

(g) $7x - 5y + 4z = 57$
$9x + 3y - 5z = 45$
$8x - 4y + 7z = 60$

(h) $4r + 3s - 2t = -21$
$5r - s - t = -15$
$3r - 2s + 5t = -16$

(i) $3x - 2y + z + 3 = 0$
$5x + 3y = 26 - 2z$
$4x + 5z = 2y - 8$

(j) $3x + 2y - 3z = -4$
$5x + y + 2z = -16$
$2x - 3y + z = 0$

(k) $4x - 3y + 6z = -9$
$2x + 4y - 3z = -10$
$3x + 2y - 4z = -11$

(l) $2x + 3y - z = 10$
$5x - 2y + 3z = 3$
$4x + 2y - 5z = 32$

3. Solve the following.

(a) $3x - 2y + 5z = 1$
$4x + 5y - 3z = 17$
$7x - 3y + 2z = 36$

(b) $4x - y + 3z = -11$
$6x - 3y - 2z = -3$
$8x - 2y + 5z = -19$

(c) $4x - 3y + 8z = 5$
$6x + 9y + 16z = 4$
$3x - 12y - 2z = 5$

(d) $x + y = 4$
$y + z = -1$
$x - z = 5$

(e) $3x + 2y = 10$
$3y - 4z = -4$
$5x - 2z = 34$

(f) $0.4x + 0.6y - 0.3z = 1.2$
$0.3x - 0.4y + 0.5z = 2.1$
$1.1x - 0.2y - 0.1z = 2.5$

(g) $4x - 2y + 3z = 0.4$
$2x - 5y + z = 1.5$
$3x - 4y - 2z = 1.9$

(h) $3w + 2x - 3y + z = 1$
$4w - x + 2y + 3z = 19$
$2w + 3x - 4y + 2z = -1$
$w - 4x + y - 4z = -7$

(i) $\dfrac{x}{2} + \dfrac{y}{4} + \dfrac{z}{3} = 9$

$\dfrac{x}{3} - \dfrac{y}{8} + \dfrac{z}{6} = 3$

$\dfrac{x}{6} + \dfrac{y}{2} + \dfrac{z}{4} = 8$

(j) $\dfrac{x}{4} + \dfrac{y}{3} - \dfrac{z}{2} = 8$

$\dfrac{x}{2} - \dfrac{y}{2} + \dfrac{z}{6} = 2$

$\dfrac{x}{6} - \dfrac{y}{3} + \dfrac{z}{3} = -2$

3.13 PROBLEM SOLVING

In this section we continue our work on problem solving by discussing two methods of proving (or disproving) statements.

Often we can show that a statement is false simply by finding an object that does not satisfy the statement. Such an object is called a counterexample.

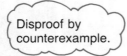

EXAMPLE 1. Prove or disprove the equation $\sqrt{a + b} = \sqrt{a} + \sqrt{b}$.

SOLUTION:

If we experiment by substituting specific numbers for a and b, we soon discover that the given equation is not usually satisfied. For instance, if we put $a = 9$ and $b = 16$, we have

$$\sqrt{a + b} = \sqrt{9 + 16} = \sqrt{25} = 5$$

whereas,

$$\sqrt{a} + \sqrt{b} = \sqrt{9} + \sqrt{16} = 3 + 4 = 7$$

This counterexample shows that, in general,

$$\sqrt{a + b} \neq \sqrt{a} + \sqrt{b}.$$

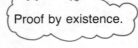

We prove an existence statement by exhibiting objects which satisfy the statement.

EXAMPLE 2. Show that there exist positive integers x, y, z, and t which satisfy the equation $x^2 + y^2 + z^2 = t^2$.

SOLUTION:

We try specific integers for x, y, and z until we find some for which $x^2 + y^2 + z^2$ is a perfect square.

Try $x = 1$, $y = 1$, $z = 1$: $x^2 + y^2 + z^2 = 1^2 + 1^2 + 1^2 = 3$, not a perfect square.

Try $x = 1$, $y = 1$, $z = 2$: $x^2 + y^2 + z^2 = 1^2 + 1^2 + 2^2 = 6$, not a perfect square.

Try $x = 1$, $y = 2$, $z = 2$: $x^2 + y^2 + z^2 = 1^2 + 2^2 + 2^2 = 9 = 3^2$, a perfect square.

Therefore, we have found a solution of the given equation in positive integers.

$$x = 1, y = 2, z = 2, \text{ and } t = 3$$

By continuing this procedure, it is possible to find other solutions.

EXERCISE 3.13

B 1. Prove or disprove the following equations.

(a) $\dfrac{1}{a + b} = \dfrac{1}{a} + \dfrac{1}{b}$

(b) $\dfrac{8 + c}{8} = 1 + \dfrac{c}{8}$

(c) $\dfrac{x}{x + y} = \dfrac{1}{1 + y}$

(d) $\sqrt{x^2} = x$

(e) $\dfrac{1 + \sqrt{x}}{1 - x} = \dfrac{1}{1 - \sqrt{x}}$

(f) $(x + y)^3 = x^3 + y^3$

(g) $\sqrt{4 + m^2} = 2 + m$

(h) $\dfrac{1}{x - 1} = \dfrac{x^2 + x + 1}{x^3 - 1}$

(i) $\sqrt[3]{a + b} = \sqrt[3]{a} + \sqrt[3]{b}$

2. Prove that there exist positive integers a, b, c, d, and e such that
$$a^2 + b^2 + c^2 + d^2 = e^2 - 7$$

3. Prove that there exists an integer n such that $\dfrac{n}{2}$ is a perfect square and $\dfrac{n}{3}$ is a perfect cube.

4. (a) Show that there exist positive integers x and y such that
$$2x + 3y = 715$$
(b) How many solutions (in positive integers) does the equation in (a) have?

MICRO MATH

SLOPE

Given two points on a line, (x_1, y_1) and (x_2, y_2), we can determine the slope of the line using the formula
$$m = \frac{y_2 - y_1}{x_2 - x_1}$$
In the BASIC language, we can represent a variable by a letter followed by a number. Therefore, it is convenient to represent x_1 as X1, x_2 as X2, y_1 as Y1, and y_2 as Y2.

The following program can be used to find the slope of a line through two given points.

NEW

```
10 REM SLOPE OF A LINE
20 PRINT "INPUT X1, Y1"
30 INPUT "X1=";X1
40 INPUT "Y1=";Y1
50 PRINT "INPUT X2, Y2"
60 INPUT "X2=";X2
70 INPUT "Y2=";Y2
80 IF X1=X2 THEN 110
90 PRINT "SLOPE =";(Y2-Y1)/(X2-X1)
   :GOTO 120
100 STOP
110 PRINT "THERE IS NO SLOPE"
120 END
```

RUN

Statement 90 in the program above should be entered on one line.

EXERCISE

1. Use the program to find the slope of the line through the given points.
(a) (3,5) and (4,7)
(b) (-2,6) and (3, -4)
(c) (-7, -9) and (-7, -4)
(d) (12, -8) and (-9, -8)
(e) (91,84) and (-57,47)
(f) (-3.42,8.79) and (-4.68,8.16)

3.14 REVIEW EXERCISE

1. Write each of the following in standard form, $Ax + By + C = 0$, where A, B, and C are integers.
(a) $2x - 3y = -3$ (b) $y = 4x - 7$
(c) $x = -6y - 8$ (d) $y - 2 = 6(x - 2)$
(e) $x - 1 = -2(y + 3)$ (f) $\frac{1}{2}x - \frac{1}{3}y = 5$

2. Solve for y.
(a) $3x + y = 4$ (b) $4x - y = 11$
(c) $2x + 4y = 7$ (d) $8x - 3y - 1 = 0$

3. Graph each of the following.
(a) $y = 4x - 5$ (b) $x + y = 11$
(c) $y = -3x + 8$ (d) $2x + y = 4$
(e) $5x - 15 = 0$ (f) $2y + 8 = 0$

4. Write an equation, in standard form, of the line through the given point and having the given slope.
(a) $(4,3)$; $m = 2$ (b) $(0,3)$; $m = 6$
(c) $(-1,2)$; $m = -3$ (d) $(-5,0)$; $m = 0$
(e) $(-4,-1)$; $m = \frac{1}{2}$ (f) $(-3,7)$; no slope

5. Write an equation, in standard form, of the line through the given points.
(a) $(2,3)$, $(4,9)$ (b) $(-2,6)$, $(-3,8)$
(c) $(-1,-3)$, $(-3,7)$ (d) $(0,2)$, $(-1,-4)$
(e) $(3,-8)$, $(2,-8)$ (f) $(4,3)$, $(4,-7)$
(g) $(2,-3)$, $(4,-4)$ (h) $(6,-7)$, $(5,-9)$

6. Determine the slope and y-intercept for each of the following.
(a) $y = 2x - 7$ (b) $3x + y = -2$
(c) $3x - y = -2$ (d) $4y + 3x = 12$
(e) $2x - 3y - 5 = 0$ (f) $x - 6y = 4$

7. Write an equation, in standard form, of the line satisfying the following conditions:
(a) through $(2,7)$ and parallel to the graph of $y = 3x + 7$.
(b) through $(-1,5)$ and perpendicular to the graph of $y = 4x - 1$.
(c) through $(-3,-6)$ and parallel to the graph of $3x + 2y - 4 = 0$.
(d) through $(4,-4)$ and perpendicular to the graph of $x - 4y + 6 = 0$.

8. Write an equation of the line through (5,6) and parallel to the x-axis.

9. Write an equation of the line through $(-4,-7)$ and parallel to the y-axis.

10. Show that the triangle with vertices (2,3), (4,7), and (6,6) is a right triangle.

11. Prove using slopes that $(-5,-2)$, $(-5,-8)$, $(-1,-8)$, and $(-1,-2)$ are the vertices of a rectangle.

12. Prove using slopes that $(-1,-3)$, (5,6), (1,5), and $(3,-2)$ are the vertices of a parallelogram.

13. Find the value of k so that the graph of $y = kx - 2$ is parallel to the graph of $3x + 2y - 6 = 0$.

14. Find the value of k so that the graph of $2x + ky - 3 = 0$ is perpendicular to the graph of $2x - y + 3 = 0$.

15. On the Mississippi Queen a banquet for 70 people costs $1900 and a banquet for 130 people costs $3100.

(a) Plot a graph of cost versus number of people.
(b) Determine an equation for this function and express it in the form $c = nm + b$ where c represents the cost and n the number of people.
(c) Use the equation to determine the cost of a banquet for 91 people.
(d) What quantity does the slope of the line represent?
(e) What meaning does the y-intercept have?

16. The dividend on 750 shares of Purple Eagle stock is $2625 per year. What is the dividend on 285 shares of the same stock?

17. In Culver City the property tax is $2220 on a property with an assessed value of $60 000.

(a) Write an equation for this direct variation function.
(b) What is the constant of proportionality?
(c) If the assessed value of a property is $53 000, what is the property tax?
(d) If the property tax is $1554, what is the property value?

18. Solve the following.
(a) $2x + 3y = 13$
   $3x - y = 3$
(b) $2x - 5y = -1$
   $x - 4y = -2$
(c) $3x + 5y = 1$
   $4x - 3y = -18$
(d) $5x - 2y = -10$
   $3x - 7y = 23$
(e) $4x + 5y = -13$
   $6x + 4y = -9$
(f) $3x - 6y = -7$
   $5x + 9y = -18$

19. Solve.
(a) $\dfrac{x}{2} - \dfrac{y}{3} = 6$

   $\dfrac{x}{4} - \dfrac{y}{6} = 3$

(b) $0.3x + 0.4y = -3.2$
   $0.5x - 0.6y = 1$

(c) $6x - 5y = 2.7$
   $5x - 4y = 2.2$

(d) $2(x - 1) - (y + 3) = -4$
   $3(x + 4) + 2(y - 1) = 1$

20. Find two numbers whose sum is 418 and whose difference is 68.

21. Nine times the larger of two numbers plus 4 times the smaller is 135. Six times the larger less 5 times the smaller is 21. Find the numbers.

22. A chemist wants to get 200 L of 37% alcohol solution by volume by mixing 40% alcohol solution with 30% alcohol solution. How many litres of each should be used?

23. Jacqueline is in the business of editing computer software. She calculated the fixed overhead costs to be $50.00 per day.
(a) If she charges $75.00 per hour, how many hours must she work per month to break-even.
(b) In a year, she made a profit of $65 000.00. How many hours of editing does that represent to the nearest one?

24. For an electric circuit, the current I varies inversely as the resistance R of the system. The following measurements were made on a system.

I (A)	R (Ω)
14	20
5	56

(a) Graph the points.
(b) Calculate the slope.
(c) Write an equation for this inverse variation.

25. The potential energy U between two planetary objects varies negatively as the inverse of the distance r separating them and directly as the product of their masses.
(a) Write an equation for this joint variation.
(b) Determine the constant of proportionality using tables of planetary data.
(c) What happens to U when r becomes
   (i) very small?
   (ii) very large?
(d) Calculate the potential energy of a satellite with a mass of 150 kg orbiting the earth at a distance of 2.0 earth radii (R_e). What does this energy represent?

3.15 CHAPTER 3 TEST

y = mx + b
slope ? def.
"y intercept"

1. Graph each of the following.
(a) $y = 3x - 7$ (b) $5x - y = 9$

2. Write an equation, in standard form, of the line satisfying the following conditions:
(a) through (3,4) with slope $m = -4$.
(b) through $(-1,7)$ and (6,0).
(c) through $(-2,-2)$ and parallel to the graph of $3x + 5y - 2 = 0$.
(d) through $(5,-1)$ and perpendicular to the graph of $x - 2y + 7 = 0$.
(e) through $(-3,-5)$ and parallel to the x-axis.
(f) through $(5,-1)$ and parallel to the y-axis.

3. Prove using slopes that (3,5), (7,5), (5,2), and (1,2) are the vertices of a parallelogram.

4. The dividend on 800 shares of Hammett stock is $3760 per year. What is the dividend on 390 shares of the same stock?

5. In Blueville, a driver must pay a fine for exceeding the speed limit. She is fined $40 for driving 60 km/h and $160 for driving 90 km/h.
(a) Plot a graph of fine versus speed.
(b) From the graph determine the fine for driving 80 km/h.
(c) Determine an equation for this function.
(d) Use the equation to determine the fine for driving 65 km/h.
(e) What quantity does the slope of the line represent?
(f) What is the legal speed limit?

6. Solve the following.
(a) $3x - 2y = 14$
    $5x + 3y = 17$

(b) $\dfrac{x}{4} + \dfrac{y}{3} = 2$

    $\dfrac{x}{2} - \dfrac{y}{5} = 1\dfrac{2}{5}$

(c) $0.4x + 0.5y = 5.9$
    $0.7x - 0.2y = 2.8$

(d) $\dfrac{x-1}{2} - \dfrac{y-3}{3} = 5$

    $4(x - 3) - 5(y + 1) = 33$

7. Seven times the larger of two numbers plus five times the smaller is 122. Eight times the larger less three times the smaller is 61. Find the numbers.

8. A chemical supply company received an order for 2000 L of 27% salt solution. To fill the order it was necessary to mix 20% salt solution with 30% salt solution. How many litres of each should be mixed?

RATIO AND PROPORTION

CHAPTER

A mathematician, like a painter or a poet, is a maker of patterns.
G.H. Hardy

DISTANCE BETWEEN TWO POINTS

The distance between $P(x_1, y_1)$ and $Q(x_2, y_2)$ is found by the formula

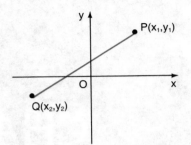

$$PQ = \sqrt{(x_2 - x_1)^2 + (y_2 - y_1)^2}$$
$$= \sqrt{(\Delta x)^2 + (\Delta y)^2}$$

The distance between $A(-3, 4)$ and $B(1, 1)$ is

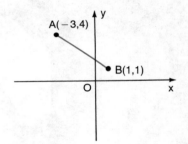

$$AB = \sqrt{(\Delta x^2) + (\Delta y)^2}$$
$$= \sqrt{(-3 - 1)^2 + (4 - 1)^2}$$
$$= \sqrt{(-4)^2 + (3)^2}$$
$$= \sqrt{25}$$
$$= 5$$

EXERCISE

1. Find the distance between each pair of points.

(a) (5,3), (1,1)
(b) (4,8), (−3,1)
(c) (−6,2), (8,0)
(d) (−3, −1), (−5,7)
(e) (4, −8), (−5,6)
(f) (−10,7), (−4, −13)
(g) (0, −4), (−5,6)
(h) (3, −11), (3,6)
(i) (−8,0), (0, −6)
(j) (2, −5), (−6, −5)
(k) (3.5,0.5), (−7.5,2.5)
(l) (0.8, −0.3), (1.1, −2.4)

(m) (−3.2,1.6), (7.8, −1.4)
(n) (1.5, −2.8), (1.3, −0.3)
(o) (h,k), (h + 3,k − 5)
(p) (a, 0), (0, b)

2. Use the distance formula to show that the triangle with vertices $A(-3,1)$, $B(1,7)$, and $C(5,1)$ is an isosceles triangle.

3. Show that the points $A(-1,3)$, $B(3,11)$, and $C(5,15)$ are collinear by showing that $AB + BC = AC$.

4. Use the distance formula to determine which of the following sets of points are collinear
(a) $(-3,-2)$, $(1,1)$, $(9,7)$
(b) $(-8,7)$, $(-4,2)$, $(5,-3)$
(c) $(-5,10)$, $(-1,4)$, $(3,-1)$
(d) $(-4,2)$, $(-2,-3)$, $(-10,17)$

5. Give the coordinates of 3 points each equidistant from $P(-1,6)$ and $Q(5,6)$.

6. If P is a point on the x-axis which is 10 units from $(3,8)$, find all the possible coordinates of P.

7. If Q is a point on the x-axis which is 13 units from $(-1,5)$, find all the possible coordinates of Q.

8. If R is a point on the y-axis which is 5 units from $(-3,-7)$, find all the possible coordinates of R.

9. Show that the points $(5,-1)$, $(2,8)$, and $(-2,0)$ lie on a circle whose centre is $(2,3)$.

10. Find a point on the y-axis which is equidistant from $(5,-5)$ and $(1,1)$.

11. Show that the point with coordinates $(5,6)$ is the midpoint of the line segment joining the points $(-11,-5)$ and $(21,17)$.

12. Find the lengths of the diagonals of a rectangle with vertices $A(-4,6)$, $B(-4,-8)$, $C(5,-8)$, and $D(5,6)$.

13. Find the perimeter of the triangle with vertices $R(6,-5)$, $S(-4,-5)$, and $T(3,-1)$.

14. Find the perimeter of the rectangle with vertices $P(-5,-2)$, $Q(-5,-9)$, $R(-1,-9)$, and $S(-1,-2)$.

15. Find the perimeter of the quadrilateral with vertices $A(-3,4)$, $B(-2,-2)$, $C(5,-1)$, and $D(2,1)$.

16. A triangle has vertices $D(-1,-1)$, $E(7,5)$, and $F(0,6)$. Is the triangle a right triangle?

17. A triangle has vertices $A(5,-6)$, $B(-4,-2)$, and $C(2,3)$. Is the triangle a right triangle?

18. Find the perimeter of the triangle with vertices $X(-a,0)$, $Y(0,b)$, and $Z(0,0)$.

19. (a) Show that the triangle with vertices $A(3,5)$, $B(-5,-7)$, and $C(3,-7)$ is a right triangle.
(b) Determine the area of the triangle.

MICRO MATH

The following program will calculate the distance between two points.

NEW
```
10 PRINT "DISTANCE BETWEEN TWO POINTS"
20 INPUT "X1=";X1
30 INPUT "Y1=";Y1
40 INPUT "X2=";X2
50 INPUT "Y2=";Y2
60 PRINT "L= ";SQR((X2-X1)↑2
   +(Y2-Y1)↑2)
70 END
```

RUN
Statement 60 in the program above should be entered on one line.

EXERCISE

1. Use the program to find the distance between each pair of points to the nearest hundredth.
(a) $(-3.75,2.46)$ and $(1.07,-0.83)$
(b) $(15.9,26.71)$ and $(-8.22,-9.66)$
(c) $(107.4,-89.23)$ and $(-17.44,93.93)$
(d) $(0.413,0.157)$ and $(-0.667,0.922)$
(e) $(-9.03,-7.46)$ and $(-3.45,-8.09)$

4.1 RATIO AND PROPORTION

The ratio of one number to another is a way of comparing two numbers by division. We sometimes use the notation $a : b$ for the ratio of a to b. For instance, the ratio of 4 to 7 is $4 : 7 = \frac{4}{7}$.

> If a, b $\in$ R and b $\neq$ 0, the ratio of a to b is
> $$a : b = \frac{a}{b}$$

If there are 18 boys and 15 girls in a mathematics class, then the ratio of boys to girls is $18 : 15$, or equivalently $6 : 5$ since

$$18 : 15 = \frac{18}{15}$$
$$= \frac{6}{5}$$

> A proportion is an equality of two ratios
> $$a : b = c : d$$
> or equivalently,
> $$\frac{a}{b} = \frac{c}{d}$$
> where b $\neq$ 0 and d $\neq$ 0.

The numbers a, b, c, d are called the first, second, third, and fourth terms respectively in the proportion. The first and fourth terms, a and d, are called the extremes, and the second and third terms, b and c, are called the means of the proportion. If the means are equal, that is, if b = c, then either b or c is called the mean proportional between a and d. For instance, in the proportion $4 : 6 = 6 : 9$, 6 is the mean proportional between 4 and 9.

means

$$a : b = c : d$$

extremes

Notice the following things about the proportion.

$$\frac{8}{12} = \frac{26}{39}$$

I. $\quad 8 \times 39 = 312$
$\quad 12 \times 26 = 312$ ⟵ The product of the extremes is equal to the product of the means.

II. $\quad \dfrac{12}{8} = \dfrac{39}{26}$ ⟵ If we turn both ratios upside down, the new proportion is still correct.

III. $\dfrac{8}{26} = \dfrac{12}{39}$ $\longleftarrow$ If we interchange the means, the new proportion is still correct.

IV. $\dfrac{39}{12} = \dfrac{26}{8}$ $\longleftarrow$ If we interchange the extremes, the new proportion is still correct.

V. $\dfrac{8 + 12}{12} = \dfrac{26 + 39}{39}$ $\longleftarrow$ If we add the denominators to the numerators, the new proportion is still correct.

Properties of Proportions

Suppose that none of the real numbers a, b, c, d is zero.

I.   If $\dfrac{a}{b} = \dfrac{c}{d}$, then $ad = bc$.   product of extremes = product of means

II. If $\dfrac{a}{b} = \dfrac{c}{d}$, then $\dfrac{b}{a} = \dfrac{d}{c}$. $\left(\dfrac{a}{b} = \dfrac{c}{d}\right)$

III. If $\dfrac{a}{b} = \dfrac{c}{d}$, then $\dfrac{a}{c} = \dfrac{b}{d}$. $\dfrac{a}{b} \diagup \dfrac{c}{d}$

IV. If $\dfrac{a}{b} = \dfrac{c}{d}$, then $\dfrac{d}{b} = \dfrac{c}{a}$. $\dfrac{a}{b} \diagdown \dfrac{c}{d}$

V.   If $\dfrac{a}{b} = \dfrac{c}{d}$, then $\dfrac{a + b}{b} = \dfrac{c + d}{d}$. $\dfrac{a + b}{b} = \dfrac{c + d}{d}$

VI. If $\dfrac{a}{b} = \dfrac{c}{d}$, then $\dfrac{a - b}{b} = \dfrac{c - d}{d}$. $\dfrac{a - b}{b} = \dfrac{c - d}{d}$

Proof of I:

Given: $\dfrac{a}{b} = \dfrac{c}{d}$

$\therefore bd \times \dfrac{a}{b} = bd \times \dfrac{c}{d}$ $\longleftarrow$ Multiply both sides by bd.

$\therefore \quad ad = bc$

Proof of V:

Adding 1 to both sides of $\dfrac{a}{b} = \dfrac{c}{d}$, we get

$$\dfrac{a}{b} + 1 = \dfrac{c}{d} + 1 \quad \text{or} \quad \dfrac{a + b}{b} = \dfrac{c + d}{d}.$$

The proofs of the other properties are left to the Exercise.

EXAMPLE 1. Solve the proportion $4 : 7 = x : 21$ for x.

SOLUTION:

$$\frac{4}{7} = \frac{x}{21}$$
$$7x = 4 \times 21 \qquad \text{Property I.}$$
$$x = \frac{4 \times 21}{7}$$
$$= 4 \times 3$$
$$= 12$$

EXAMPLE 2. Find the mean proportional between 4 and 36.

SOLUTION:
Let x be the mean proportional.

Then
$$\frac{4}{x} = \frac{x}{36}$$
$$\therefore x^2 = 4 \times 36 \qquad \text{Property I.}$$
$$= 144$$
$$\therefore x = 12 \quad \text{or} \quad -12.$$

EXERCISE 4.1

A 1. Complete the following.

(a) If $\dfrac{a}{b} = \dfrac{7}{4}$, then $\dfrac{b}{a} = \blacksquare$

(b) If $\dfrac{x}{6} = \dfrac{2}{5}$, then $5x = \blacksquare$

(c) If $\dfrac{2}{x} = \dfrac{3}{8}$, then $3x = \blacksquare$

(d) If $\dfrac{x}{y} = \dfrac{3}{7}$, then $\dfrac{x}{3} = \blacksquare$

(e) If $\dfrac{m}{n} = \dfrac{2}{9}$, then $\dfrac{m + n}{n} = \blacksquare$

(f) If $\dfrac{h}{3} = \dfrac{k}{10}$, then $\dfrac{h - 3}{3} = \blacksquare$

(g) If $\dfrac{c}{d} = \dfrac{1}{9}$, then $\dfrac{9}{d} = \blacksquare$

(h) If $\dfrac{27}{x + 1} = \dfrac{4}{y}$, then $27y = \blacksquare$

(i) If $\dfrac{r}{6} = \dfrac{7}{s}$, then $\dfrac{s}{7} = \blacksquare$

(j) If $7c = 2d$, then $\dfrac{c}{d} = \blacksquare$

(k) If $\dfrac{1}{t} = \dfrac{t}{9}$, then $t = \blacksquare$

(l) If $3x = 4y$, then $\dfrac{x}{\blacksquare} = \dfrac{y}{\blacksquare}$

B 2. Solve the following proportions for x.

(a) $3:5 = 9:x$

(b) $x:3 = 9:10$

(c) $6:17 = 2x:51$

(d) $x - 4:12 = 6:8$

(e) $\dfrac{x}{12} = \dfrac{9}{16}$

(f) $\dfrac{4}{3x} = \dfrac{15}{21}$

(g) $\dfrac{12}{38} = \dfrac{2}{x}$

(h) $\dfrac{x}{x + 2} = \dfrac{2}{3}$

(i) $\dfrac{1}{x} = \dfrac{5}{3x + 8}$

(j) $\dfrac{6 + x}{7} = \dfrac{1}{4}$

(k) $\dfrac{a}{x} = \dfrac{r}{s}$

(l) $\dfrac{1 + x}{1 - x} = \dfrac{5}{3}$

(m) $\dfrac{x + 1}{6} = \dfrac{4}{x - 1}$

(n) $\dfrac{1 + x}{2} = \dfrac{3x}{7}$

3. Find x and y in each of the following cases.

(a) $\dfrac{x}{15} = \dfrac{14}{y} = \dfrac{2}{9}$

(b) $\dfrac{x}{15} = \dfrac{y}{21} = 6$

(c) $\dfrac{2}{x} = \dfrac{4}{3y} = 10$

(d) $\dfrac{x}{8} = \dfrac{32}{x} = y$

4. If $\dfrac{a}{3} = \dfrac{b}{5}$, evaluate.

(a) $\dfrac{2a}{3b}$

(b) $\dfrac{a^2}{b^2}$

(c) $\dfrac{a + b}{2a + b}$

5. Find the value of $\dfrac{x}{y}$ in the following.

(a) $3x = 5y$

(b) $5x + 3y = 5y - 2x$

(c) $16x^2 = 25y^2$

(d) $\dfrac{x + y}{x - y} = \dfrac{4}{3}$

6. Find the mean proportionals between the following pairs of numbers.

(a) 9 and 16
(b) 4 and 25
(c) $\sqrt{2}$ and $2\sqrt{2}$
(d) 3 and 21
(e) x and x^3
(f) 3x and 12x

7. Find the fourth proportionals to the following sets of numbers.

(a) 3, 4, 7
(b) 2, 5, 8
(c) $\frac{1}{2}$, $\frac{2}{3}$, $\frac{3}{4}$
(d) a, b, c

8. Find two numbers whose sum is 50 and whose ratio is 3:7.

9. Two numbers are in a ratio of 8:5. Their sum is 117. Find the numbers.

10. The ratio of girls to boys in a class was 4:5. Then four more girls are added to the class, making equal numbers of boys and girls. What is the present size of the class?

11. A gear has 60 teeth. The ratio of teeth on this gear to the teeth on a second gear is 4:3. How many teeth are there on the second gear?

12. If a:b = 5:6, b:c = 3:7, and c:d = 5:12, find a:d.

13. Prove properties II, III, IV, and V of proportions.

14. If b is a mean proportional between a and c, prove that b = $\pm\sqrt{ac}$.

C 15. If $\dfrac{a}{b} = \dfrac{c}{d}$, use the properties of proportions to prove $\dfrac{a + b}{a - b} = \dfrac{c + d}{c - d}$.

CALCULATOR MATH

Calculators can be used to solve proportions.

EXAMPLE. Solve. 2.4:3.7 = 9.8:x

SOLUTION:
Press

9 . 8 × 3 . 7 ÷ 2 . 4 =

x = 15.1 to the nearest tenth.

EXERCISE

Solve the following using a calculator.
1. x:123 = 432:345
2. 6.7:x = 8.9:3.6
3. 67:45 = x:88
4. 12.7:21.9 = 11.4:x

4.2 MULTIPLE RATIOS

Suppose a:b = 2:3 and b:c = 3:5. We write this for short as
a:b:c = 2:3:5. These are called three-term ratios.

Since $\dfrac{a}{b} = \dfrac{2}{3}$, we have $\dfrac{a}{2} = \dfrac{b}{3}$.   Property III.

Since $\dfrac{b}{c} = \dfrac{3}{5}$, we have $\dfrac{b}{3} = \dfrac{c}{5}$.   Property III.

Therefore, $\dfrac{a}{2} = \dfrac{b}{3} = \dfrac{c}{5}$.

Let k be the common value of these three ratios.

$$\frac{a}{2} = \frac{b}{3} = \frac{c}{5} = k$$

Then a = 2k, b = 3k, and c = 5k.

In general, a three-term ratio a:b:c makes sense if b ≠ 0 and c ≠ 0.

Multiple Proportions

$$a:b:c = d:e:f$$

means there is a constant k ≠ 0 such that a = kd, b = ke, and
c = kf.

If the numbers are all nonzero, then this can be written as

$$\frac{a}{b} = \frac{b}{e} = \frac{c}{f}$$

EXAMPLE 1. What are the missing numbers
in the proportion 6:9:15 = 2:■:■?

SOLUTION:
Since 6 = 3 × 2, 9 = 3 × 3,
and 15 = 3 × 5,
∴ 6:9:15 = 2:3:5.

EXAMPLE 2. Find the values of x and y
if x:7:2 = 3:y:8.

SOLUTION:
$$\frac{x}{3} = \frac{7}{y} = \frac{2}{8}$$

$\dfrac{x}{3} = \dfrac{1}{4}$ | $\dfrac{7}{y} = \dfrac{1}{4}$

∴ $x = \dfrac{3}{4}$ | ∴ y = 7 × 4

     | = 28

EXAMPLE 3. A bag of fertilizer contains nitrogen, phosphorus, and potash in the ratio of 18:6:9.
If it contains 4 kg of nitrogen, how much phosphorus and potash does it contain?

SOLUTION:
Let the mass of phosphorus be x kg and the mass of potash y kg. Then 4:x:y = 18:6:9

∴ $\dfrac{4}{18} = \dfrac{x}{6} = \dfrac{y}{9}$

$\dfrac{2}{9} = \dfrac{x}{6}$ | $\dfrac{2}{9} = \dfrac{y}{9}$

∴ 9x = 12 | ∴ 9y = 18

∴ $x = \dfrac{12}{9}$ | ∴ y = 2

$\quad = \dfrac{4}{3}$ |

> Commercial fertilizer
> ratios are not reduced to
> lowest terms because
> they represent contents
> by percentage.

The mass of phosphorus is $1\frac{1}{3}$ kg and the mass of potash is 2 kg.

EXAMPLE 4. The sum of three numbers in the ratio 3:4:9 is 320. Find the numbers.

SOLUTION:
Let the numbers be x, y, and z.

Then $x + y + z = 320$

and $x:y:z = 3:4:9$.

∴ there is a constant k such that $\dfrac{x}{3} = \dfrac{y}{4} = \dfrac{z}{9} = k$.

∴ $x = 3k$, $y = 4k$, and $z = 9k$.

∴ $3k + 4k + 9k = 320$

$16k = 320$

$k = 20$

The numbers are 60, 80, 180.

Multiple ratios with more than three terms can also be defined. For instance,

$$1:3:5:7 = 4:12:20:28$$

and

$$\frac{1}{4} = \frac{3}{12} = \frac{5}{20} = \frac{7}{28}$$

EXERCISE 4.2

A 1. State the missing numbers in the following proportions.

(a) $1:2:3 = 3:\blacksquare:\blacksquare$ (b) $2:6:10 = 1:\blacksquare:\blacksquare$

(c) $7:3:1 = \blacksquare:6:\blacksquare$ (d) $1:3:2 = \blacksquare:\blacksquare:8$

(e) $6:\blacksquare:\blacksquare = 2:5:2$ (f) $\blacksquare:\frac{1}{4}:\blacksquare = 4:1:3$

(g) $2:\blacksquare:3 = 10:5:\blacksquare$ (h) $2:3:\blacksquare = \blacksquare:3:4$

B 2. Solve the following proportions for x, y, and z.

(a) $x:2:6 = 6:18:y$ (b) $7:x:2 = y:27:9$

(c) $2x:3y:7 = 12:1:3$

(d) $1:2:3 = x + 1:y - 2:2$

(e) $1:2:4:x = y:z:18:3$

(f) $4:7:10:13 = 10:x:y:z$

(g) $2:5:x = 6:x + y:12$

(h) $x + y:x - y:9 = 4:12:18$

3. If $a:b:c = 6:4:5$, find the value of $\dfrac{3a - b}{4b + c}$.

4. A certain brand of fertilizer contains nitrogen, phosphorus, and potash in the ratio 10:6:8. If a bag of fertilizer contains 2 kg of potash, how much nitrogen and phosphorus does it contain?

5. The M and N Nut Company advertises that its tins of mixed nuts contain peanuts, cashews, and almonds in the ratio 5:3:2 by mass.

(a) If a tin contains 250 g of cashews, find the mass of peanuts and the mass of almonds in the tin.

(b) A smaller tin contains a total mass of 500 g of nuts. Find the mass of each kind of nut in this tin.

6. Find three numbers in the ratio 7:3:2 whose sum is 72.

7. Find four numbers in the ratio 3:4:6:9 whose sum is 330.

8. The measures of the three angles of a triangle are in the ratio 1:3:5. Find the size of each angle.

C 9. If $\dfrac{a}{b} = \dfrac{c}{d} = \dfrac{e}{f}$, prove that $\dfrac{a + c + e}{b + d + f} = \dfrac{a}{b}$.

4.3 APPLYING PROPORTIONS

In this section we will solve problems using proportions.

READ

EXAMPLE 1. The Blue Jays won 55 of their first 90 games. How many games can they expect to win out of 162?

PLAN

SOLUTION:
Let w represent the number of wins after 162 games. Assuming that the ratio of wins to games after 90 games will be the same after 162 games, we write the proportion.

$E=mc^2$

$$\frac{55}{90} = \frac{w}{162}$$
$$8910 = 90w$$
$$99 = w$$

SOLVE

ANSWER

The Blue Jays can expect to win 99 games.

READ

EXAMPLE 2. Sugar consists of carbon, hydrogen, and oxygen combined in a ratio of 72:11:88 by mass. How many grams of carbon are there in 8850 g of sugar?

PLAN

SOLUTION:
Let x, y, and z represent the number of grams of carbon, hydrogen, and oxygen in 8850 g of sugar.

Then $x + y + z = 8850$

and $x:y:z = 72:11:88$

There is a constant k such that
$$\frac{x}{72} = \frac{y}{11} = \frac{z}{88} = k$$

SOLVE

$\therefore x = 72k, y = 11k,$ and $z = 88k.$

$$72k + 11k + 88k = 8850$$
$$171k = 8850$$
$$k = 50$$

$$x = 72k$$
$$= 72(50)$$
$$= 3600$$

Check.
$72k = 3600$
$11k = 550$
$88k = \underline{4400}$
8850

ANSWER

There are 3600 g of carbon in 8850 g of sugar.

EXERCISE 4.3

B 1. In a radio survey, 550 homes were called. The residents of 182 homes listened to radio station CHAT. If there are 30 000 homes in the area, in approximately how many homes will the residents listen to CHAT?

2. To estimate the number of trout in a lake, the game warden caught 85 trout, tagged them, and then released them. Later, he caught 95 trout and found tags on 7 of them. Approximately how many trout are in the lake?

3. A map is drawn so that 1 cm represents 30 km. How far apart are two cities if they are 11.4 cm apart on the map?

4. In physics, Hooke's law says that the force exerted by a spring is proportional to the amount that the spring is stretched. If a force of 70 N is needed to stretch a spring 4 cm, what force is needed to stretch the same spring 9.5 cm?

5. Propane is a combination of carbon and hydrogen in the ratio of 36:8 by mass. What mass of hydrogen is in 220 g of propane?

6. Sodium and chlorine combine in an approximate ratio of 23:36 by mass to make table salt. Approximately how many grams of chlorine are there in 1000 g of table salt?

7. Fool's gold is a combination of iron and sulphur in an approximate ratio of 56:64 by mass. Approximately how many grams of sulphur are there in 300 g of fool's gold?

8. An astronaut who has a mass of 72 kg on Earth has a mass of 12 kg on the moon. If another astronaut has a mass of 10.5 kg on the moon, what is her mass on Earth?

9. A car travelled 414 km on 46 L of gasoline. How far will the car travel on 200 L of gasoline?

10. Seven metres of steel wire has a mass of 11.9 kg. What is the mass of 53 m of the same steel wire?

11. When water freezes the ice formed has 9% more volume than the water. How much water must freeze to make 872 m^3 of ice?

12. There are three numbers in the ratio of 2:3:4. The sum of their squares is 464. Find the numbers.

13. The ratio of gravel to sand to cement in a mixture is 6:4:1. How many kilograms of each do you need for 495 kg of mix?

14. Baking soda is a combination of sodium, hydrogen, carbon, and oxygen in a ratio of 23:1:12:16 by mass. How many grams of each are there in 494 g of baking soda?

There are 25 lockers, all closed, and 25 students. The first student opens every locker. The second student closes every second locker. The third student changes the state of every third locker. (Open lockers are closed and closed lockers are opened.) The fourth student changes the state of every fourth locker. This procedure continues until the twenty-fifth student changes the state of the twenty-fifth locker. Which lockers will be open at the end of the procedure?

4.4 MIDPOINT OF A LINE SEGMENT

On a number line, the coordinate of the midpoint of a line segment is the average of the coordinates of the endpoints. For a line segment with endpoints A(2) and B(12),

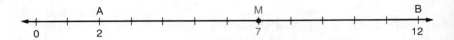

the midpoint M has coordinate $\dfrac{12 + 2}{2} = 7$.

This may be proved as follows. Consider the line segment with endpoints A(a) and B(b). If M(m) is the midpoint, then

or
$$MA = BM$$
$$m - a = b - m$$
$$2m = a + b$$

and
$$m = \frac{a + b}{2}$$

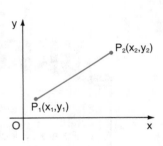

In order to determine the formula for the midpoint of a line segment in the coordinate plane a similar procedure can be used.

EXAMPLE 1. Find the coordinates of the midpoint of the line segment joining $P_1(x_1, y_1)$ and $P_2(x_2, y_2)$.

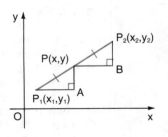

SOLUTION:
Let $P(x, y)$ be the midpoint of P_1P_2. Draw the rise and run of PP_1 and P_2P. Since $P_1A \parallel PB$, $\angle PP_1A = \angle P_2PB$ (Corresponding angles)
$$\angle A = \angle B = 90°$$
$$PP_1 = P_2P$$
$$\therefore \triangle PP_1A \cong \triangle P_2PB \qquad \text{(ASA)}$$
$$\therefore AP_1 = BP$$
$$\therefore x - x_1 = x_2 - x$$
$$2x = x_1 + x_2$$
$$x = \frac{x_1 + x_2}{2}$$

Similarly, $PA = P_2B$

and
$$y - y_1 = y_2 - y$$
$$2y = y_1 + y_2$$
$$y = \frac{y_1 + y_2}{2}$$

> If (x_1, y_1) and (x_2, y_2) are any two points in the coordinate plane, then the midpoint of the line segment joining them is
>
> $$\left(\frac{x_1 + x_2}{2}, \frac{y_1 + y_2}{2}\right)$$

EXAMPLE 2. If $(-2,8)$ and $(6,-3)$ are the endpoints of a line segment, find its midpoint.

SOLUTION:

$$x = \frac{x_1 + x_2}{2}$$
$$= \frac{-2 + 6}{2}$$
$$= 2$$

$$y = \frac{y_1 + y_2}{2}$$
$$= \frac{8 + (-3)}{2}$$
$$= \frac{5}{2}$$

∴ the midpoint is $(2, \frac{5}{2})$.

EXERCISE 4.4

B 1. Find the midpoint of the line segment joining the given pair of points.

(a) (2,2), (4,8)   (b) (3,10), (7,12)
(c) (−2,2), (−6,−8)   (d) (−7,4), (−1,9)
(e) (−3,−5), (−6,4)   (f) (10,−5), (0,−8)
(g) (0,0), (−8,10)   (h) (0,−5), (6,0)
(i) (4.2,0.4), (−1.6,−2)
(j) (3.7,−5.1), (−6.3,−4.5)

2. One endpoint of a line segment is $(-4,3)$. The midpoint is $(-3,6)$. Find the other endpoint.

3. If the midpoint of a line segment is $(-3,-5)$ and one endpoint is $(5,-6)$, find the other endpoint.

4. Find the points of division that will divide the line segment with endpoints $(-8,8)$ and $(4,-12)$ into four equal parts.

5. Find the midpoint of the line segment joining $(\sqrt{2}, \sqrt{3})$ and $(\sqrt{18}, \sqrt{75})$.

6. If the line segment joining $(-4,y)$ to $(x,-3)$ is bisected at $(1,-1)$, find the values of x and y.

7. Determine the equation of the perpendicular bisector of the line segment with endpoints $(-4,3)$ and $(2,7)$.

8. The vertices of $\triangle ABC$ are A(5,6), B(1,2), and C(9,4). Show that the line segment joining the midpoint of AB to the midpoint of AC is

(a) parallel to BC.
(b) one-half the length of BC.

9. Find the endpoints of two line segments with the midpoint $(3,-4)$.

10. Find the two points that trisect the line segment with endpoints (8,12) and (2,3).

4.5 SIMILAR TRIANGLES

Congruent triangles are equal in all respects. The triangles at the right are congruent and the corresponding sides and angles are equal.

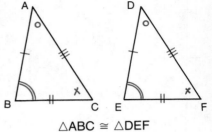

AB = DE	∠A = ∠D
BC = EF	∠B = ∠E
AC = DF	∠C = ∠F

△ABC ≅ △DEF

Similar triangles have the same shape, but not necessarily the same size. Similar triangles have the same shape because the corresponding angles are equal. The triangles at the right are similar because

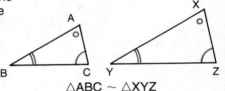

∠A = ∠X
∠B = ∠Y
∠C = ∠Z

△ABC ~ △XYZ

EXAMPLE 1. Prove that △ABC is similar to △DAE.

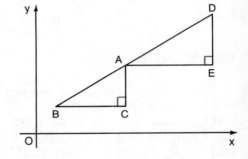

SOLUTION:
Since AC ∥ DE, ∠BAC = ∠ADE
Since BC ∥ AE, ∠ABC = ∠DAE
∠ACB = ∠DEA = 90°
Since the corresponding angles are equal, the triangles are similar and we write

△ABC ~ △DAE

EXAMPLE 2. The triangles at the right are similar.
(a) Determine the lengths of the sides of the triangles.
(b) Determine the ratio of the corresponding sides.

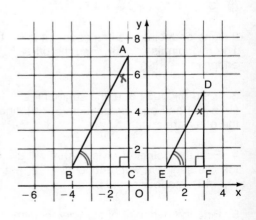

SOLUTION:

(a) △ABC

$AB = \sqrt{(\Delta x)^2 + (\Delta y)^2}$
$\quad = \sqrt{(3)^2 + (6)^2}$
$\quad = \sqrt{45}$
$\quad = 3\sqrt{5}$

$BC = \sqrt{(\Delta x)^2 + (\Delta y)^2}$
$\quad = \sqrt{(-3)^2 + (0)^2}$
$\quad = \sqrt{9}$
$\quad = 3$

$AC = \sqrt{(\Delta x)^2 + (\Delta y)^2}$
$\quad = \sqrt{(0)^2 + (6)^2}$
$\quad = \sqrt{36}$
$\quad = 6$

△DEF

$DE = \sqrt{(\Delta x)^2 + (\Delta y)^2}$
$\quad = \sqrt{(2)^2 + (4)^2}$
$\quad = \sqrt{20}$
$\quad = 2\sqrt{5}$

$EF = \sqrt{(\Delta x)^2 + (\Delta y)^2}$
$\quad = \sqrt{(-2)^2 + (0)^2}$
$\quad = \sqrt{4}$
$\quad = 2$

$DF = \sqrt{(\Delta x)^2 + (\Delta y)^2}$
$\quad = \sqrt{(0)^2 + (4)^2}$
$\quad = \sqrt{16}$
$\quad = 4$

(b) $\dfrac{AB}{DE} = \dfrac{3\sqrt{5}}{2\sqrt{5}}$ $\dfrac{BC}{EF} = \dfrac{3}{2}$ $\dfrac{AC}{DF} = \dfrac{6}{4}$

$\qquad\quad = \dfrac{3}{2}$ $\qquad\qquad\qquad\qquad\qquad\quad = \dfrac{3}{2}$

$\therefore \dfrac{AB}{DE} = \dfrac{BC}{EF} = \dfrac{AC}{DF} = \dfrac{3}{2}$

In general, the following is true.

> The corresponding sides of two similar triangles are in proportion.

EXERCISE 4.5

B 1. Prove that the pairs of triangles are similar.
  (a) △ABC ~ △CDE

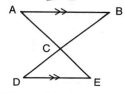

  (b) △RST ~ △RMN

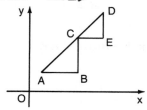

  (c) △ABC ~ △CDE

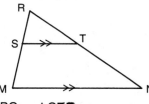

2. Find the value of x and y.

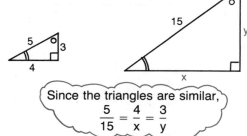

Since the triangles are similar,
$$\dfrac{5}{15} = \dfrac{4}{x} = \dfrac{3}{y}$$

3. Find the value of x and y.

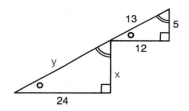

4. Find the lengths of x and y.
  (a)

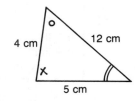

  (b)

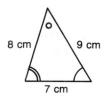

  (c)

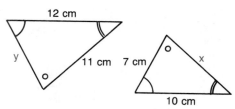

4.6 DIVISION OF A LINE SEGMENT

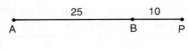

Suppose that P lies on AB and AP = 15 cm, PB = 10 cm. Then
$$\frac{AP}{PB} = \frac{15}{10} = \frac{3}{2}$$

We say that P divides the segment AB internally in the ratio 3:2.

Now suppose that P lies on AB produced and AB = 25 cm, BP = 10 cm.
Then
$$\frac{AP}{PB} = \frac{35}{10} = \frac{7}{2}$$

Here we say that P divides the segment AB externally in the ratio 7:2.

On the other hand, if P lies on BA produced and PA = 10 cm, AB = 25 cm. Then
$$\frac{AP}{PB} = \frac{10}{35} = \frac{2}{7}$$

Here P divides the segment AB externally in the ratio 2:7.

P divides AB in the ratio a:b.

$$\text{first point} \longrightarrow \frac{AP}{PB} = \frac{a}{b} \longleftarrow \text{second point}$$
$$\text{point of division}$$

If P lies on AB, the division is internal.
If P lies on AB produced, the division is external and a > b.
If P lies on BA produced, the division is external and a < b.

EXAMPLE 1. If P divides AB internally in the ratio 5:3 and AB = 12 cm, find the length of AP and PB.

SOLUTION:
Let AP = x.
Then PB = 12 − x.
$$\frac{AP}{PB} = \frac{5}{3}$$
$$\frac{x}{12 - x} = \frac{5}{3}$$
$$3x = 5(12 - x)$$
$$= 60 - 5x$$
$$8x = 60$$
$$x = 7.5$$

The length of AP is 7.5 cm.
The length of PB is 4.5 cm.

EXAMPLE 2. Find the point P(x,y) that divides the segment joining A(2,1) to B(7,6) internally in the ratio of 3:2.

SOLUTION:

Since P(x,y) is the point that divides AB internally in the ratio of 3:2 then

$$\frac{AP}{PB} = \frac{3}{2}$$

Draw the rise and run of AP and BP.

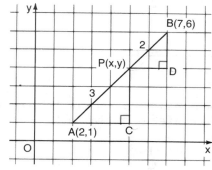

Since AC ∥ PD, ∠PAC = ∠BPD
Since PC ∥ BD, ∠APC = ∠PBD
∠C = ∠D = 90°
∴ △PAC ~ △BPD
∴ $\frac{AP}{PB} = \frac{AC}{PD} = \frac{PC}{BD} = \frac{3}{2}$

$$\frac{AC}{PD} = \frac{3}{2} \qquad \qquad \frac{PC}{BD} = \frac{3}{2}$$

$$\therefore \frac{x-2}{7-x} = \frac{3}{2} \qquad \qquad \frac{y-1}{6-y} = \frac{3}{2}$$

$$2x - 4 = 21 - 3x \qquad \qquad 2y - 2 = 18 - 3y$$

$$5x = 25 \qquad \qquad 5y = 20$$

$$x = 5 \qquad \qquad y = 4$$

∴ P(5,4) divides AB internally in the ratio of 3:2.

EXAMPLE 3. Find the point P(x,y) that divides the segment joining A(2,2) to B(5,6) externally in the ratio of 4:3.

SOLUTION:

Since P(x,y) is the point that divides AB externally in the ratio of 4:3 then

$$\frac{AP}{PB} = \frac{4}{3}$$

Draw the rise and run of AB and BP.

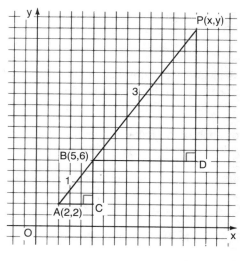

Since AC ∥ BD,
∠BAC = ∠PBD
Since BC ∥ PD,
∠ABC = ∠BPD
∠C = ∠D = 90°
∴ △BAC ~ △PBD
∴ $\frac{BA}{PB} = \frac{AC}{BD} = \frac{BC}{PD} = \frac{1}{3}$

$$\frac{AC}{BD} = \frac{1}{3} \qquad \qquad \frac{BC}{PD} = \frac{1}{3}$$

$$\frac{5-2}{x-5} = \frac{1}{3} \qquad \qquad \frac{6-2}{y-6} = \frac{1}{3}$$

$$15 - 6 = x - 5 \qquad \qquad 18 - 6 = y - 6$$

$$x = 14 \qquad \qquad y = 18$$

∴ P(14,18) divides AB externally in the ratio of 4:3.

EXERCISE 4.6

A 1. State the ratio in which P divides the line segment. State whether the division is internal or external.

(a) BC   (b) AB   (c) AC
(d) CD   (e) BD   (f) AD

```
      2    4    2    3
     •—•———•———•———•—
     A  B      P  C    D
```

2. State the ratio in which P divides the line segment. State whether the division is internal or external.

(a) RS   (b) RT   (c) ST
(d) QR   (e) TQ   (f) SQ

```
     1    5         6         4
    •—•————•—————————•—————————•
    Q R         P         S         T
```

B 3. In what ratio does P divide AB internally in the following cases.

(a) AB = 16 cm, PB = 12 cm
(b) AP = 6 cm, PB = 27 cm
(c) AB = 30 cm, AP = 22.5 cm

4. If Q divides the segment CD internally in the ratio of 2:7 and CD = 18 cm, find the lengths of CQ and QD.

5. If P divides AB externally in the ratio 6:5 and AB = 8 cm, find the lengths of BP and AP.

6. If S divides PQ externally in the ratio of 4:7 and PQ = 18 cm, find the lengths of SP and SQ.

7. For each of the following find the point P(x,y) that divides the given line segment in the given ratio.

(a) A(2,2) to B(5,3) internally in the ratio of 1:2
(b) C(−3,−5) to D(4,0) internally in the ratio of 1:4
(c) E(5,−2) to F(−4,5) internally in the ratio of 1:3
(d) G(−1,−3) to H(7,5) internally in the ratio of 4:5

(e) A(−7,3) to B(7,−1) internally in the ratio of 3:2
(f) C(6,−5) to D(−3,−1) internally in the ratio of 4:3

8. For each of the following find the point P(x,y) that divides the given line segment in the given ratio.

(a) A(−4,2) to B(4,4) externally in the ratio of 5:3
(b) C(2,1) to D(5,3) externally in the ratio of 3:1
(c) E(−6,2) to F(4,−2) externally in the ratio of 1:3
(d) G(4,−2) to H(−6,2) externally in the ratio of 2:3
(e) C(−5,−2) to D(0,−4) externally in the ratio of 2:5
(f) A(3,0) to B(−2,6) externally in the ratio of 7:2

C 9. The line segment joining (2,2) and (−7,−4) is trisected. Find the points of division.

10. The line segment joining A(−4,−5) to B(2,−2) is extended to C. If BC = 3AB, find the location of C.

11. Find the point P(x,y) that divides the line segment joining $P_1(x_1,y_1)$ to $P_2(x_2,y_2)$ internally in the ratio of a:b where a, b > 0.

MICRO MATH

NEW
```
10 REM MIDPOINT OF A LINE SEGMENT
20 INPUT "X1=";X1
30 INPUT "Y1=";Y1
40 INPUT "X2=";X2
50 INPUT "Y2=";Y2
60 PRINT "M= ("(X1+X2)/2","(Y1+Y2)/2")"
70 END
```
RUN

Use the program to find the midpoints of the following line segments.
1. A(5,7), B(−8,−9)
2. C(−6.1,−13.5), D(−21.6,15.5)
3. E(41.5,−63.8), F(−84.6,19.2)

4.7 PROBLEM SOLVING

In this section we continue our work on problem solving by discussing the method of working backward. This means that we assume the conclusion and work backward step by step until we arrive at something that is known or given. Then we may be able to reverse the steps in the argument and proceed forward from the given to the conclusion.

This procedure is commonly used in solving equations. For instance, in solving the equation $2x + 5 = 11$ we suppose that x is a number satisfying $2x + 5 = 11$ and work backward. We subtract 5 from each side of the equation and then divide each side by 2 to get $x = 3$. Since each of these steps can be reversed, we have solved the problem.

EXAMPLE. How is it possible to bring up from a river exactly 6 L of water when you only have 2 containers, a 9 L pail and a 4 L pail?

SOLUTION:

If you work this problem in the forward direction you might be lucky and discover, out of the many possibilities for proceeding, a correct solution. But it is more systematic to work backward. Imagine that we have 6 L of water in the 9 L pail, together with the full 4 L pail. We could fill the larger pail from the smaller one, leaving just 1 L in the smaller pail. Then we could empty the large pail and pour the 1 L of water into it. Finally, we could exactly fill the larger pail by adding 8 L of water using the smaller pail twice.

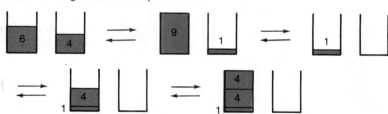

Now, by reversing this procedure, we have the solution to the problem. Start with a full 9 L pail. Use it to fill the 4 L, then empty the 4 L pail. Again fill the 4 L pail from the larger one and empty the smaller one. This will leave 1 L in the large pail. Transfer it to the small pail and fill the large one. Using the large pail to fill the small one will leave 6 L in the large pail.

EXERCISE 4.7

B 1. Al gives Bob as many cents as Bob has and Cindy as many cents as Cindy has. Bob gives Al and Cindy as many cents as each then has. Cindy gives Al and Bob as many cents as each then has. Each of them ends up with 16 cents. How many cents did each have to start with?

2. Three containers hold 19 L, 13 L, and 7 L respectively. The 19 L container is empty. The other two are full. How can you measure out 10 L using no other container?

3. If x and y are positive real numbers, prove that $\sqrt{xy} \leq \dfrac{x + y}{2}$.

4.8 REVIEW EXERCISE

1. Solve the following proportions for x.

(a) $x:4 = 9:7$

(b) $7:10 = 3:x$

(c) $\dfrac{x}{8} = \dfrac{6}{5}$

(d) $\dfrac{x}{x + 1} = \dfrac{5}{7}$

(e) $\dfrac{2 - x}{3} = \dfrac{3x}{5}$

(f) $\dfrac{x - 2}{x + 2} = \dfrac{3}{4}$

2. Solve for x and y.

(a) $\dfrac{x}{7} = \dfrac{y}{4} = \dfrac{1}{2}$

(b) $\dfrac{5}{x} = \dfrac{4}{9} = \dfrac{3}{y}$

(c) $x:y:8 = 5:7:3$

(d) $1:3:5 = 3x:2y:4$

3. The sum of three numbers in the ratio 3:4:8 is 360. Find the numbers.

4. The sum of four numbers in the ratio 2:4:6:9 is 399. Find the numbers.

5. Ammonia is a combination of nitrogen and hydrogen in the ratio of 14:3 by mass.

(a) How much nitrogen must be combined with 192 g of hydrogen to produce ammonia?

(b) How many grams of hydrogen are there in 25.5 g of ammonia?

6. A map is drawn so that 1 cm represents 50 km. How far apart are two towns if they are 6.8 cm apart on the map?

7. Oxygen and hydrogen combine in a ratio of 4:1 by mass to make heavy water.

(a) How many grams of hydrogen are there in 275 g of heavy water?

(b) How much oxygen must combine with 205 g of hydrogen to make heavy water?

8. Jerry, Joan, and Sandra contributed $2.50, $5.00, and $10.00 respectively to purchase a lottery ticket. How should the $840 000 prize money be divided?

9. Find the midpoint of the line segment joining the given pair of points.

(a) (4,9) and (8,3)

(b) (−1,2) and (−7,10)

(c) (−7,−3) and (−6,−10)

(d) (7,−13) and (7,21)

10. One endpoint of a line segment is (−6,8). The midpoint is (−2,4). Find the other endpoint.

11. Find the points of division that will divide the line segment with endpoints (−4,−6) and (10,12) into four equal parts.

12. A quadrilateral has vertices A(6,4), B(−4,2), C(−6,−6), and D(8,−2).

(a) Determine the coordinates of the midpoints of each side of the quadrilateral.

(b) Join the midpoints to form a quadrilateral. Show that the quadrilateral is a parallelogram.

13. Prove that the pairs of triangles are similar.

(a) $\triangle ABC \sim \triangle DEC$

(b) $\triangle ABC \sim \triangle ADE$

14. Find the value of x and y.

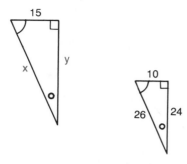

15. Find the value of s and t.

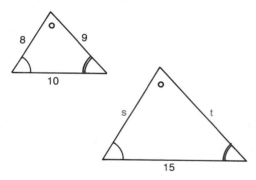

16. Find the value of m and n.

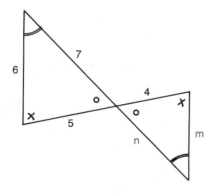

17. If P divides AB internally in the ratio 3:2 and AB = 20 cm, find the length of AP and PB.

18. If P divides AB externally in the ratio 7:3 and AB = 36 cm, find the length of AP and PB.

19. Find the point P(x,y) that divides the segment joining A(2,3) and B(9,10) internally in the ratio of 3:4.

20. Find the point P(x,y) that divides the segment joining A(1,1) and B(7,6) externally in the ratio of 3:2.

MIND BENDER

Certain pairs of two-digit numbers have the same product when both numbers are reversed. For example,

$$12 \times 42 = 21 \times 24$$
$$12 \times 63 = 21 \times 36$$
$$12 \times 84 = 21 \times 48$$
$$23 \times 96 = 32 \times 69$$
$$24 \times 84 = 42 \times 48$$
$$46 \times 96 = 64 \times 69$$

Find the eight other pairs of this nature.

Any natural number can be written as the sum of at most 4 perfect squares. For example,

$$126 = 10^2 + 5^2 + 1^2$$
$$211 = 12^2 + 7^2 + 3^2 + 3^2$$

Write each of the following as the sum of at most 4 perfect squares.

95
245
366

4.9 CHAPTER 4 TEST

1. Solve the following proportions for x.

(a) $6:x = 4:7$

(b) $\dfrac{x+1}{7} = \dfrac{3}{4}$

2. Solve for x and y.

(a) $x:y:3 = 2:3:5$

(b) $\dfrac{x}{15} = \dfrac{8}{y} = \dfrac{3}{10}$

3. Find the midpoint of the line segment joining the given pair of points.

(a) (7,9) and (1,13)

(b) $(-4, -7)$ and $(6, -11)$

4. Prove that △ABC ~ △ADE.

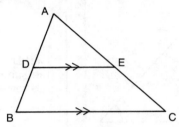

5. Find the value of x and y.

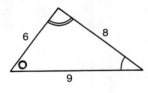 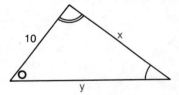

6. If P divides AB internally in the ratio 8:5 and AB = 21 cm, find the lengths of AP and PB.

7. Find the point P(x,y) that divides the line segment joining A(-2,2) and B(6,4) internally in the ratio of 3:2.

8. Find the point P(x,y) that divides the line segment joining A(-4,4) and B(4, -2) externally in the ratio of 2:3.

4.10 CUMULATIVE REVIEW FOR CHAPTERS 1 TO 4

1. Express in decimal form.

(a) $\frac{4}{5}$ (b) $\frac{3}{11}$ (c) $\frac{4}{7}$

2. Express in the form $\frac{a}{b}$.

(a) 0.315 (b) $0.\overline{4}$ (c) $0.5\overline{92}$

3. Simplify.

(a) $\frac{\sqrt{36}}{\sqrt{6}}$ (b) $\frac{\sqrt{28}}{\sqrt{7}}$ (c) $\frac{\sqrt{10x}}{\sqrt{2}}$

4. Simplify.

(a) $6\sqrt{2} + 7\sqrt{3} - 4\sqrt{2} + 5\sqrt{3}$
(b) $2\sqrt{8} + 3\sqrt{12} - 6\sqrt{18} + 2\sqrt{27}$
(c) $\sqrt{24} + \sqrt{54} - 2\sqrt{50} - 5\sqrt{18}$
(d) $3\sqrt{x} + 7\sqrt{y} - \sqrt{4y} + \sqrt{9x}$
(e) $2\sqrt{3}(1 + \sqrt{3})$
(f) $2\sqrt{5}(\sqrt{2} - \sqrt{6})$

5. Graph the following for $x \in R$.

(a) $\{\, x \mid x \leqslant 6 \,\}$
(b) $\{\, x \mid -1 \leqslant x < 2 \,\}$
(c) $\{\, x \mid x > -1 \,\} \cap \{\, x \mid x \leqslant 2 \,\}$

6. Simplify the following.

(a) $(3x^5)^2$
(b) $(5x^4)(-2x^3)$
(c) $(-2x^4)^3$
(d) $\frac{36x^5}{9x^2}$

7. Find the amount of interest on $5000 invested at 12% per annum for three years compounded semi-annually.

8. Find the present value of $4000 in three years at 8% per annum compounded quarterly.

9. The sum of the squares of two consecutive integers is 365. Find the integers.

10. Expand and simplify.

(a) $2(x - 7) - 6(x + 9)$
(b) $2x(x - 1) + x(2x + 1)$
(c) $(4x - 3)(5x + 9)$
(d) $2(x - 6)(2x + 5) - (6x - 1)(x + 3)$
(e) $(2x - 3)(x^2 - 2x - 8)$
(f) $(2x^2 - 4x - 1)(3x^2 + 5x + 1)$

11. Factor over the integers.

(a) $m^2 - m - 12$
(b) $4x^2 - 9$
(c) $6x^2 + 19x + 15$
(d) $15x^2 - 31x + 14$
(e) $28x^2 - 23x - 15$
(f) $(x - y)^2 - 16$
(g) $ax + bx + ay + by$
(h) $x^2 + 6x + 9 - m^2 - 2m - 1$

12. Divide.

(a) $\dfrac{24x + 6}{6}$

(b) $\dfrac{10x^2 - 5x + 15x}{5x}$

(c) $(x^3 - 6x^2 - 14x + 15) \div (x - 3)$
(d) $(2x^3 + 13x^2 + 13x - 3) \div (2x + 3)$
(e) $(x^3 - 4x - 3) \div (x + 1)$

13. Simplify.

(a) $\dfrac{x^2 + 5x + 6}{x^2 - 1} \times \dfrac{x^2 - 2x - 3}{x^2 - x - 6}$

(b) $\dfrac{t^2 + 9t + 20}{t^2 - 4t + 3} \div \dfrac{t^2 + 11t + 30}{t^2 + 5t - 6}$

14. Simplify.

(a) $\dfrac{1}{x + 1} + \dfrac{1}{x + 2}$

(b) $\dfrac{3}{t - 1} - \dfrac{2}{t - 2}$

(c) $\dfrac{3}{x^2 + 3x + 2} + \dfrac{1}{x^2 + 5x + 6}$

(d) $\dfrac{3}{x^2 - 4x + 4} - \dfrac{2}{x^2 + x - 6}$

15. Graph each of the following.
(a) $y = 3x + 5$
(b) $2x - 3y = 6$

16. Write an equation, in standard form, of the line through the given point and having the given slope.
(a) $(5,2)$; $m = 4$
(b) $(-3, -1)$; $m = -1$

17. Write an equation, in standard form, of the line through the given points.
(a) $(3,4)$ and $(4,6)$
(b) $(-5, -1)$ and $(-3,0)$

18. Write an equation of the line passing through $(2,3)$ and parallel to the graph of $y = 4x + 9$.

19. Write an equation of the line passing through $(-1, -5)$ and perpendicular to the graph of $2x - 3y + 4 = 0$.

20. Prove using slopes that $(-4, -1)$, $(-6, -4)$, $(-3, -4)$, and $(-1, -1)$ are the vertices of a parallelogram.

21. The dividend on 315 shares of Hillcrest stock is $866.25 per year. What is the dividend on 200 shares of the same stock?

22. Justine operates a fishing boat for tourists. A one day fishing trip for three people costs $350. A one day trip for seven people costs $550.

(a) Plot a graph of cost versus number of people.
(b) Determine an equation for this function and express it in the form $c = nm + b$ where c represents the cost and n the number of people.

(c) Determine the cost of a fishing trip for thirteen people.
(d) What quantity does the slope of the line represent?
(e) What meaning does the y-intercept have?

23. Solve the following.
(a) $3x - y = 5$
    $2x + 3y = 18$
(b) $4x - 5y = -7$
    $7x + 2y = -23$
(c) $\dfrac{x}{4} + \dfrac{y}{3} = 5, \dfrac{x}{2} - \dfrac{y}{9} = 3$
(d) $0.5x - 2y = -2$
    $0.7x + 0.3y = 3.4$

24. Find two numbers whose sum is 604 and whose difference is 118.

25. Solve the following.
(a) $2x + 3y - z = -3$
    $3x - 2y - 3z = -1$
    $4x - y + 2z = 9$
(b) $5x + 2y + 6z = 8$
    $4x - 3y + 2z = 9$
    $3x + 3y - 4z = -31$

26. A chemist wants to get 1000 L of 34% salt solution by mixing 40% salt solution with 30% salt solution. How many litres of each should be used?

27. Find the point $P(x,y)$ that divides the line segment joining $A(-1, -3)$ to $B(5,4)$ internally in the ratio of 2:1.

28. Find the point $P(x,y)$ that divides the line segment joining $A(4,4)$ to $B(10,12)$ externally in the ratio of 4:3.

29. Sandra surveyed 300 people and found that 80 of them listened to radio station CHIN. If there are 600 000 people in the city, approximately how many listen to CHIN?

QUADRATIC FUNCTIONS

CHAPTER

5

When we cannot use the compass of mathematics or the torch of experience... it is certain we cannot take a single step forward.

Voltaire

ALGEBRAIC MANIPULATION

EXERCISE

1. Simplify.

(a) $\dfrac{2}{b + 3} - \dfrac{1}{b - 3}$

(b) $\dfrac{3x - 3}{7 - x} - \dfrac{11 - 2x}{49 - x^2}$

2. Solve for each of the 3 variables.

$$\dfrac{1}{R_1} + \dfrac{1}{R_2} = \dfrac{1}{R}$$

3. Solve for each of the 6 variables.

$$\dfrac{p_1 V_1}{T_1} = \dfrac{p_2 V_2}{T_2}$$

4. Solve for x.

$$y = \dfrac{1 + x}{1 - x}$$

5. Solve for y and z.

$$x = \dfrac{y + z}{yz}$$

6. Solve for m and v.

$$E = \tfrac{1}{2}mv^2, \quad E > 0$$

7. Solve for s, b, and c.

$$m = \dfrac{(s - c)b}{(s - b)c}$$

CONTINUED FRACTIONS

EXERCISE

1. Simplify the following continued fractions by expressing each of them as a simple fraction in lowest terms.

(a) $\dfrac{1}{1 - \dfrac{1}{1 - \frac{1}{2}}}$

(b) $3 - \dfrac{4}{2 + \dfrac{5}{3 + \frac{1}{4}}}$

(c) $1 + \dfrac{1}{1 + \dfrac{1}{1 + x}}$

(d) $\dfrac{1}{1 - \dfrac{1}{1 - \frac{1}{x}}}$

(e) $1 + \dfrac{1}{1 + \dfrac{1}{1 + \frac{1}{x - 1}}}$

(f) $\dfrac{x}{1 - \dfrac{1}{x + \frac{1}{x + 1}}}$

(g) $\dfrac{\dfrac{2x}{1 + x}}{3}{1 - \frac{1}{x}}$

(h)
$$\frac{\dfrac{x+2}{x}}{x+1-\dfrac{1}{1-\dfrac{x}{x+1}}}$$

(i)
$$\frac{\dfrac{x}{x+1}}{x+2-\dfrac{\dfrac{2}{x+1}}{1-\dfrac{x}{x+1}}}$$

PIERRE LAPLACE (1749-1827)

Laplace was attracted to mathematics because of its power in solving scientific problems. His own greatest contribution was in the application of mathematics to astronomy. Using the theory of functions together with Newton's Laws, he was able to show that our solar system is stable.

It has been said that probability theory owes more to Laplace than to any other person. But even his interest in probability was inspired by the need for it in mathematical astronomy.

Laplace lived at the time of the French Revolution, but managed to survive quite well because he quickly changed his allegiance to whoever came to power. He even changed his book on mathematical astronomy to include glowing tributes to whichever side happened to be in power at the time. Laplace taught mathematics at the Military School and one of his students there was Napoleon Bonaparte. Napoleon was a great admirer of men of science. (He said, "The advancement and perfection of mathematics are intimately connected with the prosperity of the state.") When Napoleon came to power he awarded Laplace the Grand Cross of the Legion of Honour and gave him the powerful position of Minister of the Interior. So what did Laplace do when Napoleon fell from power? He signed the decree which banished Napoleon, and thrived under King Louis XVIII.

5.1 FUNCTIONS

If a car travels at a speed of 80 km/h, then after half an hour the car will have travelled 40 km. If d represents distance in kilometres and t represents time in hours, then the rule that connects t and d is d = 80t, as long as the car's speed is a constant 80 km/h. We say that distance is a function of time, because the value of d depends on the value of t. With each value of t, there is associated one value of d.

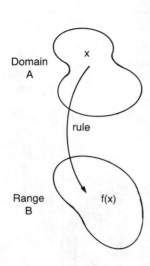

Given two sets A and B, a function f is a rule, or correspondence, that associates with each element x in A an element, called f(x), in B. A is called the domain of f. The range of f is the set of all possible values of f(x) for x in A.

You can think of a function as a machine, as shown in the figure below. If x is in the domain of the function f, then when x enters the machine it is accepted as an input and the machine produces an output f(x) according to the rule of the function. Therefore, you can think of the domain as the set of all possible inputs and the range as the set of all possible outputs.

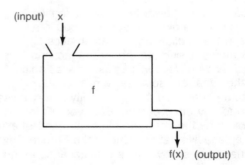

The symbol f(x) is read "f of x." It is the value of the function f at x. For example, the value of f when x = 7 is f(7). If f is given by a formula, then we find f(7) by substituting 7 for x.

EXAMPLE 1. If f(x) = 7x − 1, find (a) f(4), (b) f(−3), (c) f(0).

SOLUTION:

(a) f(4) = 7 × 4 − 1 = 27

(b) f(−3) = 7 × (−3) − 1 = −22

(c) f(0) = 7 × 0 − 1 = −1

EXAMPLE 2. If $g(x) = x + \dfrac{4}{x}$, find (a) g(2), (b) g(8), (c) g(0).

SOLUTION:

(a) $g(2) = 2 + \dfrac{4}{2} = 2 + 2 = 4$

(b) $g(8) = 8 + \dfrac{4}{8} = 8.5$

(c) g(0) is not defined because $\dfrac{4}{0}$ is not defined. This means that 0 does not belong to the domain of g.

EXAMPLE 3. In the example of the car in the opening paragraph of this section, we could write d = f(t) = 80 t. Here f(0.5) means the value of the function when t = 0.5, that is, the distance travelled after 0.5 h, and f(3) means the value of d after 3 h.

$$f(0.5) = 80 \times 0.5 = 40$$

$$f(3) = 80 \times 3 = 240$$

Since d depends on t, we call t an independent variable and d a dependent variable.

It sometimes helps to picture a function by an arrow diagram as in the figure below. Each arrow connects an element of A to an element of B. The arrow indicates that f(x) is associated with x, f(a) is associated with a, and so on.

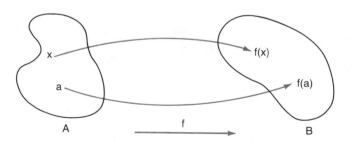

Functions are also called mappings, and we say that f maps the element x into its image f(x), and f maps the set A into the set B.

Sometimes we use arrows to name a function. In Example 1 we could have written

$$f:x \rightarrow 7x - 1 \text{ instead of } f(x) = 7x - 1.$$

This means that each real number x is mapped into its corresponding number 7x − 1.

EXAMPLE 4. Suppose that Joe is 15 years old, Brenda is 17, and Ann and John are both 16. Let f(x) be the age of x. For instance, f(Joe) = 15 and f(Ann) = 16. State the domain and range and represent f by an arrow diagram.

SOLUTION:
Domain of f: A = {Joe, Brenda, Ann, John}
Range of f: B = {15, 16, 17}

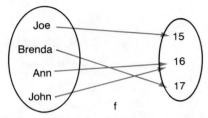

EXAMPLE 5. If f is the "squaring" function, f:x → x², and the domain is A = {−1, 0, 1, 2, 3}, find the range of f.

SOLUTION:
f(−1) = (−1)² = 1, f(0) = 0² = 0, f(1) = 1² = 1, f(2) = 2² = 4, f(3) = 3² = 9. The range is B = {0, 1, 4, 9}.

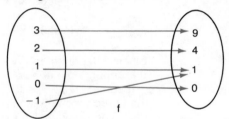

EXAMPLE 6. If f(x) = x² and the domain is R, what is the range of f?

SOLUTION:
The range consists of all squares of real numbers and the square of a real number is never negative. Therefore, the range of f consists of all non-negative real numbers: {y ∈ R | y ⩾ 0}.

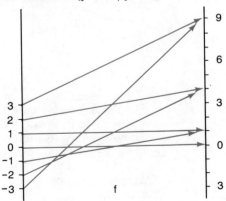

EXERCISE 5.1

A 1. If f(x) = 2 + 3x, state.

(a) f(0)   (b) f(−1)
(c) f(2)   (d) f(0.2)
(e) f(−5)   (f) f($\frac{1}{3}$)
(g) f(100)   (h) f(−10)
(i) f(a)   (j) f(b)

2. If g(x) = x² − 4, state.

(a) g(1)   (b) g(2)
(c) g(−2)   (d) g(0)
(e) g(10)   (f) g(8)
(g) g(−8)   (h) g($\sqrt{5}$)
(i) g(a)   (j) g(−a)

3. State f(3) for each of the following functions.

(a) f(x) = 2x − 5
(b) f(x) = 1 + 9x
(c) f(t) = 2 − 5t
(d) f(x) = x³
(e) f(t) = −t
(f) f:x → x² + 1
(g) f:x → $\frac{1}{3}$x − 1
(h) f(x) = 17
(i) f(x) = 8(x − 1)

4. State the domain and range of the functions represented by the following arrow diagrams.

(a)

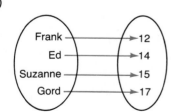

(b)

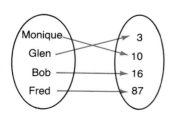

(c)

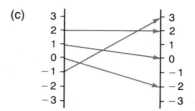

(d)

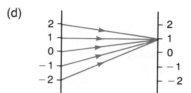

B 5. If f(x) = x² + 2x + 5, find.

(a) f(2)   (b) f(−2)
(c) f(15)   (d) f(0.3)
(e) f(−7)   (f) ff(21)
(g) f(1.4)   (h) f(0.01)

6. If g(x) = $\frac{x + 3}{2x - 5}$, find.

(a) g(1)   (b) g(11)
(c) g(−9)   (d) g($\frac{1}{2}$)
(e) g(0.2)   (f) g(100)
(g) g(−6.5)   (h) g(a)

7. If f(t) = t⁴ + t² + 1, find.

(a) f(−1)   (b) f($\sqrt{3}$)
(c) f(−4)   (d) f(1.5)
(e) f(7)   (f) f(−6)
(g) f($\sqrt[4]{2}$)   (h) f(−2.01)

8. Find f(4) for each of the following functions.

(a) f(x) = $\sqrt{x}$ (b) f(t) = $\sqrt[3]{23 + t}$
(c) f:x → (x + 2)³ (d) f:x → x⁵
(e) f(x) = x − $\frac{8}{x}$ (f) f(u) = 8 + u − u²

9. If f(x) = 3x² + x − 1, find expressions for the following.

(a) f(a)   (b) f(a + 1)

(c) f(2a) (d) f(a + b)
(e) f(-a) (f) f(-x)

10. If f(x) = 4x - 20, find x so that
(a) f(x) = 0 (b) f(x) = 16 (c) f(x) = -10

11. Patrick and Melissa Fry and their children Herman and Agnes live at 791 Main Street. Their next-door neighbour Owen Flood lives at number 793, and Patrick's mother Eleanor and father Fred live across the street at number 796. Let the domain of f be the set of these 7 people and let f(x) be the address of x on Main Street. What is the range of f? Draw an arrow diagram to represent f.

12. Given the domain A = {-2, -1, 0, 1, 2}, find the range and draw arrow diagrams for the following functions.
(a) f(x) = x² + 1 (b) g(x) = 2x

13. If the domain is R, find the ranges of the following functions.
(a) f(x) = x² + 1 (b) g(x) = 2x

14. If we take the domains of the following functions to be the largest sets for which the given expressions are meaningful, find the domain in each case.
(a) f(x) = $\frac{2 + x}{3 - x}$ (b) f(x) = $\frac{3 - x}{2 + x}$
(c) g(x) = $\sqrt{x - 2}$ (d) h(x) = $\sqrt{9 - x^2}$

15. The emergency brake cable in a car parked on a hill suddenly breaks and the car rolls down the hill. The distance, d metres, that it rolls in t seconds is given by the function d = f(t) = 0.5 t².
(a) How far has the car rolled after 10 s?
(b) When will it splash into the lake which is at the bottom of the hill 200 m away?

C16. If f(x) = x⁹ - x⁴ + $\frac{2}{x^2}$, find.
(a) f(3) (b) f(1.01)
(c) f(2.537) (d) f(-0.79)

17. If f(x) = 3x + 4 and g(x) = x² - 5, find.
(a) f(g(2)) (b) g(f(2))
(c) f(g(-1)) (d) g(f(-1))
(e) f(g(x)) (f) g(f(x))

18. If f(x) = $\frac{1}{x + 1}$ and g(x) = x + 1, find.
(a) f(g(6)) (b) g(f(6))
(c) f(g($\frac{1}{3}$)) (d) g(f($\frac{1}{3}$))
(e) f(g(x)) (f) g(f(x))

19. If f(x) = x² + x - 12, find all values of x such that
(a) f(x) = 0 (b) f(x) = 8 (c) f(x) = 30

20. Given f(x) = x² - x, find all values of t such that f(2t - 3) = f(t + 1).

21. If f(x) = 2x + 3, find $\frac{f(x_2) - f(x_1)}{x_2 - x_1}$.

5.2 GRAPHS OF FUNCTIONS

One way to picture a function is to draw an arrow diagram as we did in the preceding section. Another way is to draw its graph.

If the domain of a function f is the set A, then the graph of f is drawn by plotting all ordered pairs (x, y) such that x ∈ A and y = f(x). In other words, we plot all points in the plane such that f(x component) = y component.

> The graph of f is the set of ordered pairs {(x, y) | y = f(x), x ∈ A}.

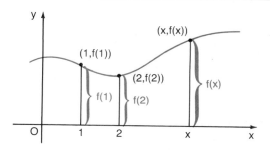

Since the y-coordinate of a point (x, y) on the graph of f is y = f(x), we can read the value of the number f(x) from the graph as the height of the graph above the point x. We can also use the graph of f to picture the domain and range of f on the x-axis and y-axis.

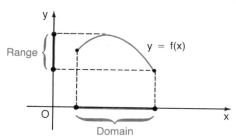

EXAMPLE 1. Draw the graph of f if f(x) = x^2 and the domain is A = {−1, 0, 1, 2, 3}.

SOLUTION:
f(−1) = 1
f(0) = 0
f(1) = 1
f(2) = 4
f(3) = 9
So the graph of f consists of the ordered pairs (−1, 1) (0, 0) (1, 1) (2, 4) (3, 9). Compare this picture with the picture in Example 5 of the preceding section. From either picture we can see at a glance that f(−1) = 1, f(3) = 9, and so on.

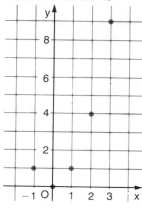

EXAMPLE 2. Draw the graph of f if $f(x) = x^2$ and the domain is R.

SOLUTION:
Here the domain consists of all real numbers and the graph $\{(x, y) \mid y = x^2, x \in R\}$ consists of infinitely many points. Obviously we cannot plot an infinite number of points individually, but we can select a few convenient values of x, calculate the corresponding values for y, and plot a few points on the graph. Then from these points we may be able to sketch the rest of the graph.

x	y
0	0
1	1
2	4
3	9
−1	1
−2	4
−3	9

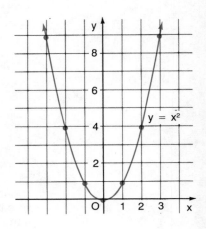

EXAMPLE 3. From the graph state f(2), f(−1), f(−3), f(0), f(1), and the domain and range of f.

SOLUTION:
$f(2) = 1$
$f(-1) = -2$
$f(-3) = -1$
$f(0) = -1$
$f(1) = 0$

The domain is $\{x \mid -4 \leqslant x \leqslant 3\}$.
The range is $\{y \mid -3 \leqslant y \leqslant 1\}$.

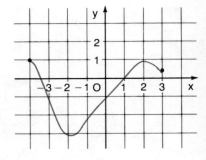

$\{x \mid -4 \leqslant x \leqslant 3\}$ is an abbreviation for $\{x \in R \mid -4 \leqslant x \leqslant 3\}$.

In Example 2, we could define the function f by the equation $y = x^2$, with the understanding that $y = f(x)$. Here are three ways of defining the same function f:

$$f(x) = x^2$$
$$f:x \rightarrow x^2$$
$$y = x^2$$

EXAMPLE 4. Sketch the graph of the function $y = 2x - 3$ if the domain is $\{x \in R\} -1 \leq x \leq 3\}$.

SOLUTION:

x	y
−1	−5
0	−3
1	−1
2	1
3	3

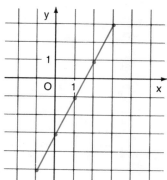

The graph is a segment of a straight line. In fact functions such as

$y = -5x + 7$, $y = \frac{3}{2}x + 1$, or in general $y = mx + b$, are called linear functions because their graphs are straight lines. These were studied in Chapter 3.

EXAMPLE 5. Temperature readings, T (in degrees Celsius), were recorded every hour starting at midnight on a day in April in Medicine Hat, Alberta, as shown in the following table. The time, x, is measured in hours from midnight.

x (hours)	0	1	2	3	4	5	6	7	8	9	10	11	12
T (°C)	6.5	6.1	5.6	4.9	4.2	4.0	4.0	4.8	6.1	8.3	10.0	12.1	14.3

x (hours)	13	14	15	16	17	18	19	20	21	22	23	24
T (°C)	16.0	17.3	18.2	18.8	17.6	16.0	14.1	11.5	10.2	9.0	7.9	7.0

Sketch the graph of the temperature as a function of time.

SOLUTION:

This is an example of a function which cannot be given by a formula. Using the given table of temperature readings, we plot the ordered pairs (0, 6.5), (1, 6.1), (2, 5.6), and so on. Then we join these points as shown in the figure below to indicate how the temperature varies between the hourly readings.

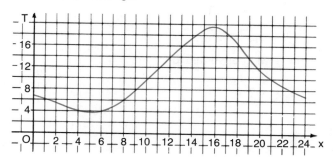

EXERCISE 5.2

A 1. From each of the given graphs state.
(a) f(0) (b) f(−1) (c) f(−2) (d) f(1) (e) f(2)
(f) the domain of f   (g) the range of f

(i)

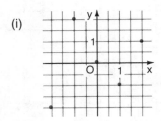

(ii)

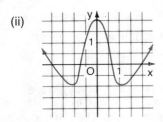

(iii)

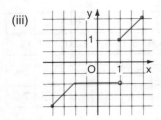

B 2. Draw the graphs of the following linear functions with the given domains A.
(a) $f(x) = 2x + 4$, $A = \{-2, 1, 0, 2\}$
(b) $g(x) = 2x + 4$, $A = R$
(c) $f(x) = 7 - 2x$, $A = \{x \in R | 1 \leqslant x \leqslant 5\}$
(d) $h(x) = \dfrac{x + 2}{3}$, $A = R$
(e) $f(t) = -t - 1$, $A = R$
(f) $y = 4x - \frac{1}{2}$, $A = R$

3. Graph the following quadratic functions with domain R.
(a) $f(x) = x^2 + x + 1$
(b) $g(x) = 1 - x^2$
(c) $h(x) = (x - 2)^2$
(d) $y = 3x^2 + 1$
(e) $y = -2x^2 + x - 5$

4. Graph the following functions.
(a) $f(x) = x^3 - 3x^2 + 1$, $x \in \{0, 1, 2, 3, 4, 5\}$
(b) $g(x) = x + \sqrt{x - 2}$, $x \in \{2, 3, 6, 11\}$
(c) $h(x) = \dfrac{1}{x}$, $\{x \in R | x \neq 0\}$
(d) $f:x \to \sqrt{x}$, $\{x \in R | x \geqslant 0\}$

5. Sketch a graph of the population of Canada as a function of time by using the following census figures.

Year	Population (in millions)
1901	5.4
1911	7.2
1921	8.8
1931	10.4
1941	11.5
1951	14.0
1961	18.2
1971	21.6
1981	24.3

6. The following table gives the number, n, of inmates admitted to penitentiaries in Canada in 1968—69 according to the grade reached in school.

Grade	n
Less than 8	1273
8	886
9	550
10	473
11	199
12	160
13 and above	85

Use this table to graph n as a function of amount of education.

C 7. Graph the following functions.
(a) $f(x) = 1.09x^3 + 0.87x^2 - 4.54x - 5.23$, $x \in R$
(b) $g(x) = \dfrac{x}{\sqrt{x^2 + 7}}$, $\{x \in R | x \geqslant 0\}$
(c) $h(x) = \sqrt[4]{x^4 + 2x^2 + 10}$, $x \in R$

5.3 QUADRATIC FUNCTIONS AND PARABOLAS

A quadratic function is a function determined by a second degree polynomial. Examples of quadratic functions are $f(x) = 2x^2$,

$g(x) = -3x^2 + 4x$, $h(x) = x^2 - 3x + 1$, and in general,

$$f(x) = ax^2 + bx + c$$

where a, b, c are constants and $a \neq 0$. The graph of f has the equation $y = ax^2 + bx + c$ and is called a parabola.

Taking the special case where $a = 1$, $b = 0$, and $c = 0$, we have the quadratic function $f(x) = x^2$. In Section 5.1 we found that this function has domain R and range $\{y \mid y \geq 0\}$. Its graph is the parabola $y = x^2$ which was sketched in Example 2 of Section 5.2, and is shown again in the figure below.

x	y
0	0
1	1
2	4
3	9
-1	1
-2	4
-3	9

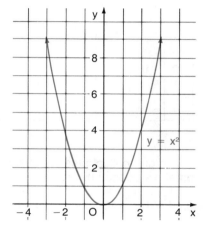

Notice that the parabola $y = x^2$ is symmetric about the y-axis since $f(x) = f(-x)$. The axis of symmetry of a parabola is the line such that the parabola is mapped onto itself by reflection in the line. In this case the axis of symmetry is the y-axis. The vertex, or turning point, of the parabola is the point where the graph intersects the axis of symmetry. In this example the vertex is (0, 0).

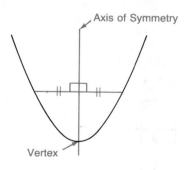

EXAMPLE 1. Sketch the graphs of the functions $y = 2x^2$, $y = \frac{1}{2}x^2$, $y = -x^2$, $y = -2x^2$, and $y = -\frac{1}{2}x^2$. Compare them to the graph of $y = x^2$.

SOLUTION:

$y = 2x^2$

x	y
0	0
1	2
2	8
3	18
−1	2
−2	8
−3	18

$y = \frac{1}{2}x^2$

x	y
0	0
1	0.5
2	2
3	4.5
−1	0.5
−2	2
−3	4.5

$y = -x^2$

x	y
0	0
1	−1
2	−4
3	−9
−1	−1
−2	−4
−3	−9

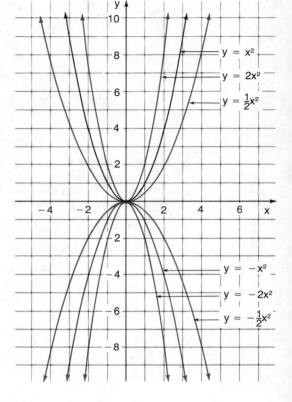

$y = -2x^2$

x	y
0	0
1	−2
2	−8
3	−18
−1	−2
−2	−8
−3	−18

$y = -\frac{1}{2}x^2$

x	y
0	0
1	−0.5
2	−2
3	−4.5
−1	−0.5
−2	−2
−3	−4.5

The graph illustrates the change in the parabola $y = ax^2$ for different values of a. If $a > 0$, the parabola opens upward. If $a < 0$, the parabola opens downward. Compare the functions to $y = x^2$.
I. If $a > 1$ or $a < -1$ there is a stretch in the y-direction.
II. If $-1 < a < 1$ there is a shrink in the y-direction.

EXAMPLE 2. Sketch the graphs of the functions $y = x^2 + 2$ and $y = x^2 - 3$. Compare them to the graph of $y = x^2$.

SOLUTION:

$y = x^2 + 2$

x	y
0	2
1	3
2	6
3	11
−1	3
−2	6
−3	11

$y = x^2 - 3$

x	y
0	−3
1	−2
2	1
3	6
−1	−2
−2	1
−3	6

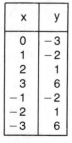

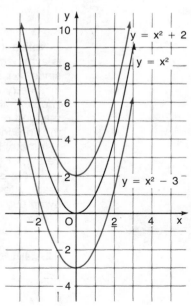

The graphs of $y = x^2 + 2$ and $y = x^2 - 3$ are congruent to $y = x^2$. All three have the same size and shape, but different positions. The graph of $y = x^2 + 2$ is the graph of $y = x^2$ shifted up 2 units. The graph of $y = x^2 - 3$ is the graph of $y = x^2$ shifted down 3 units.

EXAMPLE 3. Sketch the graphs of the following.
(a) $y = 2x^2 - 4$
(b) $y = -\frac{1}{2}x^2 + 3$

SOLUTION:

(a) The graph of $y = 2x^2 - 4$ is the graph of $y = 2x^2$ translated downward by 4 units. The vertex is $(0, -4)$, the axis of symmetry is $x = 0$, and the parabola opens upward. The range of the function is $\{y \mid y \geqslant -4\}$.

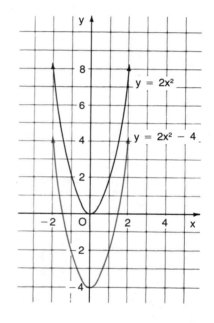

(b) The graph of $y = -\frac{1}{2}x^2 + 3$ is the graph of $y = -\frac{1}{2}x^2$ translated upward by 3 units. The vertex is $(0, 3)$, the axis of symmetry is $x = 0$, and the parabola opens downward. The range of the function is $\{y \mid y \leqslant 3\}$.

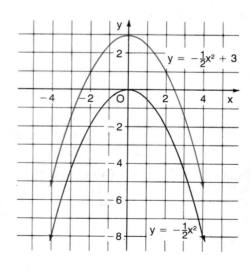

In general, if $q > 0$,

(i) the graph of $y = ax^2 + q$ is the graph of $y = ax^2$ translated upward by q units.

(ii) the graph of $y = ax^2 - q$ is the graph of $y = ax^2$ translated downward by q units.

$y = ax^2 + q$	$a > 0$	$a < 0$
vertex	$(0, q)$	$(0, q)$
axis of symmetry	$x = 0$	$x = 0$
direction of opening	upward	downward

EXERCISE 5.3

A 1. For each of the following parabolas state.
(a) the direction of the opening (up or down)
(b) the coordinates of the vertex
(c) the equation of the axis of symmetry
(i) $y = 4x^2$
(ii) $y = x^2 + 3$
(iii) $y = -\frac{1}{2}x^2$
(iv) $y = -3x^2 + 5$
(v) $y = x^2 - 6$
(vi) $y = 2x^2 + 7$
(vii) $y = -7x^2 - 7$
(viii) $y = -1.2x^2 + 3.4$
(ix) $y = -0.009x^2$

(c) $y = -3x^2$
(d) $y = -\frac{1}{3}x^2$
(e) $y = x^2 + 5$
(f) $y = x^2 - 1$
(g) $f(x) = -x^2 + 4$
(h) $f(x) = 2x^2 + 1$
(i) $f(x) = 4x^2 - 5$
(j) $g(x) = \frac{1}{4}x^2 - 2$
(k) $g(x) = -\frac{1}{2}x^2 - 3$
(l) $g(x) = 3 - \frac{1}{3}x^2$

B 2. Write an equation for a parabola with the given vertex and value for a.
(a) $(0, 0)$; $a = 4$
(b) $(0, 0)$; $a = -3$
(c) $(0, 0)$; $a = -\frac{1}{2}$
(d) $(0, 4)$; $a = 1$
(e) $(0, -3)$; $a = -2$
(f) $(0, -4)$; $a = 6$
(g) $(0, 6)$; $a = -3$
(h) $(0, 0)$; $a = -\frac{1}{3}$
(i) $(0, -7)$; $a = 4$

3. Without making a table of values, sketch the graph of each of the following functions. State the range of each function.

(a) $y = 3x^2$
(b) $y = \frac{1}{3}x^2$

C 4. Find an equation for the parabola with the given vertex and passing through the given point.
(a) vertex: $(0, 0)$; point: $(2, 12)$
(b) vertex: $(0, 0)$; point: $(-4, 4)$
(c) vertex: $(0, -2)$; point: $(1, 1)$
(d) vertex: $(0, 3)$; point: $(2, 11)$

5. Find the value of q so that the parabola $y = 2x^2 + q$ passes through the point $(-2, 3)$.

6. Find a and q so that the parabola $y = ax^2 + q$ passes through the points $(2, -3)$ and $(-4, 3)$.

5.4 GRAPHING $y = a(x - p)^2 + q$

In this section we investigate the effect of shifting parabolas to the left or right.

EXAMPLE 1. Sketch the graphs of the functions $y = (x - 2)^2$ and $y = (x + 3)^2$. Compare them to the graph of $y = x^2$.

SOLUTION:

$y = x^2$

x	y
0	0
1	1
2	4
3	9
−1	1
−2	4
−3	9

$y = (x - 2)^2$

x	y
2	0
3	1
4	4
5	9
1	1
0	4
−1	9

$y = (x + 3)^2$

x	y
−3	0
−2	1
−1	4
0	9
−4	1
−5	4
−6	9

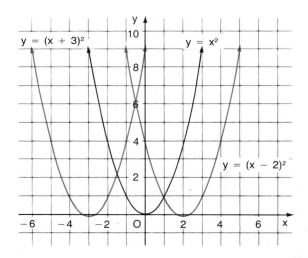

The graphs of $y = (x - 2)^2$ and $y = (x + 3)^2$ are congruent to $y = x^2$ but have different positions. The graph of $y = (x - 2)^2$ is the graph of $y = x^2$ shifted 2 units to the right. The graph of $y = (x + 3)^2$ is the graph of $y = x^2$ shifted 3 units to the left.
In general, if $p > 0$,

(i) the graph of $y = a(x + p)^2$ is the graph of $y = ax^2$ translated to the left by p units.

(ii) the graph of $y = a(x - p)^2$ is the graph of $y = ax^2$ translated to the right by p units.

EXAMPLE 2. Sketch the graphs of the following and find the vertex and the equation of the axis of symmetry.

(a) $y = 2(x - 3)^2$

(b) $y = -3(x + 2)^2$

SOLUTION:

(a) The graph of $y = 2(x - 3)^2$ is the graph of $y = 2x^2$ translated to the right by 3 units. The vertex is (3, 0). The equation of the axis of symmetry is $x = 3$.

(b) The graph of $y = -3(x + 2)^2$ is the graph of $y = -3x^2$ translated to the left by 2 units. The vertex is $(-2, 0)$. The equation of the axis of symmetry is $x = -2$.

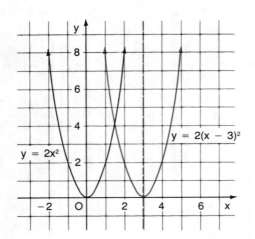

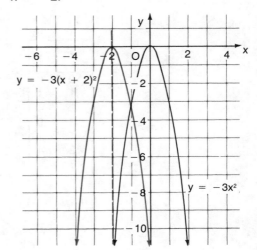

EXAMPLE 3. Sketch the graphs of the functions $y = (x - 4)^2 + 3$ and $y = (x + 3)^2 - 5$.

SOLUTION:

The graphs of $y = (x - 4)^2 + 3$ and $y = (x + 3)^2 - 5$ are congruent to $y = x^2$. The graph of $y = (x - 4)^2 + 3$ is the graph of $y = x^2$ shifted right 4 units and up 3 units. The graph of $y = (x + 3)^2 - 5$ is the graph of $y = x^2$ shifted left 3 units and down 5 units.

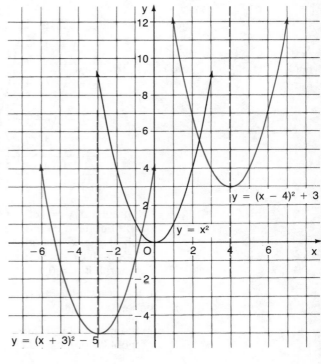

We summarize the transformations of this section and the preceding section by showing how all parabolas, $y = a(x - p)^2 + q$, are obtained from the basic function $y = x^2$.

$y = x^2$	graph is a parabola
$y = ax^2$	stretches in the y-direction if $a > 1$ or $a < -1$ shrinks in the y-direction if $-1 < a < 1$ reflects in the x-axis if $a < 0$
$y = a(x - p)^2$ $y = a(x + p)^2$	shifts p units to the right shifts p units to the left $(p > 0)$
$y = a(x - p)^2 + q$ $y = a(x - p)^2 - q$	shifts q units upward shifts q units downward $(q > 0)$

Multiply by a.

Replace x by x − p or x + p.

Add q.
Subtract q.

The geometric aspects of the parabola $y = a(x - p)^2 + q$ are summarized in the following chart.

$y = a(x - p)^2 + q$	$a > 0$	$a < 0$
vertex	(p, q)	(p, q)
axis of symmetry	$x = p$	$x = p$
direction of opening	upward	downward

EXAMPLE 4. Sketch the graph of $f(x) = 2(x + 4)^2 - 1$ and find its domain and range.

SOLUTION:
Starting with $y = x^2$, we stretch by a factor 2 in the y-direction to get $y = 2x^2$. Then we shift 4 units to the left and 1 unit downward to get $y = 2(x + 4)^2 - 1$. The vertex is $(-4, -1)$ and the equation of the axis of symmetry is $x = -4$. Since $a = 2 > 0$, the parabola opens upward. We see from the graph that the domain is R and the range is $\{y \mid y \geq -1\}$.

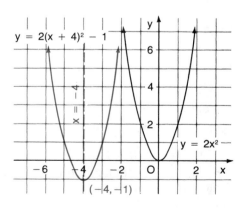

EXAMPLE 5. Find an equation for the parabola with vertex $(-3, -2)$ and passing through the point $(-2, 0)$.

SOLUTION:

Setting $(p, q) = (-3, -2)$ we substitute in the formula
$y = a(x - p)^2 + q$ to obtain
$$y = a(x - (-3))^2 + (-2)$$
$$= a(x + 3)^2 - 2$$
Since the parabola passes through the point $(-2, 0)$, we replace x by -2 and y by 0.
$$0 = a(-2 + 3)^2 - 2$$
$$0 = a - 2$$
$$a = 2$$
∴ the equation is $y = 2(x + 3)^2 - 2$.

EXERCISE 5.4

A 1. For each of the following parabolas state.

(a) the direction of the opening (up or down)
(b) the coordinates of the vertex
(c) the equation of the axis of symmetry

(i) $y = 2(x + 1)^2$
(ii) $y = -(x - 5)^2$
(iii) $y = 3(x - 4)^2 + 2$
(iv) $y = -5(x + 6)^2$
(v) $y = -\frac{1}{2}(x - 1)^2 + 6$
(vi) $y = 4(x - 3)^2$
(vii) $y = 2(x + 6)^2 - 10$
(viii) $y = -3(x - 3)^2 - 3$
(ix) $y = 3(x - \frac{1}{2})^2$
(x) $y = 0.5(x - 3)^2 - 0.7$
(xi) $y = -2(x + 5)^2 - 10$
(xii) $y = 6(x - 1)^2 + 8$

B 2. Write an equation for a parabola with the given vertex and value for a.

(a) $(5, 0)$; $a = 2$ (b) $(-4, 0)$; $a = -3$
(c) $(-6, 0)$; $a = \frac{1}{2}$ (d) $(3, 2)$; $a = 1$
(e) $(-3, 5)$; $a = -2$ (f) $(4, -3)$; $a = 4$
(g) $(-6, -7)$; $a = -5$ (h) $(-5, 0)$; $a = \frac{1}{3}$
(i) $(6, 6)$; $a = -\frac{1}{2}$ (j) $(0, 2)$; $a = -11$
(k) $(-7, 2)$; $a = -5$ (l) $(-3, -4)$; $a = 7$

3. Without making a table of values, sketch the graph of each of the following functions. State the range of each function.

(a) $y = (x + 2)^2$
(b) $y = -(x - 5)^2$
(c) $y = 2(x - 1)^2$
(d) $y = -3(x + 3)^2$

(e) $y = (x + 1)^2 + 2$
(f) $y = -(x - 4)^2 - 1$
(g) $y = 2(x + 1)^2 - 5$
(h) $y = -2(x - 3)^2 + 4$
(i) $y = -\frac{1}{2}(x - 1)^2 + 6$
(j) $f(x) = -4(x - 1)^2 + 3$
(k) $f(x) = \frac{1}{2}(x + 3)^2 + 1$
(l) $f(x) = 6(x - 2)^2 + 7$

4. Find an equation for the parabola with the given vertex and passing through the given point.

(a) vertex: $(3, 2)$; point: $(1, 6)$
(b) vertex: $(-1, -3)$; point: $(-2, -5)$
(c) vertex: $(-3, 6)$; point: $(-2, 10)$
(d) vertex: $(2, -4)$; point: $(1, -7)$
(e) vertex: $(2, -6)$; point: $(4, -4)$
(f) vertex: $(-4, 3)$; point: $(-3, 2)$

C 5. Determine the values of p or q so that the graph of the parabola will pass through the given point.

(a) $y = 2(x - 1)^2 + q$; $(2, 7)$
(b) $y = 3(x - p)^2 + 2$; $(1, 14)$
(c) $y = -3(x + 2)^2 + q$; $(-3, -7)$
(d) $y = -2(x - p)^2 + 5$; $(2, 3)$
(e) $y = -\frac{1}{2}(x - p)^2 - 4$; $(4, -6)$
(f) $y = 3(x + 2)^2 + q$; $(-3, -4)$

6. Find a and q so that the given points will lie on the parabola.

(a) $y = a(x - 1)^2 + q$; $(2, 4)$, $(3, 10)$
(b) $y = a(x + 3)^2 + q$; $(-2, 3)$, $(0, -13)$
(c) $y = a(x - 2)^2 + q$; $(1, 8)$, $(-1, 32)$

5.5 GRAPHING $y = ax^2 + bx + c$ BY COMPLETING THE SQUARE

In the previous section we graphed quadratic functions written in standard form. Functions of this type may be rewritten as follows:

$$y = 2(x - 3)^2 - 7 \longleftarrow \text{standard form}$$
$$= 2(x^2 - 6x + 9) - 7$$
$$= 2x^2 - 12x + 18 - 7$$
$$= 2x^2 - 12x + 11 \longleftarrow \text{general form}$$

The equation $y = 2x^2 - 12x + 11$ is written in the general form of a quadratic function. Many quadratic functions appear in this form, but it is easier to graph a quadratic function when it is expressed in the standard form $y = a(x - p)^2 + q$. To make the transition from general to standard we use a procedure called completing the square. In order to understand the process we first square two binomials.

$$(x + 3)^2 = x^2 + 6x + 9$$
$$(x + t)^2 = x^2 + 2tx + t^2$$

Notice that the constant term, 9 or t^2, is the square of half the coefficient of x. We use this fact when we complete the square.

EXAMPLE 1. Express $y = x^2 - 6x - 1$ in standard form.

SOLUTION:
$$y = x^2 - 6x - 1$$
We first determine what must be added to $x^2 - 6x$ to make it a square trinomial.

The square of half the coefficient of x is 9. Since we add 9 to the original function, we must also subtract 9 to keep the value of the function the same.

$$y = x^2 - 6x + 1$$
$$= x^2 - 6x + 9 - 9 + 1$$
$$= (x^2 - 6x + 9) - 9 + 1$$
$$= (x - 3)^2 - 8$$

$$\left(\frac{-6}{2}\right)^2 = 9$$

EXAMPLE 2. Sketch the graph of $y = x^2 + 8x + 11$.

SOLUTION:

We first complete the square to put the function in standard form.
$$y = x^2 + 8x + 11$$
$$= x^2 + 8x + 16$$
$$\quad - 16 + 11$$
$$= (x + 4)^2 - 5$$
The coordinates of the vertex are $(-4, -5)$ and the parabola opens upward.

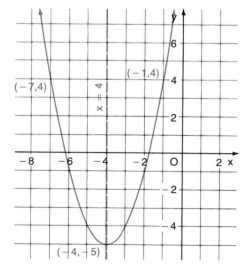

In Examples 1 and 2 we expressed quadratic functions of the form $y = x^2 + bx + c$ in standard form. We shall now consider quadratic functions of the form $y = ax^2 + bx + c$ where $a \neq 1$.

EXAMPLE 3. Express $y = 3x^2 - 12x + 7$ in standard form.

SOLUTION:

We factor the coefficient of x^2 from the first two terms.

$$
\begin{aligned}
y &= [3x^2 - 12x] + 7 \\
&= 3[x^2 - 4x] + 7 \\
&= 3[x^2 - 4x + 4 - 4] + 7 \\
&= 3[(x^2 - 4x + 4) - 4] + 7 \\
&= 3[(x - 2)^2 - 4] + 7 \\
&= 3(x - 2)^2 - 12 + 7 \\
&= 3(x - 2)^2 - 5
\end{aligned}
$$

Group the terms containing x.
Factor the coefficient of x^2.
Complete the square.

Remove square brackets.

EXAMPLE 4. (a) Sketch the graph of the function $f(x) = -3x^2 - 4x - 3$.
(b) Determine the y-intercept and plot the appropriate point on the graph.
(c) Find the range of f.

SOLUTION:

(a) The graph of f has the equation $y = -3x^2 - 4x - 3$.
We first rewrite this equation in standard form.

$$
\begin{aligned}
y &= -3x^2 - 4x - 3 \\
&= -3[x^2 + \tfrac{4}{3}x] - 3 \\
&= -3[x^2 + \tfrac{4}{3}x + \tfrac{4}{9} - \tfrac{4}{9}] - 3 \\
&= -3[(x + \tfrac{2}{3})^2 - \tfrac{4}{9}] - 3 \\
&= -3(x + \tfrac{2}{3})^2 + \tfrac{4}{3} - 3 \\
&= -3(x + \tfrac{2}{3})^2 - \tfrac{5}{3}
\end{aligned}
$$

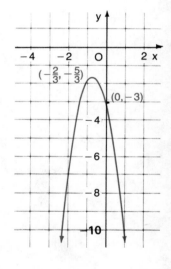

The coordinates of the vertex are $(-\tfrac{2}{3}, -\tfrac{5}{3})$ and the parabola opens downward.

(b) We determine the y-intercept by setting $x = 0$ in the original equation.

$$
\begin{aligned}
y &= -3x^2 - 4x - 3 \\
&= -3(0)^2 - 4(0) - 3 \\
&= -3
\end{aligned}
$$

∴ the y-intercept is -3.

(c) From the graph we see that the range is $\{y \mid y \leq -\tfrac{5}{3}\}$.

EXERCISE 5.5

A 1. Complete the square in each of the following.

(a) $x^2 + 8x$

(b) $x^2 + 6x$

(c) $x^2 - 12x$

(d) $x^2 - 2x$

(e) $x^2 + 10x$

(f) $x^2 - 14x$

(g) $x^2 + x$

(h) $x^2 - 3x$

(i) $x^2 - 5x$

(j) $x^2 + \frac{1}{2}x$

(k) $x^2 - \frac{6}{5}x$

(l) $x^2 - 0.8x$

(m) $x^2 + \frac{1}{4}x$

(n) $x^2 + \frac{2}{3}x$

(o) $x^2 + 0.2x$

(p) $x^2 - 1.2x$

(q) $x^2 - \frac{1}{3}x$

(r) $x^2 + 2.4x$

B 2. Without making a table of values, sketch the graph of each of the following functions. State the range of each function.

(a) $y = x^2 + 6x + 7$

(b) $y = x^2 - 4x - 1$

(c) $y = x^2 + 10x + 9$

(d) $y = x^2 - 8x$

(e) $y = x^2 + 3x$

(f) $y = x^2 - x + 1$

(g) $y = x^2 - \frac{1}{2}x - 1$

(h) $y = x^2 + \frac{2}{3}x - 2$

(i) $y = x^2 - \frac{3}{5}x$

3. Without making a table of values, sketch the graph of each of the following functions. State the range of each.

(a) $y = 2x^2 + 8x + 5$

(b) $y = 3x^2 - 6x + 4$

(c) $y = -2x^2 + 4x - 2$

(d) $y = -x^2 + 6x + 3$

(e) $y = -2x^2 + 5x + 2$

(f) $y = -3x^2 - 12x - 1$

(g) $f(x) = 5x^2 - 15x - 1$

(h) $f(x) = \frac{1}{2}x^2 + 3x - 2$

(i) $f(x) = -\frac{1}{3}x^2 - 2x + 4$

(j) $f(x) = -4x^2 + 6x$

(k) $f(x) = -0.2x^2 + 2x + 7$

(l) $f(x) = \frac{2}{3}x^2 - x + 2$

(m) $g(x) = 2x^2 + \frac{1}{2}x + 1$

(n) $g(x) = -6x^2 - x - \frac{1}{3}$

(o) $g(x) = -2x^2 - 0.8x - 2$

4. Sketch the graph of each of the following functions.

(a) $y = t^2 + 2t + 2$

(b) $y = 4t^2 - 16t + 9$

(c) $x = 2t^2 - 2t + 1$

(d) $x = 1 - t - t^2$

(e) $v = 3s^2 - 6s$

(f) $A = \frac{1}{2}r^2 + 3r + 2$

C 5. Sketch the graph of each of the following.

(a) $y + 3 = x^2 + 2x + 1$

(b) $y - 2 = 2x^2 - 6x - 1$

(c) $\frac{1}{2}y = x^2 - 3x - 1$

(d) $4y = -2x^2 + 6x - 8$

(e) $y = 3 - 4x - x^2$

(f) $y - 1 = 3x - 2x^2 + 5$

A train moving at 55 km/h meets and is passed by a train moving at 45 km/h. A passenger in the first train sees the second train take six seconds to pass. How long is the second train?

5.6 APPLICATIONS: MAXIMUM AND MINIMUM

Let us consider the function f whose graph is shown below. Notice that the point (a, b) is the lowest point on the graph of f. For this reason, the point (a, b) is called the minimum point of the graph of f. Notice also that of all the values taken on by f, the smallest value is f(a) = b. (Remember that the values of a function are the y-coordinates of points on the graph.) Therefore, the number b = f(a) is called the minimum value of the function.

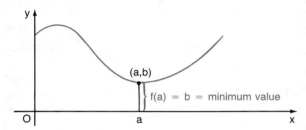

To see how to find the minimum value of a quadratic function, we consider the function
$$f(x) = 2x^2 - 4x + 5.$$
Expressed in standard form it becomes $f(x) = 2(x - 1)^2 + 3$. Since the parabola $y = 2(x - 1)^2 + 3$ opens upward, its vertex (1, 3) is the minimum point of the graph because it is the lowest point on the graph. Since f(1) = 3 and f(x) ⩾ 3 for x ∈ R, the minimum value of f is 3 and it occurs when x = 1.

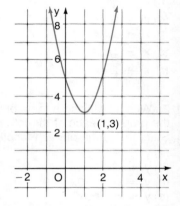

For the general function $y = a(x - p)^2 + q$, when a > 0 the parabola opens upward. Clearly $a(x - p)^2 ⩾ 0$. Therefore,
$$a(x - p)^2 + q ⩾ q$$
and the minimum occurs when $a(x - p)^2 = 0$ or x = p.

Similarly, the function
$$y = -2x^2 - 4x + 3$$
expressed in standard form becomes $y = -2(x + 1)^2 + 5$. The point (−1, 5) is called the maximum point of the graph. The maximum value of the function is 5 when x = −1. For the general function $y = a(x - p)^2 + q$, when a < 0 the parabola opens downward. Clearly $a(x - p)^2 ⩽ 0$. Therefore, $a(x - p)^2 + q ⩽ q$ and the maximum occurs when $a(x - p)^2 = 0$ or x = p.

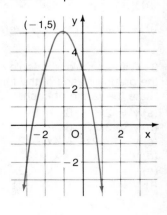

EXAMPLE 1. Determine the maximum or minimum value of the function $y = -\frac{1}{2}x^2 + 6x - 13$.

SOLUTION:
We first complete the square.

$$y = -\frac{1}{2}x^2 + 6x - 13$$
$$= -\frac{1}{2}[x^2 - 12x] - 13$$
$$= -\frac{1}{2}[x^2 - 12x + 36 - 36] - 13$$
$$= -\frac{1}{2}[(x - 6)^2 - 36] - 13$$
$$= -\frac{1}{2}(x - 6)^2 + 18 - 13$$
$$= -\frac{1}{2}(x - 6)^2 + 5$$

Since $-\frac{1}{2}(x - 6)^2 \leqslant 0$

$\therefore -\frac{1}{2}(x - 6)^2 + 5 \leqslant 5$

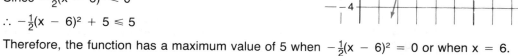

Therefore, the function has a maximum value of 5 when $-\frac{1}{2}(x - 6)^2 = 0$ or when $x = 6$.

READ

EXAMPLE 2. During the summer months Terry makes and sells necklaces on the beach. Last summer he sold the necklaces for $10 each. His sales averaged 20 per day. Considering a price increase, he took a small survey and found that for every dollar increase he would lose two sales per day. If the material for each necklace costs Terry $6, what should the selling price be to maximize profits? Determine the profit.

PLAN

SOLUTION:
Terry now sells 20 necklaces a day at $10 each. For every dollar increase, he would lose two sales per day. Expenses are $6 per necklace.
Let x be the number of dollars that Terry increases the selling price from $10.
Income from sales is (number sold) × (selling price), or

$$(20 - 2x)(10 + x)$$

Expenses are $6(20 - 2x)$.

$$\text{Profit} = \text{Income} - \text{Expenses}$$

SOLVE

$$P = (20 - 2x)(10 + x) - 6(20 - 2x)$$
$$= -2x^2 + 12x + 80$$
$$= -2[x^2 - 6x + 9 - 9] + 80$$
$$= -2[(x - 3)^2 - 9] + 80$$
$$= -2(x - 3)^2 + 18 + 80$$
$$= -2(x - 3)^2 + 98$$

Since $-2(x - 3)^2 \leqslant 0$
$\therefore -2(x - 3)^2 + 98 \leqslant 98$
The function reaches a maximum value of 98 when $-2(x - 3)^2 = 0$ or when $x = 3$.

ANSWER

The profit reaches a maximum of $98 when $x = 3$. Therefore, the selling price should be $13.

EXERCISE 5.6

B 1. Determine the maximum or minimum value of the following functions. State the value of x where each occurs.

(a) $y = 2x^2 - 12x - 7$
(b) $y = -3x^2 - 18x + 4$
(c) $y = 5x^2 - 10x + 4$
(d) $y = -x^2 + 2x - 3$
(e) $y = -\frac{1}{2}x^2 + 6x - 3$
(f) $y = \frac{1}{5}x^2 + 2x + 1$
(g) $y = \frac{1}{2}x^2 - 3x + 1$
(h) $y = -4x^2 + 2x - 1$

2. Find two numbers whose sum is 32 and whose product is a maximum.

3. Find two numbers whose difference is 6 and whose product is a minimum.

4. If a pistol bullet is fired vertically at an initial speed of 100 m/s, the height in metres after t seconds is given by $h = 100t - 5t^2$. Find the maximum height attained by the bullet.

5. A rectangular field is to be enclosed with 600 m of fencing. What dimensions will produce a maximum area?

6. A rectangular field bounded on one side by a lake is to be fenced on 3 sides by 800 m of fence. What dimensions will produce a maximum area?

7. A large car dealership has been selling new cars at $600 over the factory price. Sales have been averaging 80 cars per month. Due to inflation the $600 markup is going to be increased. The marketing manager has determined that for every $10 increase there will be one less car sold each month. What should the new markup be in order to maximize income?

8. Denise is an artist who works at a shopping centre drawing "pencil portraits." She charges $20 per portrait and she has been averaging 30 portraits per week. She decides to increase the price, but realizes that for every one dollar increase she will lose one sale per week. If materials cost her $10 per portrait, what should she set the price at in order to maximize her profit?

9.

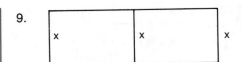

A rectangular field is to be enclosed by a fence and divided into two rectangular fields by a fence parallel to one side of the field. If 1200 m of fence are available, find the dimensions of the field giving the maximum area.

10. The TruTime Watch Company has been selling 1200 watches per week at $18 each. They are planning a price increase. A survey indicates that for every dollar increase in price there will be a drop of 40 sales per week. If it costs $10 to make each watch, what should the selling price be in order to maximize profit?

11. The effectiveness of a TV commercial depends on how many times a viewer sees it. A television advertising agency has determined that if effectiveness (e) is put on a scale from 0 to 10 where 10 is maximum positive effect, then $e = -\frac{1}{90}n^2 + \frac{2}{3}n$ where n is the number of times a viewer sees a particular commercial. Determine how many times a viewer should see a commercial to attain maximum positive effect.

12. A rocket is fired down a practice range. The height in metres after t seconds is given by
$$h = -\frac{1}{4}t^2 + 3t + 45$$
Find the maximum height attained by the rocket.

C 13. A rocket is launched vertically upward with an initial velocity v. The height, h, of the rocket at time t is equal to the height it would attain in the absence of gravity (vt) minus the free fall distance due to gravity $\left(\dfrac{gt^2}{2}\right)$.

Thus,
$$h = vt - \frac{gt^2}{2}$$

Neglecting air resistance and the variation of g with altitude, show that the rocket attains a maximum height of $\dfrac{v^2}{2g}$ at time $\dfrac{v}{g}$.

5.7 QUADRATIC REGIONS

The parabola $y = ax^2 + bx + c$ divides the plane into three regions.

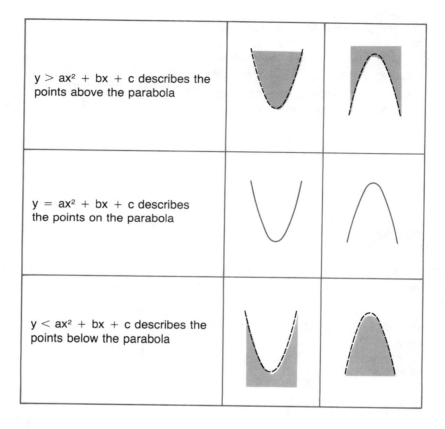

$y > ax^2 + bx + c$ describes the points above the parabola		
$y = ax^2 + bx + c$ describes the points on the parabola		
$y < ax^2 + bx + c$ describes the points below the parabola		

EXAMPLE 1. Draw the graph of
$$y \geq x^2 - 4x + 7.$$

SOLUTION:
First we express $y = x^2 - 4x + 7$ in standard form.

$$y = x^2 - 4x + 7$$
$$= x^2 - 4x + 4 - 4 + 7$$
$$= (x - 2)^2 + 3$$

The parabola has its vertex at (2, 3). The equation of the axis of symmetry is $x = 2$. The required graph is the set of points on or above the parabola.

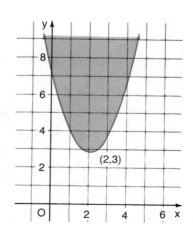

EXAMPLE 2. Graph the solution set of the system of inequalities

$$y < -x^2 - 2x + 2$$
$$\text{and } y > -\tfrac{1}{2}x - 2.$$

SOLUTION:

The solution set is the intersection of the solution sets of each inequality. For $y < -x^2 - 2x + 2$ the region is below the parabola. For $y > -\tfrac{1}{2}x - 2$ the region is above the straight line. The intersection is represented by the shaded area.

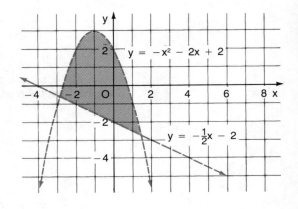

EXERCISE 5.7

B 1. Draw the graphs of the following.

(a) $y \geq 2x^2$

(b) $y < x^2 + 4$

(c) $y > -x^2 - 3$

(d) $y \leq (x + 3)^2 - 4$

(e) $y < -2(x - 1)^2 + 5$

(f) $y \geq 4(x + 2)^2 - 7$

(g) $y > x^2 + 4x + 1$

(h) $y < x^2 - 6x + 5$

(i) $y \leq -x^2 + 6x - 11$

(j) $y \geq -x^2 - 8x - 14$

(k) $y < 2x^2 - 12x + 14$

(l) $y > -2x^2 - 4x - 6$

(m) $y \geq -\tfrac{1}{3}x^2 + 2x + 3$

(n) $y < \tfrac{2}{3}x^2 - 4x + 1$

2. Graph the solution set of the following systems.

(a) $y > x^2 + 2$ and $y < x + 4$

(b) $y \geq x^2 - 4x + 6$ and $y < \tfrac{1}{2}x + 4$

(c) $y \leq -x^2 - 2x + 2$ and $y \geq x$

(d) $y > x^2 - 4x + 1$ and $y < -\tfrac{2}{3}x + 2$

(e) $y < -x^2 + 6x - 7$ and $y \geq -3$

(f) $y \leq -2x^2 + 4$ and $y > 0$

(g) $y > 2x^2 + 12x + 18$ and $y \leq -\tfrac{1}{2}x + 4$

(h) $y \geq 2x^2 - 4x + 2$ and $x \geq 1$

C 3. Graph the solution set of the following.

(a) $y \leq -x^2 + 5$ and $y \geq x^2 - 4$

(b) $y < -x^2 + 4x$ and $y \geq 2x^2 - 4x + 2$

(c) $x^2 - 1 < y < 1 - x^2$

(d) $x^2 \leq y \leq x - x^2$

(e) $4x^2 \leq y \leq x^2 + 3$

(f) $x^2 + x - 2 < y \leq -2x - x^2$

WORD LADDER

Start with the word "mine" and change one letter at a time to form a new word until you reach "coal." The best solution has the fewest steps.

m i n e

_ _ _ _

_ _ _ _

_ _ _ _

_ _ _ _

_ _ _ _

c o a l

5.8 THE GENERAL QUADRATIC FUNCTION

We shall now apply the technique of completing the square to express the general quadratic function, $y = ax + bx + c$, in standard form.

$$y = ax^2 + bx + c$$

$$= a\left[x^2 + \frac{b}{a}x\right] + c$$

$$= a\left[x^2 + \frac{b}{a}x + \frac{b^2}{4a^2} - \frac{b^2}{4a^2}\right] + c$$

$$= a\left[\left(x + \frac{b}{2a}\right)^2 - \frac{b^2}{4a^2}\right] + c$$

$$= a\left(x + \frac{b}{2a}\right)^2 - \frac{b^2}{4a} + c$$

$$= a\left(x + \frac{b}{2a}\right)^2 + c - \frac{b^2}{4a}$$

$$y = a\left(x + \frac{b}{2a}\right)^2 + \frac{4ac - b^2}{4a}$$

Comparing this equation with $y = a(x - p)^2 + q$ we conclude:

> The graph of the general quadratic function $y = ax^2 + bx + c$ is a parabola with vertex $\left(-\frac{b}{2a}, \frac{4ac - b^2}{4a}\right)$. The equation of the axis of symmetry is $x = -\frac{b}{2a}$. The parabola opens upward when $a > 0$ and downward when $a < 0$.

EXAMPLE 1. Determine the vertex and the equation of the axis of symmetry of the parabola $y = -2x^2 + 3x - 4$.

SOLUTION:
For $y = -2x^2 + 3x - 4$, $a = -2$, $b = 3$, and $c = -4$
Vertex:

$$\text{x-coordinate} = -\frac{b}{2a}$$

$$= -\frac{3}{2(-2)}$$

$$= \frac{3}{4}$$

$$\text{y-coordinate} = \frac{4ac - b^2}{4a}$$

$$= \frac{4(-2)(-4) - (3)^2}{4(-2)}$$

$$= -\frac{23}{8}$$

$\therefore$ the coordinates of the vertex are $\left(\frac{3}{4}, -\frac{23}{8}\right)$.

Axis of Symmetry:
$$x = -\frac{b}{2a}$$
$$x = \frac{3}{4}$$

EXAMPLE 2. Find a quadratic function f that satisfies the given conditions.
$$f(1) = 0, \quad f(3) = 2, \quad f(-2) = 12$$

SOLUTION:
The quadratic function is defined by

$$f(x) = ax^2 + bx + c$$

$f(1) = a(1)^2 + b(1) + c = 0$ or $a + b + c = 0$
$f(3) = a(3)^2 + b(3) + c = 2$ or $9a + 3b + c = 2$
$f(-2) = a(-2)^2 + b(-2) + c = 12$ or $4a - 2b + c = 12$

We now solve the resulting linear system.

$$a + b + c = 0 \quad ①$$
$$9a + 3b + c = 2 \quad ②$$
$$4a - 2b + c = 12 \quad ③$$

Eliminate c from ① and ②.   Eliminate c from ② and ③.

            $a + b + c = 0 \quad ①$ $9a + 3b + c = 2 \quad ②$
            $9a + 3b + c = 2 \quad ②$ $4a - 2b + c = 12 \quad ③$
Subtract $-8a - 2b = -2 \quad ④$ Subtract $5a + 5b = -10 \quad ⑤$
or $4a + b = 1 \quad ④$ or $a + b = -2 \quad ⑤$

Eliminate b from ④ and ⑤.

$$4a + b = 1 \quad ④$$
$$a + b = -2 \quad ⑤$$
Subtract. $3a = 3$
             $a = 1$

Substitute $a = 1$ in ④.

$$4a + b = 1$$
$$4(1) + b = 1$$
$$4 + b = 1$$
$$b = -3$$

Substitute $a = 1, b = -3$ in ①.

$$a + b + c = 0$$
$$(1) + (-3) + c = 0$$
$$c = 2$$

$\therefore f(x) = x^2 - 3x + 2$

EXERCISE 5.8

A 1. State the values of a, b, and c for the following quadratic functions.

(a) $y = 2x^2 + 3x - 7$
(b) $y = 2x^2 - x - 3$
(c) $y = -x^2 - 4x$
(d) $y = 3x^2 - 4$
(e) $y + 3 = x^2 + 2x$
(f) $y - 7 = 2x^2$
(g) $y = 7 - 3x + 5x^2$
(h) $y = 3x - 4x^2 - 2$
(i) $y = 5x - 3 + 7x^2$

B 2. Find the vertex and axis of symmetry of each of the following.

(a) $y = x^2 - x - 12$
(b) $y = x^2 + 2x + 3$
(c) $y = 2x^2 - 4x - 1$
(d) $y = -x^2 - 2x + 5$
(e) $y = -2x^2 + x - 5$
(f) $y = 3x^2 - x + 4$
(g) $y = \frac{1}{2}x^2 - 2x + 1$
(h) $y = -\frac{1}{4}x^2 + 3x - 1$
(i) $y = 4x^2 - 7$
(j) $y = 2x^2 - 6x$

3. Find a quadratic function that satisfies the given conditions.

(a) $f(1) = 2, f(-1) = 4, f(2) = 4$
(b) $f(1) = 0, f(3) = -2, f(-1) = 10$
(c) $f(1) = 2, f(-1) = -4, f(2) = 8$
(d) $f(0) = 2, f(-2) = 12, f(3) = 2$
(e) $f(0) = -3, f(1) = 3, f(2) = 11$
(f) $f(-1) = 5, f(0) = 4, f(1) = -1$
(g) $f(-2) = -1, f(2) = 3, f(4) = 11$
(h) $f(-2) = 12, f(1) = 6, f(2) = 16$

C 4. Determine without graphing if the following will cross the x-axis.

(a) $y = x^2 - 2x - 8$
(b) $y = 2x^2 + 4x + 3$
(c) $y = -x^2 - 7x - 12$
(d) $y = -2x^2 + 6x - 5$
(e) $y = -\frac{1}{2}x^2 + 3x - 4$
(f) $y = 3x^2 - 10x - 8$
(g) $y = x^2 + x + 1$
(h) $y = x^2 + x - 1$
(i) $y = 1 - 3x + 5x^2$

5. If $y = ax^2 + bx + c$ crosses the x-axis, does $y = -ax^2 - bx - c$ cross the x-axis? Explain.

6. If $f(x) = ax^2 + bx + c$ and $f(x) = f(-x)$ for all x, show that $b = 0$.

MICRO MATH

```
NEW
100 PRINT "THIS PROGRAM ANALYSES"
110 PRINT "THE GRAPH OF"
120 PRINT "Y = AX↑2 + BX + C"
130 PRINT "ENTER A, B, C"
140 PRINT "SEPARATED BY COMMAS"
150 INPUT A,B,C
160 PRINT
170 PRINT "THE DEFINING EQUATION IS"
180 PRINT "Y = ";A;"(X - (";-B/2/A;"))↑2
    + (";(4*A*C-B*B)/4/A;")"
190 PRINT
200 PRINT "AXIS OF SYMMETRY:"
210 PRINT "X = ";-B/2/A
220 PRINT
230 PRINT "VERTEX"
240 PRINT "(";-B/2/A;",";
    (4*A*C-B*B)/4/A;")"
250 PRINT
260 IF A<0 THEN 280
270 PRINT "OPENING UPWARD"
275 GOTO 290
280 PRINT "OPENING DOWNWARD"
290 END
RUN
```

Statements 180 and 240 in the program above should be entered on one line.

What two whole numbers, neither containing any zeros, when multiplied together equal exactly 1 000 000 000?

5.9 REVIEW EXERCISE

1. If $f(x) = 1 - 2x$ and $g(x) = x^2 + 3$, state.
(a) $f(2)$
(b) $g(2)$
(c) $f(-5)$
(d) $g(-9)$
(e) $f(0.3)$
(f) $f(\frac{1}{2})$

(g) $g(\frac{1}{2})$
(h) $g(7)$
(i) $f(a)$
(j) $g(b)$

2. Arrow diagrams are given for two functions f and g.

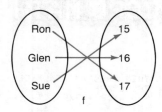

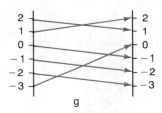

State.
(a) $f(Sue)$
(b) $f(Ron)$
(c) $g(-2)$
(d) $g(2)$
(e) domain of f
(f) range of f
(g) domain of g
(h) range of g

3. From the given graph, state.
(a) $f(0)$
(b) $f(1)$
(c) $f(-1)$
(d) $f(-2)$
(e) $f(2)$
(f) $f(-3)$
(g) domain of f
(h) range of f

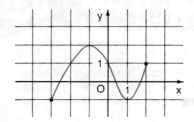

4. For each of the following parabolas state.
(a) the direction of the opening
(b) the coordinates of the vertex
(c) the equation of the axis of symmetry
(i) $y = 3x^2$
(ii) $y = x^2 - 4$
(iii) $y = -x^2 + 3$
(iv) $y = -2(x - 1)^2$
(v) $y = 2(x + 3)^2 + 4$
(vi) $y = 3(x - 4)^2 - 3$
(vii) $y = -\frac{1}{2}x^2 - 5$
(viii) $y = -3(x + 1)^2 + 2$
(ix) $y = -5(x - 4)^2 - 7$

5. Complete the square for each of the following.
(a) $x^2 - 8x$
(b) $x^2 - 12x$
(c) $x^2 - 10x$
(d) $x^2 - x$
(e) $x^2 + 3x$
(f) $x^2 - \frac{4}{3}x$
(g) $x^2 - 0.4x$
(h) $x^2 + \frac{1}{3}x$
(i) $x^2 + 5x$

6. If $f(x) = x^3 - 2x^2 + 3x - 4$ and $g(x) = \dfrac{1 - x^2}{1 + x^2}$, find.
(a) $f(2)$
(b) $g(\frac{1}{2})$
(c) $g(17)$
(d) $f(-5)$
(e) $f(12)$
(f) $g(0.1)$
(g) $f(\pi)$
(h) $g(f(1))$

7. If $f(x) = 2x^2 - 3x + 1$, find expressions for the following.
(a) $f(a)$
(b) $f(a - 1)$
(c) $f(a - b)$

(d) $f(3a)$
(e) $f(-a)$
(f) $f(-x)$

8. Draw the graphs of the following functions with the given domains A.
(a) $f(x) = 1 - 6x$ $A = R$
(b) $g(x) = 1 - 6x + x^2$ $A = \{0, 1, 2, 3, 4, 5, 6\}$.
(c) $h(x) = (x + 3)^2$ $A = \{x \in R \mid$
                                  $-6 \leq x \leq 0\}$.
(d) $f: x \rightarrow \sqrt{x - 1}$ $A = \{x \in R \mid x \geq 1\}$
(e) $g: x \rightarrow 0.5x + 1.5$ $A = R$
(f) $y = \dfrac{1}{x - 1}$ $A = \{x \in R \mid x > 1\}$

9. Sketch the graph of each of the following.
(a) $y = 2x^2$
(b) $y = -x^2 + 3$
(c) $y = 2(x - 1)^2$
(d) $y = -3(x + 1)^2$
(e) $y = -(x - 2)^2 - 1$
(f) $y = -2(x + 2)^2 + 5$
(g) $y = 2x^2 - 2$
(h) $y = 3(x - 4)^2 - 2$
(i) $y = 4(x + 3)^2 - 2$

10. Sketch the graph of each of the following.
(a) $y = 2x^2 - 8x - 1$
(b) $y = 3x^2 + 6x + 7$
(c) $y = -x^2 - 4x - 2$
(d) $y = -2x^2 + x - 1$
(e) $y = \frac{1}{2}x^2 + 3x - 2$
(f) $y = x^2 - 3x + 5$
(g) $y = -3x^2 + 12x + 5$
(h) $y = -\frac{1}{3}x^2 + x - 3$
(i) $y = \frac{2}{3}x^2 - 4x - 1$

11. Determine the maximum or minimum of each of the following functions. State the value of x where each occurs.
(a) $y = x^2 - 2x + 5$
(b) $y = -x^2 + 4x + 2$
(c) $y = 3x^2 - 6x + 1$
(d) $y = \frac{1}{2}x^2 + x + 7$
(e) $y = -2x^2 - x - 1$
(f) $y = \frac{3}{2}x^2 + 6x$

12. A rectangular field, bounded on one side by a river, is to be fenced on 3 sides by 1200 m of fence. Determine the dimensions of the field that will produce a maximum area.

13.

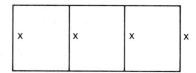

A rectangular field is to be enclosed by a fence. Two fences, parallel to one side of the field, divide the field into 3 rectangular fields. If 2400 m of fence are available, find the dimensions of the field giving the maximum area.

14. Draw the graphs of the following.
(a) $y < x^2$
(b) $y \geq x^2 - 2$
(c) $y > -x^2 + 3$
(d) $y \leq (x - 1)^2 + 4$
(e) $y \geq -2(x + 2)^2 - 1$
(f) $y < x^2 - 2x + 5$
(g) $y > 2x^2 - 6x - 2$
(h) $y \leq -x^2 + 4x - 1$
(i) $y \geq -3x^2 - 6x - 2$
(j) $y > \frac{1}{2}x^2 + x + 4$

Determine the pattern. Find the missing number.

8	7	4	5
10	9	5	6
3	2	6	7
	7	5	11

5.10 CHAPTER 5 TEST

1. If $f(x) = x^3 + x^2 - 3x$, find.
(a) $f(2)$
(b) $f(-3)$
(c) $f(0.2)$
(d) $f(a^2)$

2. Graph the following functions.
(a) $f(x) = 1 - 2x$, $\{x \in R \mid 0 \leqslant x \leqslant 3\}$
(b) $g(x) = x + \dfrac{1}{x}$, $\{x \in R \mid 0.1 \leqslant x \leqslant 10\}$

3. Without making a table of values, sketch the graph of each of the following functions and state the domain and range.
(a) $y = -(x + 4)^2$
(b) $y = \frac{1}{2}(x + 2)^2 + 1$

4. (a) Find the vertex and axis of symmetry of the parabola $y = x^2 + 10x + 16$.
(b) Sketch the graph of the parabola.

5. Find an equation for the parabola with vertex $(2, -1)$ which passes through the point $(1, 2)$.

6. Find the maximum value of the function $f(x) = 6 + 3x - x^2$.

7. Find two positive numbers whose sum is 100 and the sum of whose squares is a minimum.

8. Sketch the graph of the inequality $y \geqslant x^2 + 2x$.

QUADRATIC EQUATIONS

CHAPTER

6

I'm very well acquainted too with matters mathematical,
I understand equations, both simple and quadratical,
About binomial theorem I'm teeming with a lot o' news —
With many cheerful facts about the square of the hypotenuse.
sung by the model of a modern Major-General
in Gilbert and Sullivan's *The Pirates of Penzance*

FACTORING QUADRATIC EXPRESSIONS

EXERCISE

1. Factor.

(a) $x^2 + 7x + 12$
(b) $x^2 + 7x + 10$
(c) $y^2 - 7y + 10$
(d) $w^2 - 8w + 15$
(e) $x^2 - 2x - 8$
(f) $s^2 - 4s - 21$
(g) $x^2 + 3x - 10$
(h) $x^2 - 16$
(i) $x^2 - 25$
(j) $x^2 + 10x + 25$
(k) $x^2 - 14x + 49$
(l) $w^2 + 3w - 70$
(m) $x^2 + 2x - 15$
(n) $t^2 - t - 12$
(o) $r^2 + 2r - 24$
(p) $w^2 - 4w - 45$
(q) $t^2 - 2t + 1$
(r) $x^2 + 7x - 30$
(s) $x^2 + 11x + 28$
(t) $w^2 - 14w + 40$
(u) $x^2 + 6x - 27$
(v) $t^2 - t - 20$
(w) $x^2 + 3x - 88$
(x) $x^2 - 100$

2. Factor, if possible.

(a) $2x^2 + 7x + 3$
(b) $2x^2 - 7x + 5$
(c) $6w^2 - 7w - 3$
(d) $3w^2 - 11w - 20$
(e) $6y^2 + y - 1$
(f) $3x^2 - 3x - 4$
(g) $4x^2 + 12x + 9$
(h) $10w^2 - w - 2$
(i) $2w^2 + 9w + 10$
(j) $4x^2 - 9$
(k) $30t^2 + t - 20$
(l) $14s^2 + 41s + 15$

(m) $24x^2 - 46x + 21$
(n) $12w^2 + 29w + 15$
(o) $12t^2 - 25t + 12$
(p) $4x^2 + 20x + 25$
(q) $10 - 11x + 3x^2$
(r) $36x^2 + 1$

EQUATIONS AND INEQUALITIES

EXERCISE

1. Solve the following.

(a) $3x + 16 = x - 28$
(b) $4t - 3 = t - 45$
(c) $3(x - 4) - 6 = 5x - 12$
(d) $4(2w - 1) - (w - 5) = 11$
(e) $2(1 - 3t) - 2(4t - 5) = -2$
(f) $0 = 5 - 3(s - 5) + 4(2 - s)$
(g) $(x - 1)(x + 3) = (x + 2)(x + 1)$
(h) $4 - (w - 3) + 2(3w - 5) = 6$
(i) $\dfrac{x}{3} - \dfrac{x}{4} = \dfrac{1}{2}$
(j) $\dfrac{x + 1}{2} - \dfrac{x - 1}{4} = 5$
(k) $\dfrac{2w + 3}{3} - 1 = \dfrac{w}{2}$
(l) $5x - 3(2x - 3) + 7 = 2(1 + x)$
(m) $\dfrac{4t + 3}{2} - 2 = \dfrac{3t + 5}{5}$
(n) $0.2(x - 3) + 1 = 0.3(x + 2)$
(o) $1.2(2x - 1) - 0.2(x - 2) = 3$
(p) $\dfrac{3 - x}{4} - \dfrac{x + 1}{3} = \dfrac{x - 2}{2}$

2. Solve the following inequalities.

(a) $5x + 7 > 3x + 9$
(b) $3(w - 1) - 2 < 2w + 1$
(c) $3 - 3(t + 1) \geq 5(t - 8)$
(d) $7 - 3(x + 5) - 2(1 - 2x) \leq 2(x + 1)$
(e) $4(2 - 3t) - 5 > t - 6$
(f) $\dfrac{x}{2} + 1 < \dfrac{x}{3}$

(g) $3(2x + 1) - 2(1 - x) \leqslant 5$

(h) $\dfrac{x + 1}{3} \geqslant \dfrac{x - 2}{2}$

RADICALS

EXERCISE

1. Simplify the following.

(a) $\sqrt{2} + 3\sqrt{2} - 5\sqrt{3} + 8\sqrt{3}$

(b) $\sqrt{8} + 3\sqrt{18} - 3\sqrt{32}$

(c) $4\sqrt{27} - 5\sqrt{12} - 3\sqrt{80} - 2\sqrt{45}$

(d) $2\sqrt{90} + 5\sqrt{40} - 3\sqrt{75} + 2\sqrt{48}$

(e) $5\sqrt{63} - 2\sqrt{54} + 2\sqrt{28} - 3\sqrt{24}$

(f) $2\sqrt{68} - 5\sqrt{13} - 2\sqrt{153} - 4\sqrt{52}$

(g) $5\sqrt{363} - 2\sqrt{300} + 6\sqrt{27}$

(h) $6\sqrt{20} - 4\sqrt{125} + 8\sqrt{45} - \sqrt{500}$

2. Expand and simplify.

(a) $(3\sqrt{2} - 4)(5\sqrt{3} + 2\sqrt{2})$

(b) $(2\sqrt{3} - \sqrt{2})^2$

(c) $(5\sqrt{6} - \sqrt{3})(5\sqrt{6} + \sqrt{3})$

(d) $(2 - 3\sqrt{2})(2 + 3\sqrt{2})$

(e) $(2 + 4\sqrt{5})(\sqrt{3} - \sqrt{15})$

(f) $(6\sqrt{2} - \sqrt{5})(6\sqrt{2} + \sqrt{5})$

3. Expand and simplify.

(a) $(\sqrt{x} + 3)(\sqrt{x} - 1)$

(b) $(\sqrt{x} - 4)^2$

(c) $(2\sqrt{x} - 1)(\sqrt{x} - 3)$

(d) $(\sqrt{x + 1} + 2)(\sqrt{x + 1} + 3)$

(e) $(\sqrt{x - 5} + 1)^2$

(f) $(3\sqrt{x + 2} - 1)^2$

(g) $(1 - \sqrt{x - 3})^2$

(h) $(2 - \sqrt{x + 3})^2$

(i) $(1 + 2\sqrt{x - 1})^2$

EQUATIONS

EXERCISE

1. Solve the following for x.

(a) $x + 5 = 14$

(b) $x - 7 = -8$

(c) $4x = 12$

(d) $3x - 15 = 0$

(e) $2x + 7 = 19$

(f) $5x - 4 = 26 + 3x$

(g) $2(x + 1) - 4 = 18$

(h) $5(x - 7) - 3(x + 5) = 10$

(i) $2(x - 1) - (x + 5) = 37$

(j) $15 = 3(x - 5) + 6$

(k) $3(2x - 1) = 4(x - 6) - 5$

(l) $3(x + 6) - 2x = 1 - (2x + 3)$

(m) $4(1 - x) = 5 - (x + 5) + 11$

(n) $2(1 - 2x) - (3x + 5) = 6 - (3x - 4)$

(o) $5(x + 3) - 4 + 3(x - 1) = 8$

(p) $5x - 3(2x + 1) - 7 = 0$

(q) $3(x + 1) - 4(x - 3) - (x + 5) = 10$

(r) $5 = 2(3x - 1) - 4(x + 2) - 11$

2. Solve the following.

(a) $\dfrac{x}{2} - \dfrac{x}{3} = 5$

(b) $\dfrac{x + 1}{3} = 6$

(c) $\dfrac{x + 2}{4} - \dfrac{x - 1}{3} = 1$

(d) $\dfrac{x - 1}{2} + x = 5$

(e) $\dfrac{x - 6}{5} = \dfrac{x + 3}{4} + 2$

(f) $\dfrac{x}{4} - \dfrac{x + 1}{2} = 8$

(g) $5(x + 1) - \dfrac{x + 2}{2} = 3$

(h) $0.4(x - 8) + 3 = 4$

(i) $0.5x - 0.1(x - 3) = 4$

(j) $1.5(x - 3) - 2(x - 0.5) = 10$

(k) $1\frac{1}{2}x - \dfrac{x - 4}{3} = 7$

(l) $1.2(10x - 5) - 2(4x + 7) = 8$

(m) $(x + 2)(x - 3) - x^2 = 7$

(n) $2(x - 1)(x + 1)$
   $- (2x - 1)(x + 3) = 0$

(o) $3(x - 1)(x + 1)$
   $- 2x^2 = x(x + 1) - 6$

6.1 SOLVING QUADRATIC EQUATIONS BY GRAPHING

In this chapter we learn how to solve quadratic equations, that is, equations of the form

$$ax^2 + bx + c = 0$$

where a, b, c are real numbers and $a \neq 0$. In this first section we will look at the geometric meaning behind the solution of this equation.

For instance, suppose we are asked to solve the quadratic equation

$$x^2 - 2x - 3 = 0.$$

This means that we are required to find all real numbers x, such that $x^2 - 2x - 3 = 0$. If we consider the quadratic function $f(x) = x^2 - 2x - 3$, then we must find the numbers x, for which $f(x) = 0$. But these are just the points where the graph of f intersects the x-axis. We know from the preceding chapter that the graph of f is the parabola $y = x^2 - 2x - 3$ and we can see on the graph below that the x-intercepts of this parabola are -1 and 3. Therefore, the roots of the quadratic equation $x^2 - 2x - 3 = 0$ are -1 and 3.

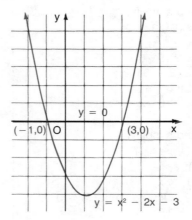

EXAMPLE. Solve $x^2 - 2x - 8 = 0$ graphically.

SOLUTION:
Draw the graph of $y = x^2 - 2x - 8$.
The graph intersects the x-axis at (4, 0) and $(-2, 0)$. Therefore, the solution set for the corresponding quadratic equation $x^2 - 2x - 8 = 0$ is $\{-2, 4\}$.

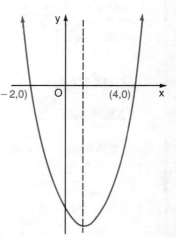

Check.

x = −2	x = 4
L.S. = x² − 2x − 8	L.S. = x² − 2x − 8
= (−2)² − 2(−2) − 8	= (4)² − 2(4) − 8
= 4 + 4 − 8	= 16 − 8 − 8
= 0	= 0
R.S. = 0	R.S. = 0

∴ the roots are −2 and 4.

In general, the roots of a quadratic equation $ax^2 + bx + c = 0$ are the zeros of the quadratic function $f(x) = ax^2 + bx + c$, or in other words, the x-intercepts of the parabola $y = ax^2 + bx + c$. The graphs below show cases where the quadratic equation has

(i) two distinct real roots, −1 and 4;

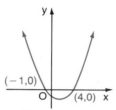

(ii) two equal real roots, 2 and 2;

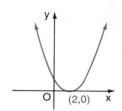

(iii) no real roots.

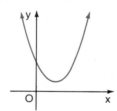

Case (iii) will be studied in Section 6.7

EXERCISE 6.1

B 1. Solve each equation by graphing. Check each root by substitution.

(a) x² + 2x − 8 = 0
(b) x² + 4x − 5 = 0
(c) x² + 6x − 7 = 0
(d) x² − x − 2 = 0
(e) x² + 4x + 3 = 0
(f) x² − 6x + 8 = 0
(g) x² + x − 20 = 0
(h) x² + 6x + 5 = 0
(i) x² − 4 = 0
(j) x² + 8x + 15 = 0

C 2. Solve graphically. Check your solutions.

(a) −x² − 2x + 3 = 0
(b) 9 − x² = 0
(c) 2x² + 3x − 2 = 0
(d) x² − 6x + 9 = 0
(e) 3x² + 5x = −2
(f) 2x² − 12x + 10 = 0

3. For the quadratic function y = x² − 4 determine the values of x so that 5 ≤ y ≤ 12.

6.2 SOLVING QUADRATIC EQUATIONS BY FACTORING

Many quadratic equations can be solved by factoring. A solution by factoring depends on the following fact.

> For any two real numbers a and b,
> ab = 0 if, and only if, a = 0 or b = 0.

This includes the possibility that both a and b = 0.

EXAMPLE 1. Solve. $x^2 + 5x + 6 = 0$

SOLUTION:
First we factor the left side of the equation.

$$x^2 + 5x + 6 = 0$$
$$(x + 2)(x + 3) = 0$$

Since the product of the two factors is zero, one or both of the factors must be 0.

$$x + 2 = 0 \qquad \text{or} \qquad x + 3 = 0$$
$$x = -2 \qquad\qquad\qquad x = -3$$

The solution is checked by substitution.

Check.

$$\begin{aligned}
x &= -2 \\
\text{L.S.} &= x^2 + 5x + 6 \\
&= (-2)^2 + 5(-2) + 6 \\
&= 4 - 10 + 6 \\
&= 0 \\
\text{R.S.} &= 0
\end{aligned}
\qquad\qquad
\begin{aligned}
x &= -3 \\
\text{L.S.} &= x^2 + 5x + 6 \\
&= (-3)^2 + 5(-3) + 6 \\
&= 9 - 15 + 6 \\
&= 0 \\
\text{R.S.} &= 0
\end{aligned}$$

The roots are -2 and -3.

EXAMPLE 2. Solve. $10x^2 - 9x = -2$

SOLUTION:
We rewrite the equation so that one side of the equation is zero.

$$10x^2 - 9x = -2$$
$$10x^2 - 9x + 2 = 0$$
$$(5x - 2)(2x - 1) = 0 \quad\longleftarrow\quad \text{Factor.}$$
$$5x - 2 = 0 \quad \text{or} \quad 2x - 1 = 0$$
$$x = \tfrac{2}{5} \qquad\qquad x = \tfrac{1}{2}$$

The solution set is $\left\{\tfrac{2}{5}, \tfrac{1}{2}\right\}$.

The following rules summarize the procedure for solving quadratic equations by factoring.

I.   Clear fractions.
II. Transform the equation so that one side is zero.
III. Divide both sides by any numerical common factors.
IV. Factor.
V.   Set each factor equal to zero and solve the resulting linear equations.

EXAMPLE 3. Solve. $10x + 24 = -38x$

SOLUTION:

$$10x^2 + 24 = -38x$$
$$10x^2 + 38x + 24 = 0$$
$$5x^2 + 19x + 12 = 0 \quad\longleftarrow\text{Divide by 2.}$$
$$(5x + 4)(x + 3) = 0$$
$$5x + 4 = 0 \quad\text{or}\quad x + 3 = 0$$
$$x = -\tfrac{4}{5} \qquad\qquad x = -3$$

The roots are $-\tfrac{4}{5}$ and -3.

EXAMPLE 4. Solve. $\dfrac{4}{x - 1} - \dfrac{3}{x + 2} = 2$

SOLUTION:

$$\frac{4}{x - 1} - \frac{3}{x + 2} = 2$$
$$4(x + 2) - 3(x - 1) = 2(x - 1)(x + 2) \quad\text{Clear fractions}$$
$$4x + 8 - 3x + 3 = 2(x^2 + x - 2) \quad\text{(multiply by } (x - 1)(x + 2)).$$
$$x + 11 = 2x^2 + 2x - 4$$
$$2x^2 + x - 15 = 0$$
$$(x + 3)(2x - 5) = 0$$
$$x + 3 = 0 \quad\text{or}\quad 2x - 5 = 0$$
$$x = -3 \qquad\qquad x = \tfrac{5}{2}$$

The roots are -3 and $\tfrac{5}{2}$.

EXERCISE 6.2

A 1. State the roots of the following quadratic equations.

(a) $(x + 3)(x - 1) = 0$
(b) $(x - 1)(x - 4) = 0$
(c) $(x + 5)(x - 4) = 0$
(d) $(w - 7)(w + 9) = 0$
(e) $(t - 11)(t - 7) = 0$
(f) $(x + 7)(x + 8) = 0$
(g) $(3x - 1)(3x + 5) = 0$
(h) $(4w - 3)(w - 5) = 0$
(i) $(2t + 5)(3t + 10) = 0$
(j) $(4s + 7)(3s + 1) = 0$
(k) $(8x + 3)(7x - 5) = 0$
(l) $(9w + 11)(3w + 14) = 0$

B 2. Solve by factoring. Check your solution.

(a) $x^2 - x - 12 = 0$
(b) $x^2 + 9x + 18 = 0$
(c) $x^2 - x - 20 = 0$
(d) $x^2 + 8x + 15 = 0$
(e) $x^2 - 4x = 77$
(f) $x^2 + 117 = -22x$
(g) $x^2 - 23x + 126 = 0$
(h) $x^2 + 8x + 16 = 0$
(i) $x^2 + 3x = 40$

3. Solve by factoring. Check your solution.

(a) $2x^2 + 3x - 2 = 0$
(b) $3x^2 + 7x + 2 = 0$
(c) $2t^2 - 7t + 5 = 0$
(d) $6x^2 - 7x + 2 = 0$
(e) $12y^2 + 29y + 15 = 0$
(f) $2x^2 + 11x - 21 = 0$
(g) $3w^2 - 4w - 32 = 0$
(h) $6x^2 + 5x - 50 = 0$
(i) $6s^2 + 11s + 5 = 0$
(j) $8x^2 + 30x + 7 = 0$
(k) $4t^2 - 11t - 45 = 0$
(l) $2w^2 - 13w - 7 = 0$
(m) $15x^2 + 19x - 10 = 0$
(n) $8w^2 - 2w - 15 = 0$
(o) $5t^2 + 23t + 24 = 0$

4. Solve by factoring.

(a) $2x^2 - 5x = 12$
(b) $10w^2 = 7w + 12$
(c) $6t^2 + 10 = 19t$
(d) $4x^2 - 18x - 10 = 0$
(e) $6x^2 + 27x + 12 = 0$

(f) $10s^2 = 17x + 20$
(g) $30w^2 + 73w + 7 = 0$
(h) $56t^2 + 14 = 65t$
(i) $5x^2 + 21x - 54 = 0$

5. Solve.

(a) $3x(x - 2) - x(x + 1) + 5 = 0$
(b) $x^2 + (x + 1)^2 = 13$
(c) $x^2 + \frac{9}{2}x - 2\frac{1}{2} = 0$
(d) $w^2 + (w + 1)^2 + (w + 2)^2 = 50$
(e) $3(x - 1)(x + 4) - 2(2x + 1)^2 = -18$
(f) $3(x - 1)(x + 2) - (x + 1)^2 = -4$
(g) $2r(r + 5) + 19 = 7(r + 3)$

6. Solve.

(a) $\dfrac{3}{x + 1} + \dfrac{4}{x + 2} = 2$

(b) $\dfrac{4}{x + 2} - 4 = \dfrac{3}{x - 3}$

(c) $\dfrac{x}{x + 1} - \dfrac{5}{x + 4} = -\dfrac{1}{6}$

(d) $\dfrac{30}{x + 15} + 1 = \dfrac{30}{x}$

(e) $\dfrac{2x}{x - 2} - 3x + 8 = 0$

(f) $\dfrac{5}{x - 1} + \dfrac{6}{x + 1} - 7 = 0$

(g) $\dfrac{1}{x + 1} = \dfrac{x - 1}{x + 5}$

(h) $2 + \dfrac{7}{x - 2} = \dfrac{1}{2x(2 - x)}$

C 7. Write a quadratic equation whose roots are the following.

(a) $3, 4$
(b) $-2, 5$
(c) $-7, -4$
(d) $\frac{1}{2}, \frac{1}{3}$
(e) $-\frac{3}{4}, -\frac{1}{5}$
(f) r, s

$2(x-2)(4x-2x^2)$

$(4x-4)(4x-2x^2)$

$16x^2 - 8x^3 - 16x + 8$

$24x^2 - 8x^3 - 16x -$

$7(4x-2x^2) = x -$

6.3 SOLVING QUADRATIC EQUATIONS— SPECIAL CASES

A quadratic equation is written in the form $ax^2 + bx + c = 0$ where a, b, and c may have any real values except $a \neq 0$. When one or more of these coefficients is 0, the resulting equations have relatively simple solutions.

CASE 1. If $c = 0$, the equation becomes $ax^2 + bx = 0$. This type can be solved by factoring and 0 is always one of the roots.

EXAMPLE. Solve. $3x^2 - 5x = 0$

SOLUTION:

$$3x^2 - 5x = 0$$
$$x(3x - 5) = 0$$
$$x = 0 \quad \text{or} \quad 3x - 5 = 0$$
$$3x = 5$$
$$x = \tfrac{5}{3}$$

$\therefore$ the roots are 0 and $\tfrac{5}{3}$.

CASE 2. If $b = 0$, the equation becomes $ax^2 + c = 0$. The roots are numerically equal but opposite in sign.

EXAMPLE. Solve. $4x^2 - 7 = 0$

SOLUTION:

$$4x^2 - 7 = 0$$
$$4x^2 = 7$$
$$x^2 = \tfrac{7}{4}$$
$$x = \pm \tfrac{1}{2}\sqrt{7}$$

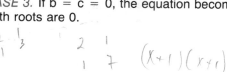

If $x^2 - d = 0$
then $x^2 = d$
and $x = \pm\sqrt{d}$

$\therefore$ the roots are $\tfrac{1}{2}\sqrt{7}$ and $-\tfrac{1}{2}\sqrt{7}$.

CASE 3. If $b = c = 0$, the equation becomes $ax^2 = 0$. In this case both roots are 0.

EXERCISE 6.3

B 1. Solve the following.

(a) $2x^2 + 7x = 0$
(b) $x^2 - 16 = 0$
(c) $5x^2 = 0$
(d) $3t^2 - 12 = 0$
(e) $3x^2 - 2x = 0$
(f) $4w^2 - 25 = 0$
(g) $-6w^2 = 0$
(h) $5t^2 = 4t$
(i) $100x^2 = 9$

(j) $\dfrac{w^2}{3} = 27$
(k) $6s - 7s^2 = 0$
(l) $25x^2 - 1 = 0$
(m) $\dfrac{3x^2 - 2x}{4} = 0$
(n) $0 = -3w^2$
(o) $\tfrac{1}{2}x^2 + \tfrac{1}{3}x = 0$

6.4 SOLVING QUADRATIC EQUATIONS BY COMPLETING THE SQUARE

Not all quadratic equations can be solved by factoring, so we must develop other methods of solution which are more generally applicable.

The principle of completing the square, which was used in graphing quadratic functions, is also used to solve quadratic equations.

The equation $x^2 - 9 = 0$ may be solved by factoring.

$$x^2 - 9 = 0$$
$$(x - 3)(x + 3) = 0$$
$$x - 3 = 0 \quad \text{or} \quad x + 3 = 0$$
$$x = 3 \qquad\qquad x = -3$$

The equation may also be solved by taking the square root of both sides if you write the equation as a perfect square equal to a constant.

$$x^2 - 9 = 0$$
$$x^2 = 9$$
$$x = \pm 3$$

This method may be used to solve equations of the following form.

$$(x + 3)^2 = 16$$

EXAMPLE 1. Solve. $(x + 3)^2 = 16$

SOLUTION:

$$(x + 3)^2 = 16$$
$$x + 3 = \pm 4$$
$$x + 3 = 4 \quad \text{or} \quad x + 3 = -4$$
$$x = 1 \qquad\qquad x = -7$$

Take the square root of both sides.

Check.

$x = 1$	$x = -7$
L.S. $= (x + 3)^2$	L.S. $= (x + 3)^2$
$= (1 + 3)^2$	$= (-7 + 3)^2$
$= 16$	$= 16$
R.S. $= 16$	R.S. $= 16$

$\therefore$ the roots are 1 and -7.

Example 1 suggests that a quadratic equation can be solved if we express it in the form

$$(x + m)^2 = d$$

or

$$x^2 + 2mx + m^2 = d$$

EXAMPLE 2. Solve. $x^2 - 6x + 4 = 0$

SOLUTION:

$x^2 - 6x + 4 = 0$	Subtract the constant term from both sides.
$x^2 - 6x + 4 - 4 = 0 - 4$	
$x^2 - 6x = -4$	Add the square of half the coefficient of x.
$x^2 - 6x + 9 = -4 + 9$	
$x^2 - 6x + 9 = 5$	Write the left side as a perfect square.
$(x - 3)^2 = 5$	
$x - 3 = \pm\sqrt{5}$	Take the square root of both sides and proceed as in Example 1.
$x - 3 = \sqrt{5}$ or $x - 3 = -\sqrt{5}$	
$x = 3 + \sqrt{5}$ $x = 3 - \sqrt{5}$	

The solution set is $\{3 + \sqrt{5}, 3 - \sqrt{5}\}$.

If the coefficient of x^2 is not 1, a preliminary step is required.

EXAMPLE 3. Solve. $2x^2 - 3x - 1 = 0$

SOLUTION:

$2x^2 - 3x - 1 = 0$	
$x^2 - \frac{3}{2}x - \frac{1}{2} = 0$	Divide by 2.
$x^2 - \frac{3}{2}x = \frac{1}{2}$	Add $\frac{1}{2}$ to each side.
$x^2 - \frac{3}{2}x + \frac{9}{16} = \frac{1}{2} + \frac{9}{16}$	Complete the square.
$(x - \frac{3}{4})^2 = \frac{17}{16}$	Simplify.
$x - \frac{3}{4} = \pm\frac{\sqrt{17}}{4}$	Take the square root.
$x = \frac{3}{4} \pm \frac{\sqrt{17}}{4}$	
$x = \frac{3 \pm \sqrt{17}}{4}$	

The solution set is $\left\{\dfrac{3 + \sqrt{17}}{4}, \dfrac{3 - \sqrt{17}}{4}\right\}$.

This method of solving quadratic equations must be understood because it leads to the derivation of the quadratic formula which will be discussed in the next section.

EXERCISE 6.4

A 1. State the value of k that makes each trinomial a perfect square.

(a) $x^2 + 6x + k$

(b) $x^2 - 8x + k$

(c) $x^2 - 10x + k$

(d) $x^2 - 2x + k$

(e) $x^2 + 4x + k$

(f) $x^2 - 12x + k$

(g) $x^2 + 18x + k$

(h) $x^2 - 22x + k$

(i) $x^2 - 3x + k$

(j) $x^2 + x + k$

(k) $x^2 - 7x + k$

(l) $x^2 + \frac{1}{2}x + k$

(m) $x^2 - \frac{2}{3}x + k$

(n) $x^2 + \frac{1}{5}x + k$

(o) $x^2 - \frac{3}{5}x + k$

B 2. Solve.

(a) $x^2 = 16$

(b) $x^2 = 36$

(c) $x^2 = 7$

(d) $4x^2 = 25$

(e) $x^2 - 49 = 0$

(f) $x^2 - 9 = 0$

(g) $x^2 - 10 = 0$

(h) $x^2 - 32 = 0$

3. Solve.

(a) $(x + 4)^2 = 9$

(b) $(x - 2)^2 = 3$

(c) $(x + 1)^2 = 7$

(d) $(x - 5)^2 = 8$

(e) $(x + 7)^2 = 27$

(f) $(x - 6)^2 = 12$

(g) $(x + \frac{1}{2})^2 = 6$

(h) $(x - \frac{1}{2})^2 = \frac{3}{4}$

(i) $(x + \frac{1}{3})^2 = \frac{5}{9}$

(j) $(x - \frac{3}{4})^2 = \frac{7}{16}$

(k) $(x + \frac{5}{2})^2 = \frac{3}{8}$

(l) $(x - \frac{4}{3})^2 = \frac{6}{27}$

4. Solve by completing the square.

(a) $x^2 - 2x - 8 = 0$

(b) $x^2 - 4x + 1 = 0$

(c) $x^2 + 6x - 2 = 0$

(d) $x^2 + 2x - 1 = 0$

(e) $x^2 + 8x + 5 = 0$

(f) $x^2 + 10x + 8 = 0$

(g) $x^2 - 3x - 4 = 0$

(h) $x^2 - 5x + 2 = 0$

5. Solve by completing the square.

(a) $2w^2 + 8w + 5 = 0$

(b) $2x^2 - 8x + 3 = 0$

(c) $3t^2 - 6t + 2 = 0$

(d) $5x^2 + 5x - 2 = 0$

(e) $2s^2 - 3s - 5 = 0$

(f) $-x^2 + 3x + 1 = 0$

(g) $\frac{1}{2}x^2 + x - 1 = 0$

(h) $3t^2 + 4t - 2 = 0$

C 6. Solve by completing the square.

(a) $x^2 + 2x = c$

(b) $x^2 = kx + 1$

(c) $x^2 + bx - 8 = 0$

(d) $ax^2 + 5x - 7 = 0$

(e) $kx^2 - 2x = k$

MIND BENDER

WORD LADDER

Start with the word "black" and change one letter at a time to form a new word until you reach "white." The best solution has the fewest steps.

b l a c k

_ _ _ _ _

_ _ _ _ _

_ _ _ _ _

_ _ _ _ _

_ _ _ _ _

_ _ _ _ _

w h i t e

6.5 THE QUADRATIC FORMULA

The method of completing the square can be used to solve the general quadratic equation $ax^2 + bx + c = 0$, $a \neq 0$.

$$ax^2 + bx + c = 0$$

$$x^2 + \frac{b}{a}x + \frac{c}{a} = 0 \qquad \text{Divide by } a.$$

$$x^2 + \frac{b}{a}x + \frac{c}{a} - \frac{c}{a} = 0 - \frac{c}{a} \qquad \text{Subtract } \frac{c}{a} \text{ from both sides.}$$

$$x^2 + \frac{b}{a}x = -\frac{c}{a}$$

$$x^2 + \frac{b}{a}x + \frac{b^2}{4a^2} = \frac{b^2}{4a^2} - \frac{c}{a} \qquad \text{Complete the square.}$$

$$\left(x + \frac{b}{2a}\right)^2 = \frac{b^2 - 4ac}{4a^2} \qquad \text{Simplify.}$$

$$x + \frac{b}{2a} = \pm\sqrt{\frac{b^2 - 4ac}{4a^2}} \qquad \text{Take the square root of both sides.}$$

$$x + \frac{b}{2a} = \pm\frac{\sqrt{b^2 - 4ac}}{2a}$$

$$x + \frac{b}{2a} - \frac{b}{2a} = -\frac{b}{2a} \pm \frac{\sqrt{b^2 - 4ac}}{2a} \qquad \text{Subtract } \frac{b}{2a} \text{ from both sides.}$$

$$x = -\frac{b}{2a} \pm \frac{\sqrt{b^2 - 4ac}}{2a}$$

$$x = \frac{-b \pm \sqrt{b^2 - 4ac}}{2a} \qquad \text{Simplify.}$$

The solution for a quadratic equation $ax^2 + bx + c = 0$ is given by the quadratic formula.

$$x = \frac{-b \pm \sqrt{b^2 - 4ac}}{2a}$$

EXAMPLE 1.   Solve $3x^2 - 5x + 2 = 0$ using the quadratic formula.

SOLUTION:

For $3x^2 - 5x + 2 = 0$, $a = 3$, $b = -5$, $c = 2$.

$$x = \frac{-b \pm \sqrt{b^2 - 4ac}}{2a}$$

$$x = \frac{-(-5) \pm \sqrt{(-5)^2 - 4(3)(2)}}{2(3)}$$

$$= \frac{5 \pm \sqrt{25 - 24}}{6}$$

$$= \frac{5 \pm \sqrt{1}}{6}$$

$$= \frac{5 \pm 1}{6}$$

Therefore,

$$x = \frac{5 + 1}{6} \quad \text{or} \quad x = \frac{5 - 1}{6}$$

$$= 1 \qquad\qquad = \tfrac{2}{3}$$

The solution set is $\{1, \tfrac{2}{3}\}$.

EXAMPLE 2. Solve. $5x^2 + 2x - 2 = 0$

SOLUTION:

For $5x^2 + 2x - 2 = 0$, $a = 5$, $\quad b = 2$, $\quad c = -2$.

$$x = \frac{-b \pm \sqrt{b^2 - 4ac}}{2a}$$

$$x = \frac{-(2) \pm \sqrt{(2)^2 - 4(5)(-2)}}{2(5)}$$

$$= \frac{-2 \pm \sqrt{4 + 40}}{10}$$

$$= \frac{-2 \pm \sqrt{44}}{10}$$

$$= \frac{-2 \pm 2\sqrt{11}}{10}$$

$$= \frac{-1 \pm \sqrt{11}}{5}$$

The roots are $\dfrac{-1 + \sqrt{11}}{5}$ and $\dfrac{-1 - \sqrt{11}}{5}$.

EXAMPLE 3. Solve. $\dfrac{t^2}{4} - \dfrac{t}{5} - 1 = 0$

SOLUTION:

$$\frac{t^2}{4} - \frac{t}{5} - 1 = 0$$

$5t^2 - 4t - 20 = 0$ ←——————— Multiply by 20 to clear fractions.

This is a quadratic equation in t with $a = 5$, $\quad b = -4$, and $\quad c = -20$.

$$t = \frac{-b \pm \sqrt{b^2 - 4ac}}{2a}$$

$$t = \frac{-(-4) \pm \sqrt{(-4)^2 - 4(5)(-20)}}{2(5)}$$

$$= \frac{4 \pm \sqrt{16 + 400}}{10}$$

$$= \frac{4 \pm \sqrt{416}}{10}$$

$$= \frac{4 \pm 4\sqrt{26}}{10}$$

$$= \frac{2(1 \pm \sqrt{26})}{5}$$

The roots are $\dfrac{2(1 + \sqrt{26})}{5}$ and $\dfrac{2(1 - \sqrt{26})}{5}$.

EXAMPLE 4. Solve. $\dfrac{2}{x} - \dfrac{3}{x + 1} = 1$

SOLUTION:

$$\frac{2}{x} - \frac{3}{x + 1} = 1$$

$2(x + 1) - 3x = x(x + 1)$ ←——————— Multiply by x(x − 1) to clear fractions.

$2x + 2 - 3x = x^2 + x$

$-x^2 - 2x + 2 = 0$

$x^2 + 2x - 2 = 0$ ←——————————— Multiply by −1.

Here $a = 1$, $b = 2$, and $c = -2$.

$$x = \frac{-b \pm \sqrt{b^2 - 4ac}}{2a}$$

Using a calculator, press

$$x = \frac{-(2) \pm \sqrt{(2)^2 - 4(1)(-2)}}{2(1)}$$

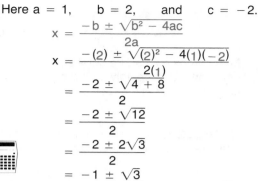

$$= \frac{-2 \pm \sqrt{4 + 8}}{2}$$

and

$$= \frac{-2 \pm \sqrt{12}}{2}$$

$$= \frac{-2 \pm 2\sqrt{3}}{2}$$

$$= -1 \pm \sqrt{3}$$

The roots are $-1 + \sqrt{3}$ and $-1 - \sqrt{3}$.

The roots may be expressed correct to 4 decimal places as follows:

$$x = -1 + \sqrt{3} \qquad\qquad x = -1 - \sqrt{3}$$
$$\doteq -1 + 1.732\ 05 \qquad\qquad \doteq -1 - 1.732\ 05$$
$$\doteq 0.7321 \qquad\qquad\qquad \doteq -2.7321$$

EXERCISE 6.5

A 1. State values for a, b, and c for each of the following.
(a) $2x^2 + 7x - 1 = 0$ (b) $4x - 7 + 5x^2 = 0$
(c) $3x^2 + 2x = 7$ (d) $9x^2 - 7 = 4x$
(e) $0 = x^2 - 7x + 1$ (f) $4 - 2x^2 = 9x$
(g) $2x^2 - 7 = 0$ (h) $5x^2 = 9x$

B 2. Solve using the quadratic formula.
(a) $x^2 + 6x + 8 = 0$ (b) $x^2 - 2x - 15 = 0$
(c) $2w^2 - 3w + 1 = 0$ (d) $10t^2 - 21t + 9 = 0$
(e) $7w^2 - 3w = 0$ (f) $5x^2 - 16 = 0$
(g) $4s^2 + 4s + 1 = 0$ (h) $x^2 - 2x - 4 = 0$
(i) $x^2 - x - 5 = 0$ (j) $x^2 + 2x - 6 = 0$
(k) $2x^2 + 8x - 3 = 0$ (l) $7x^2 - 2x - 2 = 0$

3. Solve using the quadratic formula.
(a) $2x^2 - x = 3$ (b) $6x = 2 - x^2$
(c) $3w^2 = 2w + 2$ (d) $2s^2 = 8s - 7$
(e) $2 = 3x^2 + 8x$ (f) $10x^2 - 4x - 4 = 0$
(g) $6x + 6 = 15x^2$ (h) $-x^2 - 7x - 1 = 0$
(i) $1 = 8x + 3x^2$ (j) $1 = 3t^2 + 7t$
(k) $1 = 5x^2$ (l) $4x^2 + 4x = 14$

4. Solve.
(a) $x^2 - 0.1x - 0.06 = 0$
(b) $w^2 + 2.76 = 3.5w$
(c) $t^2 + 3t - 14.56 = 0$
(d) $3.1x + 0.66 = x^2$
(e) $0.1t^2 + 0.2 = 0.45t$
(f) $0.02x^2 + 0.13x - 0.57 = 0$

5. Solve.
(a) $\frac{1}{2}x^2 - x = \frac{5}{2}$ (b) $2t(t - 1) - 3 = 0$
(c) $2(1 - x^2) - 3x(1 - x) = 7$
(d) $\frac{x^2}{3} - x - 1\frac{2}{3} = 0$ (e) $\frac{1}{2}w^2 - \frac{w}{4} - 1 = 0$
(f) $2(x - 2)(x + 1) - (x + 3) = 0$
(g) $(3x - 2)(x - 3) = (x - 4)(x - 1)$
(h) $\frac{2}{x - 1} + \frac{3}{x + 2} = 1$
(i) $\frac{3}{x + 2} + \frac{2}{x - 3} = 2$
(j) $\frac{x + 3}{2x - 1} = \frac{x - 3}{x + 4}$

6. Solve and express irrational roots correct to 4 decimal places.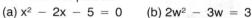
(a) $x^2 - 2x - 5 = 0$ (b) $2w^2 - 3w = 3$
(c) $\frac{x}{4} - \frac{x^2}{2} = -1$ (d) $\frac{3}{x} - \frac{4}{x + 2} = 2$
(e) $\frac{1}{2x - 1} = \frac{3}{4x^2 + 4x - 7}$
(f) $\frac{x + 2}{x - 1} = \frac{2x + 3}{x + 2}$

7. Solve for x.
(a) $\sqrt{2}x^2 - x - 3\sqrt{2} = 0$
(b) $2x^2 - \sqrt{3}x - 1 = 0$
(c) $\sqrt{2}x^2 - \sqrt{3}x - \sqrt{2} = 0$
(d) $x^2 - kx + k - 1 = 0$

6.6 SOLVING PROBLEMS USING QUADRATIC EQUATIONS

Many problems can be solved by translating the problem into an equation and then solving the equation. In this section we shall consider problems that give rise to quadratic equations.

READ

EXAMPLE 1. The sum of the squares of two consecutive even integers is 452. Find the integers.

PLAN

SOLUTION:
Let x represent the first even integer. Then x + 2 represents the next even integer. The sum of their squares is 452.

SOLVE

$$x^2 + (x + 2)^2 = 452, x \in I$$
$$x^2 + x^2 + 4x + 4 = 452$$
$$2x^2 + 4x - 448 = 0$$
$$x^2 + 2x - 224 = 0$$
$$(x - 14)(x + 16) = 0$$
$$x = 14 \quad \text{or} \quad x = -16$$

ANSWER

Since x represents the first even integer, we see that there are two solutions.
When x = 14, x + 2 = 16 and $14^2 + 16^2 = 452$.
When x = -16, x + 2 = -14 and $(-16)^2 + (-14)^2 = 452$.
The two integers are 14 and 16 or -16 and -14.

READ

EXAMPLE 2. A rectangular supermarket, 90 m by 60 m is to be built on a city block having an area of 9000 m². There is to be a uniform strip around the building for parking. How wide is the strip?

PLAN

SOLUTION:
Let the width of the strip be x metres. Then x > 0.

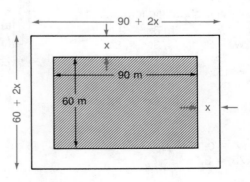

The dimensions of the city block are (90 + 2x) metres by (60 + 2x) metres. The area of the block is 9000 m².

$$(90 + 2x)(60 + 2x) = 9000$$
$$5400 + 180x + 120x + 4x^2 = 9000$$
$$4x^2 + 300x - 3600 = 0$$
$$x^2 + 75x - 900 = 0$$
$$x = \frac{-b \pm \sqrt{b^2 - 4ac}}{2a}$$
$$x = \frac{-(75) \pm \sqrt{(75)^2 - 4(1)(-900)}}{2(1)}$$
$$= \frac{-75 \pm \sqrt{9225}}{2}$$
$$= \frac{-75 \pm 15\sqrt{41}}{2} \text{ and } x > 0$$
$$\therefore x = \frac{-75 + 15\sqrt{41}}{2}$$

$$\left(\text{We reject the root } \frac{-75 - 15\sqrt{41}}{2}, \text{ since } x > 0. \right)$$

To determine the width to the nearest tenth of a metre, we evaluate $\sqrt{41}$ to the nearest hundredth and substitute.
$$x \doteq \frac{-75 + 15(6.40)}{2}$$
$$\doteq \frac{-75 + 96.00}{2}$$
$$\doteq 10.50$$
∴ the width of the strip is approximately 10.5 m.

EXAMPLE 3. A jet flew from New York to Los Angeles, a distance of 4200 km. On the return trip the speed was increased by 100 km/h. If the total trip took 13h, what was the speed from New York to Los Angeles?

SOLUTION:
Let x represent the speed from New York to Los Angeles, $x > 0$.
Then $x + 100$ represents the speed from Los Angeles to New York. Complete the DST table.

D / S T	Distance (km)	Speed (km/h)	Time (h)
N.Y. to L.A.	4200	x	$\frac{4200}{x}$
L.A. to N.Y.	4200	x + 100	$\frac{4200}{x + 100}$

The total time is 13 h.

$$\frac{4200}{x} + \frac{4200}{x + 100} = 13$$

SOLVE

$4200(x + 100) + 4200x = 13x(x + 100)$ ◄———Multiply by
$4200x + 420\,000 + 4200x = 13x^2 + 1300x$ $x(x + 100)$.
$13x^2 - 7100x - 420\,000 = 0$
$(13x + 700)(x - 600) = 0$

$$x = -\frac{700}{13} \quad \text{or} \quad x = 600$$

ANSWER

Since $x > 0$, we reject $-\frac{700}{13}$.

∴ the speed from New York to Los Angeles is 600 km/h.

EXERCISE 6.6

B 1. The product of two consecutive even integers is 224. Find the integers.

2. The product of two consecutive odd integers is 323. Find the integers.

3. The sum of a number and its square is 272. Find the number.

4. The sum of the squares of two consecutive positive integers is 481. Find the integers.

5. The sum of the squares of three consecutive integers is 434. Find the integers.

6. The difference between a number and its squares is 420. Find the number.

7. The sum of two numbers is 20 and the sum of their squares is 272. Find the numbers.

8. The sum of two numbers is 30 and their product is 209. Find the numbers.

9. The product of two consecutive even integers is 528. Find the integers.

10. The product of two consecutive positive odd integers is 195. Find the integers.

11. The sum of a number and twice its square is 300. Find the number.

12. The sum of twice a number and three times its square is 261. Find the number.

13. The sum of the squares of three consecutive even integers is 116. Find the integers.

14. The difference between three times a number and its square is 378. Find the number.

15. Two numbers differ by seven and the sum of their squares is 389. Find the numbers.

16. The hypotenuse of a right triangle is 15 cm. The sum of the other two sides is 21 cm. Find the lengths of the other two sides of the triangle.

17. The hypotenuse of a right triangle is 26 cm. The sum of the other two sides is 34 cm. Find the lengths of the other two sides of the triangle.

18. The length of a rectangle is 6 m longer than the width. If the area of the rectangle is 91 m², find the dimensions of the rectangle.

19. The hypotenuse of a right triangle is 6 cm and one side is 2 cm longer than the other. Find the length of each side to the nearest tenth of a centimetre.

20. A lidless box is constructed from a square piece of tin by cutting a 10 cm square from each corner and bending up the sides for the box. If the volume of the box is 1200 cm³, find the dimensions to the nearest tenth of a centimetre.

21. A rectangular piece of tin 40 cm by 30 cm is to be made into an open box with a base of 900 cm² by cutting equal squares from the four corners and then bending up the sides. Find, to the nearest tenth, the length of the side of the square cut from each corner.

22. A playground, which measures 60 m by 40 m, is to be doubled in area by extending each side an equal amount. By how much should each side be extended?

23. A rectangular skating rink measuring 30 m by 20 m is to be doubled in area by adding a strip at one end and a strip of the same width along one side. Find the width of the strips.

24. A local building code requires that all factories must be surrounded by a lawn. The width of the lawn must be uniform and the area of the lawn must be equal to the area of the factory. What must be the width of a lawn surrounding a rectangular factory that measures 120 m by 80 m?

25. A factory is to be built on a lot that measures 80 m by 60 m. A lawn of uniform width and equal in area to the factory must surround the factory. What dimensions must the factory have?

26. A jet flew from Montreal to San Francisco, a distance of 4000 km. On the return trip the speed was increased by 300 km/h. If the total trip took 13 h, what was the speed from Montreal to San Francisco?

27. Pete drove from Buffalo to Boston, a distance of 720 km. On the return trip he increased his speed by 10 km/h. If the total trip took 17 h, what was his speed from Boston to Buffalo?

28. Kathy drove from Dry Creek to Cactus, a distance of 250 km. She increased her speed by 10 km/h for the 360 km trip from Cactus to Spur. If the total trip took 11 h, what was her speed from Dry Creek to Cactus?

29. A jet flew from Tokyo to Bangkok, a distance of 4800 km. On the return trip the speed was decreased by 200 km/h. If the difference in the times of the flights was 2 h, what was the speed from Bangkok to Tokyo?

30. If a rock is thrown down a well, the distance, d, travelled in metres after t seconds is given by the formula
$$d = 5t^2 + vt$$
where v is the initial velocity in metres per second.
(a) How deep is the well if $v = 0$ (the rock is dropped not thrown) and the splash is heard after 3 s?
(b) How long will it take until the splash is heard if $d = 200$ m and $v = 0$?
(c) How long will it take until the splash is heard if $v = 4$ m/s and $d = 300$ m?

C 31. It took a crew $2\frac{2}{3}$ h to row 6 km upstream and back again. If the rate of flow of the stream was 3 km/h, what was the rowing rate of the crew in still water?

MIND BENDER

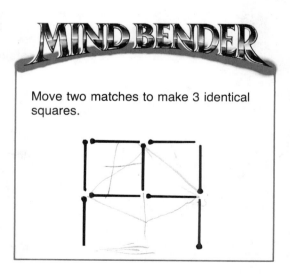

Move two matches to make 3 identical squares.

6.7 COMPLEX NUMBERS

The equation $x + 1 = 0$ has no solution in the set of whole numbers. However, it does have a solution in the set of integers. Similarly, the equation $x^2 + 1 = 0$ has no solution over the real numbers since for any real number $x^2 \geq 0$. We wish to enlarge the replacement set of x to obtain a set over which $x + 1 = 0$ does have a solution. We do this by introducing a number i, called the imaginary unit, with the property that

$$i^2 = -1$$

> The imaginary unit i is a number whose square is -1; that is,
> $$i^2 = -1$$
> or $i = \sqrt{-1}$.

The equation $x^2 + 1 = 0$ has i as a solution.

$$
\begin{aligned}
\text{L.S.} &= x^2 + 1 & \text{R.S.} &= 0 \\
&= (i)^2 + 1 \\
&= (\sqrt{-1})^2 + 1 \\
&= -1 + 1 \\
&= 0
\end{aligned}
$$

It will be necessary to combine i with real numbers so that expressions such as $4 + 3i$ and $5 - 2i$ have meaning. We want the sum of a real number and an imaginary number to be a number. This new number is called a complex number.

> A complex number is a number of the form $a + bi$, where a and b are real numbers and $i^2 = -1$.

If $b = 0$, then $a + bi = a$. In other words, a real number can also be considered to be a complex number since 5 can be written as $5 + 0i$. If $a = 0$, then $a + bi = bi$. Numbers such as 7i are called pure imaginary numbers.

Complex numbers may be summarized as follows.

Complex Numbers (C)		
$C = \{a + bi \mid a, b \text{ real numbers and } i^2 = -1\}$		
Restriction	Type	Example
$b = 0$	real	5
$a = 0$	pure imaginary	7i
$a, b \neq 0$	imaginary	$4 + 3i$

Consider the following example.

$$(2i)^2 = (2i) \times (2i) = 4i^2 = 4(-1) = -4$$

More generally, for any $x > 0$

$$(\sqrt{x}i)^2 = (\sqrt{x})^2 \times i^2 = x(-1) = -x,$$

and so we make the following definition.

> For every positive real number x, $\sqrt{-x} = i\sqrt{x}$

EXAMPLE 1. Simplify the following.
(a) $\sqrt{-81}$ (b) $\sqrt{-20}$

SOLUTION:

(a) $\sqrt{-81} = \sqrt{81} \times i$
$\qquad\quad = 9i$

(b) $\sqrt{-20} = \sqrt{20} \times i$
$\qquad\quad = 2\sqrt{5} \times i$
$\qquad\quad = 2i\sqrt{5}$

EXAMPLE 2. Simplify the following.
(a) $(3 + 2i) + (5 - 4i)$
(b) $(1 + 3i)(1 - 3i)$
(c) i^6

SOLUTION:

> We assume that the laws that hold for real numbers also apply to complex numbers.

(a) $(3 + 2i) + (5 - 4i) = (3 + 5) + (2i - 4i)$
$\qquad\qquad\qquad\qquad\quad = 8 - 2i$

(b) $(1 + 3i)(1 - 3i) = 1 - 3i + 3i - 9i^2$
$\qquad\qquad\qquad\quad = 1 - 9i^2$
$\qquad\qquad\qquad\quad = 1 - 9(-1)$
$\qquad\qquad\qquad\quad = 10$

(c) $i^6 = i^2 \times i^2 \times i^2$
$\qquad = (-1) \times (-1) \times (-1)$
$\qquad = -1$

EXAMPLE 3. Solve $x^2 - 2x + 5 = 0$, $x \in C$.

SOLUTION:
$$x^2 - 2x + 5 = 0$$
$$x = \frac{-b \pm \sqrt{b^2 - 4ac}}{2a}$$
$$x = \frac{-(-2) \pm \sqrt{(-2)^2 - 4(1)(5)}}{2(1)}$$
$$= \frac{2 \pm \sqrt{-16}}{2}$$
$$= \frac{2 \pm 4i}{2}$$
$$= 1 \pm 2i$$

$\therefore$ the roots are $1 + 2i$ and $1 - 2i$.

EXAMPLE 4. Solve $\dfrac{3}{x + 3} - \dfrac{2}{x + 2} = 1, x \in C.$

SOLUTION:

$$\frac{3}{x + 3} - \frac{2}{x + 2} = 1$$

$3(x + 2) - 2(x + 3) = (x + 3)(x + 2)$ Multiply by $(x + 3)(x + 2)$.

$3x + 6 - 2x - 6 = x^2 + 5x + 6$

$x^2 + 4x + 6 = 0$

$$x = \frac{-b \pm \sqrt{b^2 - 4ac}}{2a}$$

$$x = \frac{-(4) \pm \sqrt{4^2 - 4(1)(6)}}{2(1)}$$

$$= \frac{-4 \pm \sqrt{-8}}{2}$$

$$= \frac{-4 \pm 2i\sqrt{2}}{2}$$

$$= -2 \pm i\sqrt{2}$$

$\therefore$ the roots are $-2 + i\sqrt{2}$ and $-2 - i\sqrt{2}.$

The concept of imaginary numbers can be illustrated further by the following example.

EXAMPLE 5. Sketch the graph of $y = x^2 - 4x + 6.$

SOLUTION:

$y = x^2 - 4x + 6$

$\quad = x^2 - 4x + 4 - 4 + 6$

$\quad = (x - 2)^2 + 2$

To determine the y-intercept we let $x = 0.$

$y = x^2 - 4x + 6$

When $x = 0, y = 6.$

To determine the x-intercepts, if they exist, we let $y = 0.$

When $y = 0,$

$x^2 - 4x + 6 = 0$

$$x = \frac{4 \pm \sqrt{16 - 24}}{2}$$

$$= \frac{4 \pm \sqrt{-8}}{2}$$

$$= \frac{4 \pm 2i\sqrt{2}}{2}$$

$$= 2 \pm i\sqrt{2}$$

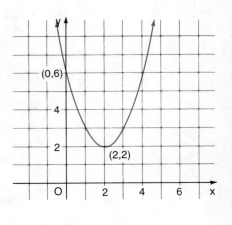

The corresponding quadratic equation has no real roots and hence the parabola has no x-intercepts.

THE ARGAND PLANE

Complex numbers can be represented geometrically in the complex plane, often called the Argand plane after Jean R. Argand who gave the representation in 1806.

The complex number $3 + 2i$ is represented by the directed line segment, or vector, from the origin to the point $(3, 2)$. The horizontal axis is the real axis, and the vertical axis is the imaginary axis. Real numbers, such as 5, are written in the form $5 + 0i$ and are represented by points on the real axis. Pure imaginary numbers such as $3i$, are written $0 + 3i$ and are represented by points on the imaginary axis.

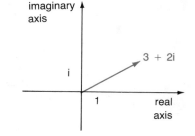

Multiplying the complex number $z = 3 + 4i$ by i,
$$(3 + 4i)i = 3i + 4i^2$$
$$= -4 + 3i$$
results in another complex number whose vector representation has the same magnitude. This new vector has been rotated 90° in a counter-clockwise direction.

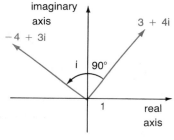

What is the result of multiplying a complex number by i^2? i^3? i^4? 2? 3? 4?

EXERCISE 6.7

A 1. Simplify.
(a) $(3 - i) + (2 + 4i)$
(b) $(2 + 5i) + (3 - 2i)$
(c) $(5 - 2i) + (3 + 4i)$
(d) $(3 + 4i) - (2 + 2i)$
(e) $(5 - 3i) - (2 + 4i)$
(f) $(3 + 4i) + (2 - 4i)$

2. Simplify the following.
(a) $\sqrt{-4}$ (b) $\sqrt{-25}$ (c) $\sqrt{-2}$
(d) $\sqrt{-100}$ (e) $\sqrt{-5}$ (f) $\sqrt{-20}$
(g) $\sqrt{-12}$ (h) $\sqrt{-18}$ (i) $\sqrt{-16}$
(j) i^2 (k) $2i^2$ (l) $(\sqrt{-3})^2$
(m) $(\sqrt{-5})^2$ (n) $\sqrt{(-5)^2}$ (o) i^3
(p) i^4 (q) $(\sqrt{-7})^2$ (r) $(5i)(2i)$

B 3. Expand and express in the form $a + bi$.
(a) $(2i - 1)(2i + 1)$ (b) $(1 - i)(1 + i)$
(c) $(1 + i)^2$ (d) $(2 + i)(3 - i)$

(e) $(i^2 - 1)^2$ (f) $(5 + 2i)(1 - i)$
(g) $3i(2i^2 - 5i + 2)$ (h) i^{100}

4. Solve for x, $x \in C$.
(a) $x^2 - 2x + 4 = 0$ (b) $x^2 - 2x + 6 = 0$
(c) $x^2 + 7 = 2x$ (d) $x^2 - 2x - 6 = 0$
(e) $2x^2 + 8x + 9 = 0$ (f) $x^2 + 9 = 0$
(g) $2x^2 + 7 = 0$ (h) $5x^2 + 2 = 2x$
(i) $0 = 3x^2 - 2x + 2$ (j) $7x^2 - 2x + 2 = 0$
(k) $2x^2 + x + 4 = 0$ (l) $3x^2 - 14x - 5 = 0$

C 5. Solve, $x \in C$.
(a) $x^2 + (x + 1)^2 + (x + 2)^2 = -1$
(b) $\dfrac{1}{x - 1} - \dfrac{2}{x - 2} = 1$
(c) $\dfrac{x - 2}{3x} = \dfrac{x - 4}{x + 2}$
(d) $\dfrac{x^2 - 2x + 1}{x^2 - 1} = \dfrac{3x - 1}{x + 2}$

6.8 THE DISCRIMINANT

Using the quadratic formula, we found the roots of $ax^2 + bx + c = 0$ to be

$$r_1 = \frac{-b + \sqrt{b^2 - 4ac}}{2a} \quad \text{and} \quad r_2 = \frac{-b - \sqrt{b^2 - 4ac}}{2a}.$$

The quantity under the radical sign, $b^2 - 4ac$, is called the discriminant of the quadratic equation. The discriminant is useful because it enables us to determine the nature of the roots of $ax^2 + bx + c = 0$ without solving the equation.

There are three possibilities for the roots of a quadratic equation.

I.   imaginary

II. real and distinct

III. real and equal

We know that the value of the discriminant must be positive, zero, or negative. We shall now consider each case.

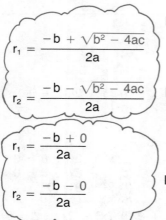

I. $b^2 - 4ac$ is negative. If $b^2 - 4ac < 0$, the expression $\pm\sqrt{b^2 - 4ac}$ represents the square root of a negative number. The equation has no real roots. The roots are imaginary.

II. $b^2 - 4ac$ is positive. If $b^2 - 4ac > 0$, the expression $\pm\sqrt{b^2 - 4ac}$ has two values, one positive and one negative. Hence, the quadratic formula will give two real numbers and there will be two real roots of the quadratic equation.

III. $b^2 - 4ac$ is zero. If $b^2 - 4ac = 0$, the expression $\pm\sqrt{b^2 - 4ac}$ has the value zero. Both roots of the quadratic equation will have the same value, $-\dfrac{b}{2a}$. The quadratic equation has equal roots.

EXAMPLE 1. Determine the nature of the roots of the following quadratic equations.
(a) $x^2 - 6x + 9 = 0$

(b) $3x^2 + 5x - 2 = 0$

(c) $2x^2 + x = -5$

SOLUTION:

(a) $b^2 - 4ac = (-6)^2 - 4(1)(9)$
$$= 36 - 36$$
$$= 0$$

∴ roots are real and equal.

(b) $b^2 - 4ac = (5)^2 - 4(3)(-2)$
$$= 25 + 24$$
$$= 49$$

∴ roots are real and distinct.

(c) $b^2 - 4ac = (1)^2 - 4(2)(5)$
$$= 1 - 40$$
$$= -39$$

∴ roots are imaginary.

EXAMPLE 2. Find the value of k such that the quadratic equation $kx^2 + (k + 8)x + 9 = 0$ has equal roots.

SOLUTION:
$$b^2 - 4ac = (k + 8)^2 - 4(k)(9)$$
$$= k^2 + 16k + 64 - 36k$$
$$= k^2 - 20k + 64$$

Since the roots are equal, the discriminant is zero.

$$k^2 - 20k + 64 = 0$$
$$(k - 4)(k - 16) = 0$$

∴ $k = 4$ or $k = 16$.

The discriminant can also be used to describe the relationship of the graph of the function $y = ax^2 + bx + c$ to the x-axis. Since the x-intercepts are found by letting $y = 0$, the discriminant determines whether the graph of $y = ax^2 + bx + c$ will intersect the x-axis

(i)   not at all;

(ii) in two distinct points;

(iii) in one point.

(i)

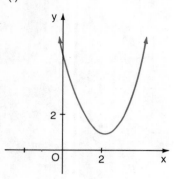

$y = x^2 - 4x + 5$
$b^2 - 4ac = -4$

(ii)

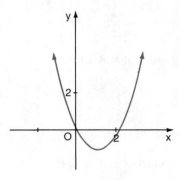

$y = x^2 - 2x$
$b^2 - 4ac = 4$

(iii)

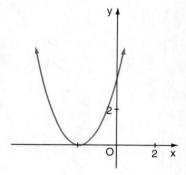

$y = x^2 + 4x + 4$
$b^2 - 4ac = 0$

The following is a summary of the three cases.

Parabola	Equation	Discriminant	Roots	Graph
$y = x^2 - 6x + 11$	$x^2 - 6x + 11 = 0$	$(-6)^2 - 4(1)(11)$ $= -8$	Imaginary	
$y = x^2 - 6x + 7$	$x^2 - 6x + 7 = 0$	$(-6)^2 - 4(1)(7)$ $= 8$	Real, Distinct	
$y = x^2 - 6x + 9$	$x^2 - 6x + 9 = 0$	$(-6)^2 - 4(1)(9)$ $= 0$	Real, Equal	

EXERCISE 6.8

A 1. If the discriminant of a quadratic equation has the given value, state the characteristics of the roots.

(a) -15 (b) 16
(c) 35 (d) -9
(e) 42 (f) 121
(g) 36 (h) 0
(i) -22 (j) 3.4
(k) -1.5 (l) 1.44

2. Assuming that a, b, and c are real numbers, how many times would the graph of $y = ax^2 + bx + c$ intersect the x-axis if

(a) the discriminant of $ax^2 + bx + c = 0$ is positive?
(b) the discriminant of $ax^2 + bx + c = 0$ is negative?
(c) the discriminant of $ax^2 + bx + c = 0$ is zero?

B 3. Find the discriminant and determine the nature of the roots of the following equations.

(a) $x^2 - 8x + 16 = 0$
(b) $x^2 - x - 5 = 0$
(c) $x^2 + 3x + 10 = 0$
(d) $x^2 + 2x + 7 = 0$
(e) $4x^2 + 9 = 12x$
(f) $x^2 - 16 = 0$
(g) $3x^2 - x + 4 = 0$
(h) $2x^2 + x = 5$
(i) $5x^2 + 7x = 0$
(j) $12x^2 - x = 6$
(k) $x^2 + 5 = 3x$
(l) $9x^2 = 5x$

4. Determine the characteristics of the roots of the following equations.

(a) $\dfrac{x^2}{2} + 4x + 4 = 0$

(b) $\frac{1}{3}x^2 + \frac{1}{2}x - 1 = 0$

(c) $\dfrac{x - 1}{2} - x^2 - 3 = 0$

(d) $(x + 1)(x - 2) = 4$
(e) $4(x^2 - 5x + 5) = -5$

(f) $\dfrac{x^2}{5} = 5$

(g) $2(x^2 - 3) = 4x$
(h) $2x^2 - \sqrt{5}x = 4$
(i) $3\sqrt{2}x^2 - x - \sqrt{2} = 0$

5. Determine the value of k that will give the indicated solution.

(a) $x^2 - 4x + k = 0$; equal roots
(b) $x^2 + 3x - 2k = 0$; imaginary roots
(c) $kx^2 - 2x + 1 = 0$; real distinct roots
(d) $3kx^2 - 3x + 1 = 0$; imaginary roots
(e) $x^2 + kx + 16 = 0$; real distinct roots
(f) $x^2 + 4kx + 1 = 0$; equal roots
(g) $(k + 1)x^2 - 2x - 3 = 0$; imaginary roots
(h) $x^2 + (k + 2)x + 2k = 0$; equal roots
(i) $x^2 + (k + 3)x + 1 = 0$; equal roots
(j) $x^2 + (k - 1)x + 1 = 0$; real distinct roots

6. Find k such that the graph of
$y = 9x^2 + 3kx + k$

(a) is tangent to the x-axis.
(b) intersects the x-axis in two points.
(c) has no intersection with the x-axis.

C 7. Under what conditions will one solution of $ax^2 + bx + c = 0$ be the reciprocal of the other?

8. Find the quadratic function whose graph passes through the points (6,0), (0,6), and (3,0).

WORD LADDER

Start with the word "four" and change one letter at a time to form a new word until you reach "five." The best solution has the fewest steps.

f o u r
_ _ _ _
_ _ _ _
_ _ _ _
_ _ _ _
_ _ _ _
f i v e

6.9 SUM AND PRODUCT OF THE ROOTS

The roots of the quadratic equation $ax^2 + bx + c = 0$ are

$$r_1 = \frac{-b + \sqrt{b^2 - 4ac}}{2a} \quad \text{and} \quad r_2 = \frac{-b - \sqrt{b^2 - 4ac}}{2a}.$$

We shall now determine expressions for $r_1 + r_2$ (sum of roots) and $r_1 \times r_2$ (product of roots) in terms of a, b, and c.

SUM

$$r_1 + r_2 = \left(\frac{-b + \sqrt{b^2 - 4ac}}{2a} \right) + \left(\frac{-b - \sqrt{b^2 - 4ac}}{2a} \right)$$

$$r_1 + r_2 = -\frac{b}{2a} + \frac{\sqrt{b^2 - 4ac}}{2a} - \frac{b}{2a} - \frac{\sqrt{b^2 - 4ac}}{2a}$$

$$= -\frac{2b}{2a}$$

$$= -\frac{b}{a}$$

PRODUCT

$$r_1 \times r_2 = \left(\frac{-b + \sqrt{b^2 - 4ac}}{2a} \right) \times \left(\frac{-b - \sqrt{b^2 - 4ac}}{2a} \right)$$

$$r_1 \times r_2 = \left(-\frac{b}{2a} + \frac{\sqrt{b^2 - 4ac}}{2a} \right) \times \left(-\frac{b}{2a} - \frac{\sqrt{b^2 - 4ac}}{2a} \right)$$

$$= \frac{b^2}{4a^2} - \frac{b^2 - 4ac}{4a^2}$$

$$= \frac{b^2 - b^2 + 4ac}{4a^2}$$

$$= \frac{c}{a}$$

$$(p + q)(p - q) = p^2 - q^2$$

If r_1 and r_2 are the roots of the quadratic equation $ax^2 + bx + c = 0$, then

$$r_1 + r_2 = -\frac{b}{a} \text{ and } r_1 \times r_2 = \frac{c}{a}.$$

EXAMPLE 1. Find the sum and product of the roots of $2x^2 - 6x - 7 = 0$.

SOLUTION:

$$r_1 + r_2 = -\frac{b}{a} \qquad\qquad r_1 \times r_2 = \frac{c}{a}$$

$$r_1 + r_2 = -\frac{(-6)}{2} \qquad\qquad r_1 \times r_2 = \frac{(-7)}{2}$$

$$= 3 \qquad\qquad\qquad = -\frac{7}{2}$$

If we divide the quadratic equation $ax^2 + bx + c = 0$ by a we have

$$x^2 + \frac{b}{a}x + \frac{c}{a} = 0$$

This equation can be written as

$$x^2 - \left(-\frac{b}{a}\right)x + \frac{c}{a} = 0$$

or

$$\boxed{x^2 - (\text{sum of roots})x + (\text{product of roots}) = 0}$$

Another way of illustrating this follows.
If the roots of a quadratic equation are r_1 and r_2 and x satisfies the equation, then

$$x = r_1 \quad \text{or} \quad x = r_2$$
$$x - r_1 = 0 \quad \text{or} \quad x - r_2 = 0$$

$$(x - r_1)(x - r_2) = 0$$
$$x^2 - r_1x - r_2x + r_1r_2 = 0$$
$$x^2 - (r_1 + r_2)x + r_1r_2 = 0$$

or $\qquad x^2 - (\text{sum of roots})x + (\text{product of roots}) = 0$

EXAMPLE 2. Determine a quadratic equation whose roots are 7 and -3.

SOLUTION:
(i) $\qquad r_1 + r_2 = 7 + (-3) = 4$
$\qquad r_1 \times r_2 = 7 \times (-3) = -21$
$\qquad x^2 - (\text{sum})x + (\text{product}) = 0$
$\qquad\quad x^2 - 4x - 21 = 0$

(ii) Since 7 and -3 are roots of a quadratic equation $(x - 7)$ and $(x + 3)$ are factors of the corresponding quadratic.
Hence
$\qquad (x - 7)(x + 3) = 0$
and
$\qquad x^2 - 4x - 21 = 0$

EXAMPLE 3. Determine a quadratic equation whose roots are the following.

(a) $2 + \sqrt{3}, 2 - \sqrt{3}$

(b) $3 - 2i, 3 + 2i$

SOLUTION:

(a) $r_1 + r_2 = (2 + \sqrt{3}) + (2 - \sqrt{3})$
$= 4$
$r_1 \times r_2 = (2 + \sqrt{3})(2 - \sqrt{3})$
$= 4 - 3$
$= 1$
$x^2 - (\text{sum})x + (\text{product}) = 0$
$x^2 - 4x + 1 = 0$

(b) $r_1 + r_2 = (3 - 2i) + (3 + 2i)$
$= 6$
$r_1 \times r_2 = (3 - 2i)(3 + 2i)$
$= 9 + 4$
$= 13$
$x^2 - (\text{sum})x + (\text{product}) = 0$
$x^2 - 6x + 13 = 0$

EXAMPLE 4. Without solving the given equation, find an equation whose roots are the reciprocals of the roots of $x^2 + 5x - 7 = 0$.

SOLUTION:

Let the roots of $x^2 + 5x - 7 = 0$ be r_1 and r_2. Then

$$r_1 + r_2 = -\frac{b}{a}$$
$$r_1 + r_2 = -5$$

and

$$r_1 \times r_2 = \frac{c}{a}$$
$$r_1 \times r_2 = -7$$

Let the roots of the new equation be R_1 and R_2.
Then $R_1 = \dfrac{1}{r_1}$ and $R_2 = \dfrac{1}{r_2}$.
The sum of the roots of the new equation is

$$R_1 + R_2 = \frac{1}{r_1} + \frac{1}{r_2}$$
$$= \frac{r_2 + r_1}{r_1 r_2}$$
$$= \frac{-5}{-7}$$
$$= \frac{5}{7}$$

The product of the roots of the new equation is

$$R_1 \times R_2 = \frac{1}{r_1} \times \frac{1}{r_2}$$
$$= \frac{1}{r_1 r_2}$$
$$= -\frac{1}{7}$$

$$x^2 - (\text{sum})x + (\text{product}) = 0$$
$$x^2 - (\tfrac{5}{7})x + (-\tfrac{1}{7}) = 0$$
$$7x^2 - 5x - 1 = 0$$

$\therefore$ the required equation is $7x^2 - 5x - 1 = 0$.

EXERCISE 6.9

A 1. State the sum and product of the roots of the following quadratic equations.
(a) $x^2 + 6x + 8 = 0$
(b) $x^2 - 3x - 4 = 0$
(c) $2z^2 - 8z + 3 = 0$
(d) $3y^2 + 4y - 3 = 0$
(e) $6t^2 - 5 = 0$
(f) $2z^2 + 7z = 0$
(g) $-3x^2 + 2x - 5 = 0$
(h) $-x^2 - 6x + 5 = 0$
(i) $3x^2 - 6x = 4$
(j) $5x^2 + 7 = 6x$
(k) $0 = 7 - 6x + 9x^2$
(l) $4x = 3x^2 - 7$

2. State a quadratic equation whose roots have the given sum and product.
(a) sum: 4; product: 3
(b) sum: -5; product: 6
(c) sum: -9; product: -7
(d) sum: 0; product: -9
(e) sum: -5; product: 0
(f) sum: -7; product: -7

B 3. Find the sum and product of the roots of the following equations.
(a) $2(x - 3) = 3x^2$
(b) $(x - 3)(x + 2) = 5$
(c) $3 - (x^2 - 2x) = 4x$
(d) $2(2x - 1)(x + 3) = x^2 + 4$

4. Write a quadratic equation having the given roots and integral coefficients.
(a) 4 and 7
(b) -6 and 3
(c) $\frac{1}{2}$ and -4
(d) $-\frac{2}{3}$ and $-\frac{1}{4}$
(e) $\frac{3}{5}$ and 0
(f) -6 and 6
(g) $1 + \sqrt{3}$ and $1 - \sqrt{3}$
(h) $3 + 2\sqrt{2}$ and $3 - 2\sqrt{2}$
(i) $2i$ and $-2i$
(j) $1 + 3i$ and $1 - 3i$
(k) $\dfrac{2 - \sqrt{5}}{2}$ and $\dfrac{2 + \sqrt{5}}{2}$
(l) $\dfrac{3 + i\sqrt{2}}{3}$ and $\dfrac{3 - i\sqrt{2}}{3}$

5. If 3 is one root of $2x^2 + bx + 3 = 0$, find the other root and the value of b.

6. Without solving the given equation, find an equation whose roots are each one less than the roots of $x^2 - 3x - 6 = 0$.

7. If -2 is one root of $5x^2 + 9x + c = 0$, find the other root and the value of c.

8. Without solving the given equation, find an equation whose roots are the negatives of the roots of $x^2 - 4x + 9 = 0$.

9. If $2 - \sqrt{2}$ is one root of $x^2 + bx + 2 = 0$, find the other root and the value of b.

10. Without solving the given equation, find an equation whose roots are the negative reciprocals of the roots of $2x^2 - 3x + 5 = 0$.

11. If $2kx^2 - (k - 6)x - 5 = 0$, find.
(a) k if the sum of the roots is 2
(b) k if the product of the roots is -7
(c) k if the sum of the roots is 0
(d) k if the reciprocal of the product of the roots is 4

12. Without solving the given equation, find an equation whose roots are the squares of the roots of $x^2 + 4x + 2 = 0$.

C 13. Show that $\dfrac{1}{r_1} + \dfrac{1}{r_2} = -\dfrac{b}{c}$.

You have 11 coins in your pocket. Their total value is more than a dollar, but you are unable to make change for a dollar. What are the values of the eleven coins?

6.10 EQUATIONS IN QUADRATIC FORM

The quadratic equation $x^4 - 5x^2 + 4 = 0$ can be solved using quadratic methods.
If we let $z = x^2$, then the equation can be rewritten as follows:

$$x^4 - 5x^2 + 4 = 0$$
$$(x^2)^2 - 5(x^2) + 4 = 0$$
$$z^2 - 5z + 4 = 0$$

This equation can be solved by factoring.

$$z^2 - 5z + 4 = 0$$
$$(z - 4)(z - 1) = 0$$
$$z = 4 \quad \text{or} \quad z = 1$$

Replace z by x^2.

$$x^2 = 4 \quad \text{or} \quad x^2 = 1$$
$$x = \pm 2 \quad \text{or} \quad x = \pm 1$$

The roots of the equations are $1, 2, -1, -2$.

We have expressed a non-quadratic equation in quadratic form. An equation is in quadratic form if it can be written as

$$a[f(x)]^2 + b[f(x)] + c = 0$$

where $a \neq 0$ and $f(x)$ is some function of x. We solve such an equation for $f(x)$ and then, if possible, solve the resulting equations for x.

EXAMPLE. Solve. $\left(x + \dfrac{2}{x}\right)^2 - 7\left(x + \dfrac{2}{x}\right) + 12 = 0$

SOLUTION:

Let $z = x + \dfrac{2}{x}$.

The given equation becomes
$$z^2 - 7z + 12 = 0$$
$$(z - 3)(z - 4) = 0$$
$$z = 3 \quad \text{or} \quad z = 4$$

Replace z by $x + \dfrac{2}{x}$.

$$x + \frac{2}{x} = 3 \quad \text{or} \quad x + \frac{2}{x} = 4$$

$$x^2 + 2 = 3x \qquad\qquad x^2 + 2 = 4x$$
$$x^2 - 3x + 2 = 0 \qquad\qquad x^2 - 4x + 2 = 0$$
$$(x - 2)(x - 1) = 0 \qquad\qquad x = \frac{4 \pm \sqrt{8}}{2}$$
$$x = 2 \quad \text{or} \quad x = 1 \qquad\qquad = \frac{4 \pm 2\sqrt{2}}{2}$$
$$= 2 \pm \sqrt{2}$$

The roots are $1, 2, 2 + \sqrt{2}, 2 - \sqrt{2}$.

EXERCISE 6.10

B 1. Solve the following equations.
(a) $x^4 - 13x^2 + 36 = 0$
(b) $x^4 + 3x^2 - 4 = 0$
(c) $\left(x + \dfrac{4}{x}\right)^2 - 9\left(x + \dfrac{4}{x}\right) + 20 = 0$
(d) $\left(x + \dfrac{1}{x}\right)^2 - 5\left(x + \dfrac{1}{x}\right) + 6 = 0$
(e) $x + 3\sqrt{x} - 10 = 0$
(f) $(x^2 + 1)^2 - 7(x^2 + 1) + 10 = 0$

2. Solve.
(a) $(x^2 - 2x)^2 - 2(x^2 - 2x) - 3 = 0$
(b) $(x^2 - 5x)^2 - 36 = 0$
(c) $(x^2 - 3x)^2 - 2(x^2 - 3x) - 8 = 0$
(d) $x^4 - 3x^2 - 4 = 0$
(e) $x^4 - 2x^2 - 24 = 0$
(f) $x^6 - 9x^3 + 8 = 0$

3. Solve.
(a) $(x^2 - 2x)^2 - 5(x^2 - 2x) = 6$
(b) $(x^2 + 2)^2 - 17(x^2 + 2) = -66$
(c) $6\left(\dfrac{1}{x - 1}\right)^2 - \left(\dfrac{1}{x - 1}\right) - 1 = 0$
(d) $2x^4 + 17x^2 - 9 = 0$
(e) $(x^2 + 1)^2 - 7(x^2 + 1) + 10 = 0$
(f) $(2x^2 + x)^2 - 7(2x^2 + x) + 6 = 0$

C 4. Solve.
(a) $x^4 - 4x^2 + 1 = 0$
(b) $x^4 - 10x^2 + 17 = 0$
(c) $\left(\dfrac{1}{2x + 1}\right)^2 + \dfrac{4}{2x + 1} + 1 = 0$
(d) $x + \dfrac{1}{x} + 2\sqrt{x + \dfrac{1}{x}} = 2$
(e) $\left(x + 1 + \dfrac{1}{x}\right)^2 + 3\left(x + 1 + \dfrac{1}{x}\right) = 1$

A UNIVERSAL MAGIC SQUARE

In a magic square, the sum of each row, each column, and each diagonal is the same. The classical magic square is a three by three square as shown below.

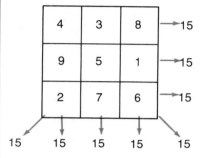

The following example is a universal four by four magic square, where you take any number as the magic sum and subtract 60 from it. This gives you a k value which can be substituted into the four spots in the square below. The result is a four by four magic square with the given magic sum.

$$k = (\text{magic sum}) - 60$$

k + 10	11	22	17
21	18	k + 9	12
15	20	13	k + 12
14	k + 11	16	19

1. Use the above model to find four by four magic squares with the following magic sums.
(a) 75   (b) 100
(c) −50   (d) 50

2. Look for other patterns in the square.

3. Why does it work?

6.11 RADICAL EQUATIONS

A radical equation is an equation in which a variable appears under a radical sign.

Before solving radical equations we shall review the operation of squaring radicals. Recall that

(i) squaring a quantity, and

(ii) determining its square root

are inverse operations, like multiplication and division. Thus, when we determine the square root of a quantity and then square the result, the quantity we obtain must be exactly the same as the one with which we started.

For example,

$$(\sqrt{5})^2 = 5$$

$$(\sqrt{x - 1})^2 = x - 1$$

EXAMPLE 1.   Square the following.

(a) $3\sqrt{x + 1}$

(b) $\sqrt{x - 1} - 3$

SOLUTION:

(a) $(3\sqrt{x + 1})^2 = 9(x + 1)$
$$= 9x + 9$$

(b) $(\sqrt{x - 1} - 3)^2 = (\sqrt{x - 1} - 3)(\sqrt{x - 1} - 3)$
$$= (\sqrt{x - 1})^2 - 6\sqrt{x - 1} + 9$$
$$= x - 1 - 6\sqrt{x - 1} + 9$$
$$= x + 8 - 6\sqrt{x - 1}$$

To solve a radical equation we use the operation that is the inverse of taking a root. However, squaring both sides of an equation does not always produce equivalent equations so all solutions must be verified.

EXAMPLE 2.   Solve. $\sqrt{x - 1} - 5 = 0$

SOLUTION:

$$\sqrt{x - 1} - 5 = 0$$
$$\sqrt{x - 1} = 5 \longleftarrow \text{Isolate the radical.}$$
$$(\sqrt{x - 1})^2 = 5^2 \longleftarrow \text{Square both sides.}$$
$$x - 1 = 25$$
$$x = 26$$

Check.

L.S. $= \sqrt{x - 1} - 5$ | R.S. $= 0$
$= \sqrt{26 - 1} - 5$ |
$= 5 - 5$ |
$= 0$ |

∴ the root is 26.

EXAMPLE 3. Solve. $x + \sqrt{x - 2} = 4$

SOLUTION:

$$x + \sqrt{x - 2} = 4$$
$$\sqrt{x - 2} = 4 - x \quad \longleftarrow \text{Isolate the radical.}$$
$$(\sqrt{x - 2})^2 = (4 - x)^2 \quad \longleftarrow \text{Square both sides.}$$
$$x - 2 = 16 - 8x + x^2$$
$$x^2 - 9x + 18 = 0$$
$$(x - 3)(x - 6) = 0$$

$x = 3$ or $x = 6$.

Check.

$x = 3$	$x = 6$
L.S. $= x + \sqrt{x - 2}$	L.S. $= x + \sqrt{x - 2}$
$= 3 + \sqrt{3 - 2}$	$= 6 + \sqrt{6 - 2}$
$= 3 + 1$	$= 6 + 2$
$= 4$	$= 8$
R.S. $= 4$	R.S. $= 4$

∴ the solution set is {3}.

Since 6 does not satisfy the original equation, it is called an extraneous root.

EXAMPLE 4. Solve. $\sqrt{4x + 5} - \sqrt{2x - 6} = 3$

SOLUTION:

$$\sqrt{4x + 5} - \sqrt{2x - 6} = 3$$

When a radical equation contains more than one radical we rewrite the equation with one radical on each side.

$$\sqrt{4x + 5} = 3 + \sqrt{2x - 6}$$
$$(\sqrt{4x + 5})^2 = (3 + \sqrt{2x - 6})^2 \quad \longleftarrow \text{Square both sides.}$$
$$4x + 5 = 9 + 6\sqrt{2x - 6} + 2x - 6$$
$$2x + 2 = 6\sqrt{2x - 6} \quad \longleftarrow \text{Isolate the radical.}$$
$$x + 1 = 3\sqrt{2x - 6} \quad \longleftarrow \text{Divide by 2.}$$
$$(x + 1)^2 = (3\sqrt{2x - 6})^2 \quad \longleftarrow \text{Square both sides.}$$
$$x^2 + 2x + 1 = 9(2x - 6)$$
$$x^2 + 2x + 1 = 18x - 54$$
$$x^2 - 16x + 55 = 0$$
$$(x - 11)(x - 5) = 0$$

$x = 11$ or $x = 5$.

Check.

$$x = 11$$
$$\text{L.S.} = \sqrt{4x + 5} - \sqrt{2x - 6}$$
$$= \sqrt{4(11) + 5} - \sqrt{2(11) - 6}$$
$$= \sqrt{49} - \sqrt{16}$$
$$= 3$$
$$\text{R.S.} = 3$$

$$x = 5$$
$$\text{L.S.} = \sqrt{4x + 5} - \sqrt{2x - 6}$$
$$= \sqrt{4(5) + 5} - \sqrt{2(5) - 6}$$
$$= \sqrt{25} - \sqrt{4}$$
$$= 3$$
$$\text{R.S.} = 3$$

$\therefore$ the solution set is $\{5, 11\}$.

EXERCISE 6.11

A 1. State the solution set of each of the following.

(a) $\sqrt{x + 1} = 3$
(b) $\sqrt{x - 1} = 4$
(c) $\sqrt{x + 2} = -3$
(d) $\sqrt{3x - 2} = 4$
(e) $\sqrt{x - 2} - 2 = 0$
(f) $\sqrt{x + 12} = 3$
(g) $\sqrt{x + 3} - 3 = 0$
(h) $\sqrt{5x + 6} = 4$
(i) $2\sqrt{x} = 3$

B 2. Solve.

(a) $2\sqrt{x - 1} - 1 = 9$
(b) $\sqrt{x + 7} - \sqrt{x} = 1$
(c) $\sqrt{x - 5} + \sqrt{x + 4} = 9$
(d) $\sqrt{x + 2} - \sqrt{x + 5} = 3$
(e) $\sqrt{3x - 5} + 1 = \sqrt{3x}$
(f) $\sqrt{x + 1} = 1 + \sqrt{x - 4}$
(g) $\sqrt{x} - \sqrt{x - 5} = 5$
(h) $\sqrt{x} - \sqrt{x - 16} = 2$

3. Solve.

(a) $\sqrt{x + 2} = x$
(b) $5\sqrt{x - 6} = x$
(c) $\sqrt{x + 4} + 8 = x$
(d) $\sqrt{x - 1} - x = -$
(e) $7 - x = -\sqrt{x - 1}$
(f) $x + \sqrt{x + 7} = 5$
(g) $x = 3 + \sqrt{x - 1}$
(h) $\sqrt{14 - 10x} + 3 = x$

4. Solve.

(a) $\sqrt{2x - 1} + \sqrt{x - 1} = 1$
(b) $\sqrt{3x - 2} - 1 = \sqrt{2x - 3}$
(c) $\sqrt{x + 1} + \sqrt{3x + 1} = 2$

(d) $\sqrt{x^2 - 4x + 3} + \sqrt{x^2 - 2x + 2} = 1$
(e) $\sqrt{3x - 2} - 2\sqrt{x} = 1$
(f) $2\sqrt{x + 6} + \sqrt{2x + 10} = 2$

C 5. Solve.

(a) $\sqrt[3]{x - 1} = 3$
(b) $\sqrt[3]{x - 2} = 2$
(c) $\sqrt{x + 1} + \dfrac{2}{\sqrt{x + 1}} = \sqrt{x + 6}$
(d) $\sqrt{x - 7} + \sqrt{x} = \dfrac{21}{\sqrt{x - 7}}$
(e) $\sqrt{x + 2} + \sqrt{x - 1} = \sqrt{4x + 1}$
(f) $\sqrt{x^2 + 4x + 4} - \sqrt{x^2 + 3x} = 1$
(g) $\sqrt{x} + \sqrt{3} = \sqrt{x + 3}$
(h) $\dfrac{1}{1 - x} + \dfrac{1}{1 + \sqrt{x}} = \dfrac{1}{1 - \sqrt{x}}$

6. Solve.

(a) $x = \sqrt{2x + 1}$
(b) $\sqrt{2x + 2} = x$

MIND BENDER

24 teams enter a basketball tournament. Each team plays until is loses. How many games are required to produce a winner?

6.12 QUADRATIC INEQUALITIES

The solution of quadratic inequalities $ax^2 + bx + c > 0$ and $ax^2 + bx + c < 0$ is best understood if we consider the graph of $y = ax^2 + bx + c$. The next two examples illustrate this.

EXAMPLE 1. Sketch the graph of $y = x^2 - 4x - 5$ and indicate where

(a) $x^2 - 4x - 5 = 0$
(b) $x^2 - 4x - 5 > 0$
(c) $x^2 - 4x - 5 < 0$

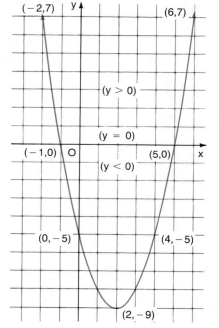

SOLUTION:
We complete the square to sketch the graph of $y = x^2 - 4x - 5$.

$$y = x^2 - 4x - 5$$
$$= x^2 - 4x + 4 - 4 - 5$$
$$= (x - 2)^2 - 9$$

(a) The function is zero $(x^2 - 4x - 5 = 0)$ when $x = 5$ or $x = -1$.
(b) The function is positive $(x^2 - 4x - 5 > 0)$ when $x > 5$ or $x < -1$.
(c) The function is negative $(x^2 - 4x - 5 < 0)$ when $x < 5$ and $x > -1$, that is, when $-1 < x < 5$.

EXAMPLE 2. Use a graph to find x for which $-x^2 + 4x - 2 \geqslant 0$.

SOLUTION:
We first sketch the graph of $y = -x^2 + 4x - 2$.

$$y = -x^2 + 4x - 2$$
$$= (x^2 - 4x + 4 - 4) - 2$$
$$= -(x - 2)^2 + 4 - 2$$
$$= -(x - 2)^2 + 2$$

When $y = 0$,

$$-x^2 + 4x - 2 = 0$$
$$x^2 - 4x + 2 = 0$$
$$x = \frac{4 \pm \sqrt{16 - 8}}{2}$$
$$= 2 \pm \sqrt{2}$$

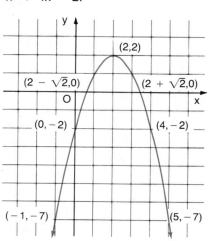

From the graph we see that $-x^2 + 4x - 2 \geq 0$, $(y \geq 0)$, for all points (x, y) which are on the parabola and also on or above the x-axis. Thus, $-x^2 + 4x - 2 \geq 0$ if $2 - \sqrt{2} \leq x \leq 2 + \sqrt{2}$.

The solution set for a quadratic inequality may also be found algebraically.

EXAMPLE 3. Solve. $x^2 - 2x - 15 > 0$

SOLUTION:
First we factor.
$$x^2 - 2x - 15 > 0$$
$$(x - 5)(x + 3) > 0$$
The product of the two factors must be positive. Hence, both factors must be positive or both factors must be negative. This means that we have two cases to consider.

CASE I—Both Positive

$$x - 5 > 0 \quad \text{and} \quad x + 3 > 0$$
$$x > 5 \quad \text{and} \quad x > -3$$
The intersection of the solution sets of these inequalities is

$$\{x \mid x > 5\}.$$

CASE II—Both Negative

$$x - 5 < 0 \quad \text{and} \quad x + 3 < 0$$
$$x < 5 \quad \text{and} \quad x < -3$$
The intersection of the solution sets of these inequalities is

$$\{x \mid x < -3\}.$$

$\therefore$ the solution set of the given inequality is $\{x \mid x > 5\} \cup \{x \mid x < -3\}$ whose graph is

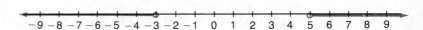

EXAMPLE 4. Solve. $2x^2 + x - 6 < 0$

SOLUTION:
$$2x^2 + x - 6 < 0$$
$$(2x - 3)(x + 2) < 0$$
The product of the two factors must be negative. Hence, one of the factors must be positive and the other one must be negative. There are two cases to consider.

CASE I

$$2x - 3 > 0 \quad \text{and} \quad x + 2 < 0$$
$$x > \tfrac{3}{2} \quad \text{and} \quad x < -2$$
There is no intersection for these inequalities.

$$2x - 3 < 0 \quad \text{and} \quad x + 2 > 0$$
$$x < \tfrac{3}{2} \quad \text{and} \quad x > -2$$

The intersection of the solution sets of these inequalities is

$$\{x \mid -2 < x < \tfrac{3}{2}\}.$$

∴ the solution set of the given inequality is $\{x \mid -2 < x < \tfrac{3}{2}\}$ whose graph is

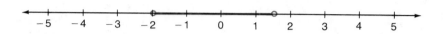

EXERCISE 6.12

B 1. Sketch the graph of $y = x^2 - 2x - 3$ and indicate where

(a) $x^2 - 2x - 3 = 0$
(b) $x^2 - 2x - 3 > 0$
(c) $x^2 - 2x - 3 < 0$

2. Sketch the graph of $y = -x^2 + 4x - 3$ and indicate where

(a) $-x^2 + 4x - 3 = 0$
(b) $-x^2 + 4x - 3 > 0$
(c) $-x^2 + 4x - 3 < 0$

3. Graph the solution set for each of the following.

(a) $x^2 + 6x + 5 > 0$
(b) $x^2 - 6x + 8 < 0$
(c) $x^2 + 2x - 15 < 0$
(d) $x^2 - 4 > 0$
(e) $x^2 + 2x > 0$
(f) $x^2 - 4x + 4 < 0$
(g) $x^2 + x - 12 \geqslant 0$
(h) $x^2 - 7x + 12 \leqslant 0$
(i) $x^2 + 9x + 20 < 0$
(j) $x^2 + 5x - 14 \geqslant 0$

4. Graph the solution set for each of the following.

(a) $x^2 + 7x > -12$
(b) $x^2 + 6 > 5x$
(c) $x^2 - 5x + 4 \geqslant 0$
(d) $x^2 - x - 12 \leqslant 0$
(e) $x^2 - 9 \leqslant 0$
(f) $x^2 - 5x \geqslant 0$
(g) $2x^2 + 5x - 3 > 0$
(h) $2x^2 - 13x + 15 \leqslant 0$
(i) $2x^2 - 5x < 12$
(j) $2x^2 - 7x + 5 > 0$

5. Graph the solution set.

(a) $2x^2 + 3x - 15 < 2x$
(b) $x^2 - 2x - 2 \geqslant 0$
(c) $x^2 - 2x - 1 > 0$
(d) $\dfrac{x^2}{2} - \tfrac{3}{2}x - 2 < 0$
(e) $x^2 - \dfrac{x - 1}{3} \leqslant 15$
(f) $x^2 + (x + 1)^2 + (x + 2)^2 > 14$
(g) $\dfrac{x^2 + 4}{2} < \dfrac{2x(x - 1)}{3}$
(h) $\dfrac{4}{x - 2} \geqslant 3$

6.13 REVIEW EXERCISE

1. State the roots of the following equations.
(a) $(x - 6)(x + 7) = 0$
(b) $(x + 5)(x + 11) = 0$
(c) $(w - 10)(w + 3) = 0$
(d) $(2x + 1)(x - 4) = 0$
(e) $(3t - 4)(2t - 5) = 0$
(f) $(5w + 7)(3w + 11) = 0$
(g) $x(3x - 4) = 0$
(h) $(5w + 12)w = 0$
(i) $(4x + 5)^2 = 0$

2. Simplify the following.
(a) $\sqrt{-9}$
(b) $\sqrt{-8}$
(c) $3i^2$
(d) i^7
(e) $(\sqrt{-6})^2$
(f) $(2i)(7i)$

3. If the discriminant of a quadratic equation has the given value, state the characteristics of the roots.
(a) 15
(b) 0
(c) -13
(d) 25
(e) 20

4. State the solution set of each of the following.
(a) $\sqrt{x + 2} = 3$
(b) $\sqrt{x - 1} = 4$
(c) $\sqrt{x + 2} = 1$
(d) $2\sqrt{x} = 5$

5. State the sum and product of the roots of the following quadratic equations.
(a) $x^2 - 5x + 5 = 0$
(b) $2x^2 + 3x - 7 = 0$
(c) $7x^2 - 6x = 0$
(d) $13x^2 + 8 = 0$

6. State a quadratic equation whose roots have the given sum and product.
(a) sum: 6; product: -2
(b) sum: -7; product: 6
(c) sum: 0; product: -5
(d) sum: -2; product: 0

7. Solve by factoring.
(a) $2x^2 - 7x - 30 = 0$
(b) $9x^2 - 4 = 0$
(c) $6t^2 - 25t + 4 = 0$
(d) $10x^2 + 27x + 18 = 0$
(e) $2w^2 - 7w + 5 = 0$
(f) $2x^2 + 16x + 30 = 0$
(g) $18t^2 + 24t - 10 = 0$
(h) $14w^2 - 17w + 5 = 0$
(i) $16x^2 + 34x - 15 = 0$

8. Solve.
(a) $5x^2 + 14x = 0$
(b) $x^2 - 5x + 3 = 0$
(c) $6t^2 + 13t + 7 = 0$
(d) $2w^2 + 5 = 0$
(e) $2(x^2 + 3) - 4x = 5$
(f) $2x^2 - 3x = -6$
(g) $(x + 1)^2 - 3(x + 1) = 0$
(h) $(2z + 5)(z + 2) = 6$

9. Solve.
(a) $\dfrac{x + 1}{2} + \dfrac{x - 1}{3} = x^2$
(b) $w^2 + w + 1 = 0$
(c) $\dfrac{2}{x - 1} + \dfrac{3}{x + 2} = 2$
(d) $x^2 + 6 = -2x$
(e) $\dfrac{x + 1}{x - 2} = \dfrac{2x + 3}{x - 3}$
(f) $2t^2 + t + 3 = 0$
(g) $2w(3w - 4) + 6 = 0$
(h) $\frac{2}{3}x^2 + x = 2$

10. Solve.
(a) $(x - 3)(x - 4) - (2x + 1)(x - 5) = 8$
(b) $x^2 + 0.1x - 0.2 = 0$
(c) $\dfrac{3x + 1}{x} - \dfrac{x}{2} = 6$
(d) $\dfrac{3}{x} - \dfrac{4}{x - 3} = 2$
(e) $0.1x^2 + 0.3x = 1$
(f) $1.2x^2 = 0.3x + 0.2$

11. The sum of the squares of two consecutive integers is 145. Find the integers.

12. The sum of the squares of three consecutive integers is 302. Find the integers.

13. The sum of a number and three times its square is 200. Find the number.

14. A rectangular building 100 m by 80 m is to be surrounded by a lawn of uniform width. The area of the lawn must be equal to the area of the building. Find the width of the lawn to the nearest tenth of a metre.

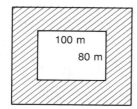

15. Julie drove from Dog's Nest to Bullet, a distance of 270 km. For the 300 km trip from Bullet to Dust she increased her speed by 10 km/h. If the total trip took 6 h, what was her speed from Bullet to Dust?

16. Use the discriminant to determine the nature of the roots of the following.

(a) $2x^2 + x - 3 = 0$
(b) $9w^2 + 4 = 12w$
(c) $2x^2 - x + 4 = 0$
(d) $3x^2 = -5x - 2$
(e) $2(x^2 - 1) + 3x = -4$
(f) $(x - 1)(x - 2) = -3x$

17. Determine the value of k that will give the indicated solution.

(a) $x^2 - 3x + k = 0$; equal roots
(b) $2x^2 + x - 2k + 1 = 0$; imaginary roots
(c) $2kx^2 + x^2 - x - 3 = 0$; real distinct roots

18. Write a quadratic equation having the given roots and integral coefficients.

(a) -5 and -8
(b) $\frac{1}{3}$ and $-\frac{1}{4}$
(c) $2 + 2\sqrt{3}$ and $2 - 2\sqrt{3}$
(d) $2 + 3i$ and $2 - 3i$

19. If 4 is one root of $3x^2 - 11x + c = 0$, find the other root and the value of c.

20. Without solving the given equation, find an equation whose roots are two more than the roots of $x^2 + 2x - 5 = 0$.

21. Solve the following equations.

(a) $x^4 - 10x^2 + 9 = 0$
(b) $(x^2 - 2x)^2 - 11(x^2 - 2x) + 24 = 0$
(c) $\left(x + \dfrac{1}{x}\right)^2 - 6\left(x + \dfrac{1}{x}\right) + 8 = 0$
(d) $(x^2 - 3x)^2 - 16 = 0$

22. Solve.

(a) $\sqrt{x + 1} + \sqrt{x - 2} = 3$
(b) $\sqrt{x - 3} = 2 - \sqrt{x + 5}$
(c) $\sqrt{x - 2} = 4 - x$
(d) $x = 10 - \sqrt{x + 2}$
(e) $\sqrt{x + 6} = 1 + \sqrt{x - 1}$
(f) $\sqrt{x - 3} + \sqrt{x} = \sqrt{2}$

23. Graph the solution set for each of the following.

(a) $x^2 - x - 12 \leq 0$
(b) $x^2 - 6x > 0$
(c) $x^2 - 3x < 10$
(d) $x^2 + 8x + 15 \geq 0$
(e) $2w^2 - 5w - 3 > 0$
(f) $3x^2 \leq 10x + 8$

6.14 CHAPTER 6 TEST

1. Solve the following equations by factoring.
(a) $x^2 - 8x + 15 = 0$
(b) $2x^2 - 7x = 4$

2. Solve the following equations.
(a) $3x^2 + x - 1 = 0$
(b) $x^2 + 2x + 2 = 0$

3. Solve. $\dfrac{1}{x-1} + \dfrac{2}{x+1} = 3$

4. The hypotenuse of a right triangle is 3 cm. The sum of the other two sides is 4 cm. Find the lengths of the other two sides of the triangle.

5. Express the complex number $(2 + 3i)(4 - i)$ in the form $a + bi$.

6. Use the discriminant to determine the nature of the roots of the following equations.
(a) $x^2 + 6x + 47 = 0$
(b) $3x^2 + 9x + 5 = 0$

7. For what values of k does the equation $2x^2 + kx + 1 = 0$ have equal roots?

8. Without solving the given equation, find an equation whose roots are each three less than the roots of $x^2 + 2x - 7 = 0$.

9. The sum of the squares of three consecutive integers is 245. Find the integers.

THE CIRCLE

CHAPTER

7

My joy, my grief, my hope, my love, did all within this circle move!
Edmund Waller

DISTANCE BETWEEN TWO POINTS

$$\sqrt{(x_2 - x_1)^2 + (y_2 - y_1)^2}$$

EXERCISE

1. Calculate the distance between each pair of points.
(a) (3, 7) and (−2, −9)
(b) (−5, −4) and (8, −1)
(c) (0, −5) and (11, −7)
(d) (−2, 8) and (−2, −9)
(e) (12, 8) and (−13, 8)
(f) (−6, −6) and (−11, −11)
(g) (0, 0) and (14, −3)

MIDPOINT OF A LINE SEGMENT

$$\left(\frac{x_1 + x_2}{2}, \frac{y_1 + y_2}{2}\right)$$

EXERCISE

1. Find the midpoint of the line segment joining the following pairs of points.
(a) (8, 6) and (2, 2)
(b) (−3, 9) and (−7, 1)
(c) (−5, −1) and (−13, −15)
(d) (16, −5) and (−1, 8)
(e) (0, 0) and (−7, −7)
(f) (−1, −4) and (20, −17)
(g) (−13, 34) and (0, −6)

SLOPE

$$m = \frac{y_2 - y_1}{x_2 - x_1}$$

EXERCISE

1. Determine the slope of the line segment joining the following pairs of points.
(a) (5, 5) and (8, 8)

(b) (3, −7) and (−5, −7)
(c) (−4, −9) and (−1, −11)
(d) (0, −12) and (0, 8)
(e) (23, 12) and (−14, 8)
(f) (−1, −2) and (−3, −4)
(g) (5, 9) and (−6, −6)

EQUATIONS OF LINES

EXERCISE

1. Determine an equation of the line passing through the point (3, 6) with slope m = 2.

2. Determine an equation of the line through (−4, −7) with slope m = −4.

3. Determine an equation of the line through (5, 6) and (7, 8).

4. Determine an equation of the line through (−3, 7) and (−5, 8).

5. Determine an equation of the line through (4, 3) and parallel to the line y = 3x − 8.

6. Determine an equation of the line through (−5, −7) and perpendicular to y = 2x + 8.

7. Determine an equation of the line through (0, 0) and parallel to the line segment with endpoints (3, 4) and (−3, 8).

8. Determine an equation of the line through (0, 0) and perpendicular to the line segment with endpoints (−1, −1) and (3, 7).

LINEAR SYSTEMS

EXERCISE

1. Solve the following systems of linear equations.

(a) 2x + y = 10
    4x − y = 8

(b) x − 2y = −17
    5x − 4y = −43

(c) 4x − 6y = −27
    x + 2y = −5

(d) 2x + 3y = −6
    3x − 5y = −9

THE PYTHAGOREAN THEOREM

In the right triangle

$$a^2 = b^2 + c^2$$

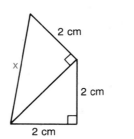

EXERCISE

Calculate the value of x in each of the figures correct to two decimal places.

1.

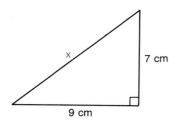

2.

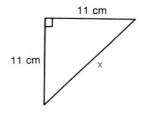

3.

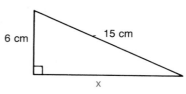

4.

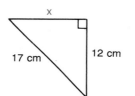

5.

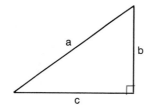

6.

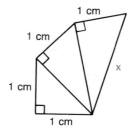

7.

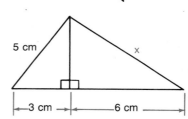

7.1 THE EQUATION OF A CIRCLE, CENTRE (0, 0)

A locus is a set of points that obey a rule or satisfy a given condition.
    In order to find the equation of a locus, we go through the following three steps:
I. Name a point, such as P(x, y), to represent a typical point in the locus.
II. State the locus in geometric form.
III. Express the law of the locus in algebraic form.
These three steps are illustrated in the following example.

EXAMPLE 1. Find an equation of the locus of a point equidistant from A(2, 5) and B(6, 9).

SOLUTION:

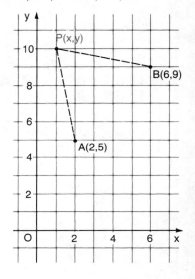

I. Let P(x, y) be any point equidistant from A(2, 5) and B(6, 9).
II. P is equidistant from A and B if, and only if,
    AP = BP.
III. Using the distance formula

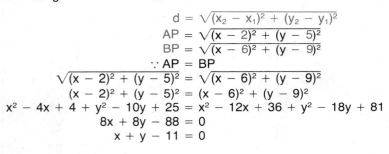

$$d = \sqrt{(x_2 - x_1)^2 + (y_2 - y_1)^2}$$
$$AP = \sqrt{(x - 2)^2 + (y - 5)^2}$$
$$BP = \sqrt{(x - 6)^2 + (y - 9)^2}$$
$$\because AP = BP$$
$$\sqrt{(x - 2)^2 + (y - 5)^2} = \sqrt{(x - 6)^2 + (y - 9)^2}$$
$$(x - 2)^2 + (y - 5)^2 = (x - 6)^2 + (y - 9)^2$$
$$x^2 - 4x + 4 + y^2 - 10y + 25 = x^2 - 12x + 36 + y^2 - 18y + 81$$
$$8x + 8y - 88 = 0$$
$$x + y - 11 = 0$$

$\therefore$ an equation of the locus is $x + y - 11 = 0$.

It turns out that the locus is the right bisector of AB.

The locus of points in the plane that are a fixed distance from a fixed point is called a circle. The fixed distance is the radius and the fixed point is called the centre.

EXAMPLE 2. Find an equation of the circle having centre O(0, 0) and radius 4.

SOLUTION:

Let P(x, y) be any point on the circle. P is on the circle if, and only if, PO = 4, that is,

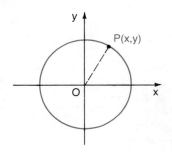

$$\sqrt{(x - 0)^2 + (y - 0)^2} = 4$$
$$(x - 0)^2 + (y - 0)^2 = 16$$
$$x^2 + y^2 = 16$$

Since these steps can be reversed, a point belongs to the circle if, and only if, its coordinates satisfy $x^2 + y^2 = 16$.
∴ an equation of the circle is $x^2 + y^2 = 16$.

EXAMPLE 3. Find an equation of the circle with centre O(0, 0) and radius r.

SOLUTION:
Let P(x, y) be any point on the circle. Then PO $= r$.

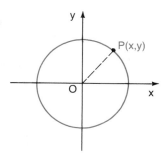

$$\sqrt{(x - 0)^2 + (y - 0)^2} = r$$
$$(x - 0)^2 + (y - 0)^2 = r^2 \quad \text{Both sides}$$
$$x^2 + y^2 = r^2 \quad \text{are positive.}$$

Conversely, if P(x, y) is any point whose coordinates satisfy

then
$$x^2 + y^2 = r^2$$
$$\sqrt{x^2 + y^2} = r$$
$$\therefore \sqrt{(x - 0)^2 + (y - 0)^2} = r$$
$$\therefore PO = r$$

and P(x, y) is a point of the circle.

> An equation of the circle with centre (0, 0) and radius r is
> $$x^2 + y^2 = r^2$$

EXERCISE 7.1

A 1. State an equation of a circle with centre (0, 0) and the given radius.

(a) 5   (b) 9   (c) 11
(d) $\sqrt{3}$ (e) $2\sqrt{5}$ (f) m

2. For each equation, give the radius of the circle.

(a) $x^2 + y^2 = 36$ (b) $x^2 + y^2 = 64$
(c) $x^2 + y^2 = 144$ (d) $x^2 + y^2 - 16 = 0$
(e) $x^2 + y^2 = 24$ (f) $x^2 + y^2 = 1$

B 3. Find an equation of the locus of each of the following.

(a) points 5 units above the x-axis
(b) points 2 units to the left of the y-axis
(c) points 5 units from the origin
(d) points 7 units from A(2, 5)
(e) points equidistant from B(0, 5) and C(0, −1)
(f) points equidistant from D(−3, 2) and E(5, −6)

4. For each of the following state an equation of the circle with centre (0, 0) and

(a) passing through (3, 4).
(b) passing through (−4, 2).
(c) having an x-intercept of 6.
(d) having a y-intercept of 2.

5. Find an equation of a circle with centre (0, 0) and passing through the point of intersection of $3x - 2y = 2$ and $4x + 3y = -20$.

6. Determine which of the following define a circle with centre (0, 0) and state the radius.

(a) $x^2 + y^2 + 16 = 0$
(b) $x^2 - 20 = y^2$
(c) $4x^2 + 4y^2 = 36$
(d) $2x^2 + 3y^2 = 24$
(e) $x^2 - y^2 = 1$
(f) $100 - x^2 - y^2 = 0$

7.2 THE GENERAL EQUATION OF A CIRCLE

In this section we use the distance formula to determine the equations of circles whose centres are not the origin.

EXAMPLE 1. Find an equation of the circle having centre C(2, 5) and radius 4.

SOLUTION:
Let P(x, y) be any point on the circle. P is on the circle if, and only if, CP = 4.

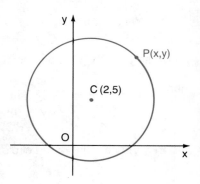

$$\sqrt{(x - 2)^2 + (y - 5)^2} = 4$$
$$(x - 2)^2 + (y - 5)^2 = 16$$

Since these steps can be reversed, a point belongs to the circle if, and only if, its coordinates satisfy $(x - 2)^2 + (y - 5)^2 = 16$.

∴ an equation of the circle is
$(x - 2)^2 + (y - 5)^2 = 16$.

We can use the method of Example 1 to find the general form of the equation of the circle having centre C(h, k) and radius r.

Let P(x, y) be any point on the circle. P is on the circle if, and only if, CP = r.

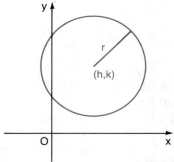

$$\sqrt{(x - h)^2 + (y - k)^2} = r$$
$$(x - h)^2 + (y - k)^2 = r^2$$

An equation of the circle with centre (h, k) and radius r is
$$(x - h)^2 + (y - k)^2 = r^2$$

EXAMPLE 2. Find the equation of the circle with centre (−3, 5) and radius 6.

SOLUTION:
Using the formula
$$(x - h)^2 + (y - k)^2 = r^2$$
$$(x - (-3))^2 + (y - 5)^2 = 6^2$$

∴ $(x + 3)^2 + (y - 5)^2 = 36$ is an equation of the circle.

EXAMPLE 3. Find the centre and radius of the circles defined by the following equations.

(a) $x^2 + y^2 - 6x = 7$

(b) $x^2 + y^2 + 4x - 10y - 7 = 0$

SOLUTION:

(a) We wish to express the equation in the form $(x - h)^2 + (y - k)^2 = r^2$.

$$x^2 + y^2 - 6x = 7$$
$$x^2 - 6x + y^2 = 7$$

Complete the square.

$$x^2 - 6x + 9 + y^2 = 7 + 9$$

Factor.

$$(x - 3)^2 + (y - 0)^2 = 16$$

∴ the centre is (3, 0) and the radius is 4.

(b) $x^2 + y^2 + 4x - 10y - 7 = 0$

$$x^2 + 4x + y^2 - 10y = 7$$

Complete the square.

$$x^2 + 4x + 4 + y^2 - 10y + 25$$
$$= 7 + 4 + 25$$

Factor.

$$(x + 2)^2 + (y - 5)^2 = 36$$

∴ the centre is (−2, 5) and the radius is 6.

EXERCISE 7.2

A 1. State the coordinates of the centre and the radius of the following circles.

(a) $x^2 + y^2 = 49$
(b) $x^2 + (y - 2)^2 = 25$
(c) $(x + 3)^2 + y^2 = 36$
(d) $(x - 5)^2 + (y + 4)^2 = 100$
(e) $2(x - 2)^2 + 2(y + 1)^2 = 32$
(f) $(x - a)^2 + (y - b)^2 = c^2$

2. State the equation of each of the following circles.

(a) centre (2, 5); radius $\sqrt{7}$
(b) centre (−2, 3); radius 5
(c) centre (−3, −4); radius 5
(d) centre (5, −2); radius 3
(e) centre (0, 3); radius 6
(f) centre (−4, 0); radius $4\sqrt{2}$
(g) centre (−3, −8); radius 10
(h) centre (a, b); radius c
(i) centre (a, −b); radius c

B 3. Find the equation of each of the following circles.

(a) centre (0, 0); passing through (−3, 4)

(b) centre (2, 5); passing through (2, 8)
(c) centre (−3, −2); passing through (−3, 8)
(d) centre (0, 5); passing through (3, −4)
(e) centre (−3, 0); passing through (0, 4)

4. Find the coordinates of the centre and the length of the radius of each of the following circles.

(a) $x^2 + y^2 + 6x - 27 = 0$
(b) $x^2 + y^2 - 4y - 5 = 0$
(c) $x^2 + y^2 + 10x - 8y + 16 = 0$
(d) $x^2 + y^2 - 2x - 2y - 3 = 0$
(e) $x^2 + y^2 - 4x + 2y - 4 = 0$
(f) $x^2 + y^2 - 10x + 12y - 3 = 0$
(g) $x^2 + y^2 + 5x - 3y + \frac{1}{2} = 0$

5. Find the equation of each of the following circles.

(a) having a diameter with endpoints A(2, −5) and B(2, 5)
(b) having a diameter with endpoints A(2, −5) and B(8, 3)

7.3 CHORDS AND TANGENTS

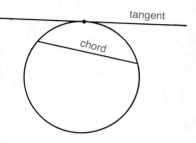

A chord is a line segment having its endpoints on the circle. A tangent is a line that intersects the circle in only one point.

In this section you will require the following formulas from your previous work.

Equations of Lines	Midpoint
$y - y_1 = m(x - x_1)$ $y = mx + b$ $y = m(x - a)$ $\dfrac{x}{a} + \dfrac{y}{b} = 1$	$\left(\dfrac{x_1 + x_2}{2}, \dfrac{y_1 + y_2}{2}\right)$
Distance Between Two Points	**Slope**
$d = \sqrt{(x_2 - x_1)^2 + (y_2 - y_1)^2}$	$m = \dfrac{y_2 - y_1}{x_2 - x_1}$

EXAMPLE. Given the circle defined by $x^2 + y^2 = 25$ and the points A(4, 3) and B(3, −4),

(a) show that AB is a chord of the circle.
(b) find the equation of the chord AB.
(c) find the equation of the line through (0, 0) and perpendicular to AB.
(d) show that the perpendicular from the centre of the circle to the chord bisects the chord.

SOLUTION:
(a) A(4, 3) lies on the circle since

$$x^2 + y^2 = (4)^2 + (3)^2$$
$$= 25$$

B(3, −4) lies on the circle since

$$x^2 + y^2 = (3)^2 + (-4)^2$$
$$= 25$$

Since A and B are on the circle, AB is a chord of the circle.
(b) The slope of AB is

$$m_{AB} = \frac{3 - (-4)}{4 - 3}$$
$$= 7$$

Use the slope, 7, and the point A(4, 3) to find the equation of AB.

$$y - y_1 = m(x - x_1)$$
$$y - 3 = 7(x - 4)$$
$$y - 3 = 7x - 28$$
$$7x - y = 25$$

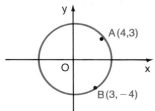

(c) The slope of lines perpendicular to AB is $-\frac{1}{7}$. Use the slope, $-\frac{1}{7}$, and the point (0, 0) to find the equation of the line through (0, 0) and perpendicular to AB.

$$y - y_1 = m(x - x_1)$$
$$y - 0 = -\frac{1}{7}(x - 0)$$
$$y = -\frac{1}{7}x$$
$$x + 7y = 0$$

(d) Solve the two equations to find the point of intersection.

$$7x - y = 25 \qquad ①$$
$$x + 7y = 0 \qquad ②$$

①
② × 7
Subtract.

$$7x - y = 25$$
$$7x + 49y = 0$$
$$-50y = 25$$
$$y = -\frac{1}{2}$$

Substitute $y = -\frac{1}{2}$ in ②.

$$x + 7\left(-\frac{1}{2}\right) = 0$$
$$x - \frac{7}{2} = 0$$
$$x = \frac{7}{2}$$

∴ the chord AB and the perpendicular from the centre of the circle intersect at $\left(\frac{7}{2}, -\frac{1}{2}\right)$.

The midpoint of AB is

$$\left(\frac{4 + 3}{2}, \frac{3 + (-4)}{2}\right) = \left(\frac{7}{2}, -\frac{1}{2}\right)$$

∴ the perpendicular from the centre of the circle to the chord AB bisects the chord.

EXERCISE 7.3

B 1. Given the circle defined by $x^2 + y^2 = 25$ and points A(3, 4) and B(−4, 3),

(a) show that AB is a chord of the circle.
(b) find the equation of the right bisector of AB.
(c) show that the centre of the circle lies on the right bisector of AB.

2. Given the circle defined by $x^2 + y^2 = 40$ and the points C(2, 6) and D(6, −2),

(a) show that CD is a chord of the circle.
(b) find the midpoint, M, of CD.
(c) find the slope of OM.
(d) find the slope of CD.
(e) show that CD ⊥ OM.

3. Given the circle $x^2 + y^2 = 16$ and the point P(5, 9) and T, the point of contact of a tangent from P,

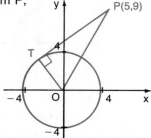

(a) show that P lies in the exterior of the given circle.
(b) find the lengths of OT and OP.
(c) find the length of the tangent PT.

4. Find the length of the tangent to the given circle from the given point.
(a) $x^2 + y^2 = 25$; P(6, 8)
(b) $x^2 + y^2 = 9$; P(7, −6)
(c) $x^2 + y^2 = 16$; P(0, 10)
(d) $x^2 + y^2 = 40$; P(5, −5)
(e) $x^2 + y^2 = 50$; P(−8, 10)
(f) $x^2 + y^2 = 1$; P(8, 1)
(g) $(x − 5)^2 + y^2 = 36$; P(8, −8)
(h) $(x + 3)^2 + (y − 2)^2 = 25$; P(8, 8)

5. Given the circle defined by $x^2 + y^2 = 25$ and the point T(−3, 4),

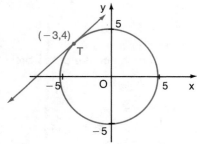

(a) show that T is a point of the circle.
(b) find the slope of OT.
(c) find the slope of the tangent at T, perpendicular to OT.
(d) find an equation of the tangent to the circle at T.

6. Find an equation of the tangent to the given circle at the given point of contact.
(a) $x^2 + y^2 = 13$; T(2, 3)
(b) $x^2 + y^2 = 34$; T(3, 5)
(c) $x^2 + y^2 = 74$; T(−5, −7)

(d) $x^2 + y^2 = 20$; T(2, −4)
(e) $x^2 + y^2 = 36$; T(0, 6)
(f) $x^2 + y^2 = 34$; T(−5, 3)
(g) $(x − 2)^2 + (y − 3)^2 = 20$; T(4, 7)
(h) $(x + 3)^2 + (y − 4)^2 = 25$; T(0, 0)

7. Given the circle defined by $x^2 + y^2 = 36$ and the points C(6, 0) and D(0, −6), show that the perpendicular from the centre of the circle to the chord CD bisects the chord.

8. Given the circle defined by $x^2 + y^2 = 25$ and the points A(0, 5) and B(−4, 3) on the circle,
(a) find the equation of the perpendicular bisector of the chord AB.
(b) show that the perpendicular bisector of the chord AB is a diameter of the circle.

C 9. Given the circle defined by $x^2 + y^2 = r^2$ and the points P(a, b) and Q(−b, a) on the circle,
(a) find the equation of the right bisector of the chord PQ.
(b) show that the right bisector of the chord PQ is a diameter of the circle.

10. Given the circle defined by $x^2 + y^2 = r^2$ and the points P(p, q) and Q(q, −p) on the circle,
(a) find the midpoint, M, of PQ.
(b) find the slope of OM.
(c) find the slope of PQ.
(d) show that OM ⊥ PQ.

11. Given the circle defined by $x^2 + y^2 = r^2$ and the point P(a, b) in the exterior of the circle, T is the point of contact on the circle of a tangent from P.
(a) Find the lengths of OT and OP.
(b) Find the length of the tangent PT and write a formula for the length of a tangent from a point.

12. Given the circle defined by $x^2 + y^2 = r^2$ and the point T(a, b) on the circle,
(a) find the slope of OT.
(b) find the slope of the tangent at T, perpendicular to OT.
(c) find an equation of the tangent to the circle at T.

7.4 LINEAR-QUADRATIC SYSTEMS

In this section we shall consider the intersection of two loci, where one locus is represented by a linear equation and the other by a quadratic. The point(s) of intersection of the loci satisfy both equations in the system.

Substitution is one method of solving a system of linear equations in two variables. This method is illustrated in the following example.

EXAMPLE 1. Find the point of intersection of the lines
$$2x + y = 7$$
$$3x - 4y = 5.$$

SOLUTION:

$$2x + y = 7 \quad ①$$
$$3x - 4y = 5 \quad ②$$

Solve for y in ①.

$$y = 7 - 2x$$

Substitute in ②.

$$3x - 4(7 - 2x) = 5$$
$$3x - 28 + 8x = 5$$
$$11x = 33$$
$$x = 3$$

Substitute.
$$y = 7 - 2x$$
$$= 7 - 2(3)$$
$$= 1$$

$\therefore x = 3$ and $y = 1$.

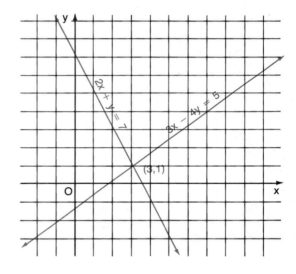

You may use the substitution method to solve a system consisting of a linear and a quadratic equation. You do this by solving the linear equation for one of the variables and then substituting into the quadratic equation.

EXAMPLE 2. Find the points of intersection of the line
$$x - y = 7 \quad ①$$
and the circle
$$x^2 + y^2 = 25 \quad ②$$
and illustrate with a graph.

SOLUTION:
To do this we must solve the linear quadratic system.
$$x + y = 7 \quad ①$$
$$x^2 + y^2 = 25 \quad ②$$
From the linear equation ①
$$y = 7 - x$$

Substitute in the quadratic equation ②.

$$x^2 + (7 - x)^2 = 25$$
$$x^2 + 49 - 14x + x^2 = 25$$
$$2x^2 - 14x + 24 = 0$$
$$x^2 - 7x + 12 = 0$$
$$(x - 3)(x - 4) = 0$$
$$x - 3 = 0 \quad \text{or} \quad x - 4 = 0$$
$$x = 3 \qquad\qquad x = 4$$

Substituting in the linear equation,

$x = 3$	$x = 4$
$y = 7 - 3$	$y = 7 - 4$
$= 4$	$= 3$
$\therefore (x, y) = (3, 4)$	$\therefore (x, y) = (4, 3)$

∴ there are two solutions
(3, 4) and (4, 3).

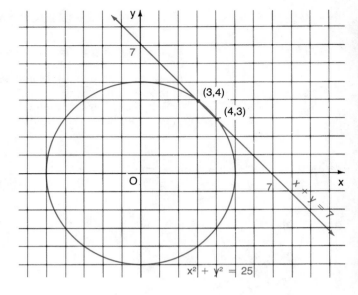

Steps in Solving a Linear-Quadratic System

I. Solve the linear equation for one of the variables.

II. Substitute in the quadratic equation.

III. Solve the quadratic equation.

IV. Substitute all values obtained in III. into the linear equation to find all solutions.

EXERCISE 7.4

B Solve and illustrate with a diagram.

1. $y = x^2 + x, \ y = x + 1$

2. $y = x^2 - 3, \ y = x - 1$

3. $y = x^2 + 4x - 5, \ y = x - 7$

4. $y = x^2 + 3x - 5, \ y = 4x + 1$

5. $y = x^2 - x - 2, \ y = x - 3$

6. $x^2 + y^2 = 25, \ x + y = 5$

7. $x^2 + y^2 = 25, \ y = 3x - 5$

8. $2x + 3y = 6, \ 4x^2 + 9y^2 = 36$

9. $y = x^2, \ y = 2x + 3$

10. $x^2 + y^2 = 25, \ 3x - 4y = 25$

11. $y^2 = 4x, \ x - 2y = -3$

12. $y^2 + 2x = 16, \ x + y = 9$

13. $x^2 - y^2 = 8, \ y = x + 2$

14. Find the value(s) of k making $y = 4x + k$ tangent to the circle $x^2 + y^2 = 25$.

15. Find the value of k making $x + y = k$ a tangent to the parabola $y^2 = 4x$.

7.5 QUADRATIC-QUADRATIC SYSTEMS

In this section we shall find the points of intersection of two quadratic equations.

EXAMPLE. Find the points of intersection of the circle
$$x^2 + y^2 = 25 \qquad ①$$
and the parabola
$$x^2 + 4y = 25 \qquad ②.$$

SOLUTION:
From equation ②

$$x^2 = -4y + 25$$

Substitute $x^2 = -4y + 25$ into equation ①.

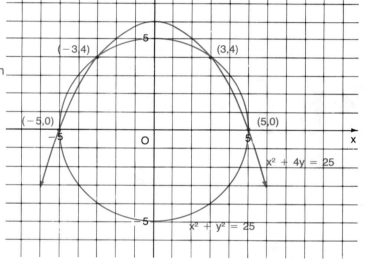

$$(-4y + 25) + y^2 = 25$$
$$y^2 - 4y = 0$$
$$y(y - 4) = 0$$
$$y = 0 \quad \text{or} \quad y - 4 = 0$$
$$y = 4$$

Substituting $y = 0$ and $y = 4$ in turn into equation ②,

$$y = 0$$
$$x^2 + 4(0) = 25$$
$$x^2 = 25$$
$$x = \pm 5$$

$$y = 4$$
$$x^2 + 4(4) = 26$$
$$x^2 = 9$$
$$x = \pm 3$$

∴ the solution set is
$\{(-5, 0), (5, 0), (-3, 4), (3, 4)\}$.

EXERCISE 7.5

B Solve the following systems and illustrate with a diagram.

1. $x^2 + y^2 = 25$, $4x + y^2 = 25$

2. $x^2 + y^2 = 1$, $y = x^2 - 1$

3. $x^2 + y^2 = 25$, $xy = 12$

4. $4x^2 + y^2 = 17$, $xy = 2$

5. $x^2 + y^2 = 10$, $9x^2 + y^2 = 18$

6. $x^2 + y^2 = 6$, $y = x^2$

7. $y = x^2$, $y = -x^2 + 2$

8. $y = 5 - x^2$, $x^2 + y^2 = 25$

9. $x^2 + y^2 = 9$, $\dfrac{x^2}{9} + \dfrac{y^2}{6} = 1$

10. $ax^2 + by^2 = 1$, $bx^2 - ay^2 = 1$

7.6 PROBLEM SOLVING

In this section we discuss the procedure called "proof by contradiction." When we argue by contradiction, we assume that the conclusion is not true and eventually something contradicts either what is given, the indirect method, or what is known to be true, reductio ad absurdum. For instance, in proving that $\sqrt{2}$ was irrational in Section 1.2, we first assumed that $\sqrt{2}$ was rational and proceeded to derive a contradiction.

The following example is a famous proof about prime numbers from Euclid's *Elements*. Recall that a prime is an integer whose only factors are itself and 1. Thus, the numbers

$$2, 3, 5, 7, 11, 13, 17, 19, 23, 29, 31,...$$

are primes because they cannot be factored as products of smaller numbers.

READ

EXAMPLE. Prove that there are infinitely many primes.

SOLUTION:

PLAN

Using the method of proof by contradiction, we assume the statement is false, that is, we assume that there is only a finite number of primes,

$$2, 3, 5, 7, 11, 13, ..., P$$

SOLVE

where P is the supposed largest prime. To get a contradiction, we consider the number N obtained by multiplying all these primes together and adding 1.

$$N = (2 \times 3 \times 5 \times 7 \times ... \times P) + 1$$

ANSWER

Observe that if N is divided by any of the primes 2, 3, ..., P, it leaves the remainder 1. Therefore, N is not divisible by any prime number. This means that N has no factors other than itself and 1. In other words, N is a prime number. But N is obviously larger than P. This contradicts our assumption that P is the largest prime. Since we have arrived at a contradiction, our original assumption must be wrong. Therefore, there must be an infinite number of primes.

EXERCISE 7.6

B 1. Two hundred people are at a party and among any set of 4 people there is at least one person who knows the other 3. There are 3 people who don't know each other. Prove that the other 197 people know everyone at the party. (It is assumed that if A knows B, then B knows A.)

2. Prove that $\sqrt{3}$ is an irrational number.

3. Is it possible to write numbers using each of the ten digits exactly once so that the sum of the numbers is equal to 100?

7.7 REVIEW EXERCISE

1. Determine the coordinates of the centre and the radius of the following circles.
(a) $x^2 + y^2 = 49$
(b) $x^2 + y^2 = 64$
(c) $(x - 2)^2 + (y - 1)^2 = 16$
(d) $(x + 3)^2 + (y + 5)^2 = 9$
(e) $(x - 7)^2 + (y + 1)^2 = 25$
(f) $(x - 2)^2 + y^2 = 81$
(g) $x^2 + (y + 5)^2 = 100$

2. Write an equation for the following circles.
(a) centre (0, 0); radius 4
(b) centre (0, 0); radius $\sqrt{5}$
(c) centre $(-3, 5)$; radius 7
(d) centre $(4, -2)$; radius $\sqrt{10}$
(e) centre (0, 0); passing through (2, 7)
(f) centre (3, 3); passing through (0, 0)

3. Find the equation of each of the following circles.
(a) centre $(-3, 4)$; passing through (5, 2)
(b) centre (0, 3); passing through $(-1, -6)$
(c) centre $(-4, 0)$; passing through $(-5, 4)$
(d) centre $(-3, -3)$; passing through (0, 0)

4. Find the coordinates of the centre and the radius of the following circles.
(a) $x^2 + y^2 + 8x - 6y - 11 = 0$
(b) $x^2 + y^2 - 2x + 8y + 13 = 0$
(c) $x^2 + y^2 - 6x - 8y = 0$

5. Given the circle defined by $x^2 + y^2 = 20$ and the points $A(-2, 4)$ and $B(4, 2)$,
(a) show that AB is a chord of the circle.
(b) find an equation of the chord AB.
(c) find an equation of the line through (0, 0) and perpendicular to AB.
(d) show that the perpendicular from the centre of the circle to the chord bisects the chord.

6. Find the length of the tangent to the given circle from the given point.
(a) $x^2 + y^2 = 16$; P(5, 4)
(b) $x^2 + y^2 = 25$; P$(-3, -7)$
(c) $x^2 + y^2 = 1$; P$(-5, 4)$

7. Find an equation of the tangent to the given circle at the given point.
(a) $x^2 + y^2 = 26$; P(1, 5)
(b) $x^2 + y^2 = 45$; T$(-3, 6)$

8. Solve and illustrate with a diagram.
(a) $y = x^2 + 2x$, $y = x + 2$
(b) $x^2 + y^2 = 16$, $x + y = 4$
(c) $x^2 + y^2 = 25$, $3x + 4y = 25$
(d) $x^2 + 2y^2 = 9$, $x - y - 3 = 0$

9. Solve the following systems and illustrate with a diagram.
(a) $y = x^2$, $x^2 + y^2 = 20$
(b) $x^2 + y^2 = 20$, $4x^2 + 5y^2 = 80$

10. Find the equation of each of the following circles.
(a) centre $(2, -5)$ and tangent to the x-axis
(b) centre $(-4, 2)$ and tangent to the y-axis
(c) having a diameter with endpoints $(4, -7)$ and $(-2, 9)$

11. The sum of two numbers is 22 and the sum of their squares is 250. Find the numbers.

In the following multiplication, all of the digits from 0 to 9 have been used once. Complete the multiplication.

7.8 CHAPTER 7 TEST

1. Determine the coordinates of the centre and the radius of the following circles.
(a) $x^2 + y^2 = 81$
(b) $(x + 4)^2 + (x - 5)^2 = 36$

2. Write an equation for the following circles.
(a) centre (0, 0); radius 3
(b) centre (-2, 1); radius 6

3. Find the coordinates of the centre and the radius of the circle defined by the following equation.

$$x^2 + y^2 - 8x + 4y - 11 = 0$$

4. Find the length of the tangent from P(-5, 4) to the circle $x^2 + y^2 = 16$.

5. Find an equation of the tangent to the circle defined by $x^2 + y^2 = 169$ at the point (-5, 12).

6. Solve.
$$x^2 + y^2 = 25$$
$$4x - 3y = 0$$

7. Solve.
$$x^2 + y^2 = 25$$
$$y = x^2 - 5$$

FUNCTIONS
AND
RELATIONS

CHAPTER

The flower of modern mathematical thought... the notion of a function.

Thomas J. McCormack

FUNCTIONS

EXERCISE

1. If $f(x) = x^3 - 2x^2 + x - 4$, find.
(a) $f(0)$ (b) $f(2)$
(c) $f(-1)$ (d) $f(\frac{1}{2})$
(e) $f(0.1)$ (f) $f(a)$
(g) $f(2x)$ (h) $f(-x)$

2. If $g(x) = 4x - x^4$, find.
(a) $g(1)$ (b) $g(-2)$
(c) $g(-3)$ (d) $g(\frac{1}{2})$
(e) $g(\sqrt{2})$ (f) $g(0.2)$
(g) $g(-x)$ (h) $g(2x)$

3. From each of the given graphs, state.
(a) $f(1)$ (b) $f(-1)$
(c) $f(0)$ (d) $f(2)$
(e) the domain of f   (f) the range of f

(i)

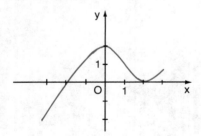

(ii)

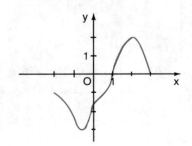

LINEAR FUNCTIONS

EXERCISE

1. Find the equation of the form $y = mx + b$ for the line
(a) with slope 2 and y-intercept 5.
(b) with slope $-\frac{1}{2}$ and y-intercept -6.
(c) with slope -3, through $(1,2)$.
(d) with slope 4, through $(1,\frac{1}{2})$.
(e) through $(6,2)$ and $(-1,-1)$.
(f) through $(1,-2)$ and $(6,3)$.

2. Draw the graphs of the following linear functions.
(a) $f(x) = x + 1$
(b) $F(x) = 1 - x$
(c) $g(x) = 2x - 3$
(d) $G(x) = 1 + \frac{1}{2}x$
(e) $h(x) = 2 + 3x$
(f) $H(x) = -\frac{1}{3}(x + 2)$

QUADRATIC FUNCTIONS

EXERCISE

1. Write each of the following quadratic functions in standard form and then sketch the graph. State the domain and range of each function.
(a) $y = x^2 + 2$
(b) $y = 1 - x^2$
(c) $f(x) = x^2 + 2x$
(d) $f(x) = x - x^2$
(e) $g(x) = x^2 - 8x + 4$
(f) $g(x) = x^2 + 5x + 10$
(g) $h(x) = -2x^2 + 6x + 7$
(h) $h(x) = 3x^2 + 6x + 8$

TRANSFORMATIONS

EXERCISE

1. Find the images of the point (2,6) under the following transformations.

(a) $(x,y) \rightarrow (x + 3,y)$,
   horizontal translation

(b) $(x,y) \rightarrow (x,y + 4)$,
   vertical translation

(c) $(x,y) \rightarrow (x,2y)$,
   vertical stretch

(d) $(x,y) \rightarrow (x,\frac{1}{4}y)$,
   vertical shrink

(e) $(x,y) \rightarrow (x,-y)$,
   reflection in x-axis

(f) $(x,y) \rightarrow (-x,y)$,
   reflection in y-axis

(g) $(x,y) \rightarrow (y,x)$,
   reflection in the line
   $y = x$

2. Find the images of the point $(4,-3)$ under the following transformations.

(a) $(x,y) \rightarrow (x - 5,y)$
(b) $(x,y) \rightarrow (x,y + 8)$
(c) $(x,y) \rightarrow (x,3y)$
(d) $(x,y) \rightarrow (x,\frac{1}{2}y)$
(e) $(x,y) \rightarrow (x,-y)$
(f) $(x,y) \rightarrow (-x,y)$
(g) $(x,y) \rightarrow (y,x)$

3. Find the images of the points
(a) (3,6)
(b) $(-5,2)$
under the following transformations.
(i)   shift 6 units upward
(ii)   shift 3 units downward
(iii) shift 2 units upward and 3 units to the right
(iv)   shift 5 units to the left
(v)   vertical stretch by a factor of 3
(vi)   vertical shrink by a factor of $\frac{1}{2}$

(vii) reflection in the y-axis
(viii) reflection in the line $y = x$
(ix)   reflection in the x-axis
(x)   reflection in the x-axis and a vertical stretch by a factor of 2

4. Draw the triangle with vertices (0,0), (3,3), (4,1), and show how it is transformed by the following transformations.

(a) reflection in the y-axis
(b) upward translation by 3 units
(c) vertical shrink by a factor of $\frac{1}{3}$
(d) translation to the left by 6 units
(e) reflection in the x-axis
(f) downward translation by 4 units
(g) vertical stretch by a factor of 2
(h) reflection in the line $y = x$
(i) shift to the right 5 units, up 3 units, shrink by a factor of $\frac{1}{2}$, and reflect in the x-axis

5. Draw a quadrilateral with vertices (0,0), (0,2), $(-4,4)$, $(-6,-2)$, and show how it is transformed by the following transformations.

(a) translation to the left 3 units and down 8 units
(b) vertical shrink by a factor of $\frac{1}{2}$
(c) vertical stretch by a factor of 2
(d) reflection in the y-axis
(e) reflection in the x-axis
(f) reflection in the line $y = x$
(g) translation to the right 2 units, followed by reflection in the y-axis

Put the numbers from 1 to 9 in the spaces to make the statements true.

$$\blacksquare \div \blacksquare + \blacksquare = 9$$
$$\blacksquare \times \blacksquare + \blacksquare = 9$$
$$\blacksquare \times \blacksquare - \blacksquare = 9$$

8.1 TYPES OF FUNCTIONS

Functions were defined and studied in Sections 5.1 and 5.2. In this section we look at some of the types of functions that occur most frequently.

I. LINEAR FUNCTIONS

In Chapter 3, we studied functions f defined by an equation of the form

$$f(x) = mx + b$$

Such a function is called a linear function because its graph is the straight line $y = mx + b$ with slope m and y-intercept b. For instance, the graph of the linear function $f(x) = 2x - 1$ is shown in the figure below.

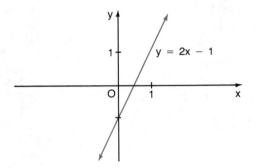

A special case of the linear function occurs when the slope is $m = 0$. The function given by $f(x) = b$, where b is a given number, is called a constant function because all its values are the same number, namely b. Its graph is a horizontal line. For example, the graph of the constant function $f(x) = 2$ is shown in the figure below.

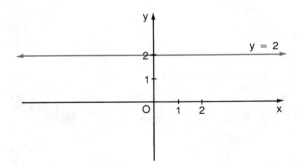

II. QUADRATIC FUNCTIONS

In Chapter 5, we studied quadratic functions which are functions defined by an equation of the form

$$f(x) = ax^2 + bx + c \qquad (a \neq 0)$$

We found that the graph is a parabola which opens upward if $a > 0$ and downward if $a < 0$. Illustrations are given in the figures below.

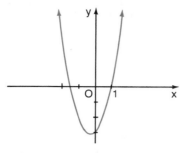

Graph of $f(x) = 2x^2 + x - 3$

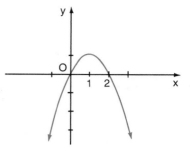

Graph of $f(x) = 2x - x^2$

III. POLYNOMIAL FUNCTIONS

A function P is called a polynomial function if it has the form

$$P(x) = a_n x^n + a_{n-1} n^{n-1} + \ldots + a_2 x^2 + a_1 x + a_0$$

where $a_n, a_{n-1}, \ldots, a_2, a_1, a_0$ are real numbers. If $a_n \neq 0$, then P has degree n. For instance, a polynomial function of degree 1 is a linear function and a polynomial function of degree 2 is a quadratic function. The function

$$P(x) = 5x^3 - 4x^2 + x - 2$$

is a polynomial function of degree 3, also called a cubic function, and

$$Q(x) = 2x^6 - x^4 + \tfrac{3}{5}x^2 + x - \sqrt{3}$$

has degree 6.

EXAMPLE 1. Sketch the graph of the function $f(x) = x^3$.

SOLUTION:

The given function is a polynomial of degree 3. We graph it as usual by plotting some representative points and joining them as follows.

x	y
0	0
$\frac{1}{2}$	$\frac{1}{8}$
1	1
2	8
$-\frac{1}{2}$	$-\frac{1}{8}$
-1	-1
-2	-8

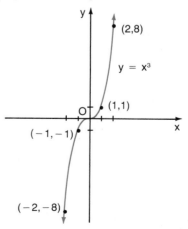

Notice from the graph that the domain and range of f are both R.

IV. RATIONAL FUNCTIONS

A rational function is a ratio of polynomials. For instance, the functions f, g, and h defined by

$$f(x) = \frac{x + 1}{x^2 + 1}, \qquad g(x) = \frac{2}{x - 3}, \qquad h(x) = \frac{2x^4 - 5x}{x^3 + 3x^2 + 2}$$

are all rational functions. In determining the domain of a rational function we use the following convention.

> If a formula is given for f(x) but no domain is given, then the domain of f is assumed to be the set of all values of x for which the given expression for f(x) is meaningful.

EXAMPLE 2. Find the domain of g if $g(x) = \dfrac{2}{x - 3}$.

SOLUTION:
The expression $\dfrac{2}{x - 3}$ makes sense except when the denominator is zero; $x - 3 = 0$ when $x = 3$. Therefore, the domain of g consists of all real numbers except for 3. In set notation we write the domain as $\{\, x \in R \mid x \neq 3 \,\}$.

V. ROOT FUNCTIONS

If n is a positive integer, the function f given by $f(x) = \sqrt[n]{x}$ is called a root function.

EXAMPLE 3. Find the domain and sketch the graph of the square root function $f(x) = \sqrt{x}$.

SOLUTION:
We recall from Chapter 1 that the expression $\sqrt{x}$ is only defined when $x \geq 0$. Thus, the domain of f is $\{\, x \mid x \geq 0 \,\}$. To graph f we plot points whose y-coordinates are found using a calculator.

x	y
0	0
0.5	0.71
1	1
2	1.41
3	1.73
4	2
5	2.24

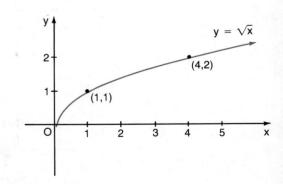

VI. FUNCTIONS WITH SEVERAL DEFINING EQUATIONS

Some functions cannot be described by a single simple formula. The next two examples illustrate functions which are described by different formulas in different parts of their domains.

EXAMPLE 4. A function f is defined by

$$f(x) = \begin{cases} 0 & \text{if } x \leq 0 \\ x & \text{if } 0 < x \leq 1 \\ 2 - x & \text{if } 1 < x \leq 2 \\ 0 & \text{if } x > 2 \end{cases}$$

Evaluate f(0.5), f(1.8), f(2.4), and sketch the graph of f.

SOLUTION:

Remember that a function is a rule. For this function the rule says: "First look at the value of the input x. If $x \leq 0$ or $x > 2$, then the value of f(x) is 0. If $0 < x \leq 1$, then the value of f(x) is x. If $1 < x \leq 2$, then the value of f(x) is $2 - x$."

Since $0 < 0.5 \leq 1$, we have f(0.5) = 0.5.
Since $1 < 1.8 \leq 2$, we have f(1.8) = 2 − 1.8 = 0.2.
Since $2.4 > 2$, we have f(2.4) = 0.

To draw the graph of f, notice that if $0 < x \leq 1$ then f(x) = x. Therefore, the part of the graph of f between x = 0 and x = 1 coincides with the line y = x which has slope 1 and passes through the origin. If $1 < x \leq 2$, then f(x) = 2 − x; therefore, between x = 1 and x = 2 the graph of f coincides with the line y = 2 − x, which has slope −1 and joins the points (1,1) and (2,0). To the left of 0 and to the right of 2, the graph of f coincides with y = 0 which is the x-axis.

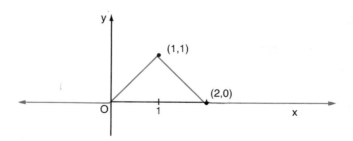

This function f is called piecewise linear because the pieces of its graph are line segments.

EXAMPLE 5. Postage rates for first-class letters in the country of Urbania are stated as follows: "Ten cents for the first 25 g or less, five cents for each additional 25 g or fraction thereof." If x is the mass of a letter in grams, and y = f(x) is the cost of postage, draw a graph of f.

SOLUTION:

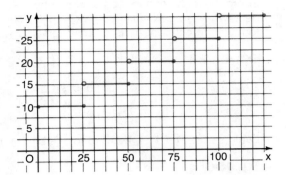

$$f(x) = \begin{cases} 10 & \text{if } 0 \leq x \leq 25 \\ 15 & \text{if } 25 < x \leq 50 \\ 20 & \text{if } 50 < x \leq 75 \\ \text{and so on.} \end{cases}$$

This is an example of a step function.

EXERCISE 8.1

A 1. State whether the given function is constant, linear, quadratic, polynomial, rational, or a root function.

(a) $f(x) = x^2 + 2x - 2$
(b) $g(x) = -1$
(c) $h(x) = \sqrt[4]{x}$
(d) $F(x) = 17x + 6$
(e) $G(x) = \dfrac{1 + x + x^2}{1 - x + x^2}$
(f) $H(t) = 12t^4 - t + 37$
(g) $R(t) = 1 - 2t$
(h) $S(x) = \sqrt{2}$
(i) $y = 1 + x^5$
(j) $y = \dfrac{x + 3}{2x + 4}$

(e) $f(x) = \dfrac{3}{x(x - 1)}$
(f) $f(x) = \dfrac{x}{(x - 4)(x + 3)}$
(g) $g(x) = \dfrac{4}{x^2 + 6x + 5}$
(h) $g(x) = \dfrac{x^2}{x^2 - 1}$
(i) $h(x) = \dfrac{x^2}{x^2 + 1}$
(j) $h(x) = \dfrac{x^2 + x + 1}{x^2 - 3x + 2}$
(k) $y = \sqrt[3]{x}$
(l) $y = \sqrt[4]{x}$
(m) $y = \sqrt{x - 1}$
(n) $y = \dfrac{1}{\sqrt{x + 1}}$

B 2. Find the domains of the following functions.

(a) $f(x) = 2x + 7$
(b) $f(x) = x^4 - x^2 + 1$
(c) $f(x) = \dfrac{1}{x - 4}$
(d) $f(x) = \dfrac{x - 2}{x + 2}$

3. Sketch the graphs of the following functions.

(a) $f(x) = 4 - 3x, \ -1 \leq x \leq 6$
(b) $f(x) = -1$
(c) $f(x) = \pi$
(d) $f(x) = 2x - 1, \ -2 \leq x \leq 2$
(e) $f(x) = x^3 - 3x^2$

(f) $f(x) = x^4$
(g) $y = 2x^2 - x^4$
(h) $y = x^5$
(i) $y = \sqrt[3]{x}$
(j) $y = \sqrt[4]{x}$
(k) $y = \sqrt{x - 1}$
(l) $y = \sqrt{2 - x}$

4. Sketch the graphs of the following functions.

(a) $f(x) = \begin{cases} 1 & \text{if } x < 2 \\ 3 & \text{if } x \geqslant 2 \end{cases}$

(b) $f(x) = \begin{cases} -1 & \text{if } 0 \leqslant x < 1 \\ 1 & \text{if } 1 < x \leqslant 2 \end{cases}$

(c) $f(x) = \begin{cases} 1 - x & \text{if } x \leqslant 1 \\ 2x - 2 & \text{if } x > 1 \end{cases}$

(d) $f(x) = \begin{cases} 2x + 1 & \text{if } -1 \leqslant x \leqslant 0 \\ x + 1 & \text{if } 0 < x \leqslant 1 \end{cases}$

(e) $f(x) = \begin{cases} -1 & \text{if } x \leqslant -1 \\ x & \text{if } -1 < x < 1 \\ 1 & \text{if } x \geqslant 1 \end{cases}$

(f) $f(x) = \begin{cases} x & \text{if } 0 \leqslant x < 1 \\ x - 1 & \text{if } 1 \leqslant x < 2 \\ x - 2 & \text{if } 2 \leqslant x \leqslant 3 \end{cases}$

5. A long distance call costs \$2.50 for the first 3 min and \$0.30 for each additional minute (or part of a minute).
(a) Draw a graph of cost versus time.
(b) Determine the equations that define this function.
(c) How much will it cost to talk for 7 min?
(d) How long did you talk if the cost was \$4.60?

6. A taxi company charges \$2.00 for the first 0.4 km (or part) and \$0.25 for each additional 0.2 km (or part).
(a) Draw the graph of cost as a function of distance.
(b) Determine the equations that define this function.
(c) How much will it cost to go 15 km by taxi?

C 7. Find the domains of the following functions.

(a) $f(x) = \dfrac{1}{8x^3 + 6x^2 - 9x}$

(b) $f(x) = \dfrac{x^2 + 1}{x^2 + x - 1}$

(c) $f(x) = \sqrt{x^2 - 2x}$

(d) $f(x) = \sqrt[4]{2 - x - x^2}$

8. Sketch the graph of the function f defined by

$$f(x) = \begin{cases} -x - 2 & \text{if } x \leqslant -2 \\ x + 2 & \text{if } -2 < x < -1 \\ x^2 & \text{if } -1 \leqslant x \leqslant 1 \\ 2 - x & \text{if } 1 < x < 2 \\ x - 2 & \text{if } x \geqslant 2 \end{cases}$$

9. Sketch the graph of the function g defined by

$$g(x) = \begin{cases} -5 & \text{if } x < -3 \\ 4 - x^2 & \text{if } -3 \leqslant x \leqslant 2 \\ 2x - 4 & \text{if } 2 < x \leqslant 4 \\ 4 & \text{if } x > 4 \end{cases}$$

10. Let [x] be the largest integer which is less than or equal to x, $x \in R$. For example, $[2.6] = 2$, $[2] = 2$, $[\pi] = 3$, $[-2.6] = -3$.
(a) Draw the graph of the function f given by $f(x) = [x]$, $-5 \leqslant x \leqslant 5$.
(b) Is f continuous?

WORD LADDER

Start with the word "eye" and change one letter at a time to form a new word until you reach "lid." The best solution has the fewest steps.

e y e
_ _ _
_ _ _
l i d

8.2 SYMMETRY

The graphs of f(x) = x² and f(x) = x³ in the figures below, are both symmetric. The one on the left is symmetric in the y-axis, that is, the graph is mapped onto itself when reflected in the y-axis. The one on the right is symmetric about the origin, that is, the graph is mapped onto itself when rotated through 180° about the origin. Notice that in either case we need only plot the graph of the function for x ⩾ 0. The rest of the graph is then obtained by symmetry. Therefore, our work is cut in half.

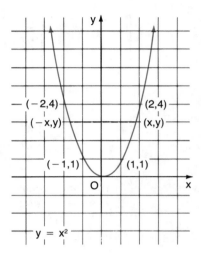
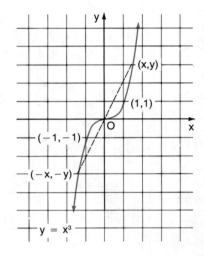

(x,y)→(−x,y) for reflection in the y-axis.

(x,y)→(−x,−y) for rotation of 180° about the origin.

Notice that if the graph of f is symmetric in the y-axis and (x,y) is on the graph, then so is (−x,y). Thus, f(x) = f(−x). Functions which satisfy this condition are called even functions.

Likewise if the graph of f is symmetric about the origin and (x,y) is on the graph, then so is (−x,−y). Thus, f(−x) = −f(x). Functions which satisfy this condition are called odd functions.

An even function f satisfies f(−x) = f(x) for all x in its domain. The graph of an even function is symmetric in the y-axis.
An odd function f satisfies f(−x) = −f(x) for all x in its domain. The graph of an odd function is symmetric about the origin.

EXAMPLE. Determine whether each of the following functions is even, odd, or neither.
(a) f(x) = x⁵   (b) g(x) = 1 − x²   (c) h(x) = x² + x

SOLUTION:

(a) $f(-x) = (-x)^5$
$= (-1)^5x^5$
$= -x^5$
$= -f(x)$

∴ f is odd.

(b) $g(-x) = 1 - (-x)^2$
$= 1 - x^2$
$= g(x)$

∴ g is even.

(c) $h(-x) = (-x)^2 + (-x)$
$= x^2 - x$
$h(-x) \neq h(x)$
$h(-x) \neq -h(x)$

∴ h is neither even nor odd.

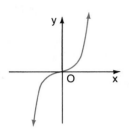

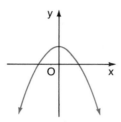

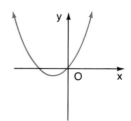

EXERCISE 8.2

A 1. State whether the functions whose graphs are given are even, odd, or neither.

(a)

(b)

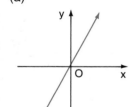

(c)

(d)

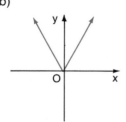

(e)

(f)

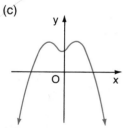

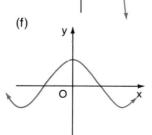

B 2. (a) Find $f(-x)$ for each of the following functions.
(b) Determine whether f is even, odd, or neither.

(i) $f(x) = 3x^2 + 1$
(ii) $f(x) = x^3 + 2x$
(iii) $f(x) = x^3 + 2x^2$
(iv) $f(x) = 1 + x$

(v) $f(x) = \dfrac{1}{x}$

(vi) $f(x) = x^4 + \dfrac{1}{x^2}$

(vii) $f(x) = x^2 + \dfrac{1}{x}$

(viii) $f(x) = \dfrac{x}{1 + x^2}$

3. (a) Determine whether f is odd or even.
(b) Sketch the graph of f for $x \geq 0$.
(c) Use symmetry to sketch the entire graph of f.

(i) $f(x) = x^4$
(ii) $f(x) = x^3 - 3x$

(iii) $f(x) = \dfrac{1}{x^3}$

(iv) $f(x) = \dfrac{1}{x^2 + 1}$

8.3 VERTICAL AND HORIZONTAL SHIFTS

In this section and the next, we shall see how transformations can help us graph functions. We first look at vertical and horizontal shifts.

In Section 5.3 we saw that, if $c > 0$, the graph of $y = x^2 + c$ is just the parabola $y = x^2$ translated upward by c units, and the graph of $y = x^2 - c$ is the graph of $y = x^2$ translated downward by c units. In a similar fashion, as illustrated by the figure below, we have the following general rule.

VERTICAL SHIFTS

> Let $c > 0$.
> The graph of $y = f(x) + c$ is the graph of $y = f(x)$ translated upward by c units.
> The graph of $y = f(x) - c$ is the graph of $y = f(x)$ translated downward by c units.

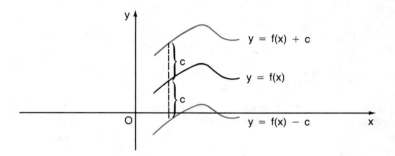

We saw in Section 5.4 that, if $c > 0$, then the graph of $y = (x + c)^2$ is the parabola $y = x^2$ translated c units to the left, and the graph of $y = (x - c)^2$ is the parabola $y = x^2$ translated c units to the right. Similarly, we have the following rule for general functions.

HORIZONTAL SHIFTS

> Let $c > 0$.
> The graph of $y = f(x + c)$ is the graph of $y = f(x)$ translated to the left by c units.
> The graph of $y = f(x - c)$ is the graph of $y = f(x)$ translated to the right by c units.

This is true because if we let $g(x) = f(x - c)$, then the value of g at x is the same as the value of f at $x - c$ (c units to the left of x).

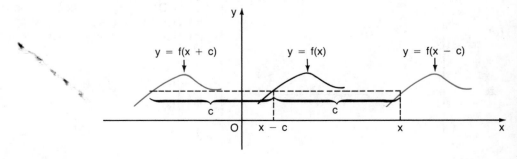

EXAMPLE. Sketch the graphs of the following functions.
(a) $y = x^3 + 2$
(b) $y = x^3 - 1$
(c) $y = (x + 3)^3$
(d) $y = (x - 4)^3$

SOLUTION:

In each case we start with the graph of $y = x^3$, which we obtained in Example 1 of Section 8.1. To draw the graph of
(a) $y = x^3 + 2$, we shift the graph of $y = x^3$ 2 units upward.
(b) $y = x^3 - 1$, we shift the graph of $y = x^3$ 1 unit downward.
(c) $y = (x + 3)^3$, we shift the graph of $y = x^3$ 3 units to the left.
(d) $y = (x - 4)^3$, we shift the graph of $y = x^3$ 4 units to the right.

$y = x^3$

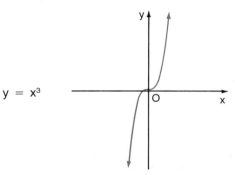

(a) $y = x^3 + 2$

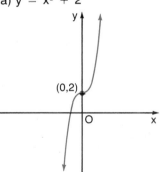

(b) $y = x^3 - 1$

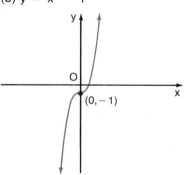

(c) $y = (x + 3)^3$

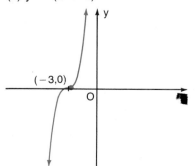

(d) $y = (x - 4)^3$

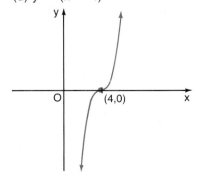

EXERCISE 8.3

A 1. If the graph of f is given, describe how the graphs of the following functions are obtained.

(a) $y = f(x) + 8$

(b) $y = f(x) - 3$

(c) $y = f(x - 3)$

(d) $y = f(x + 6)$

(e) $y = f(x) + \frac{1}{2}$

(f) $y = f(x + \frac{1}{2})$

(g) $y = f(x - 2)$

(h) $y = f(x) - 2$

(i) $y = 1 + f(x)$

(j) $y = f(5 + x)$

B 2. The graphs of two functions $y = f(x)$ are given.

(a)

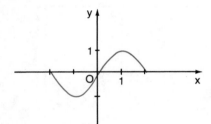

(b)

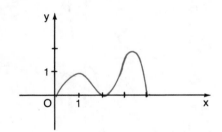

In each case draw the graphs of the following functions.

(i) $y = f(x) + 2$ (ii) $y = f(x) - 4$

(iii) $y = f(x + 2)$ (iv) $y = f(x - 4)$

3. The graph of the function $y = g(x)$ is given.

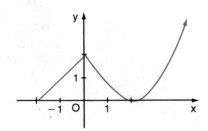

Draw the graphs of the following functions.

(a) $y = g(x - 3)$

(b) $y = g(x) + 3$

(c) $y = g(x) - 1$

(d) $y = g(x + 5)$

(e) $y = g(x - 2) + 1$

(f) $y = g(x + 2) - 3$

4. Use shifts to graph the following functions starting from the graph of $y = \sqrt{x}$.

(a) $y = \sqrt{x} - 2$

(b) $y = \sqrt{x - 2}$

(c) $y = 4 + \sqrt{x}$

(d) $y = \sqrt{4 + x}$

5. Use shifts to graph the following functions starting from the graph of $y = x^4$.

(a) $y = x^4 + 1$

(b) $y = (x + 2)^4$

(c) $y = x^4 - 3$

(d) $y = (x - 3)^4$

(e) $y = (x + 1)^4 + 1$

(f) $y = (x - 1)^4 - 1$

8.4 STRETCHING AND REFLECTING

In this section we look at the effect of stretching and reflecting on the graph of a function. In Section 5.3 we found that if we start with the parabola $y = x^2$, then to get the parabola $y = ax^2$, we have to stretch in the y-direction if $a > 1$ and shrink in the y-direction if $0 < a < 1$. If $a < 0$, we have to combine a reflection with a stretching or shrinking.

Likewise, for a general function f, the graph of $y = a f(x)$ is the graph of $y = f(x)$ stretched by a factor of a in the vertical direction if $a > 1$. If $0 < a < 1$, we shrink instead of stretch. The graph of $y = -f(x)$ is the graph of $y = f(x)$ reflected in the x-axis because the point (x,y) is replaced by the point $(x, -y)$.

STRETCHING
AND
REFLECTING

The graph of $y = a f(x)$ is obtained from the graph of $y = f(x)$ by	
stretching in the y-direction	if $a > 1$
shrinking in the y-direction	if $0 < a < 1$
reflecting in the x-axis	if $a = -1$
shrinking and reflecting	if $-1 < a < 0$
stretching and reflecting	if $a < -1$

EXAMPLE 1. Given the graph of $y = f(x)$, draw the graphs of the following functions.

(a) $y = 2f(x)$ (b) $y = \frac{1}{2}f(x)$

(c) $y = -f(x)$ (d) $y = -2f(x)$

(e) $y = -\frac{1}{2}f(x)$

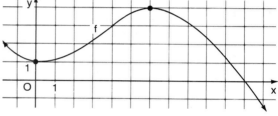

SOLUTION:
Using the rules for stretching and reflecting, we sketch the graphs as shown at the right.

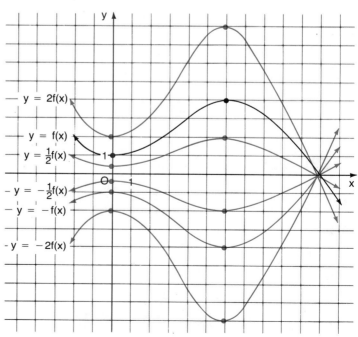

EXAMPLE 2. Use the transformations to draw the graph of the function $y = 2\sqrt{x + 3} - 1$.

SOLUTION:
(i) We start with the graph of $y = \sqrt{x}$ which we obtained in Example 3 of Section 8.1.
(ii) We shift it 3 units to the left to get the graph of $y = \sqrt{x + 3}$.
(iii) Then we stretch it vertically by a factor of 2 to get the graph of $y = 2\sqrt{x + 3}$.
(iv) Finally we shift it 1 unit downward to get the graph of $y = 2\sqrt{x + 3} - 1$.

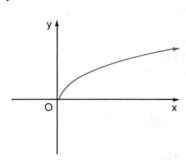

(i) $y = \sqrt{x}$

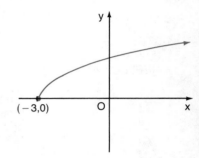

(ii) $y = \sqrt{x + 3}$

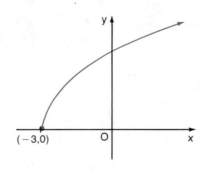

(iii) $y = 2\sqrt{x + 3}$

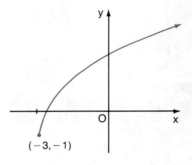

(iv) $y = 2\sqrt{x + 3} - 1$

EXERCISE 8.4

A 1. If the graph of f is given, describe how the graphs of the following functions are obtained.

(a) $y = 4f(x)$
(b) $y = -f(x)$
(c) $y = \frac{1}{4}f(x)$
(d) $y = -3f(x)$
(e) $y = -\frac{1}{5}f(x)$
(f) $y = 2f(x)$
(g) $y = 2f(x) + 1$
(h) $y = -f(x) + 6$
(i) $y = 4f(x + 1)$
(j) $y = 3f(x - 2)$
(k) $y = \frac{1}{2}f(x) - 5$
(l) $y = -\frac{1}{3}f(x) + 4$

B 2. The graphs of two functions $y = f(x)$ are given.

(a)

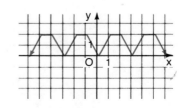

(b)

In each case draw the graphs of the following functions.

(i) $y = \frac{1}{2}f(x)$ (ii) $y = -2f(x)$

(iii) $y = 3f(x)$ (iv) $y = 3f(x) + 1$

(v) $y = -f(x)$ (vi) $y = 2 - f(x)$

3. The graphs of 2 functions $y = f(x)$ are given.

(a)

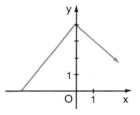

(b)

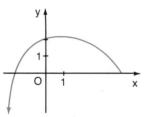

In each case draw the graphs of the following functions.

(i) $y = \frac{1}{2}f(x)$ (ii) $y = -\frac{1}{2}f(x)$

(iii) $y = 2f(x)$ (iv) $y = 1 - \frac{1}{2}f(x)$

4. Use transformations to graph the following functions starting from the graph of $y = \sqrt{x}$.

(a) $y = -\sqrt{x}$ (b) $y = 2\sqrt{x}$

(c) $y = -2\sqrt{x}$ (d) $y = 2\sqrt{x} + 1$

(e) $y = \frac{1}{2}\sqrt{x} - 1$ (f) $y = \frac{1}{2}\sqrt{x - 1}$

(g) $y = 3\sqrt{x + 2}$ (h) $y = 3\sqrt{x + 2} + 2$

5. Use transformations to graph the following functions starting from the graph of $y = x^3$.

(a) $y = -x^3$ (b) $y = \frac{1}{2}x^3$

(c) $y = \frac{1}{2}x^3 + 1$ (d) $y = \frac{1}{2}(x + 1)^3$

(e) $y = (x - 1)^3 + 2$ (f) $y = 1 - (x + 2)^3$

MICRO MATH

We can find the area under the graph of
$$y = x^2, 0 \leqslant x \leqslant 1$$
by dividing the region into rectangles.

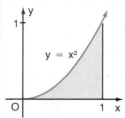
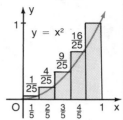

For 5 rectangles the area is

$$\frac{1}{5}\left(\frac{1}{25}\right) + \frac{1}{5}\left(\frac{4}{25}\right) + \frac{1}{5}\left(\frac{9}{25}\right) + \frac{1}{5}\left(\frac{16}{25}\right) + \frac{1}{5}\left(\frac{25}{25}\right)$$

$$= \frac{1}{125} + \frac{4}{125} + \frac{9}{125} + \frac{16}{125} + \frac{25}{125}$$

$$= \frac{1 + 4 + 9 + 16 + 25}{125}$$

$$= \frac{55}{125} = 0.44$$

If we consider n rectangles, then the area under the curve is given by

$$\frac{1}{n}\left(\frac{1}{n^2}\right) + \frac{1}{n}\left(\frac{4}{n^2}\right) + \frac{1}{n}\left(\frac{9}{n^2}\right) + \cdots + \frac{1}{n}\left(\frac{n^2}{n^2}\right)$$

$$= \frac{1 + 4 + 9 + \cdots + n^2}{n^3}$$

This equation leads us to write the following program.

```
NEW
10 PRINT "AREA UNDER Y = X↑2"
20 PRINT "NUMBER OF RECTANGLES?"
30 INPUT N
40 S = 0
50 FOR I = 1 TO N
60 A = I↑2
70 S = S + A
80 NEXT I
90 S = S/N↑3
100 PRINT "AREA IS "; S
110 END
RUN
```

Run the program for the following number of triangles.

1. 5   2. 10

3. 100   4. 1000

What is a rational approximation for the area under $y = x^2, 0 \leqslant x \leqslant 1$?

8.5 RELATIONS

The statements

> "Allan is married to Sharon,"
>
> "George is the brother of Felix,"
>
> "Juliet loves Romeo,"
>
> "2 is less than 5,"
>
> "7 is a factor of 21,"

all involve the idea of a relationship. In each case the relation is given by the words in red. A relation connects the members of two sets.

Consider the relation "drinks" which connects the sets A = {Henry, Sue} and B = {cola, root beer, orange}. Suppose that Henry drinks only cola and orange, and Sue drinks only root beer and orange. Then the relation "drinks" can be described by the set of ordered pairs

{(Henry, cola), (Henry, orange), (Sue, root beer), (Sue, orange)}.

A relation is a set of ordered pairs. The domain of the relation is the set of all first components. The range of the relation is the set of all second components.

In the above example the domain is A and the range is B.

EXAMPLE 1. A = B = {Mary, Joan, Bill, Brian, Peter}. Mary knows Bill, Bill knows Joan, Joan knows Brian, Brian knows Mary, and Peter knows both Mary and Joan. Express the relation "knows" as a set of ordered pairs. What are the domain and the range?

SOLUTION:
{(Mary, Bill), (Bill, Joan), (Joan, Brian), (Brian, Mary), (Peter, Mary), (Peter, Joan)}
Domain = A
Range = {Mary, Joan, Bill, Brian}

EXAMPLE 2. (a) List the relation "is less than" on the set A = {10, 9, 3, 7, 4} as a set of ordered pairs.
(b) State the domain and range.

SOLUTION:
(a) {(3,4), (3,7), (3,9), (3,10), (4,7), (4,9), (4,10), (7,9), (7,10), (9,10)}
(b) Domain = {3, 4, 7, 9}
Range = {4, 7, 9, 10}

We can picture a relation by drawing its graph, that is, by plotting the ordered pairs which make up the relation. For instance the figure at the right shows the graph of the relation in Example 2.

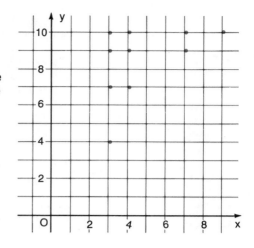

EXAMPLE 3. Graph the relation {(x,y) | y < x + 2, x ∈ R, y ∈R}. (This relation could be written simply as y < x + 2.)

SOLUTION:

We already know how to graph the line y = x + 2. We want to plot all ordered pairs (x,y) such that y < x + 2. If y = x + 2 for points on the line, then y < x + 2 for points below the line because the value of y decreases as we go below the line. Let us check this: (0,0) is obviously a point below the line and we verfiy that (0,0) satisfies the inequality y < x + 2.

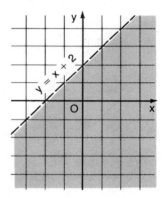

> L.S. y = 0
> R.S. x + 2 = 0 + 2
> = 2
> L.S. < R.S.
∴ y < x + 2

Thus, the graph of y < x + 2 is the shaded region below the line y = x + 2.

Functions and relations both have graphs. What is the connection between functions and relations? We can consider any function f to be a special kind of relation because when we graph f we are plotting the set of ordered pairs (x, f(x)). It is a special set of ordered pairs since for each first component x there is only one second component, namely f(x).

Not every relation is a function because in some relations a given first component can have more than one second component. The relation in Example 1 is not a function because Peter knows both Mary and Joan. The relation in Example 2 is not a function because, for instance, the ordered pairs (3,4) and (3,7) both occur. Likewise the relation in Example 3 is not a function.

In order to tell if a given graph represents a function or a relation we use the following test.

> **Vertical Line Test**
> If it is possible to draw a vertical line which intersects a graph in more than one point, then the graph does not represent a function.

This test works because if $x = a$ intersects the graph in two points (a,b) and (a,c), then we have two ordered pairs with the same first component.

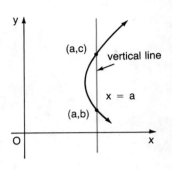

EXAMPLE 4. Which of the following graphs represent functions?

(a) 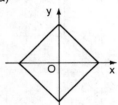 (b) 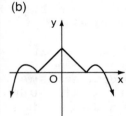 (c) 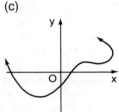 (d)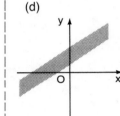

SOLUTION:
In (b) every vertical line intersects the graph only once. But in (a), (c), and (d) the indicated vertical lines intersect the graphs more than once. Thus, only (b) represents a function.

(a) 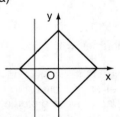 (b) 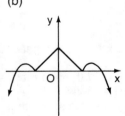 (c) 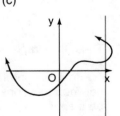 (d)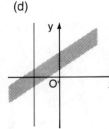

EXAMPLE 5. Graph the relation $x^2 + y^2 = 25$.

SOLUTION:
Remember that $x^2 + y^2 = 25$ is short for

Unless otherwise stated, we assume that $x \in R$, $y \in R$.

$$\{(x,y) \mid x^2 + y^2 = 25\}.$$

The graph of this relation is the circle with radius 5 and centre (0,0). It is not a function. Note that if we try to solve the equation $x^2 + y^2 = 25$ for y we get

$$y^2 = 25 - x^2$$
$$y = \pm\sqrt{25 - x^2}$$

This represents two functions:
The function $y = \sqrt{25 - x^2}$ represents the top half of the circle.
The function $y = -\sqrt{25 - x^2}$ represents the bottom half of the circle.

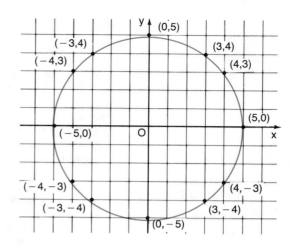

EXERCISE 8.5

A 1. State the domain and range of the following relations. Which of them are functions?

(a) {(elephant, peanut), (monkey, banana), (elephant, banana)}
(b) {(3,17), (4,0), (5,1), (6,0)}
(c) {(π,1), (π,2), (π,3)}

2. State the domain and range of the relations whose graphs are given. Which of them represent functions?

(a) 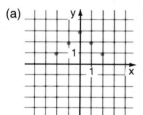 (b)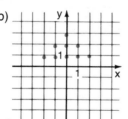

(c) 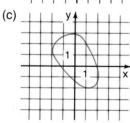 (d)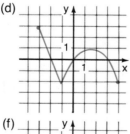

(e) 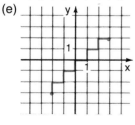 (f)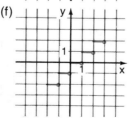

B 3. The five children in the Smith family are Joe, Jane, Judy, Jim, and John, who were born in that order.
(a) Express the relation "is younger than" as a set of ordered pairs. State the domain and range of this relation.
(b) Express the relation "is a sister of" as a set of ordered pairs. State the domain and range of this relation.

4. (a) If A = {2, 3, 4, 5, 6, 7, 8, 9}, express the relation "is a factor of" as a set of ordered pairs.
(b) Graph this relation.

5. Graph the following relations.
(a) $\{(x,y) \mid y \geq x\}$ (b) $\{(x,y) \mid y > x + 1\}$
(c) $\{(x,y) \mid y \geq 2\}$ (d) $\{(x,y) \mid x < 5\}$
(e) $\{(x,y) \mid y < 2x - 3\}$
(f) $\{(x,y) \mid y \leq 3 - 2x\}$
(g) $\{(x,y) \mid x + y < 1\}$
(h) $\{(x,y) \mid 1 < x + y < 2\}$

6. Graph the following relations. Which are functions?
(a) $x = y^2$ (b) $y = x^2$ (c) $y > x^2$
(d) $x^2 + y^2 = 1$ (e) $x + y = 1$

C 7. (a) Graph the following relations using the same axes and a large scale.
     (i) $x^2 + y^2 = 1$ (ii) $x^4 + y^4 = 1$
     (iii) $x^6 + y^6 = 1$ (iv) $x^8 + y^8 = 1$
(b) What shape does the graph of $x^{2n} + y^{2n} = 1$ approach as n becomes large?

8.6 INVERSES OF RELATIONS AND FUNCTIONS

Consider the relation "is a factor of." If A = {2, 3, 4, 5}, and B = {4, 5, 6}, it consists of the set of ordered pairs

$$F = \{(2,4), (2,6), (3,6), (4,4), (5,5)\}.$$

For instance, (2,6) belongs to F because 2 is a factor of 6. Here the domain is A and the range is B.

Now consider the relation "is a multiple of." It consists of the set of ordered pairs

$$M = \{(4,2), (6,2), (6,3), (4,4), (5,5)\}.$$

Here the domain is B and the range is A.

Notice that each ordered pair in M can be obtained by changing the order of the components in F. We say that M is the inverse of F and we write M = F⁻¹. The domain of M is equal to the range of F, and the range of M is equal to the domain of F.

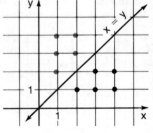

$2^{-1} = \frac{1}{2}$ but F⁻¹ does not mean $\frac{1}{F}$.

> We find the inverse F⁻¹ of any relation F by interchanging the components (that is, by reversing the order of the components). If (a,b) ∈ F, then (b,a) ∈ F⁻¹.

EXAMPLE 1. If F = {(2,1), (3,1), (4,1), (3,2), (4,2)} graph both F and F⁻¹ and state their domains and ranges.

SOLUTION:
F⁻¹ = {(1,2), (1,3), (1,4), (2,3), (2,4)}

Domain of F: {2, 3, 4} Range of F: {1, 2}

Domain of F⁻¹: {1, 2} Range of F⁻¹: {2, 3, 4}

The graph of F is shown in black.

The graph of F⁻¹ is shown in red.

You can see that in Example 1 the graph of F⁻¹ could be obtained by reflecting the graph of F in the line y = x. If you placed a mirror along the line y = x, then the graph of F⁻¹ would be the reflection, or mirror image, of the graph of F.

In fact this is true for any relation F because if (a,b) ∈ F, then (b,a) ∈ F⁻¹ and the point (b,a) is obtained from (a,b) by reflecting in the line y = x.

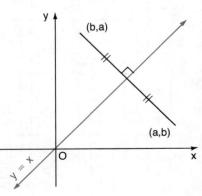

> The graph of any inverse relation F^{-1} is the reflection of the graph of F in the line $y = x$.

We recall from the preceding section that any function f can be considered as a relation. Therefore, it has an inverse relation f^{-1}. This inverse relation, however, may or may not be a function.

EXAMPLE 2. The graph of the function $f(x) = x^2$ is the parabola $y = x^2$. Draw the graph of its inverse relation.

SOLUTION:

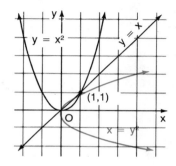

First we graph f as in Chapter 5 (shown in black). Then we reflect in the line $y = x$ to get the graph of f^{-1} (shown in red).
Note that the inverse relation f^{-1} is not a function because it fails the Vertical Line Test.

To find the equation of the inverse relation, we simply interchange x and y in the equation $y = x^2$ (because we interchanged x and y to obtain the graph of f^{-1}). Therefore, the equation of the inverse relation is $x = y^2$. This is also a parabola, but the axis of symmetry is now horizontal.

In general we can adapt the rules of Chapter 5 to graph relations with equations of the form $x = ay^2 + by + c$, $a \neq 0$. As in Example 2, the graphs will all be parabolas with horizontal axes of symmetry. For such parabolas,

(i) an "upward translation" becomes a "translation to the right";

(ii) a "downward translation" becomes a "translation to the left";

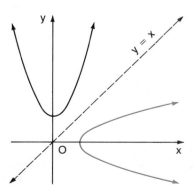

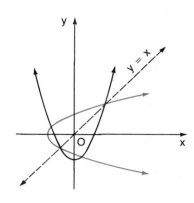

(iii) a "translation to the right" becomes an "upward translation";

(iv) a "translation to the left" becomes a "downward translation";

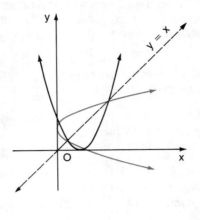 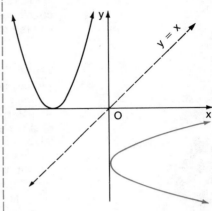

EXAMPLE 3. Sketch the graph of $x = 2y^2 - 4y + 5$.

SOLUTION:
We first complete the square to put the relation in standard form.

$$
\begin{aligned}
x &= 2y^2 - 4y + 5 \\
&= 2[y^2 - 2y] + 5 \\
&= 2[y^2 - 2y + 1 - 1] + 5 \\
&= 2[(y - 1)^2 - 1] + 5 \\
&= 2(y - 1)^2 - 2 + 5 \\
&= 2(y - 1)^2 + 3
\end{aligned}
$$

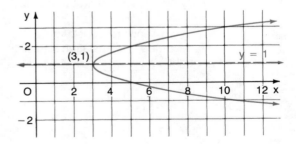

The coordinates of the vertex are (3,1) and the equation of the axis of symmetry is $y = 1$.

We have seen in Example 2 that the inverse of a function is not necessarily a function. However, it is often possible to restrict the domain of a function so that its inverse will be a function.

The inverse of the function $f(x) = x^2 + 1$ is not a function because the graph of f^{-1} is the parabola $x = y^2 + 1$. Let us restrict the domain to the non-negative numbers by considering the function

$$g(x) = x^2 + 1, \qquad x \geq 0.$$

The graph of g^{-1}, obtained by reflection in the line $y = x$, is shown in the figure at the right.

We see by the Vertical Line Test that g^{-1} is a function.

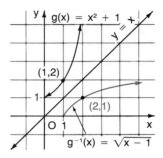

To find an expression for g^{-1} we write the original function g in the form

$$y = x^2 + 1, \qquad x \geq 0$$

Then we interchange x and y.

$$x = y^2 + 1, \qquad y \geq 0$$

Finally, we solve for y.

$$y^2 = x - 1$$
$$y = \pm\sqrt{x - 1}$$

But $y \geq 0$, so we reject the negative sign and we obtain

$$g^{-1}(x) = \sqrt{x - 1}.$$

From the above we see that the procedure for finding the inverse function of f (if it has one) is as follows.

> I. Write $y = f(x)$.
> II. Interchange x and y.
> III. Solve the equation for y.
> IV. The resulting equation is $y = f^{-1}(x)$.

EXAMPLE 4. Find the inverse function of $f(x) = 2x + 5$.

SOLUTION:

Write.	$y = 2x + 5$
Interchange x and y.	$x = 2y + 5$
Solve for y.	$2y = x - 5$
	$y = \dfrac{x - 5}{2}$

The inverse function is $f^{-1}(x) = \dfrac{x - 5}{2}$.

If f has an inverse function f^{-1}, then the general rule for obtaining f^{-1} by interchanging x and y can be stated as follows.

> $f^{-1}(y) = x \qquad$ if $\qquad f(x) = y$

Thus, f^{-1} reverses the effect of f. If f has domain A and range B, then f^{-1} has domain B and range A. In terms of arrow diagrams, we simply reverse the direction of the arrows to get f^{-1} from f.

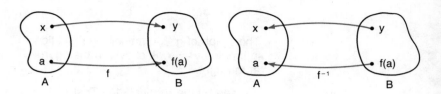

EXAMPLE 5. (a) If $f(x) = x^3 - 2$ and f has domain A = {−2, −1, 0, 1, 2}, draw an arrow diagram for f.
(b) Is f^{-1} a function? If so, draw an arrow diagram for f^{-1}.
(c) State the values of $f^{-1}(-3)$ and $f^{-1}(6)$.
(d) State the domain and range of f^{-1}.

SOLUTION:
(a) We compute the values of f in order to draw the arrow diagram.
$$f(-2) = (-2)^3 - 2 = -10$$
$$f(-1) = (-1)^3 - 2 = -3$$
$$f(0) = 0^3 - 2 = -2$$
$$f(1) = 1^3 - 2 = -1$$
$$f(2) = 2^3 - 2 = 6$$

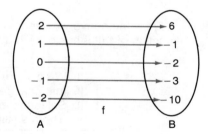

(b) Since the 5 values are all different, f^{-1} is a function.

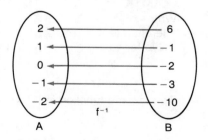

(c) $f^{-1}(-3) = -1$, $f^{-1}(6) = 2$
(d) The domain of f^{-1} is B = {−10, −3, −2, −1, 6}. The range of f^{-1} is A = {−2, −1, 0, 1, 2}.

EXERCISE 8.6

A 1. (a) State the ordered pairs of the inverses of the following relations.
(b) State the domain and range of each inverse.
 (i) {(0,2), (0,3), (4,6), (8,2)}
 (ii) {(1, −2), (2, −1), (3,0), (1,1)}
 (iii) {(−1,1), (−2,2), (−3,3)}
 (iv) {(3,$\frac{1}{3}$), (4,$\frac{1}{4}$), (3,3), (4,4)}

2. Arrow diagrams are given for two functions f and g.

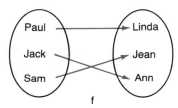

f

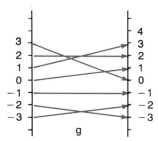

g

(a) Are f⁻¹ and g⁻¹ both functions?
(b) State.
 (i) f(Jack)   (ii) f⁻¹(Linda)
 (iii) f⁻¹(Ann)   (iv) f(Sam)
 (v) g(−3)   (vi) g(−1)
 (vii) g⁻¹(0)   (viii) g⁻¹(−3)
 (ix) g⁻¹(3)   (x) g(1)
 (xi) g(g⁻¹(3))   (xii) g⁻¹(g(0))

B 3. In each part of this question graph the relation and its inverse using the same axes but different colours.
(a) {(−4,1), (−3,1), (−2,1), (−2,2), (−1,2), (−1,3), (−1,4)}
(b) {(−2,0), (−1,1), (0, −1), (0,2), (1,0)}
(c) {(x,y) | x = y²}
(d) {(x,y) | y = 2x}
(e) {(x,y) | x + y = 0}
(f) {(x,y) | y > 1 − x}
(g) {(x,y) | x² + y² = 1}

4. Draw arrow diagrams for each of the following functions f with domain A. State whether or not f⁻¹ is a function. If it is, draw an arrow diagram for f⁻¹.
(a) f(x) = x⁴ − 4, A = {0, 1, 2, 3, 4}
(b) f(x) = x⁴ − 4, A = {−2, −1, 0, 1, 2}
(c) f(x) = 17 − x³, A = {−2, −1, 0, 1, 2}
(d) f(x) = the largest prime factor of x, A = {2, 3, 4, 5, 6, 7, 8, 9}

5. Graph the following functions f and their inverses f⁻¹. In which cases if f⁻¹ a function?
(a) f(x) = x + 1, x ∈ R
(b) f(x) = 2, x ∈ R
(c) f(x) = $\frac{1}{x}$, x ≠ 0
(d) f(x) = x² + 1, x ∈ R
(e) f(x) = √x, x ⩾ 0

6. Find formulas for the inverses of the following functions.
(a) f(x) = 5x + 7
(b) f(x) = 2 − 3x
(c) f(x) = x² − 5, x ⩾ 0
(d) f(x) = √x
(e) f(x) = ∛x
(f) f(x) = (x + 1)³
(g) f(x) = $\frac{1}{x + 1}$
(h) f(x) = 1 + x⁴, x ⩾ 0

C 7. Without making a table of values, sketch the graph of each of the following. State the domain of each.
(a) x = 2y²   **(b)** x = −y²
(c) x = y² − 4   **(d)** x = −y² + 3
(e) x = (y + 2)²   **(f)** x = −2(y − 3)²
(g) x = 2(y + 1)² + 1   **(h)** x = −(y − 2)² − 4

8. Without making a table of values, sketch the graph of each of the following.
(a) x = y² + 4y   **(b)** x = y² − 6y + 7
(c) x = 2y² + 4y   **(d)** x = −y² + 8y
(e) x = 2y² − 4y − 1   **(f)** x = −2y² + 8y + 1
(g) x = 3y² − 12y − 2
(h) x = −4y² − 16y + 3
(i) x = $\frac{1}{2}$y² + y + 2
(j) x = 2y² + y − 3

8.7 APPLICATIONS

Nature herself exhibits to us measurable and observable quantities in definite mathematical dependence; the conception of a function is suggested by all the processes of nature where we observe natural phenomena varying according to distance or time.

T.J. Merz

In this section we shall graph some of the functions which occur in science or in everyday situations.

EXAMPLE. You put some ice cubes in a glass, fill the glass with cold water and then let the glass sit on a table. The temperature T of the water (in degrees Celsius) is a function of the time t (in seconds) that the glass has been sitting on the table. Sketch a graph of T as a function of t.

SOLUTION:

The cold water gradually becomes colder as the ice melts, and so T decreases until T = 0 is reached. After a while all the ice melts and T increases until it levels off at room temperature.

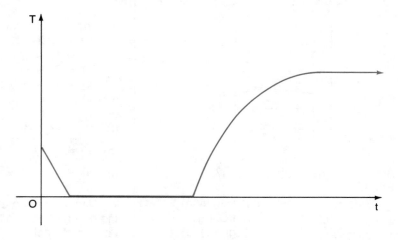

EXERCISE 8.7

B 1. The price P (in dollars) of a new house on a given lot depends on the area A (in square metres) of floor space in the house. Sketch a graph of P as a function of A.

2. You fill an electric kettle with cold water and plug in the kettle. After the kettle boils, you then pull the plug out. The temperature T of the water (in degrees Celsius) is a function of the time t (in seconds) that has passed since the kettle was plugged in. Sketch a graph of T as a function of t and state the domain and range of this function.

3. The number n of swim suits sold in a store depends on the time of year. Let t be the time measured in weeks (from the beginning of the year). Sketch a graph of n as a function of t.

4. Your height h (in centimetres) depends on your age A (in years). Sketch a graph of h as a function of A.

5. The price P (in cents) of a bottle of ketchup depends on the volume V (in millilitres) of ketchup in the bottle. Sketch a graph of P as a function of V.

6. A train goes from Toronto to Windsor and stops for 2 min in Dundas, 5 min in Brantford, 2 min in Woodstock, 10 min in London, and 5 min in Chatham. The distance d (in kilometres) travelled by the train depends on the time t (in minutes) since the train left Toronto. Sketch a graph of d as a function of t.

7. A biologist observes that the number of bacteria in a laboratory dish doubles every day. If N is the number of bacteria (in millions) and N = 1 after the first day, sketch a graph of N as a function of time t (in days). Can you give a formula for N as a function of t?

8. A farmer finds that the value V of his crop (in dollars) depends on the amount x of fertilizer (in cubic metres) that he uses. Sketch a graph of V as a function of x.

9. On a sunny day the length ℓ (in metres) of the shadow of a telephone pole depends on the time of day t (in hours). Sketch a graph of ℓ as a function of t. State the domain and range of this function.

10. Nancy bought a 1 kg bag of sugar and used it only for her morning cup of coffee in which she put 10 g of sugar. If S is the amount of sugar remaining (in grams) after t days, sketch a graph of S as a function of t. State the domain and range of this function.

11. When you blow up a round balloon, the diameter D (in centimetres) of the balloon depends on the number N of breaths that you have blown into it. Sketch a graph of D as a function of N. Can you give a formula for D as a function of N?

C 12. The mass m (in kilograms) of a given kind of fish depends on its length ℓ (in centimetres). Sketch a graph of m as a function of ℓ. Can you give a formula for m as a function of ℓ?

THE PERILS OF SUBTRACTION

The subtraction of two numbers which are close to each other can be a dangerous operation on a calculator because of the loss of significant digits. As an illustration, use your calculator to compute

$$8721\sqrt{3} - 10\,681\sqrt{2}$$

Your answer will probably not be very accurate because the numbers $8721\sqrt{3}$ and $10\,681\sqrt{2}$ agree to 8 significant digits which, after subtraction, become zeros before the first nonzero digit. To make matters worse, the formerly small errors in the square roots are now more crucial.

Nonetheless, we can get around the problem, in this case, as follows. First, show that

$$8721\sqrt{3} - 10\,681\sqrt{2} =$$

$$\frac{1}{8721\sqrt{3} + 10\,681\sqrt{2}}$$

This expression enables us to avoid the loss of significant digits. Use it and your calculator to show that

$$8721\sqrt{3} - 10\,681\sqrt{2} \approx 0.000\,033\,101$$

8.8 REVIEW EXERCISE

1. State whether the given function is constant, linear, quadratic, polynomial, rational, or a root function.

(a) $f(x) = 1 + 2x + x^2 + x^3 - 2x^4$
(b) $f(x) = x^6$
(c) $f(x) = \sqrt[6]{x}$
(d) $f(x) = 2x + 3$
(e) $y = 3x^2 - 2x + 4$
(f) $y = \dfrac{x^3 + x}{x^2 + x + 1}$
(g) $y = 10$

2. Describe how the graph of the given function can be obtained from the graph of f.

(a) $y = f(x + 2)$
(b) $y = f(x) + 2$
(c) $y = 4f(x)$
(d) $y = -f(x)$
(e) $y = -4f(x)$
(f) $y = \frac{1}{3}f(x)$
(g) $y = f(x - 1)$
(h) $y = 2f(x) - 1$

3. State the domain and range of the relations whose graphs are given. Which of them represent functions?

(a)

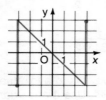

(b)

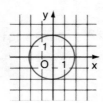

(c)

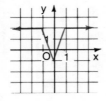

(d)

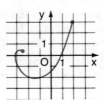

(e)

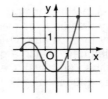

(f)

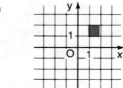

4. Find the domains of the following functions.

(a) $f(x) = x^3 - x^2 + x - 1$
(b) $g(x) = \sqrt{x + 2}$
(c) $h(x) = \dfrac{x + 1}{2x + 1}$
(d) $s(x) = \dfrac{3x - 2}{x^2 - x}$

5. Sketch the graphs of the following functions.

(a) $f(x) = 5$
(b) $f(x) = 1 + 2x,\ -1 \leqslant x \leqslant 1$
(c) $g(x) = x^2,\ -2 \leqslant x \leqslant 1$
(d) $g(x) = x^3 - 4x$

(e) $h(x) = \begin{cases} x & \text{if } x \leqslant -1 \\ 2x + 3 & \text{if } x > -1 \end{cases}$

(f) $h(x) = \begin{cases} -1 & \text{if } x < 0 \\ x - 1 & \text{if } 0 \leqslant x \leqslant 3 \\ 2 & \text{if } x > 3 \end{cases}$

6.

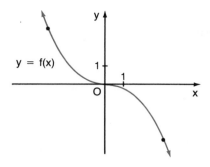

y = f(x)

Given the graph of f, draw the graphs of the following functions.

(a) $y = \frac{1}{3}f(x)$

(b) $y = f(x + 1)$
(c) $y = f(x) - 2$
(d) $y = -2f(x)$
(e) $y = f(x - 3) + 1$
(f) $y = 1 - f(x)$

7. (a) Sketch the graph of $y = x^4$.
(b) Use the graph in part (a) and transformations to stretch the graphs of the following functions.

    (i) $y = (x + 2)^4$
    (ii) $y = (x - 1)^4$
    (iii) $y = -x^4$
    (iv) $y = 2x^4 - 1$

8. Determine whether the following functions are even, odd, or neither.
(a) $f(x) = 3 - 2x^2$
(b) $g(x) = 3x - 2x^2$
(c) $h(x) = 3x - 2x^3$

9. (a) If $A = \{2, 4, 6, 8, 10, 12\}$, express the relation "is a multiple of."
(b) Graph the relation.

10. Graph the following relations.
(a) $y \geqslant 1 - x$
(b) $x \geqslant 2$
(c) $x = -y^2$
(d) $x < -y^2$
(e) $0 < y < 1$
(f) $x^2 + y^2 = 4$
(g) $x^2 + y^2 < 4$
(h) $x^2 + y^2 > 4$

11. Graph the following functions f and their inverses f^{-1}. In which cases is f^{-1} a function?
(a) $f(x) = 2x - 3$ $(x \in R)$
(b) $f(x) = \sqrt{1 - x^2}$ $(-1 \leqslant x \leqslant 1)$
(c) $f(x) = \sqrt{x - 1}$ $(x \geqslant 1)$
(d) $f(x) = 0$ $(x \in R)$

12. Find formulas for the inverses of the following functions.

(a) $f(x) = \frac{1}{6}(x - 7)$

(b) $f(x) = 3x^2 + 7, x \geqslant 0$
(c) $f(x) = \sqrt[4]{x - 7}$

(d) $f(x) = \dfrac{2x - 1}{x + 3}$

13. When you turn on a hot water faucet the temperature T of the water (in degrees Celsius) depends on the time t (in seconds) since the tap was turned on. Sketch a graph of T as a function of t.

8.9 CHAPTER 8 TEST

1. Find the domains of the following functions.

(a) $f(x) = \dfrac{x^2}{x + 6}$

(b) $g(x) = \sqrt{2x - 3}$

2. Sketch the graph of the function f defined by

$$f(x) = \begin{cases} 1 & \text{if } x < 0 \\ 1 - x & \text{if } x \geq 0 \end{cases}$$

3. Describe how the graph of the given function can be obtained from the graph of f.

(a) $y = f(x) - 5$

(b) $y = f(x - 5)$

(c) $y = -2f(x)$

(d) $y = \frac{1}{2}f(x) + 3$

4. (a) Sketch the graph of $y = x^3$.

(b) Sketch the graph of $y = (x + 2)^3 + 1$.

5. (a) If $A = \{2, 4, 6, 8, 10\}$ and $B = \{3, 5, 7, 9\}$, express the relation "is greater than" as a set of ordered pairs.

(b) State the domain and range and graph this relation.

6. (a) Graph the relation $x = y^2$. Is it a function?

(b) Graph $x = (y - 1)^2 + 3$.

7. Find a formula for the inverse of the function $f(x) = \sqrt{2x + 1}$.

1. If $f(x) = 3x^2 + 2x - 5$, find.
(a) $f(6)$
(b) $f(-6)$
(c) $f(\frac{1}{2})$
(d) $f(-x)$
(e) $f(\pi)$
(f) $f(a + 2)$

2. Without making a table of values, sketch the graphs of the following functions.
(a) $y = -x^2$
(b) $y = (x - 5)^2$
(c) $y = x^2 - 5$
(d) $y = 3x^2$
(e) $y = (x + 1)^2 - 3$
(f) $y = \frac{1}{2}x^2 + 2$
(g) $y = x^2 - 2x$
(h) $y = 4x^2 + 4x + 5$

3. (a) Solve the equation $x^2 + x - 6 = 0$ by factoring.
(b) Graph the function $y = x^2 + x - 6$.

4. (a) Find the roots of the equation $x^2 + x - 1 = 0$ correct to one decimal place.
(b) Use part (a) to help graph the function $y = x^2 + x - 1$.

5. (a) Find the vertex and axis of symmetry of the parabola $y = x^2 - 6x + 10$.
(b) Sketch the graph of the parabola.

6. Solve the following equations.
(a) $x^2 + 2x - 6 = 0$
(b) $x^2 + x + 1 = 0$
(c) $2x^2 - 3x + 4 = 0$
(d) $3x^2 + 8x + 2 = 0$
(e) $2x^2 - 3x + 1 = 0$
(f) $x^2 - 10x + 26 = 0$

7. Find the minimum value of the function $g(x) = 4x^2 + x + 3$.

8. A hockey team plays in an arena with a seating capacity of 15 000 spectators. With ticket prices at $12, average attendance at a game has been 11 000. A market survey indicates that for each dollar that ticket prices are lowered, average attendance will increase by 1000. How should the owners of the team set ticket prices so as to maximize their revenue from ticket sales?

9. Solve the following equations.
(a) $x + \dfrac{1}{x} = 3$
(b) $x^4 - 6x^2 + 8 = 0$
(c) $x + \sqrt{x - 1} = 2$

10. A plane flew from Toronto to Edmonton, a distance of 2000 km. On the return trip the speed was increased by 60 km/h. The total trip took 9 h. What was the speed from Toronto to Edmonton?

11. Use the discriminant to determine the nature of the roots of the following equations.
(a) $4x^2 - 28x + 49 = 0$
(b) $3x^2 + 4x + 5 = 0$
(c) $2x^2 + 8x - 3 = 0$
(d) $4x^2 + 8x + 5 = 0$
(e) $9x^2 + 48x + 64 = 0$
(f) $3x^2 - 5x + 3 = 0$

12. For what values of k will the equation $kx^2 + x + 1 = 0$ have real distinct roots?

13. Solve. $x^2 - x - 12 < 0$

14. Find equations for the following circles.
(a) centre $(0,0)$; radius 6
(b) centre $(2,3)$; radius 1
(c) centre $(-1,5)$; radius $\sqrt{2}$
(d) centre $(2,-4)$; radius 3
(e) centre $(-6,0)$; radius 10
(f) centre $(-1,-2)$; passing through the origin

15. Find the centre and radius of the circle with equation $x^2 + y^2 + 4x - 2y = 11$.

16. Find the centre and radius of the circle with equation $x^2 + y^2 - 12x - 6y + 40 = 0$.

17. Find the midpoint of the line segment joining the points $(-1,6)$ and $(3,-10)$.·

18. A circle has equation $x^2 + y^2 = 61$.
(a) Find an equation of the tangent to the circle at the point $(5,6)$.
(b) Find the length of the tangent to the circle from the point $(10,12)$.

19. Given the circle $x^2 + y^2 = 34$ and the points $A(-3,5)$ and $B(5,3)$,
(a) show that AB is a chord of the circle.
(b) find an equation of the chord AB.
(c) find an equation of the line through the origin which is perpendicular to AB.
(d) show that the line in part (c) bisects the chord.

20. Find the domains of the following functions.

(a) $y = \sqrt{x - 2}$

(b) $f(x) = \dfrac{1}{x(x + 1)}$

(c) $f(x) = \dfrac{3}{x^2 + 4x + 3}$

(d) $g(x) = \dfrac{2}{x^2 - 9}$

21. The graph of the function $y = f(x)$ is given.

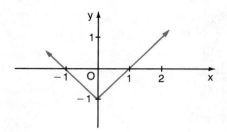

Draw the graphs of the following functions.
(a) $y = f(x + 2)$ (b) $y = f(x) - 4$
(c) $y = f(x - 3)$ (d) $y = f(x + 3) + 1$

22. Graph the following relations.
(a) $\{(x,y) \mid y \leqslant x\}$
(b) $\{(x,y) \mid y < x - 1\}$
(c) $\{(x,y) \mid y > 3\}$
(d) $\{(x,y) \mid y \leqslant -2\}$
(e) $\{(x,y) \mid x + y > 1\}$
(f) $\{(x,y) \mid y \geqslant 2 - 3x\}$

23. Graph the relation and its inverse using the same axes but different colours.
(a) $\{(-4,0), (-2,2), (0,-2), (0,4)\}$
(b) $\{(x,y) \mid y = 3x\}$
(c) $\{(x,y) \mid x = y^2 + 2\}$

SEQUENCES AND SERIES

CHAPTER

9

Wherever there is number, there is beauty.
Proclus

EXPONENTS

Recall the Laws of Exponents from Chapter 1.

$a^m \times a^n = a^{m+n}$

$a^m \div a^n = a^{m-n}$

$(a^m)^n = a^{mn}$

$a^0 = 1$

$a^{-n} = \dfrac{1}{a^n}$

$a^{\frac{1}{n}} = \sqrt[n]{a}$

$(ab)^n = a^n b^n$

$\left(\dfrac{a}{b}\right)^n = \dfrac{a^n}{b^n}$

$a^{\frac{p}{q}} = \sqrt[q]{a^p} = (\sqrt[q]{a})^p$

EXERCISE

Simplify.

1. (a) $3a^5 \times 2a^3$
(b) $12x^7 \div 3x^4$
(c) $(2y^4)^3$
(d) $(3a^5b^2)^0$

2. (a) $3x^{-5} \times 5x^4$
(b) $\sqrt[4]{16x^4y^8}$
(c) $(a^4b^6)^{\frac{3}{2}}$
(d) $(3a^0b^3)^{\frac{1}{3}}$

3. (a) $\left(\dfrac{x^3}{y^0}\right)^2$
(b) $\dfrac{a^0}{b^0}$

(c) $(x^3)^{\frac{2}{3}}$
(d) $\sqrt[5]{32a^5b^{10}}$

4. (a) $\dfrac{2^{-1} + 3^{-1}}{2^{-1} - 3^{-1}}$

(b) $\dfrac{3(5^2 - 1)}{5 - 1}$

(c) $\dfrac{2(1 - 3^3)}{1 - 3}$

(d) $4^{\frac{3}{2}} - 2^0 + \left(\dfrac{1}{4}\right)^{-\frac{1}{2}}$

(e) $\left(\dfrac{3}{2}\right)^{-2} + \dfrac{4}{9}$

(f) $5^0 - 9^{\frac{1}{2}} - \left(\dfrac{1}{81}\right)^{-\frac{1}{4}}$

(g) $\dfrac{3^{-1}}{3^{-2} + 3^{-3}}$

(h) $\dfrac{2^{-1} \times 3^{-1}}{2^{-1} + 3^{-1}}$

(i) $4^{\frac{1}{2}} \times \dfrac{2^{-1}}{2^{-2}}$

5. Solve the following equations.
(a) $3^x = 9$
(b) $2^x = \dfrac{1}{4}$
(c) $3^{-x} = 27$
(d) $8^x = 64$
(e) $4^x = 64$
(f) $2^x = 64$
(g) $\left(\dfrac{3}{2}\right)^x = \dfrac{8}{27}$
(h) $2 \times 3^x = 162$
(i) $4^x = 8^5$
(j) $3^{x+1} = 81$
(k) $5^x = 1$
(l) $0.5^x = 0.0625$

LINEAR SYSTEMS

EXERCISE

1. Solve the systems of equations.
(a) $3x + 2y = -7$
$4x - 5y = 6$
(b) $8x + 3y = 43$
$2x + 7y = 17$
(c) $4x - y = -8$
$5x + 2y = -23$
(d) $6x - 7y = -60$
$x + 9y = 51$
(e) $\dfrac{x}{2} + \dfrac{y}{3} = 7$

$\dfrac{x}{4} - \dfrac{y}{9} = 1$
(f) $5x - 3y = 5$
$8x + 2y = 1.2$
(g) $\dfrac{x}{3} + \dfrac{y}{5} = 0$

$\dfrac{x}{6} - \dfrac{y}{2} = 6$
(h) $y = 3x - 7$
$y = -2x - 2$

FUNCTIONS

EXERCISE

1. If $f(x) = 3x + 2$, evaluate.
(a) f(4)   (b) f(10)
(c) f(−5)   (d) f(0)

2. If $t(x) = 4x - 7$, evaluate.
(a) t(6)   (b) t(32)
(c) t(−7)   (d) t(−11)

3. If $f(x) = 2^x$, evaluate.
(a) f(3)   (b) f(5)
(c) f(−2)   (d) f(0)

4. If $t(x) = (-3)^x$, evaluate.
(a) t(2)   (b) t(4)
(c) t(−1)   (d) t(−3)

5. If $f(x) = \dfrac{x + 1}{2}$, evaluate.
(a) f(7)   (b) f(41)
(c) f(−13)   (d) f(−67)

6. If $t(x) = \dfrac{t + 3}{t - 1}$, evaluate.
(a) t(5)   (b) t(15)
(c) t(−7)   (d) t(−17)

7. If $f(x) = \dfrac{x^2 + 1}{x}$, evaluate.
(a) f(5)   (b) f(9)
(c) f(−7)   (d) f(−12)

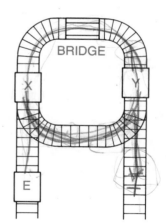

BRIDGE

Use the engine, E, to switch the positions of boxcars X and Y and return the engine to its original position. The bridge is wide enough for the engine, but not wide enough for either boxcar.
Each end of the engine may be used for pushing or pulling.
The two boxcars may be joined to each other.

9.1 SEQUENCES

The following are examples of sequences.

2, 4, 6, 8, 10, ...
3, 9, 27, 81, ...
100, 50, 25, 12.5, ...
7, 8, 3, 4, 11, 13, ...

A sequence need not follow a particular pattern. The sequence 7, 8, 3, 4, 11, 13, ... was taken from a table of random numbers. Most sequences that mathematicians are interested in do follow a pattern.

Gail bought a stereo paying $100 as a down payment, and then $40 every month until paid. We can express her payment plan as shown in the following table:

Month	1st	2nd	3rd	...	nth
Total Amount Paid	$100	$140	$180	...	

This table represents a function whose domain is a subset of the natural numbers.

> **Sequence**
> A sequence is a function whose domain is a subset of the natural numbers. The values in the range of the function are called the terms of the sequence.

The table can then be written:

n	1	2	3	...	n
$f(n)$	100	100 + 40	100 + 80	...	$100 + (n - 1)40$

We use three dots to indicate that there are more terms than have been written. When we use the three dots we must be careful not to assume terms that are not there. For example, if we write

1, 2, 3, 4, ...

we cannot assume that the next term is 5 unless we know more about the sequence.

The sequence defined by

$$f(n) = n + (n - 1)(n - 2)(n - 3)(n - 4)(n^2 + 1)$$

has 1, 2, 3, 4 as its first four terms. However, the fifth term is 629.

The terms of a sequence are often named using a single letter

$$t_1, t_2, t_3, t_4, ..., t_n$$

where t_n is the nth term or general term of the sequence. Relating this notation to our function notation

$$t_1 = f(1), t_2 = f(2), t_3 = f(3), ..., \text{ and } t_n = f(n)$$

EXAMPLE 1. Find the first five terms, given:
(a) $t_n = 2n + 1$
(b) $t_n = n^2 - 1$
(c) $f(n) = 3 - 2n$
(d) $f{:}x \rightarrow 2x$

SOLUTION:
(a) $t_n = 2n + 1$
$t_1 = 2(1) + 1 = 3$
$t_2 = 2(2) + 1 = 5$
$t_3 = 2(3) + 1 = 7$
$t_4 = 2(4) + 1 = 9$
$t_5 = 2(5) + 1 = 11$

(b) $t_n = n^2 - 1$
$t_1 = 1^2 - 1 = 0$
$t_2 = 2^2 - 1 = 3$
$t_3 = 3^2 - 1 = 8$
$t_4 = 4^2 - 1 = 15$
$t_5 = 5^2 - 1 = 24$

(c) $f(n) = 3 - 2n$
$f(1) = 3 - 2(1) = 1$
$f(2) = 3 - 2(2) = -1$
$f(3) = 3 - 2(3) = -3$
$f(4) = 3 - 2(4) = -5$
$f(5) = 3 - 2(5) = -7$

(d) $f{:}x \rightarrow 2x$
$f{:}1 \rightarrow 2(1) = 2$
$f{:}2 \rightarrow 2(2) = 4$
$f{:}3 \rightarrow 2(3) = 6$
$f{:}4 \rightarrow 2(4) = 8$
$f{:}5 \rightarrow 2(5) = 10$

It is sometimes more convenient to describe the pattern of a sequence in terms of getting from one term to the next, rather than stating a general term. Given the terms of a sequence

$$1, 4, 7, 10, 13, ...$$

we can express the pattern using a recursion formula:

$$t_1 = 1$$
$$t_{n+1} = t_n + 3, \qquad n \in N$$

This means that the first term, $t_1 = 1$, and that we add 3 to the nth term to get the $(n + 1)$th term.

EXAMPLE 2. Find the first five terms determined by the following recursion formulas.

(a) $t_1 = 3$

    $t_n = t_{n-1} - 2, \qquad n > 1$

(b) $t_1 = -2$

    $t_{n+1} = t_n + (2n - 1)$

(c) $t_1 = 1, \qquad t_2 = 1$

    $t_n = t_{n-1} + t_{n-2}, \qquad n > 2$

SOLUTION:

(a) $t_1 = 3$

$n = 2, t_2 = t_1 - 2 = 3 - 2 = 1$
$n = 3, t_3 = t_2 - 2 = 1 - 2 = -1$
$n = 4, t_4 = t_3 - 2 = -3 - 2 = -5$
$n = 5, t_5 = t_4 - 2 = -5 - 2 = -7$

$\therefore$ the first five terms are $3, 1, -1, -5, -7$.

(b) $t_1 = -2$

$n = 1, t_2 = t_1 + (2(1) - 1) = -2 + 1 = -1$
$n = 2, t_3 = t_2 + (2(2) - 1) = -1 + 3 = 2$
$n = 3, t_4 = t_3 + (2(3) - 1) = 2 + 5 = 7$
$n = 4, t_5 = t_4 + (2(4) - 1) = 7 + 7 = 14$

$\therefore$ the first five terms are $-2, -1, 2, 7, 14$.

(c) $t_1 = 1, \qquad t_2 = 1$

$t_n = t_{n-1} + t_{n-2}, \qquad n > 2$
$n = 3, t_3 = t_2 + t_1 = 1 + 1 = 2$
$n = 4, t_4 = t_3 + t_2 = 2 + 1 = 3$
$n = 5, t_5 = t_4 + t_3 = 3 + 2 = 5$

$\therefore$ the first five terms are $1, 1, 2, 3, 5$.

This is called the Fibonacci sequence, named after the Italian mathematician, Leonardo Fibonacci. The Fibonacci sequence is often seen in patterns involving growth of leaves on a stem, or the seed spirals in a sunflower.

EXERCISE 9.1

A 1. State the first five terms of the sequence whose nth term is given.

(a) $t_n = 2n$ (b) $t_n = 1 - 2n$
(c) $t_n = 2^n$ (d) $t_n = 2^{n-1}$
(e) $t_n = n^2$ (f) $t_n = 3(2^n)$
(g) $t_n = 1 + 2n$ (h) $t_n = 1 + 2(n - 1)$
(i) $t_n = \dfrac{1}{n}$ (j) $t_n = \dfrac{n + 1}{n}$
(k) $t_n = 2^{-n}$ (l) $t_n = \dfrac{n - 1}{2^n}$

B 2. State a possible rule that determines the following terms. Use your rule to find the next three terms.

(a) 5, 10, 15, 20, ...   (b) 5, 25, 125, ...
(c) 2, 4, 6, 8, ...   (d) 1, 3, 5, 7, ...
(e) 1, 4, 9, 16, ...   (f) 1, 3, 9, 27, ...
(g) $1, \frac{1}{2}, \frac{1}{3}, \frac{1}{4}, ...$ (h) $1, \frac{1}{2}, \frac{2}{3}, \frac{3}{4}, ...$
(i) $2, -1, \frac{1}{2}, -\frac{1}{4}, ...$ (j) a, 2a, 3a, 4a, ...
(k) $a, ar, ar^2, ar^3, ...$ (l) $1, 1 + d, 1 + 2d, ...$

3. List the first five terms of the sequences determined by each of the following.

(a) $t_n = 3n$ (b) $t_n = 2n - 5$
(c) $t_n = (n + 1)(n - 1)$ (d) $t_n = 3^n$
(e) $t_n = 3^{n-1}$ (f) $t_n = 3^n - 1$
(g) $t_n = (-1)^n$ (h) $t_n = (-1)^{n+1}3n$
(i) $t_n = 2n - 3$ (j) $t_n = \dfrac{n - 1}{n + 1}$
(k) $t_n = \dfrac{n}{2}(n + 1)$ (l) $t_n = \dfrac{n(n - 1)}{2n + 1}$

4. List the first five terms of the sequences determined by the following functions.

(a) $f(n) = 2n - 1$ (b) $f(n) = 2n$
(c) $f(k) = (-2)^k$ (d) $f(k) = \dfrac{2k}{2k - 1}$
(e) $f(n) = \dfrac{1}{n}$ (f) $f(k) = 2^{-k}$
(g) $f{:}n \rightarrow 5n - 3$ (h) $f{:}k \rightarrow k^2$
(i) $f{:}k \rightarrow 2^k$ (j) $f{:}n \rightarrow \dfrac{n - 1}{n + 1}$
(k) $f{:}k \rightarrow \dfrac{k}{2}(k - 1)$ (l) $f{:}n \rightarrow \dfrac{n^2 - 1}{n}$

5. Find a general term that determines the following sequences, then list the next three terms.

(a) 1, 2, 3, 4, ...   (b) 4, 3, 2, 1, ...
(c) $2, -1, \frac{1}{2}, -\frac{1}{4}, ...$ (d) $\frac{1}{8}, \frac{1}{4}, \frac{1}{2}, 1, ...$
(e) 4, 8, 16, 32, ...   (f) 4, 1, -2, -5, ...
(g) 1, -1, 1, -1, ...   (h) 2, 6, 10, 14, ...
(i) 2, 6, 18, 54, ...   (j) $1, x, x^2, x^3, ...$
(k) 3a + b, 2a + 2b, a + 3b, ...
(l) $a^3b, a^2b^2, ab^3, b^4, ...$

C 6. Write the first five terms of the sequence whose first term is 3 and every other term is 5 less than twice the preceding term.

7. Find an expression for the nth term of a sequence whose first term is 5 and every other term is 4 more than the preceding term.

8. A sequence has the first term 2 and every other term is 3 more than the preceding term.
(a) Find the first five terms.
(b) Find the nth term.
(c) Find.
   (i) t_{25}
   (ii) t_{1000}

9. A sequence has the first term 3 and every other term is 5 less than the preceding term.
(a) Find the first five terms.
(b) Find the nth term.
(c) Find.
   (i) t_{50}
   (ii) t_{500}

10. Find the first five terms determined by the following recursion formulas.

(a) $t_1 = 3$
$t_{n+1} = t_n + 2$

(b) $t_1 = 1$
$t_{n+1} = 2t_n + 4$

(c) $t_1 = 3$
$t_n = t_{n-1} - 2, n > 1$

(d) $t_1 = 0$
$t_n = t_{n-1} + 2, n > 1$

(e) $t_1 = 2$
$t_n = t_{n-1} + 2n, n > 1$

(f) $t_1 = 1, t_2 = 1$
$t_{n+2} = t_{n+1} + t_n$

9.2 ARITHMETIC SEQUENCES

Sequences such as 2, 5, 8, 11, ... where the difference between consecutive terms is constant are called arithmetic sequences. Each term of the sequence is formed by adding a fixed quantity to the preceding term. The arithmetic sequence is defined by a linear function. The above sequence is defined by the linear function

$$f(n) = 3n - 1$$

Computing the first four terms,

$$t_1 = 3(1) - 1 = 2$$
$$t_2 = 3(2) - 1 = 5$$
$$t_3 = 3(3) - 1 = 8$$
$$t_4 = 3(4) - 1 = 11$$

We can write the terms of the sequence as follows.

$$2 \quad\quad 5 \quad\quad 8 \quad\quad 11 \quad ...$$
$$\updownarrow \quad\quad \updownarrow \quad\quad \updownarrow \quad\quad \updownarrow$$
$$2, 2 + 1(3), 2 + 2(3), 2 + 3(3), ...$$

The general arithmetic sequence is

$$a, a + d, a + 2d, a + 3d, ...$$

where a is the first term and d is the common difference.

$$t_1 = a$$
$$t_2 = a + d$$
$$t_3 = a + 2d$$
$$.$$
$$.$$
$$.$$

$$\boxed{t_n = a + (n - 1)d}$$

EXAMPLE 1. Find t_{10} and t_n for the arithmetic sequence
2, 6, 10, 14, ...

SOLUTION:
$a = 2$ and $d = 6 - 2 = 4$

$$t_n = a + (n - 1)d$$

$$t_{10} = 2 + 9(4) \quad\quad t_n = 2 + (n - 1)4$$
$$= 2 + 36 \quad\quad\quad\quad = 2 + 4n - 4$$
$$= 38 \quad\quad\quad\quad\quad\quad = 4n - 2$$

$\therefore t_{10} = 38$ and $t_n = 4n - 2$.

EXAMPLE 2. How many terms are there in the arithmetic sequence
$$-5, -2, 1, 4, \ldots, 103?$$

SOLUTION:
The sequence is arithmetic with
$$a = -5, \quad d = 3, \quad \text{and} \quad t_n = 103.$$

$$t_n = a + (n - 1)d$$
$$103 = -5 + (n - 1)3$$
$$103 = -5 + 3n - 3$$
$$111 = 3n$$
$$37 = n$$

∴ the sequence has 37 terms.

EXAMPLE 3. In an arithmetic sequence $t_8 = 130$ and $t_{12} = 166$. Find the first three terms and t_n.

SOLUTION:
$$t_n = a + (n - 1)d$$

$$t_{12} = 166 \qquad a + 11d = 166$$
$$t_8 = 130 \qquad a + 7d = 130$$

Subtraction.
$$4d = 36$$
$$d = 9$$

Substitute in t_8.

$$a + 7(9) = 130$$
$$a + 63 = 130$$
$$a = 67$$

∴ the first three terms are 67, 76, and 85.

$$t_n = a + (n - 1)d$$
$$t_n = 67 + (n - 1)9$$
$$= 67 + 9n - 9$$
$$= 9n + 58$$

∴ $t_n = 9n + 58$

The terms between any two given terms of an arithmetic sequence are called arithmetic means between the given terms.

EXAMPLE 4. Insert five arithmetic means between 5 and 29.

SOLUTION:
The sequence will have seven terms.

$$5, t_2, t_3, t_4, t_5, t_6, 29$$

$$t_n = a + (n - 1)d$$

Where $a = 5, \quad n = 7, \quad \text{and} \quad t_n = 29.$

Substitute.
$$29 = 5 + (7 - 1)d$$
$$24 = 6d$$
$$4 = d$$

∴ the terms of the sequence are 5, 9, 13, 17, 21, 25, 29.

EXERCISE 9.2

A 1. Which of the following are successive terms of an arithmetic sequence? State the values of a and d for those that are arithmetic.

(a) 1, 5, 10, 15, ...
(b) 1, 6, 11, 16, ...
(c) 20, 16, 12, 8, ...
(d) 2, 4, 8, 16, ...
(e) $-2, -5, -8, -11, ...$
(f) $1, \frac{1}{2}, \frac{1}{4}, \frac{1}{8}, ...$
(g) $2, 2\frac{1}{2}, 2\frac{3}{4}, 3\frac{1}{4}, ...$
(h) 1, 1.6, 2.2, 2.8, ...
(i) 5, 4.3, 3.7, 3.0, ...
(j) $2x^2, 3x^2, 4x^2, ...$
(k) $x, x^2, x^3, x^4, ...$
(l) $a, a + b, a + 2b, ...$

2. State the first five terms of the arithmetic sequence, given:

(a) $a = 1, d = 4$
(b) $a = 3, d = 5$
(c) $a = -8, d = -3$
(d) $a = x, d = y$
(e) $a = x + 1, d = x + 2$
(f) $a = 5, d = x$
(g) $t_1 = 5m, d = 3m$
(h) $t_1 = 3, d = x + 1$
(i) $t_1 = 5 + x, d = 3$
(j) $t_1 = 2a, d = -a$
(k) $t_1 = 0.5, d = 1.5$
(l) $t_1 = 3a, d = 2 - a$

B 3. Find the indicated terms for each of the following arithmetic sequences.

(a) t_{10} and t_{44} for 8, 10, 12, ...
(b) t_{16} and t_{51} for 10, 14, 18, ...
(c) t_7 and t_{100} for 10, 17, 24, ...
(d) t_{15} and t_{71} for 6, 0, -6, ...
(e) t_5 and t_{62} for $-12, -8, -4$, ...
(f) t_7 and t_{93} for $a, a + 2b, a + 4b$, ...
(g) t_{11} and t_{101} for x, 3x, 5x, ...
(h) t_8 and t_{105} for $a + b, a, a - b$, ...
(i) t_{22} and t_n for 4, 7, 10, ...
(j) t_{51} and t_n for $-11, -5, 1$, ...
(k) t_{30} and t_n for 2, 9, 16, ...
(l) t_{24} and t_n for $a, a + 6, a + 12$, ...

4. Find a, d, and t_n for the following arithmetic sequences.

(a) $t_5 = 16, t_8 = 25$
(b) $t_{12} = 52, t_{22} = 102$
(c) $t_{50} = 140, t_{70} = 180$
(d) $t_2 = -12, t_5 = 9$
(e) $t_7 = 37, t_{10} = 22$
(f) $t_5 = -20, t_{18} = -53$
(g) $t_{13} = -177, t_{22} = -207$
(h) $t_7 = 3 + 5x, t_{11} = 3 + 23x$

5. Find the number of terms in each of the following arithmetic sequences.

(a) 3, 5, 7, ..., 129
(b) $-1, 2, 5, ..., 164$
(c) $-29, -24, -19, ..., 126$
(d) 61, 55, 49, ..., -119
(e) 5, 5.5, 6, ..., 87
(f) $-53, -49, -45, ..., 51$
(g) x, x + 2, x + 4, ..., x + 256
(h) p + 3q, p + 7q, p + 11q, ..., p + 111q

6. How many multiples of 5 are there from 25 to 750 inclusive?

7. How many multiples of 7 are there from -56 to 560 inclusive?

8. How many multiples of 6 are there between 65 and 391?

9. When money is lent at simple interest rates, the amounts required to pay off the loan at the end of each year are the terms of an arithmetic sequence.

Year	Amount
Now	P
1	P + Pi
2	P + 2Pi
3	P + 3Pi
...	...
n	P + nPi

Where P represents the principal, and i the interest rate.

(a) If $1000 is lent at 9% per annum simple interest, find the amount at the end of 1, 2, 3, 4, and n years. Find a and d for the sequence.
(b) Find the amount required to repay a loan of $2500 at 7% per annum simple interest after 12 a.

10. Find t_{37} of an arithmetic sequence with $t_5 = 11$, $t_{18} = 65$.

11. Find x so that x, $\frac{1}{2}x + 7$, and $3x - 1$ are three terms of an arithmetic sequence.

12. Find x so that 2x, $3x + 1$, and $x^2 + 2$ are three terms of an arithmetic sequence.

13. Find the common difference of the sequence determined by $t_n = 5n + 4$.

14. Given $t_n = 2n + 5$, find.
(a) the first five terms
(b) t_k
(c) t_{k-1}
(d) $t_k - t_{k-1}$

15. (a) Insert one arithmetic mean between 11 and 17.
(b) Insert two arithmetic means between 5 and 23.
(c) Insert seven arithmetic means between 15 and 39.
(d) Insert six arithmetic means between 9 and -45.
(e) Insert three arithmetic means between $x + 2y$ and $4x + 14y$.

16. The arithmetic mean of two numbers is 9 and the sum of their squares is 180. Find the numbers.

17. Five fence posts are to be equally spaced between two corner posts that are 42 m apart. How far apart should the five line posts be installed?

18. A management trainee is hired at a salary of $15 000 with half-yearly raises of $375 until the maximum salary of $28 500 is reached. How long will it take to reach the maximum salary?

19. A small car depreciates $1500 the first year, and $600 each year thereafter. How long will it take a $5200 car to depreciate to $1300?

20. The gas company charges a basic monthly rate plus a certain amount per unit of consumption. Consumption of 40 units gives a bill of $67.20, and 73 units gives a bill of $116.70. Find the basic monthly rate and the cost per unit.

21. Colleen purchased a rare painting for $15 000. The painting increased in value each year by 10% of the original price. What was the value of the painting after ten years?

22. At the end of the third week there were 870 members in the video club. At the end of the seventh week there were 1110 members. The increases each week were arithmetic.

(a) How many joined the first week?
(b) How many members were in the club after 11 weeks?

You have 12 identical coins, one of which is counterfeit. The counterfeit coin is either lighter or heavier than the rest. Using a simple balance 3 times, find the counterfeit coin.

9.3 GEOMETRIC SEQUENCES

Sequences such as

$$3, 6, 12, 24, \ldots$$

$$\frac{6}{3} = \frac{12}{6} = \frac{24}{12} = \ldots = 2$$

where the ratio of consecutive terms is constant are called geometric sequences.

Each term of the sequence is found by multiplying the preceding term by a fixed quantity.

The geometric sequence is defined by an exponential function. The above sequence is defined by the exponential function

$$f(n) = 3 \times 2^{n-1}.$$

Computing the first four terms:

$$t_1 = 3 \times 2^{1-1} = 3$$
$$t_2 = 3 \times 2^{2-1} = 6$$
$$t_3 = 3 \times 2^{3-1} = 12$$
$$t_4 = 3 \times 2^{4-1} = 24$$
$$\vdots$$

The general geometric sequence is $a, ar, ar^2, \ldots$ where a is the first term and r is the common ratio.

$$t_1 = a$$
$$t_2 = ar$$
$$t_3 = ar^2$$
$$\vdots$$

$$\boxed{t_n = ar^{n-1}}$$

EXAMPLE 1. Find t_5 and t_n for the geometric sequence 2, 6, 18, ...

SOLUTION:

$$t_n = ar^{n-1}$$

$a = 2, \qquad r = 3$

$$t_5 = ar^4 \qquad\qquad t_n = ar^{n-1}$$
$$\quad = 2 \times 3^4 \qquad\qquad = 2(3)^{n-1}$$
$$\quad = 162$$

$\therefore t_5 = 162$ and $t_n = 2(3)^{n-1}$.

EXAMPLE 2. How many terms are there in the geometric sequence 3, 6, 12, ..., 768?

SOLUTION:

The sequence is geometric: $a = 3$, $\quad r = 2$, $\quad t_n = 768$

$$t_n = ar^{n-1}$$
$$768 = 3(2^{n-1})$$
$$256 = 2^{n-1}$$
$$2^8 = 2^{n-1}$$
$$n - 1 = 8$$
$$n = 9$$

$\therefore$ the sequence has 9 terms.

EXAMPLE 3. In a geometric sequence of real numbers $t_5 = 1875$ and $t_7 = 46\ 875$. Find t_1, t_2, and t_n.

SOLUTION:

$$t_7 = 46\ 875 \rightarrow ar^6 = 46\ 875$$
$$t_5 = 1875 \quad \rightarrow ar^4 = 1875$$

Divide.

$$\frac{ar^6}{ar^4} = \frac{46\ 875}{1875}$$
$$r^2 = 25$$
$$r = \pm 5$$

This indicates that there are two possible solutions.

(i) $r = 5$	(ii) $r = -5$
Substitute in t_5.	Substitute in t_5.
$a(5)^4 = 1875$	$a(-5)^4 = 1875$
$625a = 1875$	$625a = 1875$
$a = 3$	$a = 3$
$\therefore t_1 = 3$, $t_2 = 15$, and $t_n = 3(5)^{n-1}$.	$\therefore t_1 = 3$, $t_2 = -15$, and $t_n = 3(-5)^{n-1}$.

The terms between any two given terms of a geometric progression are called geometric means between the two terms.

EXAMPLE 4. Insert three geometric means between 5 and 80.

SOLUTION:

$a = 5$, $\quad t_5 = 80$, $\quad$ and $\quad t_n = ar^{n-1}$

$$80 = 5r^4$$
$$16 = r^4$$

$$r^2 = 4 \qquad \text{or} \qquad r^2 = -4$$
$$r = \pm 2 \qquad\qquad\quad r = \pm 2i$$

We reject these roots since only real values are considered.

$\therefore r = \pm 2$

Hence there are two sequences with terms:
(i) 5, 10, 20, 40, 80
(ii) 5, -10, 20, -40, 80
Unless otherwise stated, we shall assume that we are working with sequences of real numbers.

EXERCISE 9.3

A 1. Which of the following are successive terms of a geometric sequence? State the values of a and r for those that are geometric.
(a) 1, 4, 9, 16, ...
(b) 1, 2, 4, 8, ...
(c) 5, 10, 15, 20, ...
(d) 3, 12, 24, 72, ...
(e) 32, 16, 8, 4, ...
(f) 64, -16, 4, -1, ...
(g) x, x^3, x^5, x^7, ...
(h) x, x^3, x^6, x^9, ...
(i) x, $-x^2$, x^3, $-x^4$, ...
(j) $2ax$, $2x$, $2\dfrac{x}{a}$, ...
(k) $3a^5b$, a^4b^2, $\frac{1}{3}a^3b^3$, ...
(l) 1, $2x$, $3x^2$, $4x^3$, ...

(h) $3x^{13}$, $3x^{12}$, $3x^{11}$, ..., 3
(i) 1458, 486, 162, ..., 2
(j) $\dfrac{1}{x}$, $\dfrac{1}{x^2}$, $\dfrac{1}{x^3}$, ..., $\dfrac{1}{x^{11}}$
(k) $\dfrac{5}{x}$, $\dfrac{1}{x^2}$, $\dfrac{1}{5x^3}$, ..., $\dfrac{1}{625x^6}$
(l) $2x$, 2, $\dfrac{2}{x}$, ..., $\dfrac{2}{x^{17}}$

4. Find a, t, and t_n for the following geometric sequences.
(a) $t_3 = 36$, $t_4 = 108$
(b) $t_5 = 48$, $t_8 = 384$
(c) $t_2 = 28$, $t_4 = 448$
(d) $t_3 = 64$, $t_8 = 2$
(e) $t_4 = -9$, $t_5 = -3$
(f) $t_2 = 12$, $t_4 = 192$
(g) $t_5 = 12$, $t_9 = 108$
(h) $t_5 = 3$, $t_{14} = 1536$
(i) $t_6 = 486$, $t_9 = 2250$
(j) $t_3 = 5x^6$, $t_{10} = 5x^{20}$
(k) $t_4 = 8x^3$, $t_9 = 256x^8$
(l) $t_3 = 32k^8$, $t_7 = 2k^4$

B 2. Find the terms indicated for each of the following geometric sequences.
(a) t_5 and t_n for 2, 4, 8, ...
(b) t_6 and t_n for 1, 5, 25, ...
(c) t_6 and t_k for 3, 6, 12, ...
(d) t_6 and t_k for 32, 16, 8, ...
(e) t_5 and t_n for 2, -4, 8, ...
(f) t_7 and t_n for 64, -32, 16, ...
(g) t_8 and t_n for 81, -27, 9
(h) t_6 and t_n for $\frac{1}{2}$, $\frac{1}{4}$, $\frac{1}{8}$, ...
(i) t_{10} and t_n for $2x$, $4x^2$, $8x^3$, ...
(j) t_8 and t_n for 1, $\dfrac{x}{2}$, $\dfrac{x^2}{4}$, ...
(k) t_{25} and t_{50} for $\dfrac{1}{x^4}$, $\dfrac{1}{x^2}$, 1, ...
(l) t_{20} and t_{60} for $3x^{10}$, $-3x^9$, $3x^8$, ...

3. How many terms are there in the following geometric sequences.
(a) 4, 12, 36, ..., 972
(b) 3, 6, 12, ..., 768
(c) 2, -4, 8, ..., 512
(d) $\frac{1}{2}$, $\frac{1}{4}$, $\frac{1}{8}$, ..., $\dfrac{1}{1024}$
(e) $\frac{1}{25}$, $\frac{1}{5}$, 1, ..., 625
(f) $\frac{2}{81}$, $\frac{4}{27}$, $\frac{8}{9}$, ..., 6912
(g) $2x^2$, $2x^3$, $2x^4$, ..., $2x^{16}$

5. Find the number of terms in the following geometric sequences.
(a) $t_1 = 8$, $r = 1.5$, $t_n = 40.5$
(b) $t_1 = 567$, $r = \frac{1}{3}$, $t_n = \frac{7}{9}$
(c) $t_1 = 6$, $r = 2$, $t_n = 1536$
(d) 5, 35, 245, ..., 588 245
(e) 3, 6, 12, ..., 96
(f) 64, 32, 16, ..., 0.125
(g) $\frac{1}{4}$, $\frac{1}{2}$, 1, ..., 32
(h) b, ab, a^2b, ..., $a^{12}b$

6. Find x so that $2x$, $x + 5$, and $x - 7$ are consecutive terms of a geometric sequence.

7. Find y so that $4y + 1$, $y + 4$, and $10 - y$ are consecutive terms of a geometric sequence.

8. A car depreciates 30% every year. Find the value of a car five years old if the original price was $9000.00.

9. When money is lent and compound interest is charged, the amount required to repay the loan at the end of each year forms a geometric sequence.

Year	Amount
Now	P
1	$P(1 + i)$
2	$P(1 + i)^2$
3	$P(1 + i)^3$
...	...
n	$P(1 + i)^n$

Where P represents the principal and i the annual rate of interest.
If $300 is lent at 9% per annum compounded annually, show the amount at the end of 1, 2, 3, and n years.

10. Find the amount of $500 invested for 4 a at 8% compounded annually.

11. A virus reproduces by dividing into two, and after a growth period by dividing again. How many virus cells will be in a system starting with a single virus cell after ten divisions?

12. (a) Graph the sequence defined by $t_n = 3(2)^{n-1}$ for $1 \leqslant n \leqslant 5$.
(b) What type of growth is illustrated here?

13. In a certain region, the number of highway accidents increased by 20% per year over a four year period. How many accidents were there in 1984 if there were 5120 in 1980?

14. The population of Satellite City increased from 12 000 to 91 125 over a five year period. Find the annual rate of increase assuming the increase was geometric.

15. A house worth $80 000 sold for $106 480 three years later. Find the annual rate of increase if the value of the house increased geometrically.

C16. (a) Insert three geometric means between 5 and 12 005.
(b) Insert three geometric means between 27 and 2187.
(c) Insert four geometric means between 48 and $1\frac{1}{2}$.
(d) Insert five geometric means between 1458 and 2.

17. If a, x, and b are consecutive terms of a geometric sequence, then x is called the geometric mean of a and b.
(a) Find the geometric mean of 2 and 8.
(b) Find the geometric mean of 5 and 180.
(c) Find the geometric mean of m and n.

18. Show that the sequence defined by $t_{n+1} = 2t_n$, $t_1 = x + y$ is a geometric sequence and state the values of a and r.

19. In case of disaster, St. Mary's General Hospital has a fan out system for calling in staff where each person makes two calls. How many people are called in the sixth level of calls if the person who initiates the first call is considered the first level?

20.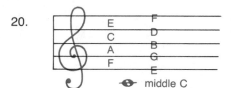

The notes of the musical scale are based on frequency of vibration. A, above middle C, is 440 Hz, that is, 440 vibrations, or cycles, per second. The frequency of a note just above is given by

$$t_{n+1} = 2^{\frac{1}{12}} t_n$$

where t_n is the given note and t_{n+1} the note just above.
(a) Find the frequency of C, 3 notes above A.
(b) Find the frequency of F, 4 notes below A.
(c) Find the ratio of the frequencies $\dfrac{t_{13}}{t_1}$.

9.4 SERIES

We have studied sequences such as

1, 3, 9, 27, 81, 243.

If we add the terms of a sequence, the resulting sum is called a series. The series that corresponds to this sequence is

1 + 3 + 9 + 27 + 81 + 243.

A series is a sum of the terms of a sequence.
Given the sequence t_1, t_2, t_3, ..., t_n, the nth partial sum of the corresponding series is

$$S_n = t_1 + t_2 + t_3 + ... + t_n.$$

If a series has a finite number of terms it has a sum which can be found. For example,

2 + 4 + 6 + 8 + 10 = 30

Series can be described using the Greek letter Σ (sigma). For example, the sum

2 + 4 + 6 + 8 + 10

can be written

$$\sum_{n=1}^{5} 2n$$

which is read

"the sum of 2n from n = 1 to n = 5."

EXAMPLE 1. Find the sums of the following series.

(a) $\sum_{n=1}^{6} n$

(b) $\sum_{n=1}^{5} (2n - 1)$

SOLUTION:

(a) $\sum_{n=1}^{6} n = 1 + 2 + 3 + 4 + 5 + 6$

$\qquad = 21$

(b) $\sum_{n=1}^{5} (2n - 1)$

$= [2(1) - 1] + [2(2) - 1] + [2(3) - 1] + [2(4) - 1] + [2(5) - 1]$
$= 1 + 3 + 5 + 7 + 9$
$= 25$

EXAMPLE 2. Write the following sums explicitly as series.

(a) $\sum_{i=1}^{5} ix$

(b) $\sum_{i=1}^{6} x^i$

(c) $\sum_{k=1}^{5} kx^k$

SOLUTION:

(a) $\sum_{i=1}^{5} ix = 1x + 2x + 3x + 4x + 5x$

(b) $\sum_{i=1}^{6} x^i = x^1 + x^2 + x^3 + x^4 + x^5 + x^6$

(c) $\sum_{k=1}^{5} kx^k = 1x^1 + 2x^2 + 3x^3 + 4x^4 + 5x^5$

EXERCISE 9.4

A 1. Write the following sums explicitly as a series.

(a) $\sum_{i=1}^{4} (2i + 1)$ (b) $\sum_{j=1}^{6} 2j$

(c) $\sum_{k=1}^{5} \frac{1}{k}$ (d) $\sum_{n=1}^{5} (-1)^n$

(e) $\sum_{i=1}^{4} 2^i$ (f) $\sum_{i=1}^{5} \frac{1}{2i - 1}$

(g) $\sum_{n=2}^{5} (2n - 1)$ (h) $\sum_{k=1}^{6} k(k + 1)$

B 2. Write each series in expanded form and find the sum.

(a) $\sum_{n=1}^{5} (2 - n)$ (b) $\sum_{n=1}^{4} (-1)^n n^2$

(c) $\sum_{k=1}^{5} (-1)^k(k^2 - 1)$ (d) $\sum_{i=1}^{5} (3i - 1)$

(e) $\sum_{k=1}^{5} 1 - k$ (f) $\sum_{i=1}^{4} (1 - j)^2$

(g) $\sum_{n=2}^{6} (2 - n)$ (h) $\sum_{k=3}^{7} (-1)^k k^2$

(i) $\sum_{i=0}^{4} 2^i$ (j) $\sum_{n=1}^{5} \frac{1}{n + 1}$

(k) $\sum_{k=2}^{6} \frac{3}{k - 1}$ (l) $\sum_{i=0}^{5} \frac{2i}{i + 1}$

3. Use the summation sign to write each series.

(a) $2 + 4 + 6 + 8 + 10 + 12$
(b) $1 + 4 + 9 + 16 + ... + n^2$
(c) $3 + 5 + 7 + 9 + 11$
(d) $1 + 2 + 4 + 8 + ... + 2^{n-1}$
(e) $1 + 4 + 7 + 10 + ... + (3n - 2)$
(f) $3 + 6 + 12 + ... + 3(2)^{n-1}$
(g) $5 + 5x + 5x^2 + 5x^3$
(h) $3x + 4x + 5x + 6x$
(i) $x - x^2 + x^3 - x^4 + x^5$
(j) $a + 2a^2 + 3a^3 + 4a^4$
(k) $\frac{1}{2} + \frac{1}{3} + \frac{1}{4} + \frac{1}{5}$
(l) $\frac{1}{x} + \frac{2}{x^2} + \frac{3}{x^3} + \frac{4}{x^4}$

What is the only word in the English language that contains the letter sequence GNT?

9.5 ARITHMETIC SERIES

To the finite sequence 2, 4, 6, 8, 10 corresponds the finite series

$$2 + 4 + 6 + 8 + 10.$$

In this series we write $S_5 = 30$ where the symbol S_5 denotes the sum of the first five terms of the series. In this section we shall develop and use a formula to find the sum of a series.

Carl Friedrich Gauss, when only eight years old, used the following method to sum the natural numbers from 1 to 100. Letting S_{100} represent the sum of the first 100 natural numbers, we write out the series first explicitly and again in reverse:

Add.

$$
\begin{aligned}
S_{100} &= 1 + 2 + 3 + \ldots + 99 + 100 \\
S_{100} &= 100 + 99 + 98 + \ldots + 2 + 1 \\
\hline
2S_{100} &= 101 + 101 + 101 + \ldots + 101 + 101 \\
&= 100(101) \\
S_{100} &= \tfrac{100}{2}(101) \\
&= 5050
\end{aligned}
$$

This method of pairing terms can be used to find the sums of other series as in Example 1.

EXAMPLE 1. Find the sum of 300 terms of the series
$$1 + 5 + 9 + 13 + \ldots .$$

SOLUTION:
In order to use Gauss' method, we first find t_{300} from the corresponding sequence.

$$t_n = a + (n - 1)d$$

$$a = 1, \quad d = 4, \quad n = 300$$

$$
\begin{aligned}
t_{300} &= 1 + (300 - 1)4 \\
&= 1197
\end{aligned}
$$

$$
\begin{aligned}
S_{300} &= 1 + 5 + 9 + \ldots + 1193 + 1197 \\
S_{300} &= 1197 + 1193 + 1189 + \ldots + 5 + 1 \\
\hline
2S_{300} &= 1198 + 1198 + 1198 + \ldots + 1198 + 1198 \\
&= 300(1198) \\
S_{300} &= \tfrac{300}{2}(1198) \\
&= 179\ 700
\end{aligned}
$$

We now use the method illustrated in Example 1 to derive a formula for the sum of the general arithmetic series.

The general arithmetic sequence

$$a, (a + d), (a + 2d), ..., (t_n - d), t_n$$

has n terms with first term n and last term t_n. The corresponding arithmetic series is

Reverse.

Add.

$$
\begin{array}{l}
S_n = a \qquad\quad + (a + d) + (a + 2d) + ... + (t_n - d) + t_n \\
S_n = t_n \qquad\quad + (t_n - d) + (t_n - 2d) + ... + (a + d) + a \\
\hline
2S_n = (a + t_n) + (a + t_n) + (a + t_n) \ + ... + (a + t_n) + (a + t_n) \\
\quad\ = n[a + t_n]
\end{array}
$$

$$\boxed{S_n = \frac{n}{2}[a + t_n]} \qquad (1)$$

Since $t_n = a + (n - 1)d$, we can substitute for t_n in formula (1):

$$S_n = \frac{n}{2}[a + a + (n - 1)d]$$

$$\boxed{S_n = \frac{n}{2}[2a + (n - 1)d]} \qquad (2)$$

EXAMPLE 2. Find the sum of the first 50 terms of the given arithmetic series.

(a) $5 + 8 + 11 + 14 + ...$

(b) $-10 - 12 - 14 - ...$

SOLUTION:
(a) $a = 5, \qquad d = 3, \qquad n = 50$

$$S_n = \frac{n}{2}[2a + (n - 1)d]$$
$$S_{50} = \frac{50}{2}[2(5) + (50 - 1)3]$$
$$= 25[10 + 147]$$
$$= 25 \times 157$$
$$= 3925$$

(b) $a = -10, \qquad d = -2, \qquad n = 50$

$$S_n = \frac{n}{2}[2a + (n - 1)d]$$
$$S_{50} = \frac{50}{2}[2(-10) + (50 - 1)(-2)]$$
$$= 25[-20 - 98]$$
$$= 25[-118]$$
$$= -2950$$

EXAMPLE 3. Find the sum of the arithmetic series
$3 + 7 + 11 + \ldots + 483$.

SOLUTION:
Before we can use either formula (1) or formula (2) we must determine n, the number of terms.

$$t_n = a + (n - 1)d$$

$$a = 3, \quad d = 4, \quad t_n = 483$$

$$483 = 3 + (n - 1)4$$
$$483 = 4n - 1$$
$$484 = 4n$$
$$121 = n$$

Using formula (1),

$$S_n = \frac{n}{2}(t_1 + t_n)$$

$$S_{121} = \frac{121}{2}(3 + 483)$$

$$= \frac{121 \times 486}{2}$$

$$= 29\ 403$$

Using formula (2),

$$S_n = \frac{n}{2}[2a + (n - 1)d]$$

$$S_{121} = \frac{121}{2}[2(3) + (121 - 1)4]$$

$$= \frac{121 \times 486}{2}$$

$$= 29\ 403$$

Although both formulas gave the required answer in this example, the form that is used depends on the given data.

EXERCISE 9.5

A 1. Find the sum of the arithmetic series given.
(a) $a = 4, t_n = 9, n = 6$
(b) $a = 5, t_n = 29, n = 9$
(c) $a = 7, t_n = -22, n = 12$
(d) $a = -4, t_n = 17, n = 8$
(e) $a = 0, t_n = 64, n = 16$
(f) $a = 3x, t_n = 21x, n = 10$

B 2. Find the sum of the following series using Gauss' method as in Example 1.
(a) $2 + 4 + 6 + 8 + \ldots + 2000$
(b) $1 + 3 + 5 + 7 + \ldots + 1999$
(c) $1 + 2 + 3 + \ldots + 1000$
(d) $1 + 2 + 3 + \ldots + n$
(e) $3 + 6 + 9 + \ldots$ to 150 terms
(f) $5 + 10 + 15 + \ldots$ to 200 terms

3. Find the sum of the first 1000 terms of the following series.
(a) $1 + 4 + 7 + \ldots$ (b) $10 + 8 + 6 + \ldots$
(c) $5 + 8 + 11 + \ldots$ (d) $0 - 2 - 4 \ldots$

4. Find the indicated sums for the following.
(a) S_{15} of $5 + 9 + 13 + \ldots$
(b) S_{20} of $20 + 25 + 30 + \ldots$
(c) S_{14} of $-14 - 8 - 2 + \ldots$
(d) S_{21} of $-2 + 6 + 14 + \ldots$
(e) S_{50} of $50 + 48 + 46 + \ldots$
(f) S_{15} of $20 + 15 + 10 + \ldots$
(g) S_{50} of $\frac{1}{2} + \frac{3}{2} + \frac{5}{2} + \ldots$
(h) S_{61} of $\frac{1}{2} + \frac{3}{4} + 1 + \ldots$

5. Find the sums of the following series.
(a) $4 + 8 + 12 + \ldots + 400$
(b) $5 + 10 + 15 + \ldots + 265$
(c) $100 + 90 + 80 + \ldots - 100$
(d) $52 + 47 + 42 + \ldots - 48$
(e) $-17 - 10 - 3 + \ldots + 74$
(f) $2 - 5 - 12 - \ldots - 222$
(g) $\frac{5}{2} + \frac{11}{2} + \frac{17}{2} + \ldots + \frac{53}{2}$
(h) $\frac{1}{2} + \frac{1}{4} + 0 - \ldots - \frac{11}{2}$

6. Find the sums of the following series.

(a) $\sum_{n=1}^{100} (2n - 1)$

(b) $\sum_{i=1}^{2000} 2i$

(c) $\sum_{n=1}^{80} (3n - 2)$

7. Find the number of terms in the following arithmetic series.
(a) $78 = 1 + 2 + 3 + ...$
(b) $1830 = 3 + 7 + 11 + ...$
(c) $1250 = 15 + 20 + 25 + ...$
(d) $-350 = 10 + 8 + 6 + ...$
(e) $-120 = -30 - 26 - 22 - ...$
(f) $-345 = 5 + 1 - 3 - ...$

8. A pile of logs is formed by first laying 12 logs side by side, then piling others on top forming a prism tapering to one log at the top. How many logs are there in the pile?

9. A student is offered a job to last 20 h. The pay is $3.25 for the first hour, $3.50 for the second hour, $3.75 for the third hour, and so on, or a straight $6/h for the 20 h. Which system pays more?

10. An experimental theatre has 25 seats in the front row and one additional seat in each following row. How many seats are there if the theatre has 25 rows of seats?

11. An insurance broker earned $24 000 per year and had increases of $800 per year for the next five years. What was this person's total income over the five year period?

12. (a) Find a formula for
$S_n = 1 + 2 + 3 + ... + n$.
(b) Find
(i) S_{100}
(ii) S_{1000}
(iii) S_{2000} for this series.

13. Find a formula for the sum of
(a) the first n odd natural numbers.
(b) the first k even natural numbers.

14. A 12 h clock strikes the same number as the hour from 07:00 to 20:00 inclusive. How many times does the clock strike in one day?

15. In one type of billiards, numbered balls (1 to 15) are used and the player gets the number of points equal to the number on the ball he sinks.
(a) How many points are on the table at the start of the game?
(b) What is the least number of balls required to win if order is not important?
(c) How many balls must you sink to win if you must shoot the balls in order beginning with the 1?

C 16. Find an expression for the sum of n terms of a series with $t_n = 3n - 2$.

17. Find an expression for the sum of n terms of a series with $t_n = 1 - 4n$.

18. Find the first five terms of the arithmetic series with $t_{12} = 35$ and $S_{20} = 610$.

19. A ball rolls down an inclined track and gains speed. If the distance the ball travels is 3 cm in the first second, 6 in the second, 9 in the third, and so on, find.
(a) the distance the ball travels in the 15 s
(b) the total distance travelled in 15 s

20. Boxes are stored in a warehouse. The stack is 4 boxes wide, and 20 boxes long at the bottom. Each layer is one box shorter than the previous layer but the same width. How many boxes are there if the top layer is four boxes long?

MIND BENDER

Augustus de Morgan, who lived in the 19th century, said "I was x years old in the year x^2." When was he born?

9.6 GEOMETRIC SERIES

The sum of the terms of the finite geometric sequence 1, 3, 9, 27, 81, 243, 729 is the finite geometric series

$$1 + 3 + 9 + 27 + 81 + 243 + 729.$$

We can find the sum of this series directly by addition, or by the following procedure.

$$S_7 = 1 + 3 + 9 + 27 + 81 + 243 + 729$$
$$3 \times S_7 = \quad\; 3 + 9 + 27 + 81 + 243 + 729 + 2187$$

Subtract top from bottom.

$$2S_7 = -1 \qquad\qquad\qquad\qquad\qquad\qquad\qquad + 2187$$
$$2S_7 = 2186$$
$$S_7 = \frac{2186}{2}$$
$$= 1093$$

This method is used to develop a formula for the sum of the general geometric series.
For the general geometric series,

$$S_n = a + ar + ar^2 + ... + ar^{n-1}$$

We now develop a formula for the sum of n terms of the geometric series.

$$S_n = a + ar + ar^2 + ... + ar^{n-1}$$
$$rS_n = \quad\;\; ar + ar^2 + ... + ar^{n-1} + ar^n$$

Subtract.

$$S_n - rS_n = a \qquad\qquad\qquad\qquad\qquad - ar^n$$
$$(1 - r)S_n = a(1 - r^n)$$
$$S_n = \frac{a(1 - r^n)}{1 - r}, \quad r < 1$$

or

$$S_n = \frac{a(r^n - 1)}{r - 1}, \quad r > 1$$

We have stated two versions of the formula to avoid negative denominators.

EXAMPLE 1. Find S_{10} for the series $1 + 2 + 4 + ...$

SOLUTION:

$a = 1, \quad r = 2, \quad n = 10$

$$S_n = \frac{a(r^n - 1)}{r - 1}$$
$$S_{10} = \frac{1(2^{10} - 1)}{2 - 1}$$
$$= \frac{1024 - 1}{1}$$
$$= 1023$$

EXAMPLE 2. Find the sum of the series
$$5 + 15 + 45 + ... + 10\,935.$$

SOLUTION:

$$S_n = \frac{a(r^n - 1)}{r - 1}$$

where $a = 5$, $r = 3$,   and n is unknown.

$$t_n = ar^{n-1}$$
$$10\,935 = 5(3)^{n-1}$$
$$3^{n-1} = \frac{10\,935}{5}$$
$$3^{n-1} = 2187$$
$$3^{n-1} = 3^7$$
$$n - 1 = 7$$
$$n = 8$$

$$S_8 = \frac{5(3^8 - 1)}{3 - 1}$$
$$= \frac{5(6561 - 1)}{2}$$
$$= \frac{5(6560)}{2}$$
$$= 16\,400$$

EXERCISE 9.6

B 1. Find the indicated sum for the following series.
(a) $S_8 = 10 + 20 + 40 + ...$
(b) $S_5 = 2 + 6 + 18 + ...$
(c) $S_6 = 3 + 15 + 75 + ...$
(d) $S_8 = 2 - 6 + 18 - ...$
(e) $S_6 = 256 + 128 + 64 + ...$
(f) $S_6 = 972 + 324 + 108 + ...$

2. Find the sum of the following series.
(a) $1 + 2 + 4 + ... + 256$
(b) $1 + 3 + 9 + ... + 2187$
(c) $2 - 4 + 8 - ... + 512$
(d) $5 - 15 + 45 - ... + 3645$
(e) $243 + 81 + 27 + ... + \frac{1}{27}$
(f) $2700 + 270 + 27 + ... + 0.0027$

3. Every person has two natural parents, four grandparents, and so on into the ancestral past. What is the total number of direct ancestors in six generations?

4. An emergency measures organization uses a fan-out system to alert staff. The executive officer who initiates the action makes four calls. Each person in turn makes four calls, and so on. How many people have been alerted after the fifth level of calls if the executive officer is considered the first level?

5. A superball bounces to $\frac{3}{4}$ of its initial height when dropped on dry pavement. If the ball is dropped from a height of 16 m,
(a) how high does it bounce after the fifth bounce?
(b) how far does the ball travel by the time it hits the ground for the sixth time?

6. In a certain town the number of accidental deaths decreased 20% per year over the last five years. How many people died accidentally during this period if there were 52 accidental deaths five years ago?

7. In a popular lottery the first prize money was $10 000. Each succeeding ticket paid $\frac{1}{2}$ as much as the ticket before it. How much was paid out in prizes if six tickets were drawn?

C 8. Evaluate the following expressions.

(a) $\displaystyle\sum_{n=1}^{4} (2^n + 3^n)$

(b) $\displaystyle\sum_{n=1}^{6} (2^n - 3^{n-1})$

(c) $\displaystyle\sum_{n=1}^{5} (3^n + 2^{-n})$

(d) $\displaystyle\sum_{n=1}^{1000} 5^n$

(e) $\displaystyle\sum_{n=1}^{96} \frac{1}{2^n}$

(f) $\displaystyle\sum_{n=1}^{5} (2n - 1)$

9.7 THE INFINITE GEOMETRIC SERIES

If we take a sheet of paper and cut it in half, we have $\frac{1}{2} + \frac{1}{2}$.
We repeat this process with one of the halves.

$$\frac{1}{2} + \frac{1}{4} + \frac{1}{4} = 1$$

Repeating again,

$$\frac{1}{2} + \frac{1}{4} + \frac{1}{8} + \frac{1}{8} = 1.$$

Continuing this process forever,

$$\frac{1}{2} + \frac{1}{4} + \frac{1}{8} + \frac{1}{16} + \frac{1}{32} + \dots$$

This is called an infinite geometric series with $a = \frac{1}{2}$ and $r = \frac{1}{2}$.
From our example, we say that the sum of this infinite series is

$$\frac{1}{2} + \frac{1}{4} + \frac{1}{8} + \frac{1}{16} + \frac{1}{32} + \dots = 1.$$

We can find the sum of any infinite geometric series
$a + ar + ar^2 + \dots + ar^{n-1} + \dots$ when $-1 < r < 1$.

$$S_n = \frac{a(1 - r^n)}{1 - r}, \qquad -1 < r < 1$$

For $-1 < r < 1$, as n increases, r^n decreases in value.
As n becomes larger and larger, r^n approaches zero.
Therefore, S_n approaches $\dfrac{a}{1 - r}$. Hence for an infinite number of
terms, the sum of the series is

$$S = \frac{a}{1 - r}$$

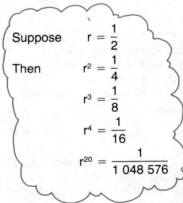

Suppose $r = \dfrac{1}{2}$

Then $r^2 = \dfrac{1}{4}$

$r^3 = \dfrac{1}{8}$

$r^4 = \dfrac{1}{16}$

$r^{20} = \dfrac{1}{1\ 048\ 576}$

EXERCISE 9.7

B 1. Find the sum of the following series.

(a) $1 + \frac{1}{3} + \frac{1}{9} + \frac{1}{27} + \dots$

(b) $2 + 1\frac{1}{2} + 1\frac{1}{8} + \frac{27}{32} + \dots$

(c) $5 + 2\frac{1}{2} + 1\frac{1}{4} + \frac{5}{8} + \dots$

2. Use the infinite series formula to express $0.\overline{2}$ as a common fraction.

3. A ball is dropped from a height of 1 m and it bounces to one-half its previous height. Find the total distance travelled by the ball.

9.8 PROBLEM SOLVING

In this section we continue our work on problem solving by discussing network theory. Network theory has applications to electrical circuitry and economics. It was originated over two hundred years ago by Leonard Euler.

The diagram below shows how a river loop divides the city of Königsberg into four areas, marked A, B, C, and D. The townspeople knew that the seven bridges could not all be crossed in a continuous walk without recrossing one of the bridges, but no one knew the explanation. Euler used network theory to solve the bridges of Königsberg problem.

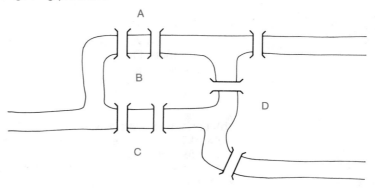

Before solving problems using network theory, we must know the meaning of order as it applies to vertices. The order of a vertex in a network is the number of lines (or arcs) which end at the vertex. In the following figures each vertex has been labelled with its order.

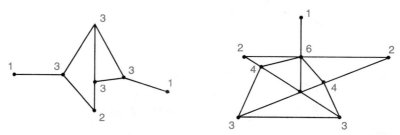

A network is called traversible if it can be drawn in one stroke (that is, without lifting the pencil) and without retracing any lines.

The following are the rules for determining whether a network is traversible.

I. If all vertices are even, the network is traversible.

II. If a network has two odd vertices, it is traversible. One odd vertex must be the starting point and the other odd vertex the ending point.

III. If a network has more than two odd vertices it is not traversible.

EXAMPLE 1. Determine whether the following networks are traversible. If a network is traversible indicate a path by numbering its lines in order.

(a)

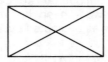

(b)

SOLUTION:

(a) Label each vertex with its order. Since there are more than 2 vertices with odd order, this network is not traversible.

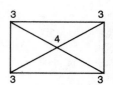

(b) Label each vertex with its order. This network is traversible since there are only 2 vertices with odd order. To find a path, we start and end at the odd vertices. One such path is indicated by numbering the lines in order.

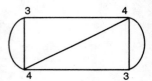

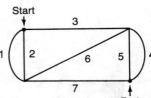

EXAMPLE 2. Solve the bridges of Königsberg problem.

SOLUTION:

Represent the four areas of the city as points.

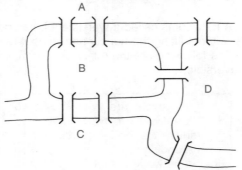

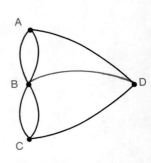

Since there are three odd vertices, the network is not traversible.

EXERCISE 9.8

B 1. Copy each of the following figures and label each vertex with its order.

(a)

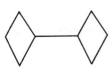

(b)

(c)

(d)

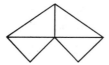

(e)

(f)

2. Determine whether the following networks are traversible. If a network is traversible, indicate a path by numbering its lines in order.

(a)

(b)

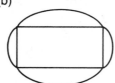

(c)

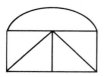

(d)

(e)

(f)

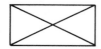

(g)

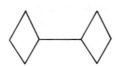

(h)

(i)

3. Is it possible to find a path which goes through each door of the house once and only once? If the answer is yes, draw a path that will do it.

(a)

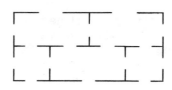

(b)

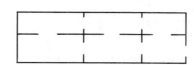

(c)

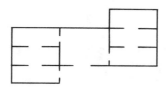

(d)

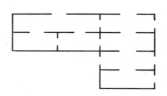

9.9 REVIEW EXERCISE

1. State the first five terms of the sequence whose nth term is given.

(a) $t_n = 3n$

(b) $t_n = 2 - 2n$

(c) $t_n = n^2$

(d) $t_n = 2^n$

(e) $t_n = \dfrac{2}{n}$

(f) $t_n = 3 + 2(n - 3)$

2. State a possible general term for each of the following sequences. Use your general term to find the next three terms.

(a) 5, 9, 13, 17, 21, ...

(b) 3, 9, 27, 81, ...

(c) 128, 64, 32, 16, 8, ...

(d) 51, 46, 41, 36, 31, ...

(e) 1, −2, 4, −8, 16, ...

3. List the first five terms of the sequences defined by the following functions.

(a) $f(n) = 3n - 4$

(b) $f(n) = 7 - 5n$

(c) $f(n) = 3^n$

(d) $f(n) = n^3$

(e) $f:n \to 5n - 8$

(f) $f:n \to \dfrac{n - 2}{n + 2}$

4. Find the first five terms determined by the following recursion formulas.

(a) $t_1 = 4$, $t_n = t_{n-1} + 7$, $n > 0$

(b) $t_1 = 3$, $t_{n+1} = t_n - 3$, $n > 0$

(c) $t_1 = 2$, $t_{n+1} = t_n + 3n$, $n > 0$

5. Write the first three terms of the arithmetic sequence, given:

(a) $a = 4$, $d = 3$

(b) $a = -2$, $d = 2$

(c) $a = -7$, $d = -1$

(d) $a = 3$, $d = x$

6. Find the indicated terms for each of the following arithmetic sequences.

(a) t_n and t_8 for 4, 6, 8, 10, ...

(b) t_n and t_{11} for 20, 17, 14, 11, ...

(c) t_n and t_{15} for −7, −9, −11, ...

(d) t_n and t_{21} for 3, 9, 15, ...

(e) t_n and t_{19} for −1, −7, −13, ...

7. Find a, d, and t_n for each of the following arithmetic sequences.

(a) $t_{10} = 6$ and $t_{18} = 14$

(b) $t_9 = 39$ and $t_{15} = 63$

(c) $t_{12} = -8$ and $t_{20} = -16$

(d) $t_6 = -11$ and $t_{12} = -29$

8. Find the number of terms in each of the following arithmetic sequences.

(a) 8, 12, 16, ..., 100

(b) 52, 46, 40, ..., −8

(c) −4, −9, −14, ..., −154

(d) $x + 1$, $x + 6$, $x + 11$, ..., $x + 241$

9. How many multiples of 5 are there from 35 to 495 inclusive?

10. How many multiples of 8 are there from 12 to 236 inclusive?

11. Find t_n of an arithmetic sequence with $t_4 = -4$ and $t_{20} = -52$.

12. Find the terms indicated for each of the following geometric sequences.

(a) t_n and t_5 for 2, 4, 8, ...

(b) t_n and t_6 for 27, 9, 3, ...

(c) t_n and t_4 for 1, −3, 9, ...

13. Find a, r, and t_n for each of the following geometric sequences.

(a) $t_4 = 24$ and $t_7 = 192$

(b) $t_2 = -6$ and $t_5 = -162$

(c) $t_2 = -12$ and $t_4 = -192$

14. Find the number of terms in the following geometric sequences.

(a) 2, 4, 8, ..., 1024

(b) 16 384, 4096, 1024, ..., 1

15. Write each series in expanded form and find the sum.

(a) $\displaystyle\sum_{n=1}^{6} \frac{n + 1}{n}$ (b) $\displaystyle\sum_{k=2}^{6} \frac{k - 2}{k}$ (c) $\displaystyle\sum_{i=1}^{5} 2^{i-1}$

16. Find the sum of the first 500 terms of the following arithmetic series.
(a) $1 + 4 + 7 + 10 + ...$
(b) $60 + 58 + 56 + 54 + ...$

17. Find the indicated sums of the following arithmetic series.
(a) S_{40} of $3 + 8 + 13 + 18 + ...$
(b) S_{50} of $50 + 46 + 42 + 38 + ...$
(c) S_{80} of $-20 - 17 - 14 - 13 ...$

18. Find the sums of the following arithmetic series.
(a) $6 + 9 + 12 + ... + 126$
(b) $100 + 90 + 80 + ... - 220$

19. Find the indicated sums of the following geometric series.
(a) S_8 of $2 + 4 + 8 + 16 + ...$
(b) S_6 of $1 - 3 + 9 - 27 + ...$
(c) S_6 of $512 + 256 + 128 + 64 + ...$

20. Find the sums of the following geometric series.
(a) $5 + 10 + 20 + ... + 1280$
(b) $1 + 3 + 9 + ... + 2187$

21. Find t_4 and S_6 for the following geometric series.
(a) $t_3 = 6$ and $t_6 = 48$
(b) $t_2 = 10$ and $t_5 = 1250$

22. If $t_4 = 24$ and $t_7 = 192$, find the sum of the first four terms if the sequence is
(a) arithmetic.
(b) geometric.

23. There is a legend that the inventor of chess chose the following as his reward. One grain of wheat on the first square, two grains on the second square, four on the third, eight on the fourth, and so on for all sixty-four squares on the chess board. Find an expression for the amount of wheat required to fulfill his request.

24. Tom purchased a rare stamp for $9000. Tom was guaranteed that the stamp would increase in value by 15% each year based on the original value. What is the value of the stamp if Tom owns it for five years?

25. A theatre has 32 rows of seats. The first row has 24 seats. Each of the remaining rows increases by 2 seats. How many seats are there in the theatre?

26. A jogger runs 300 m in the first minute. The distance covered decreases by 20 m in each succeeding minute. What distance will the jogger cover in the seventh minute?

MIND BENDER

1×1

1 square

2×2

5 squares in all

3×3

14 squares in all

Determine a method to find the number of squares in all for any $n \times n$ diagram of squares.

9.10 CHAPTER 9 TEST

1. Write the first three terms of the arithmetic sequences, given:
   (a) $a = 4$, $d = 7$ (b) $a = -4$, $d = -2$

2. Write the first three terms of the geometric sequences, given:
   (a) $a = 2$, $r = -2$ (b) $a = -1$, $r = 3$

3. Find the indicated terms for the arithmetic sequence 2, 5, 8, ...
   (a) t_{21} (b) t_{33} (c) t_n

4. Find the indicated terms for the geometric sequence 1, 3, 9, ...
   (a) t_5 (b) t_7 (c) t_n

5. Find a, d, and t_{22} for the following arithmetic sequence.
   $$t_8 = 26 \text{ and } t_{14} = 50$$

6. Find a, r, and t_7 for the following geometric sequence.
   $$t_4 = -24 \text{ and } t_6 = -96$$

7. Find the number of terms in the following arithmetic sequence.
   $$5, 9, 13, ..., 365$$

8. Find the indicated sum of the arithmetic series.
   $$S_{100} \text{ of } 3 + 9 + 15 + 21 + ...$$

9. Find the indicated sum of the geometric series.
   $$S_6 \text{ of } 1 + 2 + 4 + 8 + ...$$

10. Find the sum of the arithmetic series.
    $$60 + 56 + 52 + 48 + ... - 100$$

11. Find the sum of the geometric series.
    $$3 - 6 + 12 - 24 ... - 1536$$

INTEREST
AND
ANNUITIES

CHAPTER

He who is unfamiliar with mathematics remains more or less a
stranger to our time.

E. Dillmann

PERCENT

EXERCISE

1. Express as decimals.
(a) 15%   (b) 25%
(c) 18%   (d) 2.5%

2. Express as percents.
(a) 0.35   (b) 0.375
(c) 0.625   (d) 1.125
(e) $\frac{1}{4}$ (f) $\frac{5}{8}$
(g) $\frac{7}{8}$ (h) $1\frac{3}{8}$

3. Calculate.
(a) 10% of 35   (b) 15% of 60
(c) 20% of 55   (d) 25% of 32
(e) 7% of 50   (f) 7% of 515
(g) 110% of 80   (h) 107% of 25

4. (a) What percent of 32 is 8?
(b) What percent of 18 is 18?
(c) What percent of 24 is 32?
(d) What percent of 32 is 24?
(e) What percent of 65 is 13?
(f) What percent of 250 is 10?

5. (a) 24 is 50% of what number?
(b) 32 is 10% of what number?
(c) 55 is 12% of what number?
(d) 75 is 18% of what number?
(e) 125 is 125% of what number?
(f) 88 is 110% of what number?
(g) 57 is 19% of what number?

6. Which is greater?
(a) 10% of 35 or 35% of 10
(b) 15% of 20 or 25% of 10
(c) 30% of 18 or 60% of 10
(d) 40% of 30 or 25% of 50

7. (a) Find 18% of $49.95.
(b) $2.50 is what percent of $20.00?
(c) What is 107% of $29.50?
(d) What amount is $29.95 increased by 15%?

(e) What amount is $69.50 decreased by 15%?
(f) $55 is 110% of what amount?
(g) $74.98 is 107% of what amount?

8. $33.50 is increased by 10% and then increased again by an additional 15%.

(a) What is the new amount?
(b) What single percentage increase will result in the same amount?

MARK-UP AND DISCOUNT

Selling price = Cost price + Mark-up

Sale price = Selling price − Discount

EXERCISE

1. The cost price of a shirt is $28.50, and the mark-up is 30%. What is the selling price?

2. The invoice price of a car is $12 035.00, and the mark-up is 15%. What is the selling price of the car?

3. A car dealer sells a car for $17 200.00. The profit on this sale is $2000.00. What is the percentage mark-up?

4. In a special promotion sale, a car dealer had a straight mark-up of $400.00 per car. What was the percentage mark-up for cars that sold for the following amounts.
(a) Buick $21 545.00
(b) Buick $14 624.00
(c) Pontiac $13 000.00
(d) Camaro $19 495.00
(e) Pontiac $18 450.00

5. The regular selling price of a shirt is $29.95, and it is on sale at a discount of 20%. What is the sale price of the shirt?

6. There is a 12% reduction in the price of a car that regularly sells for $21 500.
(a) What is the discount?
(b) What is the sale price?

7. The sale price of a tie is $15.96 following a 20% discount. What is the regular selling price of the tie?

8. A sweater is on sale at 30% off the regular selling price. It cost $67.37 to purchase the sweater including a 7% sales tax. What is the regular selling price?

COMMISSION

EXERCISE

1. The real estate commission for multiple listing service is usually set at 6% of the sale price of the property. Of this, 3% is for the office and agent that lists the property, 1.5% is for the office that sells the property, and the balance is the commission for the agent who sells the property. A house has been sold for $129 500.00.
(a) What is the total real estate commission?
(b) What is the selling agent's commission?
(c) How much did the house owner receive after paying the commission?

2. Miss Greening receives $125.00 each week plus 1.5% of sales as a broker. What is her total income over a 4 week period if sales were $4200.00, $3965.00, $4825.00, and $4400.00?

3. On a sale of $35 000.00, Mr. Labelle's commission was $525.00. What percentage commission was he paid?

4. Robert Leclerc receives $475.00 per week or 2% of sales, whichever is greater. What was his pay for each of the weeks with the following sales?
(a) sales of $27 000.00
(b) sales of $22 125.00
(c) sales of $25 500.00
(d) sales of $28 750.00
(e) sales of $30 500.00

5. Jennifer Davis receives 1% commission on sales up to $12 000.00, and 1.5% after that. What is Jennifer's commission on sales of $16 250.00?

6. In a used car sale, the salesmen were paid a flat rate of $50.00 per car. Find the percentage commission for the salesman for each of the following cars sold.
(a) Ford $12 000
(b) Plymouth $9 500
(c) Buick $14 500
(d) Chevette $4 025

SIMPLE INTEREST

$$I = Prt \qquad A = P + I$$
where I : interest
         r : annual rate of interest
         t : time in years
         P : principal
         A : amount

EXERCISE

1. Express as a fraction of a year.
(a) 6 months   (b) 8 months
(c) 30 d   (d) 90 d
(e) 400 d   (f) 18 months

2. Calculate the simple interest and amount on each of the following.
(a) $500 at 11% per annum for 3 a
(b) $1200 at 10.5% per annum for 2 a
(c) $2500 at 11.5% per annum for 3 a
(d) $5600 at 10% per annum for 6 months
(e) $10 500.00 at 9.5% per annum for 9 months

3. Calculate the simple interest and amount.
(a) $1000 at 10% per annum for 30 d
(b) $2500 at 9.5% per annum for 60 d
(c) $1500 at 11% per annum for 90 d
(d) $3600 at 8% per annum for 120 d

10.1 AMOUNT OF A LUMP SUM

When a lump sum of money is borrowed from a bank, or other lending institution, interest is charged for the use of the money. If the interest is payable at regular intervals during the duration of the loan, and it is not actually paid but added on to the principal for the next interest period, it is called compound interest.

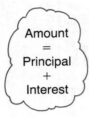

Amount
=
Principal
+
Interest

EXAMPLE 1. If $100 is invested at 8% per annum compounded annually, show how the amount grows over a term of
(a) n years
(b) 20 a

SOLUTION:
After 1a, $100 accumulates to

$$A = \underbrace{100}_{\text{principal}} + \underbrace{100(0.08)}_{\text{interest}} = 100(1 + 0.08)$$

$$= 100(1.08)$$

This amount, $100(1.08) now becomes the principal for the following year's investment. After 2a, we have

$$A = \underbrace{100(1.08)}_{\text{principal}} + \underbrace{100(1.08)(0.08)}_{\text{interest}}$$

$$= 100(1.08)(1 + 0.08)$$
$$= 100(1.08)(1.08)$$
$$= 100(1.08)^2$$

We continue this process until we have $100(1.08)^n$ for n years. It is convenient to present the data on a time line.

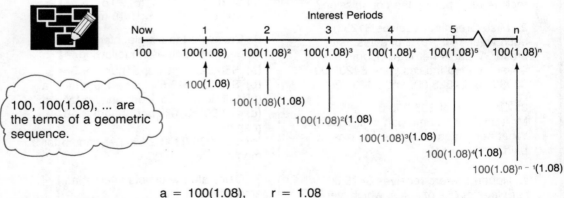

Interest Periods

| Now | 1 | 2 | 3 | 4 | 5 | n |

100   100(1.08)   100(1.08)²   100(1.08)³   100(1.08)⁴   100(1.08)⁵   100(1.08)ⁿ

100(1.08)

100(1.08)(1.08)

100(1.08)²(1.08)

100(1.08)³(1.08)

100(1.08)⁴(1.08)

100(1.08)ⁿ ⁻ ¹(1.08)

100, 100(1.08), ... are the terms of a geometric sequence.

$a = 100(1.08), \quad r = 1.08$

(a) For n years,
$$t_n = ar^{n-1}$$
$$t_n = 100(1.08)(1.08)^{n-1}$$
$$= 100(1.08)^n$$
A formula is $t_n = 100(1.08)^n$

(b) For n = 20,
$$t_{20} = 100(1.08)^{20}$$
$$= 100(4.660\ 957\ 1)$$
$$\doteq 466.10$$
The money accumulates to $466.10.

$3000(1.055)^{10}$

Press

3 0 0 0 ×
1 · 0 5 5
y^x 1 0 =

Display `5124.4334`

From tables:

$5\frac{1}{2}\%$	n
1.619 09	9
1.708 14	10
1.802 09	11
1.901 21	12

In these sections, a calculator was used to 8 figure accuracy. For added convenience, five place tables are provided in the Appendix. Calculations using the five place tables are generally accurate only to four figures.

EXAMPLE 2. Find the amount of $3000 invested for 5 a at 11% per annum compounded semi-annually.

SOLUTION:

The 11% per annum compounded semi-annually for 5 a is equivalent to 5.5% per conversion period (in this case 6 months) for 10 conversion periods (twice a year for 5 a). For this geometric sequence,

$a = 3000(1.055)$, $\quad r = 1.055$, $\quad n = 10$.

$$t_n = ar^{n-1}$$
$$t_{10} = 3000(1.055)(1.055)^{10-1}$$
$$t_{10} = 3000(1.055)^{10}$$
$$= 3000(1.708\ 145)$$
$$\doteq 5124.43$$

The amount after 5 a would be $5124.43.

per annum means per year

We generalize the results of Example 2 in the following formula for Amount of a Lump Sum.

$$A = P(1 + i)^n$$
where A is the amount,
P is the principal,
i is the rate of interest per conversion period,
n is the total number of conversion periods.

EXAMPLE 3. Find the amount of $5000 invested for 10 a at 9% compounded semi-annually.

SOLUTION:

At 9% compounded semi-annually, $i = 0.045$.
For 10 a compounded semi-annually, $n = 20$.

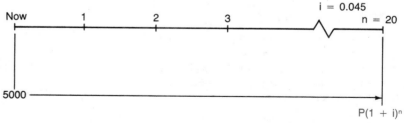

$$A = P(1 + i)^n$$
$$A = 5000(1 + 0.045)^{20}$$
$$= 5000(2.411\ 711\ 4)$$
$$= 12\ 058.57$$

The amount would be $12 058.57.

A 1. (a) Find n and i for a 10% per annum semi-annual rate for 7 a.
(b) Find n and i for an 8% per annum quarterly rate for 10 a.
(c) Find the per annum rate and conversion period if there are 10 conversion periods in 5 a and i = 4.5%.
(d) Find the per annum rate if the rate per conversion period is 2% compounded quarterly.

B 2. Find the amount of each of the following investments.
(a) $5000 at 10% per annum compounded semi-annually for 5 a.
(b) $7500 at 9% per annum compounded semi-annually for 3 a.
(c) $2000 at 8% per annum compounded quarterly for 4 a.

3. Albert Catello invests $5000 for 5 a at 9% per annum compounded semi-annually.
(a) What is his investment worth after 5 a?
(b) How much interest has been earned?

4. Find the amounts that $1000 invested at 12% per annum for 3 a will grow to if the interest is compounded
(a) monthly.
(b) quarterly.
(c) semi-annually.

5. Find the length of time to the nearest half year that it would take a sum of money to double if it is invested at
(a) 8% semi-annually.
(b) 9% semi-annually.
(c) 10% semi-annually.

6. Bob Brown invests $5000 in an account that pays 9% per annum compounded semi-annually. How long must he wait (to the nearest half year) in order to have $8400 for the purchase of a new car?

7. Marie Dubois is saving money for a vacation. On June 1, she invests $800 at 10% compounded semi-annually, and on Dec. 1, she invests an additional $700 at the same rate. How much does she have the following Dec. 1?

C 8. Ken Jones borrowed $10 000 at 9% per annum semi-annually using his insurance policy for collateral. He then reinvested the $10 000 in second mortgages at 16% per annum compounded quarterly. What profit did he make over 5 a?

9. Debbie Lubinski borrows money at 8% per annum compounded quarterly and reinvests it at 12% per annum compounded monthly. Find the annual rate of return by considering an investment of $1 for 1 a.

10. The sum of $20 000 was invested five years ago at 11% compounded semi-annually. Since then, interest rates have dropped and today the principal and interest were reinvested at 9% per annum compounded semi-annually for an additional 5 a. What will the investment be worth in 5 a?

MICRO MATH

This program calculates simple interest and amount.

NEW

```
10 PRINT "SIMPLE INTEREST"
20 INPUT "PRINCIPAL     :  P =";P
30 INPUT "ANNUAL RATE   :  R =";R
40 INPUT "NUMBER OF DAYS: T =";T
50 I=P*R/100*T/365
60 A=P+I
70 PRINT "THE INTEREST IS $ ";I
80 PRINT "THE AMOUNT IS $";A
90 END
```

RUN

EXERCISE

1. Find the simple interest and the amount.
(a) $2500 for 42 d at 12.5% per annum
(b) $3600 for 90 d at 11.25% per annum
(c) $10 525 for 30 d at 15% per annum
(d) How many days will it take an amount of money to double at 12% per annum?

10.2 PRESENT VALUE OF A LUMP SUM

The present value of a lump sum is the sum of money that must be invested now, at a given rate of interest, to produce a desired amount at a later date.

EXAMPLE 1. What principal invested now, at 10% per annum compounded semi-annually, will produce an amount of $9000 for the purchase of a new car in 3 a?

SOLUTION:
The principal to be invested is called the present value, PV. We can put this information on a time line.

Time: n = 6 interest periods   i = 0.05

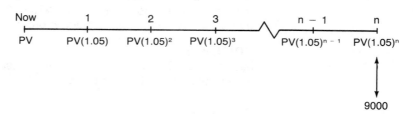

9000

The last term of the sequence, $PV(1.05)^n$, represents the value of the investment after 6 interest periods.

This amount is $9000.

$$A = 9000, \quad i = 0.05, \quad P = PV, \quad n = 6$$

$$A = P(1 + i)^n$$
$$9000 = PV(1 + 0.05)^6$$
$$9000 = PV(1.05)^6$$
$$\frac{9000}{(1.05)^6} = PV$$
$$\frac{9000}{1.340\ 095\ 6} = PV$$
$$6715.938\ 8 = PV$$

$6715.94 should be invested today to give $9000 in 3 a.

The results of Example 1 can be generalized to the formula for Present Value of a Lump Sum.

$$PV = \frac{A}{(1 + i)^n}$$

where PV is the present value,
A is the amount to be achieved,
i is the interest rate per conversion period,
n is the number of conversion periods.

A table of values for $\frac{1}{(1 + i)^n}$ is provided in the Appendix.

EXAMPLE 2. Mr. and Mrs. Conrad have sold their condominium for $70 000 and moved into an apartment closer to their jobs. When they retire in 12 a they would like to have $100 000 for the purchase of a retirement home. How much of the $70 000 should they invest in bonds that pay 9% semi-annually in order to have $100 000 in 12 a?

SOLUTION:
Mr. and Mrs. Conrad should invest the present value of $100 000 at 9% per annum for 12 a. In this case,

$$A = 100\ 000, \quad i = 0.045, \quad \text{and} \quad n = 24.$$

$$PV = \frac{A}{(1 + i)^n}$$

$$PV = \frac{100\ 000}{(1.045)^{24}}$$

$$= \frac{100\ 000}{2.876\ 013\ 8}$$

$$= 34\ 770.35$$

$$\frac{100\ 000}{(1.045)^{24}}$$

Press

[1][0][0][0][0][0][÷]
[1][.][0][4][5]
[yˣ][2][4][=]

Display 34770.347

From tables:

$4\frac{1}{2}$	n
0.363 35	23
0.347 70	24
0.332 73	25

They should invest $34 770.35 to produce the desired $100 000 in 12 a.

EXERCISE 10.2

B 1. Find the present value of each of the following amounts.

(a) $15 000 in 10 a at 9% compounded semi-annually

(b) $2500 in 5 a at 8% compounded quarterly

(c) $500 in 2 a at 12% compounded monthly

(d) $1575 in 3 a at 12% compounded quarterly

(e) $25 000 in 5.75 a at 10% compounded quarterly

2. How much money should be invested today at 10% compounded semi-annually in order to have $8750 in 7 a for the purchase of a new automobile?

3. Ellen Boucher has a paid-up policy that will pay her $20 000 at age 65. What is the value of the policy at age 58? Money is worth 5% per annum.

4. Find the present values of $1000 due in 3 a at 12% per annum if the interest is compounded

(a) monthly.

(b) quarterly.

(c) semi-annually.

5. George Fontaine has signed a promissory note to pay $1500 on Dec. 1. What is the value of the note on April 1 in the same year if money is worth 1% per month on the unpaid balance?

6. Ian and Vera Adams wish to provide for their new baby's education. How much should they invest on the day the child is born in order to have $16 000 on the child's 18th birthday if money is worth 9% per annum compounded semi-annually?

7. Heather White has won a lottery and wishes to set up educational funds for her two children. How much must she invest today at 10% compounded semi-annually in order to have $10 000 each for Colin in 7 a and Dorothy in 12 a if money is worth 9% per annum compounded semi-annually?

8. Louis Vanderman owes $1000 in 1 a and $3000 in 2.5 a. How much money is required to retire both debts if money is worth 11% per annum compounded semi-annually?

C 9. What principal invested for the next 5 a at 10% semi-annually and for the following 4 a at 12% per annum compounded quarterly will amount to $20 000 in 9 a?

10. A station wagon sells today for $19 400. In 3 a, the cost of a new model will increase by 16%. How much should you invest today at 9% compounded semi-annually in order to buy the new model in 3 a with a $7500 trade-in allowance?

WORD LADDER

Start with the word "one" and change one letter at a time to form a new word until you reach "two." The best solution has the fewest steps.

o n e
_ _ _
_ _ _
_ _ _
_ _ _
_ _ _
_ _ _
_ _ _
t w o

10.3 AMOUNT OF AN ANNUITY

An annuity is a sum of money paid as a series of regular equal payments. Although the word annuity suggests annual or yearly, payments can be made semi-annually, quarterly, or monthly. Payments are usually made at the end of the payment interval unless otherwise stated. The amount of an annuity is the sum of the amounts of the individual payments invested from the time of the first payment until the last payment is made, including all the interest.

EXAMPLE 1. Joanne Tate deposits $500 into a high interest bearing savings account every Dec. 15 and June 15 for 8 a. How much will she have in the account at the time of the last payment if interest is earned at 8% per annum compounded semi-annually?

SOLUTION:
We are asked to find the amount of an annuity of 16 semi-annual payments of $500 each at 8% per annum compounded semi-annually, so that $i = 0.04$.
The last payment receives no interest.
The second last payment is in for one conversion period.
The first payment is in for 15 conversion periods.
The problem can be illustrated on a time line.

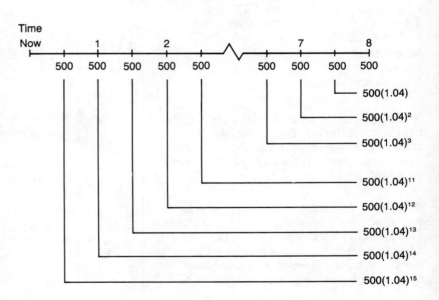

The amount of the annuity is the sum of the amount of the 16 payments.

$$A = 500 + 500(1.04) + 500(1.04)^2 + ... + 500(1.04)^{15}$$

which is a geometric series with
$a = 500,$ $r = 1.04,$ and $n = 16.$

$$S_n = \frac{a(r^n - 1)}{r - 1}$$

$$S_{16} = \frac{500(1.04^{16} - 1)}{1.04 - 1}$$

$$= \frac{500(1.872\,981\,2 - 1)}{0.04}$$

$$= \frac{500(0.872\,981\,2)}{0.04}$$

$$= 10\,912.27$$

The amount of the annuity after 8 a is $10 912.27.

The regular annuity payment is called the periodic rent, R. As in the study of lump sums, i is the interest rate per conversion period and n is the number of payments. We shall assume that the conversion period is equal to the time between payments.

EXAMPLE 2. How much money should be invested every month at 12% per annum compounded monthly in order to have $5000 in 18 months?

SOLUTION:
12% per annum compounded monthly gives i = 0.01.

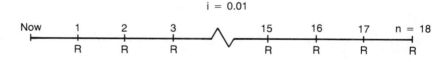

The annuity is
$$R + R(1.01) + R(1.01)^2 + ... + R(1.01)^{17} = 5000$$
where a = R,   r = 1.01,   n = 18,   and   A = 5000.

$$S_n = \frac{a(r^n - 1)}{r - 1}$$

$$5000 = \frac{R(1.01^{18} - 1)}{1.01 - 1}$$

$$5000 = \frac{R(1.196\,147\,5 - 1)}{0.01}$$

$$\frac{5000 \times 0.01}{0.196\,147\,5} = R$$

$$R = 254.91$$

The monthly investment should be $254.91.

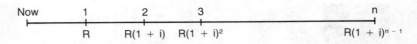

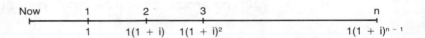

The general annuity
$$R + R(1 + i) + R(1 + i)^2 + ... + R(1 + i)^{n-1}$$
is a geometric series with
$a = R,$ $r = (1 + i),$ having n terms.

$$S_n = \frac{a(r^n - 1)}{r - 1}$$

$$A = \frac{R((1 + i)^n - 1)}{(1 + i) - 1}$$

$$= \frac{R((1 + i)^n - 1)}{i}$$

Solving this formula for R,
$$Ai = R((1 + i)^n - 1)$$
so that,

$$R = \frac{Ai}{(1 + i)^n - 1}$$

To find a general expression for the amount of an annuity, we consider n payments of \$1 each at a rate i per payment interval.

The amount of this annuity is given by the series
$$1 + 1(1 + i) + 1(1 + i)^2 + 1(1 + i)^3 + ... \text{ to n terms}$$
This quantity is represented by the symbol $S_{\overline{n}|i}$ and is evaluated in the annuity table in the Appendix for various values of i and n.

$$S_n = \frac{a(r^n - 1)}{(r - 1)}$$

$$S_{\overline{n}|i} = \frac{1[(1 + i)^n - 1]}{(1 + i) - 1}$$

$$= \frac{(1 + i)^n - 1}{i}$$

$S_{\overline{n}\|i} = \dfrac{(1 + i)^n - 1}{i}$	The amount of an annuity of n payments of \$1 at i% per interval.
$RS_{\overline{n}\|i} = \dfrac{R[(1 + i)^n - 1]}{i}$	The amount of an annuity of n payments of \$R at i% per payment interval.

EXERCISE 10.3

B 1. Find the amount of each of the following annuities.

(a) 20 semi-annual payments of $500 each into an account that pays 8% per annum compounded semi-annually

(b) 16 quarterly payments of $300 each into an account that pays 10% per annum compounded quarterly

(c) 36 monthly payments of $50 each into an account that pays 12% per annum compounded monthly

(d) 10 semi-annual payments of $1000 into an account that pays 10% compounded semi-annually

2. Find the periodic rent in each of the following annuities.

(a) 40 semi-annual payments amounting to $20 000 at 10% compounded semi-annually

(b) an amount of $8000 in 6 a if money is worth 12% per annum compounded quarterly

(c) 24 monthly payments amounting to $3200 at 12% per annum compounded monthly

(d) an amount of $12 000 in 4 a if money is worth 12% per annum compounded every two months

3. Tom Vogel bought a small tractor for his lawn care business. He expects this machine to last for 3 a, and then he will buy a new one for about $21 000. How much must he invest at 1% per month, each month in order to meet this expense in 3 a?
This type of investment plan to meet a future expense is called a sinking fund.

4. The ABC Arena Company wishes to establish a sinking fund to replace the ice making equipment in 12 a. How much should be deposited every 6 months into an account that pays 9% per annum compounded semi-annually in order to have $142 000 for the replacement in 12 a?

5. Mrs. Benton makes provision for her own pension. She invests $1000 every 6 months starting 6 months before her 35th birthday, into an account that pays 9% per annum compounded semi-annually. How much will she have on her 50th birthday?

6. Jones Printing has a small press worth $230 000 with a useful life expectancy of 20 a. How much should the company invest semi-annually in a sinking fund that pays 9% per annum compounded semi-annually to meet this expense in 20 a?

7. The Direct Route Courier Service purchased a new van for $24 700. In 3 a the van will have a trade-in value of $9000 and will be traded in on a similar vehicle expected to cost $28 200. How much should the company invest semi-annually in a sinking fund that pays 10% per annum compounded semi-annually to meet this expense in 3 a?

8. Find the amounts of the following annuities after the last payment has been made.

(a) 12 semi-annual payments of $500 at 8% per annum compounded semi-annually

(b) 36 quarterly payments of $200 at 12% per annum compounded quarterly

(c) 24 semi-annual payments of $1000 at 9% per annum compounded semi-annually

(d) 36 monthly payments of $100 at 18% per annum compounded monthly

C 9. Mrs. Shuster has purchased a new car for $16 200. She realizes that due to driving conditions, she will have to replace the car in 4 a. In that time, new car prices will increase by 40%, and her present car will depreciate 70%. There will be a 7% retail sales tax on the difference between the trade-in value and the new car price. How much should she invest every 3 months in a sinking fund in order to purchase a new car with cash in 4 a if money is worth 12% per annum compounded quarterly?

10. Anne and Todd add to the family allowances received for their son Jason so that they save $50 per month for his education. Every 6 months the funds are deposited into an account that pays 10% per annum compounded semi-annually. How much will be in the account when Jason is 18 if his parents started saving on the day he was born?

10.4 PRESENT VALUE OF AN ANNUITY

An annuity is a sum of money that is received or paid in instalments at regular intervals.

The present value of an annuity is the principal which must be invested now at a given rate of interest in order to provide a given periodic rent. We find the present value of the annuity by finding the present values of all of the lump sums.

EXAMPLE 1. How much money must be invested now at 9% per annum compounded semi-annually to provide an annuity of 10 payments of $200 every 6 months, the first payment being in 6 months.

SOLUTION:
We can illustrate the problem on a time line.

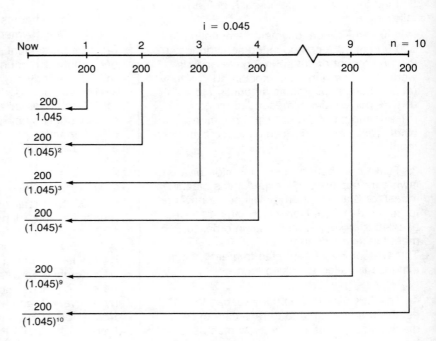

The present value of the annuity is
$$PV = \frac{200}{1.045} + \frac{200}{(1.045)^2} + \frac{200}{(1.045)^3} + \ldots + \frac{200}{(1.045)^{10}}$$
which is a geometric series with
$$a = \frac{200}{1.045}, \quad r = \frac{1}{1.045}, \quad \text{and} \quad n = 10.$$
$$S_n = \frac{a(1 - r^n)}{1 - r}$$

$$PV = \frac{200}{1.045}\left(\frac{1 - \dfrac{1}{1.045^{10}}}{1 - \dfrac{1}{1.045}}\right)$$

$$= \frac{200(1 - 0.643\,927\,7)}{1.045(1 - 0.956\,94)}$$

$$= \frac{200(0.356\,072\,3)}{1.045(0.043\,062\,201)}$$

$$\doteq 1582.54$$

Investing $1582.54 now at 9% per annum compounded semi-annually will provide an annuity of $200 every 6 months for 5 a.

EXAMPLE 2. Rod Carter plans to retire at age 57, and to receive a cash payment of $50 000 from a profit sharing plan. The total sum received will be used to set up an annuity with an insurance company at 8% per annum compounded semi-annually until age 65. How large is each payment if Rod is to receive two equal payments per year?

SOLUTION:
Let each payment in dollars be R.

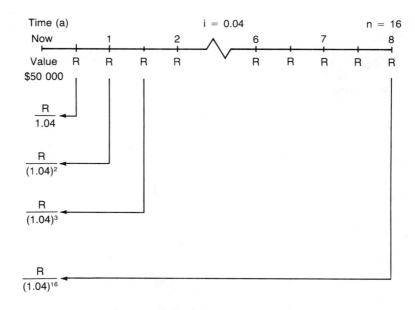

The present value, $50 000, is the sum of the present values:

$$50\,000 = \frac{R}{1.04} + \frac{R}{(1.04)^2} + \frac{R}{(1.04)^3} + \ldots + \frac{R}{(1.04)^{16}}$$

which is a geometric series with

$S_{16} = 50\,000, \qquad a = \dfrac{R}{1.04}, \qquad r = \dfrac{1}{1.04}, \qquad$ and $\qquad n = 16.$

$$S_n = \frac{a(1 - r^n)}{1 - r}$$

$$50\,000 = \frac{\dfrac{R}{1.04}\left(1 - \dfrac{1}{1.04^{16}}\right)}{1 - \dfrac{1}{1.04}}$$

$$50\,000 = \frac{\dfrac{R}{1.04}\,(1 - 0.533\,908\,2)}{1 - 0.961\,538\,5}$$

$$= \frac{\dfrac{R}{1.04}\,(0.466\,091\,8)}{0.038\,461\,539}$$

$$\frac{50\,000 \times 1.04 \times 0.038\,461\,539}{0.466\,091\,8} = R$$

$$R \doteq 4291.00$$

Calculations are simplified if we reverse the series.

$$50\,000 = \frac{R}{1.04^{16}} + \frac{R}{1.04^{15}} + \frac{R}{1.04^{14}} + \ldots + \frac{R}{1.04}$$

$r = 1.04$

$$S_n = \frac{a(r^n - 1)}{r - 1}$$

$$50\,000 = \frac{R}{1.04^{16}}\left(\frac{1.04^{16} - 1}{1.04 - 1}\right)$$

$$50\,000 = R\left(\frac{1 - \dfrac{1}{1.04^{16}}}{0.04}\right)$$

$$50\,000 = R\left(\frac{1 - 0.533\,908\,2}{0.04}\right)$$

$$50\,000 = R\left(\frac{0.466\,091\,8}{0.04}\right)$$

$$\frac{50\,000 \times 0.04}{0.466\,091\,8} = R$$

$$4291.00 = R$$

Each semi-annual payment is $4291.00.

EXERCISE 10.4

B 1. Find the present value of each of the following.

(a) 20 payments of $300 each at 8% per annum compounded semi-annually

(b) 24 payments of $1000 each at 12% per annum compounded quarterly

(c) 36 payments of $75 each at 12% per annum compounded monthly

(d) 10 payments of $5000 each at 10% per annum compounded semi-annually

2. Find the periodic rent of each of the following.

(a) 40 semi-annual payments of 8% per annum with a present value of $10 000

(b) 12 quarterly payments at 12% per annum with a present value of $8000

(c) 24 monthly payments at 12% per annum with a present value of $5600

(d) 15 semi-annual payments at 10% per annum with a present value of $9000

3. Find the present value of $200/month for 2 a beginning in 1 month if interest is earned at 2% per month.

4. Richard Banks has won $5000 to attend university. If he invests the money in a small second mortgage at 12% per annum compounded monthly, how much can he draw monthly for the next 3 a starting 1 month later?

5. Tom Sullivan retires from the Acme Steel Company with his choice of $50 000 cash or 30 equal half-yearly payments earning interest at 9% per annum compounded semi-annually. If the first payment is made 6 months after he retires, how large is each payment?

6. W. "Bill" Harvey pays into an annuity with interest at 9% per annum compounded semi-annually that will pay him $5000 every 6 months for 10 a starting at age 65. Find the present value of the annuity on his 65th birthday.

C 7. An annuity pays $3000 every 6 months for 10 a from an account that pays 9% per annum compounded semi-annually. Find the present value of the annuity if the first payment is due

(a) now.

(b) in 6 months.

8. Find the monthly payment required to pay off an automobile worth $10 125 including taxes, if the down payment was $2500 and the balance was financed over 2 a at 12% per annum compounded monthly.

9. Mr. and Mrs. Yates win $250 000 in a lottery on Mr. Yates' 45th birthday. How much should they invest as a lump sum today in order to have 10 payments of $10 000 each every 6 months starting 6 months after Mr. Yates' 60th birthday? The money is invested at 10% per annum compounded semi-annually.

10. Tim Stevens signs a professional baseball contract that will pay him $125 000 per year for 5 a plus a bonus for signing in the form of an annuity of $25 000 per year for 20 a to follow. Money is worth 6% per annum, compounded annually. Find.

(a) the present value of the annuity at the time of signing

(b) the total cost of the contract to the club

(c) Compare the total cost of the contract to the club with the total amount of money Tim Stevens will receive.

A small boat is floating in a bathtub. Which will raise the level of the water in the tub more, dropping a penny into the tub or into the boat?

10.5 DEFERRED ANNUITIES

When the first payment of an annuity is not due for some time, we say that the annuity is deferred.

EXAMPLE 1. Find the present value of an annuity of ten quarterly payments of $500. Interest is earned at 10% per annum compounded quarterly. The first payment is deferred one year.

10% compounded quarterly means 10% per annum compounded quarterly.

SOLUTION:
If the annuity had not been deferred the first payment would have been due one quarter from "Now." The deferment of one year makes the first payment due 5 quarters from "Now."

$$i = (\tfrac{1}{4})(10\%)$$
$$= 2.5\%$$

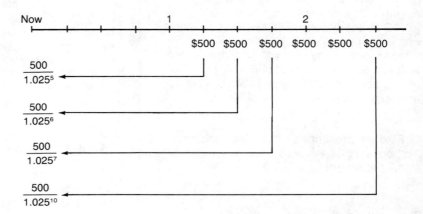

From the diagram it can be seen that present value of the deferred annuity,

$$PV = \frac{500}{1.025^5} + \frac{500}{1.025^6} + \frac{500}{1.025^7} + ... + \frac{500}{1.025^{10}}$$

$$a = \frac{500}{1.025^5}, \qquad r = \frac{1}{1.025}, \qquad n = 6, \qquad \text{and} \qquad S_n = PV.$$

$$S_n = \frac{a(1 - r^n)}{1 - r}$$

$$PV = \frac{\dfrac{500}{1.025^5}\left(1 - \dfrac{1}{1.025^6}\right)}{1 - \dfrac{1}{1.025}}$$

$$\doteq \frac{500\,(1 - 0.862\ 296\ 9)}{1.025^5\,(1 - 0.975\ 609\ 8)}$$

$$\therefore PV \doteq 2495.05$$

The present value is $2495.05.

EXAMPLE 2. Find the amount of an annuity of 4 semi-annual payments of $1200 if the repayment is deferred to 3 a after the last rent payment. Interest is paid at 8% per annum compounded semi-annually.

SOLUTION:

$$i = (\tfrac{1}{2})(8\%)$$
$$= 4\%$$

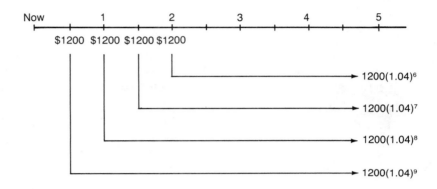

From the diagram it can be seen that,
the deferred amount of an annuity is
$$A = 1200(1.04)^6 + 1200(1.04)^7 + 1200(1.04)^8 + 1200(1.04)^9$$
$$a = 1200(1.04)^6, \qquad r = 1.04, \qquad n = 4, \qquad \text{and} \qquad S_n = A.$$

$$S_n = \frac{a(r^n - 1)}{r - 1}$$

$$A = \frac{1200(1.04)^6(1.04^4 - 1)}{1.04 - 1}$$

$$\doteq \frac{1200(1.265\ 319)(1.169\ 858\ 6 - 1)}{0.04}$$

$$\doteq 6447.76$$

The amount is $6447.76.

EXAMPLE 3. Mrs. Wilson invested $10 000 at 12% per annum compounded monthly. She wishes to withdraw the money in 24 equal monthly payments, the first payment deferred for one year. How much will she receive in each payment?

SOLUTION:

$$i = \frac{1}{12}(12\%)$$
$$= 1\%$$

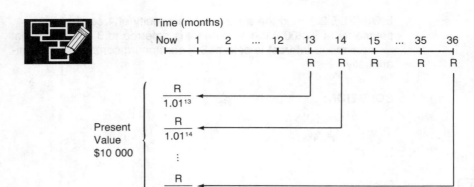

From the diagram it can be seen that, present value of the deferred annuity,

$$PV = \frac{R}{1.01^{13}} + \frac{R}{1.01^{14}} + \dots + \frac{R}{1.01^{36}}$$

$$S_n = 10\ 000, \qquad a = \frac{R}{1.01^{13}}, \qquad n = 24, \qquad \text{and} \qquad r = 0.01.$$

$$S_n = \frac{a(1 - r^n)}{1 - r}$$

$$10\ 000 = \frac{\dfrac{R}{1.01^{13}}\left(1 - \dfrac{1}{1.01^{24}}\right)}{1 - \dfrac{1}{1.01}}$$

$$10\ 000 = \frac{R\ (1 - 0.787\ 566\ 2)}{1.01^{13}\ (1 - 0.990\ 099)}$$

$$R = 530.44$$

$$R = \frac{10\ 000(1.01^{13})(0.009\ 901)}{(0.212\ 243\ 38)}$$

Mrs. Wilson receives $530.44 each month.

EXERCISE 10.5

A 1. Describe what is meant by the following terms:

(a) annuity
(b) present value of an annuity
(c) amount of an annuity
(d) deferred annuity

2. A deferred annuity of 12 quarterly payments is purchased on November 1, 1978. If the annuity is deferred for 5 months, when will the first payment be received?

3. (a) Will deferring an annuity increase or decrease its present value? Why?
(b) When purchasing an annuity with a large sum, will deferring the annuity increase or decrease the rental payments?

B 4. An annuity of 6 semi-annual payments is deferred for 2 a. If the semi-annual rent is $750 and interest is paid at 12% compounded semi-annually, find the present value.

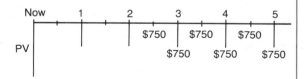

5. Find the present value of an annuity of $250 per month, the first payment is 6 months from now, and it is to continue for 12 payments. Interest is paid at a rate of 1.5% per month.

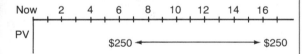

6. $5000 is invested at 12% compounded quarterly and withdrawn in 8 equal quarterly payments. If the first withdrawal is made 2 a after the investment date, how large is each payment?

7. A debt of $2500 is to be repaid by 24 equal monthly payments. The first monthly payment is due 3 months from the date of purchase. If interest is charged at 1.5% per month, how large is each payment?

8. Find the present value of an annuity of 20 quarterly payments of $500, the first payment to be made in one year. The money is invested at 10% compounded quarterly.

9. An annuity of 30 semi-annual payments of $1000 is allowed to gather interest for an additional 2 a before being cashed in. If it has earned interest at 7% compounded semi-annually, how much is it worth?

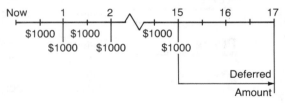

10. (a) Find the amount of an annuity of 36 monthly payments of $75, interest is paid at 1% per month.
(b) Find the amount of the same annuity if it is allowed to accumulate interest for an additional year.

C 11. Mr. Kowalchuk wishes to invest while his earnings are high, so that he and his wife can take a trip when he retires. He wants $10 000 when he is 65, and makes 20 equal semi-annual payments into an annuity with the last payment on his 55th birthday. If the investment earns 8% compounded semi-annually, how large must each payment be?

12. (a) Find the present value of an annuity of three annual payments of $500 which is deferred for 3 a, by finding the value at year 3 and then using compound interest tables to find the present value of this sum. Interest is paid at 6% per annum.
(b) Repeat the solution using the method described in this section as a check.

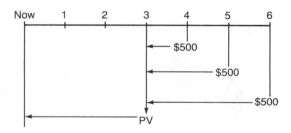

10.6 PROBLEMS SOLVED AS ANNUITIES

The problems that follow are related to annuities. In solving them, we apply the following steps in our problem solving model.

READ

Determine whether you are given the periodic rent, the present value, a simple annuity, or a deferred annuity.

PLAN

Draw a time payment diagram. Determine whether you are required to find a periodic rent payment, the present value, or the amount of an annuity.
Decide on the appropriate formula.

SOLVE

State the formula and show the substitution.
Perform the calculations using:
(i) tables,
(ii) a calculator,
(iii) a computer.

ANSWER

Check for gross error. Conclude with a statement.

EXAMPLE. Mr. Armstrong purchased a boat for $4500. He paid $1000 down and financed the remainder at 1.5% per month on the unpaid balance over 3 a. What are the equal monthly payments?

SOLUTION:
Let each payment in dollars be R. The amount financed is
$4500 − $1000 = $3500, and i = 1.5%.

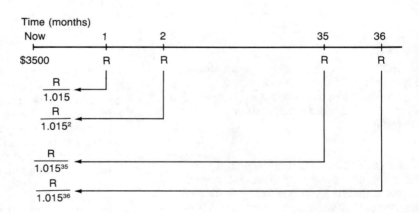

From the diagram we see that we are given the present value, are required to find the periodic rent, and are dealing with a simple annuity.

$$3500 = \frac{R}{1.015} + \frac{R}{1.015^2} + \frac{R}{1.015^3} + \dots + \frac{R}{1.015^{36}}$$

$$S_n = 3500, \quad a = \frac{R}{1.015}, \quad r = \frac{1}{1.015}, \quad n = 36$$

$$S_n = \frac{a(1 - r^n)}{1 - r}$$

$$3500 = \frac{\dfrac{R}{1.015}\left(1 - \dfrac{1}{1.015^{36}}\right)}{1 - \dfrac{1}{1.015}}$$

$$3500 \doteq \frac{R\,(1 - 0.585\ 089\ 7)}{1.015\,(1 - 0.985\ 221\ 7)}$$

$$\therefore R \doteq 126.53$$

The monthly payment will be \$126.53.

Checking for gross error.
$125 \times 36 = 4500$

EXERCISE 10.6

B 1. A stereophonic sound system costs \$565 with no down payment. The account is paid in 18 equal monthly instalments with interest of 1.5% per month on the unpaid balance. Interest starts immediately with payments due in one month. How large is each payment?

2. A new automobile costs \$9500 with an allowance of \$2200 on a trade-in as down payment. If the difference is paid in 36 equal monthly payments with interest at 1% per month on the unpaid balance, find the monthly payments. The first payment is due in one month.

3. A piano is advertised at \$200 down and \$30 per month for 24 months. If interest is being charged at 1.5% per month, what is the cash price?

4. A stereo is sold for \$150 down and \$20 per month for 12 months. If interest is being charged at 2% per month, what is the cash price?

5. A fur coat costs \$2000. If a down payment of \$500 is made and the remainder financed over 12 months at 1% per month service charge on the unpaid balance, what are the monthly payments?

6. A portable stereo cassette tape recorder costs \$75 cash, or \$10 down with interest on the unpaid balance at 2% per month, financed over 10 months. Find the monthly payment, the total amount paid, and the total interest charged.

7. Mr. Firth and Mr. Lasky both buy a new car every second year and pay an average of \$4000 plus trade-in.
(a) Mr. Firth borrows the money for his car and pays the loan in 24 monthly payments, starting the month after the car is purchased. If the money is borrowed at 1% per month on the unpaid balance, what are his monthly payments?
(b) Mr. Lasky pays cash for his car and starts one month later to make monthly payments into an account paying 1% per month compounded monthly, so that in 24 months he will have \$4000 for his new car. How much must he invest monthly?
(c) What is the difference in their total payments for the 2 a?

8. Acme Construction Company replaces its power shovel every 5 a and invests in a sinking fund at 10% compounded quarterly. To provide the funds, the company expects to pay \$25 000 plus trade and starts making quarterly deposits 3 months after the purchase of the latest shovel. If the last payment is made when the new shovel is bought, how large must each quarterly payment be?

9. (a) At age 45, Mr. McDermitt invests \$20 000 to be repaid in 20 semi-annual payments starting on his 55th birthday. If the money is invested at 11% compounded semi-annually, how large will each payment be?
(b) How large are the payments if he waits until his 60th birthday?

10.7 EFFECTIVE ANNUAL RATE: LUMP SUM

When interest is added to the principal, the effective rate of interest is greater than the advertised, or nominal rate. For example, a nominal rate of 10% compounded semi-annually would have interest added to the principal every six months. Considering a deposit of $100,

$$A = P(1 + i)^n$$

$$A = 100(1 + 0.05)^2$$

$$= 100 \times 1.1025$$

$$= 110.25$$

The amount at the end of the year would be $110.25 instead of $110.00 due to the semi-annual calculation. When it was stated that the interest rate was 10%, it was only nominally correct. The effective annual rate of interest is 10.25%. We use this method to derive a formula for the effective annual rate.

Let the effective annual rate be i.

Let the nominal (advertised) rate be j.

Let the number of conversion periods in a year be m.

The amount of $1 for one year will be

$$A = 1\left(1 + \frac{j}{m}\right)^m$$

$$= \left(1 + \frac{j}{m}\right)^m$$

Subtracting 1, the principal at the beginning of the year, we have the actual interest for the year

$$\left(1 + \frac{j}{m}\right)^m - 1$$

which is the effective rate.

> For a lump sum, the effective annual rate of interest for a nominal rate j, converted m times per year is
>
> $$i = \left(1 + \frac{j}{m}\right)^m - 1$$

EXAMPLE. Find the effective rate when the advertised rate of interest is 9.5% compounded monthly.

SOLUTION:
Using the formula

$$i = \left(1 + \frac{j}{m}\right)^m - 1$$

$j = 0.095$ and $m = 12$.

$$i = \left(1 + \frac{0.095}{12}\right)^{12} - 1$$

$$= (1.007\ 916\ 7)^{12} - 1$$

$$= 1.099\ 248 - 1$$

$$= 0.099\ 248$$

∴ the effective annual rate of interest is 9.925%.

EXERCISE 10.7

B 1. Find the effective annual rate of interest on each of the following advertised rates.
   (a) 9% semi-annually
   (b) 12% quarterly
   (c) 15% semi-annually
   (d) 10% quarterly
   (e) 8% quarterly

2. Find the difference in interest between 8% compounded annually, and 8% compounded semi-annually.

3. Compare the effective rates of interest for 1% per month, and 12% compounded monthly.

4. What is the additional interest cost to a bank that normally pays 9% annually if 9% semi-annually was paid on deposits totalling $3 500 000?

5. Find the effective rate of interest when 10% is compounded
   (a) annually.
   (b) semi-annually.
   (c) quarterly.
   (d) monthly.

6. An automobile dealer advertises cars sold with no down payment and an annual rate of 8.9% interest compounded monthly on the price of the car. What is the effective annual rate of interest?

7. The Acme Auto Parts Company charges 15% interest compounded monthly on overdue accounts. What is the effective annual rate of interest on an account of $18 925 which is 7 months overdue?

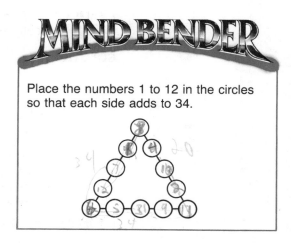

MIND BENDER

Place the numbers 1 to 12 in the circles so that each side adds to 34.

10.8 EFFECTIVE RATE FOR AN INTERVAL: ANNUITIES

In this section we derive a formula to determine the rate of interest for the interval p by considering an amount of 1. Where the interest is converted m times per year, the amount at the end of 1 a would be

$$\left(1 + \frac{j}{m}\right)^m$$

Letting the interest for the period, p, be i_p, the amount for p periods, or 1 a, would be

$$(1 + i_p)^p$$

Equating these, we have

$$(1 + i_p)^p = \left(1 + \frac{j}{m}\right)^m$$

Taking the pth root of each side,

$$(1 + i_p)^{\frac{p}{p}} = \left(1 + \frac{j}{m}\right)^{\frac{m}{p}}$$

$$1 + i_p = \left(1 + \frac{j}{m}\right)^{\frac{m}{p}}$$

$$i_p = \left(1 + \frac{j}{m}\right)^{\frac{m}{p}} - 1$$

The interest rate for the interval p of an amount converted m times per year at a nominal rate j is

$$i_p = \left(1 + \frac{j}{m}\right)^{\frac{m}{p}} - 1$$

EXAMPLE 1. Find the rate for a 6 month period if payments are made twice yearly and interest is at the rate of 10% per annum.

SOLUTION:

$$i_p = \left(1 + \frac{j}{m}\right)^{\frac{m}{p}} - 1$$

Where j = 0.10,   m = 1,   and   p = 2.

$$i_p = \left(1 + \frac{0.1}{1}\right)^{\frac{1}{2}} - 1$$

$$= (1.1)^{\frac{1}{2}} - 1$$
$$= 1.048\ 808\ 8 - 1$$
$$= 0.048\ 808\ 8$$

∴ the effective interest rate is 4.88% for 6 months.

EXAMPLE 2. Find the effective rate when the payments are made quarterly with interest at 9% compounded semi-annually.

SOLUTION:

$$i_p = \left(1 + \frac{j}{m}\right)^{\frac{m}{p}} - 1$$

Where $j = 0.09$, $\qquad m = 2$, $\qquad$ and $\qquad p = 4$.

$$i_p = \left(1 + \frac{0.09}{2}\right)^{\frac{2}{4}} - 1$$

$$= (1.045)^{\frac{2}{4}} - 1$$
$$= 1.022\ 252\ 4 - 1$$
$$= 0.022\ 252\ 4$$

∴ the effective rate of interest is 2.23% for 3 months.

EXERCISE 10.8

B 1. What is the effective interest rate for a 3 month period if payments are made quarterly, and the nominal rate of interest is 8% per annum?

2. Determine the rate for a 6 month period if payments are made twice yearly and interest is at the nominal rate of 12% per annum.

3. What is the effective rate if payments are made monthly and the nominal rate is 12% per annum?

4. Compare the effective rates of interest when payments are made quarterly with interest at 11% compounded
(a) annually.
(b) semi-annually.

5. What is the effective quarterly rate of interest if payments are made every three months and the nominal rate is 12% per annum compounded semi-annually?

6. What is the effective rate of interest if payments are made every three months and the nominal rate of interest is 8% per annum compounded quarterly?

7. Samantha Charette makes payments every three months on a special mortgage arrangement. Interest is charged at 10.5% per annum compounded semi-annually. What is the effective rate of interest for a three month interval?

8. Fay and Charles purchase a new three wheel vehicle and pay for it with a trade-in followed by monthly payments. They will make monthly payments on the loan and pay interest at a rate of 12% per annum compounded semi-annually. What is the rate of interest charged for a one month period?

9. Find the effective rate of interest when monthly payments are made and the nominal rate of interest is 11.5% per annum compounded semi-annually.

10.9 ANNUITIES IN PERPETUITY

An annuity in perpetuity is one which continues forever. These investments are useful in setting up scholarship funds to ensure that the award continues. For example, if we were to invest $10 000 at 10% per annum, we could set up an annuity that pays $1000 per annum.

EXAMPLE. How much money should be invested today at 11% per annum compounded semi-annually in order to set up an annuity in perpetuity of $5000? The first payment is due in one year.

SOLUTION:

$$PV = \frac{5000}{1.055^2} + \frac{5000}{1.055^4} + \frac{5000}{1.055^6} + \dots$$

For this infinite series,

$$S = \frac{a}{1 - r}$$

$$a = \frac{5000}{1.055^2} \qquad r = \frac{1}{1.055^2}.$$

$$PV = \frac{\dfrac{5000}{1.055^2}}{1 - \dfrac{1}{1.055^2}}$$

$$= \frac{5000}{1.055^2 - 1}$$

$$= \frac{5000}{1.113\ 025 - 1}$$

$$= 44\ 238$$

$44 238.00 should be invested.

EXERCISE 10.9

B 1. How much money must be invested now in order to set up a scholarship of $1000 per annum continuing forever, if the funds are invested at 12% per annum compounded semi-annually.

2. What is the present value of an annuity in perpetuity of $500 if the funds are invested at 10% per annum compounded semi-annually? The first payment takes place in one year.

3. How much should be invested today in an account that pays 15% per annum compounded semi-annually in order to set up an annuity in perpetuity of $500 for a citizenship award.

4. The sum of $15 250 was collected to set up a memorial scholarship that would continue on forever. The funds are invested in a plan that pays 10% per annum, and equal payments will begin next year. Find the value of each payment.

5. The Scriven Prize of $10 000 is awarded annually for literary achievement. The fund is set up as an annuity in perpetuity where the original funds continue to earn interest at 8% per annum compounded semi-annually. What would the prize be if the interest was at 12% per annum compounded monthly?

10.10 COMPUTER PROGRAMS FOR INTEREST AND ANNUITIES

The following programs can be used to solve problems in this chapter. All statements should be entered on one line.

1. Amount of a lump sum.
NEW
```
10 PRINT "AMOUNT OF A LUMP SUM"
20 PRINT "A = P(1 + I)↑N"
30 INPUT "PRINCIPAL";P
40 INPUT "RATE/CONVERSION PERIOD";I
50 INPUT "CONVERSION PERIODS";N
60 A = P*(1 + I)↑N
70 PRINT "THE AMOUNT IS"
80 PRINT "& =";P;" (1 + ";I;")↑";N
90 PRINT "A =";A
100 END
```
RUN

2. Present value of a lump sum.
NEW
```
10 PRINT "PRESENT VALUE OF A LUMP SUM"
20 PRINT "A = P (1 + I)↑N"
30 INPUT "PRINCIPAL";P
40 INPUT "RATE/CONVERSION PERIOD";I
50 INPUT "CONVERSION PERIODS";N
60 A=P/(1+I)↑N
70 PRINT "THE AMOUNT IS"
80 PRINT "A =";P;"/ (1 + ";I;")↑";N
90 PRINT "A =";A
100 END
```
RUN

3. Amount of an annuity.
NEW
```
10 PRINT "AMOUNT OF AN ANNUNITY"
20 PRINT "A = R((1 + I)↑N-1)/I"
30 INPUT "PERIODIC RENT";R
40 INPUT "RATE/CONVERSION PERIOD";I
50 INPUT " CONVERSION PERIODS";N
60 A=R*((1+I)↑N-1)/I
70 PRINT "THE AMOUNT IS"
80 PRINT "A =";R;"((1 + ";I;)
   ↑";N "- 1)/"I
90 PRINT "A =";A
100 END
```
RUN

4. Present value of an annuity.
NEW
```
10 PRINT "PRESENT VALUE OF AN ANNUITY"
20 PRINT "A = R/(1 + I)↑N*((1 + I)↑N - 1)/I
30 INPUT "PERIODIC RENT";R
40 INPUT "RATE/CONVERSION PERIOD";I
50 INPUT "CONVERSION PERIODS";N
60 A=R/(1+I)↑:*((1+I)↑N-1)/I
70 PRINT "THE AMOUNT IS"
80 PRINT "A =" R"/(1+"I")↑"N"
   *((1+"I")↑"N"-1)/"I
90 PRINT "A =";A
100 END
```
RUN

5. Periodic payment for an amount of an annuity.
NEW
```
10 PRINT "PERIODIC PAYMENT"
20 PRINT "R = A*I/((1+I)↑N-1)"
30 INPUT "AMOUNT";A
40 INPUT "INTEREST RATE";I
50 INPUT "NUMBER PAYMENTS";N
60 R=A*I/((1+I)↑N-1)
70 PRINT "THE PAYMENTS ARE"
80 PRINT "R="A"*"I"/((1+"I")↑"N"-1)"
90 PRINT "R =";R
100 END
```
RUN

6. Periodic payment for a present value of an annuity.
NEW
```
10 PRINT "PERIODIC PAYMENT"
20 PRINT "R=V*I*(1+I)↑N/((1+I)↑N-1)"
30 INPUT "PRESENT VALUE";V
40 INPUT "INTEREST RATE";I
50 INPUT "NUMBER PAYMENTS";N
60 R=V*I*(1+I)↑N/((1+I)↑N-1)
70 PRINT "THE PAYMENTS ARE"
80 PRINT "R="V"*"I"*(1+"I")↑"N"/"
85 PRINT "((1+"I")↑"N"-1)"
90 PRINT "R =";R
100 END
```
RUN

7. Effective annual rate of a loan.
NEW
```
10 PRINT "EFFECTIVE ANNUAL RATE"
20 PRINT "R = 2*N1*I/(A*(N2 + 1))"
30 INPUT "PAYMENTS/YEAR N1";N1
40 INPUT "TOTAL PAYMENTS N2";N2
50 INPUT "TOTAL INTEREST I";I
60 INPUT "PRINCIPAL AMOUNT A";A
70 PRINT "THE RATE IS"
80 R=2*N1*I/A/(N2+1)
90 PRINT "THE EFFECTIVE ANNUAL RATE IS"
100 PRINT "R = 2*"N1"*"/"A"/("N2"+1)"
110 PRINT "R ="R
120 END
```
RUN

These programs can be made reiterative by inserting the following three lines in each program ahead of the END statement.

```
□□ PRINT "ANOTHER QUESTION?"
□□ INPUT "Y" OR "N";Z$
□□ IF Z$="Y" THEN 10
```

10.11 REVIEW EXERCISE

1. Find the compound amount of the following.
(a) $3500 at 9% for 3 a, compounded semi-annually
(b) $100 at 1% per month compounded monthly for 10 months
(c) $1200 at 10% per annum compounded semi-annually for 18 months
(d) $2400 at 9% per annum compounded semi-annually for 36 months
(e) $4500 at 9% per annum for 12 a compounded semi-annually

2. Find the present value of the following.
(a) $1000 due in 3 a at 8% per annum compounded semi-annually
(b) $3000 due in 2 a at 10% per annum compounded semi-annually
(c) $1200 due in 2 a at 8% per annum compounded semi-annually
(d) $1200 due in 2 a at 8% per annum compounded quarterly
(e) $500 due in 24 months at 1% per month compounded monthly
(f) $100 per month paid for 3 months the first payment made 6 months from now. Money is worth $1\frac{1}{2}$% per month compounded monthly.

3. Find the compound amount of the following.
(a) $500 for 3 a at 8% per annum compounded annually
(b) $100 for 2 a at 10% per annum compounded semi-annually
(c) $1500 for 4 a at 8% per annum compounded annually
(d) $325 for 48 months at 12% per annum compounded quarterly
(e) $600 for 3 a at 13% per annum compounded quarterly

4. In how many years will $100 double if it is invested at 10% per annum compounded semi-annually?

5. What is the smallest rate of interest to the nearest $\frac{1}{10}$% that must be charged for $650 to triple in 18 a, if the interest is to be compounded semi-annually?

6. What sum of money should a family invest on the birth of a child to provide $600 at the age of 18 and $1000 at the age of 20? The money is to be invested at 7% per annum compounded semi-annually.

7. In how many years will $500 double if it is invested at 8% per annum compounded quarterly?

8. Which has the greater present value and by how much?
(i) $600 due in 5 a at 8% per annum compounded quarterly
(ii) $600 due in 6 a at 7% per annum compounded annually

9. What sum of money paid 5 a from today is equivalent to payments of $1500 paid 2 a from now, and $3500 paid 7 a from now, if money is worth 8% per annum compounded semi-annually?

10. Find the present value of 5 annual payments of $500 if the first payment is made 1 a from now and money is worth 8% compounded semi-annually.

11. Find the present value of 6 semi-annual payments of $100 if the first payment is made 2 a from now and money is worth 9% compounded semi-annually.

12. $50 per month is placed in an account for 6 months. Interest at 1% per month is added to the account at the end of each month after the first deposit.
(a) How much is in the account at the time of the last deposit?
(b) One month after the last deposit, withdrawals of $50 per month are made for 6 months. How much is in the account after the last withdrawal?

13. Find the amount of each of the following annuities.
(a) 20 semi-annual payments of $500 each into an account that pays 12% per annum compounded semi-annually

(b) 16 quarterly payments of $300 each into an account that pays 10% per annum compounded quarterly
(c) 36 monthly payments of $50 each into an account that pays 16% per annum compounded quarterly
(d) 10 semi-annual payments of $2000 into an account that pays 11% per annum compounded semi-annually.

14. Find the periodic rent in each of the following annuities.
(a) 40 semi-annual payments amounting to $50 000 at 10% per annum compounded semi-annually
(b) an amount of $24 000 in 6 a if interest is paid at 12% per annum compounded quarterly
(c) 24 monthly payments amounting to $7500 at 12% per annum compounded monthly
(d) an amount of $25 000 in 4 a if interest is at 12% per annum compounded every 2 months

15. Find the present value of each of the following annuities.
(a) 20 payments of $300 each at 12% per annum compounded semi-annually
(b) 24 payments of $2000 each at 12% per annum compounded quarterly
(c) 36 payments of $175 each at 12% per annum compounded monthly

16. Find the periodic rent of each of the following.
(a) 40 semi-annual payments at 10% per annum with a present value of $10 000
(b) 12 quarterly payments at 12% per annum with a present value of $12 000
(c) 24 monthly payments at 18% per annum with a present value of $6000

17. An investment plan where funds are invested in order to meet a future expense is called a sinking fund. The Patine Arena Company wishes to establish a sinking fund to replace the ice-making equipment in 12 a. How much money should be invested every 6 months into an account that pays 11% per annum compounded semi-annually in order to have $85 000 to replace the equipment in 12 a?

18. Janice Amyot bought a van for her courier business. She expects the van to last for 3 a, and then she will buy a new one for about $25 000. How much must she invest every month in a plan that pays 12% per annum compounded monthly in order to meet this expense in 3 a?

19. Jean Lacroix won a $5000 scholarship to attend university. He can invest the money for 3 a at 12% per annum compounded monthly. How much can he draw monthly from this investment starting 1 month later?

20. Mark Santini retires from the Acme Steel Company with his choice of $100 000 cash, or 30 equal half-yearly payments earning interest at 13% per annum compounded semi-annually. If the first payment is made 6 months after he retires, how large is each payment?

21. Betty Jones Printing has a small press worth $500 000 with a useful life expectancy of 20 a. How much money should the company invest semi-annually in a sinking fund that pays interest at a rate of 11% per annum compounded semi-annually in order to meet this expense in 20 a?

22. Rose and Ralph Lee pay into an annuity with interest at 13% per annum compounded semi-annually. The annuity will pay them $8000 every 6 months for 10 a. Find the present value of the annuity on the day of the first payment.

23. Shelly deposits $100 every month for 2 a into an account that pays 12% per annum compounded monthly. She plans to withdraw the money in 12 equal payments over the next year.
(a) How much money is in the account after 2 a?
(b) What is the monthly payment she will receive?

24. Find the monthly payment required to pay off a car worth $18 725 including taxes if the trade-in was worth $5275 and the balance was financed over 3 a at 18% per annum compounded monthly.

10.12 CHAPTER 10 TEST

1. Find the amount of $8000 invested for 5 a at 11% per annum compounded semi-annually.

2. What lump sum should be invested today in an account that pays 1% per month compounded monthly in order to have $1000 in 3 a?

3. How much money should be invested every month at 12% per annum compounded monthly in order to have $10 000 in 2.5 a?

4. What amount of money should be invested now at 11% per annum compounded semi-annually in order to provide an annuity of $2000 every 6 months for 5 a?

5. How much money will there be in 20 months if $50 is deposited every month in an account that pays 9% per annum compounded monthly?

6. Margot Tilley invested $4000 in an account that pays 1% per month, and she wants to draw 8 equal monthly payments while she is at university. Find the value of each payment.

7. Find the present value of an annuity of eight semi-annual payments of $3000. Interest is earned at 11% per annum compounded semi-annually. The first payment is deferred 2 a.

8. A synthesizer is advertised at $500 down and $100 per month for 18 months. If interest is charged at 1.5% per month compounded monthly, what is the cash price of the synthesizer?

MATHEMATICS
OF
INVESTMENT

CHAPTER

11

The greatest mathematicians, as Archimedes, Newton and Gauss,
always united theory and applications in equal measure.

Felix Klein

REVIEW AND PREVIEW TO CHAPTER 11

PERCENTS AND MONEY

EXERCISE

1. Find the following.
(a) 10% of $25 000
(b) 11.5% of $75 000
(c) 12% of $68 000
(d) 10.25% of $62 000

2. (a) What percent of $65 is $26?
(b) What percent of $4500 is $500?
(c) What percent of $72 000 is $24 000?
(d) What percent of $66 000 is $22 000?

3. (a) $50 000 is 10% of what amount?
(b) $65 000 is 15% of what amount?
(c) $135 000 is 12% of what amount?
(d) $84 000 is 11% of what amount?

4. Jackie Hames receives 2.5% commission on the first $100 000 of gross sales and 3% after that each month. Calculate her commission for each of the following months.
(a) January $97 000
(b) February $118 000
(c) March $105 000
(d) April $124 000

5. Using the formula $I = Prt$, find the simple interest on the following amounts as indicated.
(a) $20 000 at 12% for 1 a
(b) $65 000 at 11.5% for 4 a
(c) $55 000 at 10% for 30 d
(d) $125 000 at 12% for 90 d

6. Find the real estate commission paid on the following sales if the rate is 6%.
(a) Bungalow $ 85 900
(b) Two-storey $145 000
(c) Split-level $125 000
(d) Business $250 000

INVESTING MONEY

EXERCISE

1. Find the amount of the following lump sums.
(a) $500 for 3 a at 10% compounded semi-annually
(b) $2500 for 2 a at 12% compounded monthly
(c) $65 000 for 10 a at 12% compounded quarterly

2. Find the present value of the following.
(a) $1000 due in 3 a at 9% compounded semi-annually
(b) $5000 due in 4 a at 10% compounded quarterly
(c) $16 000 due in 18 a at 12% compounded quarterly

3. Find the amount of the following annuities.
(a) $200 per month for 24 months at 12% per annum compounded monthly
(b) $500 every 6 months for 3 a in an account that pays 11% compounded semi-annually

4. Find the present value of the following annuities.
(a) $100 per month for 3 a at 1% per month
(b) $500 every 6 months for 36 months in an account that pays 11% compounded semi-annually
(c) $1000 per annum for 10 a in an account that pays 13%

5. How much should Derek deposit every month starting today in order to have $15 000 in 3 a, if the money earns 1% per month?

INTERPOLATION

A difficulty arises in some questions when we are unable to approximate the answer using normal methods. If the problem does not require a high degree of accuracy, we can interpolate using the numbers we have. Interpolation is similar to averaging two numbers to get an "in-between" value.

EXAMPLE. Using interpolation, find the compound amount of $1000 invested at $9\frac{1}{2}\%$ for 5a compounded annually, given:

$1000 at 9% for 5 a: $1000(1.09)^5 = 1538.62$
$1000 at 10% for 5 a: $1000(1.10)^5 = 1610.51$

SOLUTION:
From the given information, $1538.62 at 9% is too low, and $1610.51 at 10% is too high.

Let the amount by which the $9\frac{1}{2}\%$ amount exceeds $1538.62 be $x. Then the $9\frac{1}{2}\%$ amount is $(1538.62 + x)$. We order the figures and set up the question as follows.

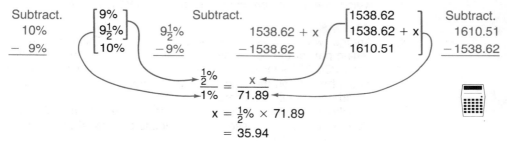

Subtract.
$$
\begin{array}{c}
10\% \\
-\ 9\% \\
\end{array}
\qquad
\begin{bmatrix}
9\% \\
9\frac{1}{2}\% \\
10\%
\end{bmatrix}
\qquad
\begin{array}{c}
9\frac{1}{2}\% \\
-9\%
\end{array}
$$

Subtract.
$$
\begin{array}{c}
1538.62 + x \\
-1538.62
\end{array}
\qquad
\begin{bmatrix}
1538.62 \\
1538.62 + x \\
1610.51
\end{bmatrix}
\qquad
\begin{array}{c}
1610.51 \\
-1538.62
\end{array}
$$
Subtract.
$$ 1610.51 - 1538.62 $$

$$\frac{\frac{1}{2}\%}{1\%} = \frac{x}{71.89}$$

$$x = \tfrac{1}{2}\% \times 71.89$$

$$= 35.94$$

The amount is $1538.62 + $35.94 = $1574.56.
Checking by direct calculation, $1000(1.095)^5 = 1574.24$. In this way, we can interpolate between two known values to find an unknown value. In our example above, we are accurate within $0.32, or to the nearest dollar. Interpolation is more accurate if the spread between the known values is narrow.

EXERCISE

1. Find the unknown quantity using interpolation.

(a)
5	13
7	?
9	21

(b)
12	61
15	?
18	85

2. Use interpolation to find the unknown quantity.

$9\frac{1}{4}\%$	$778.17
10%	?
11%	$842.53

3. An investment at 8% pays $925.47, while the same investment at 10% pays $1071.79. Use interpolation to calculate the amount this investment would pay at $9\frac{1}{4}\%$.

4. A sum of money invested at 7.5% will amount to $830 over a period of time. The same sum over the same period of time, invested at 8.25% will amount to $871. What interest rate will cause the same sum to amount to $857 over the same period of time?

11.1 KINDS OF INVESTMENTS

When money is invested, we usually think of some sort of bond, stock, or mortgage that will increase the value of the investment. To illustrate the difference between these kinds of investments, suppose a company, Condor Computer, wishes to build a second assembly plant. To do this, the company must raise $2 000 000 from outside sources. The following are three ways Condor Computer could raise money to build the second plant.

BONDS

The company may decide to issue bonds to the public. To raise $2 000 000, Condor Computer could issue 2000 bonds with a value of $1000 each. A person buying a bond would get interest payments plus their $1000 back when the bond reaches maturity. Bondholders are investors who do not want to take the risk of owning part of the company like shareholders do. Bondholders receive a predetermined amount of interest from their investment and their money is eventually returned. Bondholders become the company's creditors. If the company goes out of business, creditors are paid before owners, which means that bondholders are paid before stockholders.

STOCKS

Another way to raise money is to issue stocks. This would involve selling part of the company. To raise $2 000 000, the company would issue 200 000 shares with an initial value of $10 each. These shares would be sold on the stockmarket. The people that buy the stock would become part owners of the company and are called stockholders or shareholders. They provide money in exchange for a share in the profits, losses, and business decisions with respect to the company.

MORTGAGES

The $2 000 000 could be borrowed, for example from a bank, using the existing plant as security. This is called a mortgage loan. The bank is said to hold the mortgage. Condor Computer would then have to repay the $2 000 000 plus interest in a specified time. If the loan was not repaid, the holder of the mortgage, in this case the bank, has the right to foreclose. This means that the bank can take the plant used as security and sell it for whatever it can get to recover the principal and any interest owing from the loan.

There are many variations on the above ways for an individual to invest money, or for a company to raise money. In this chapter we shall focus our attention on bonds, stocks, and mortgages.

BONDS

When a government, a government agency, or a company requires money to operate, or for expansion, funds may be obtained by issuing bonds. A bond is a loan by the purchaser to the government or company that issues the bond. The bond is simply a promissory note, printed on quality paper in $1000 or other denominations in order to be easily available for resale. Canada Savings Bonds are also available in denominations of $300 and $500.

Every bond displays five pieces of information:

I.   The amount or face value of the bond
II. The interest rate or bond rate of the bond
III. How the interest is paid
IV. The maturity date
V.   The name of the government or company that is issuing the bond

The total annual interest is expressed as a percentage of the value of the bond and is called the coupon rate. For example, a 9% coupon bond would pay 9% of $1000, or $90 per annum until maturity, when the bond is redeemed for $1000.

Government Bonds

Bonds can be issued by the federal, provincial, or municipal governments. Government of Canada Bonds, issued and guaranteed by the federal government, are considered to be the safest long-term debt securities available. Provincial and municipal bonds are considered more risky than federal bonds, but they still have a high quality rating. A provincial government has the power to limit the amount of money that a municipality raises by issuing bonds. Provincial and municipal bonds generally have interest rates slightly higher than federal bonds due to the difference in risk. In general, the greater the risk, the higher the interest rate.

Agencies and Utilities

Bonds can also be issued by Crown corporations such as Air Canada and Canadian National Railways or by provincial utilities such as Ontario Hydro or Hydro Quebec. These bonds and the issuing organizations are listed in the *Bank of Canada Review*. Generally, the interest rates on these types of bonds are slightly above comparable rates for government bonds.

Corporate Bonds

While small businesses borrow money from individual lenders or banks, large corporations can issue bonds. Corporate bonds generally have longer terms to maturity than direct loans. The money raised by a corporation through a bond issue is usually used to pay for capital expenditures such as a new plant or equipment. There is a wide variety of corporate bonds available to the purchaser.

Risk

When money is invested by lending it to an individual, a corporation, a government or one of its agencies, or utilities, there is always an element of risk involved. Whenever there is an uncertainty about what will happen next, there is a risk. In the investment world, the investor is encouraged to take greater risks by being offered higher interest rates. Some examples of risks are:

I. Inflation risk: How does the rate of inflation affect the value of the investment as compared to the amount of return?

II. Financial risk: For any organization, how does the amount of debt affect the risk?

Face Value

The face value, or maturity value of a bond is the amount that is to be paid to the bondholder, in addition to interest, when the bond matures. The face value is also referred to as the bond's par value, or principal value. For example, a twenty-year, $1000 bond with a coupon rate of 11% sold on January 1, 1986, will pay $110 per annum until January 1, 2006 when the bondholder will receive the face value of $1000 on maturity. Prior to maturity, the bond may be more or less than the face value depending on the money market, and the market value.

Market Value

The market value of a bond is the value for which the bond may be bought or sold in the market. The market value is determined by the market interest rate on comparable securities. If the market interest rate is greater than the coupon rate, then the bond will sell at a discount. If the market interest rate is less than the coupon rate, then the bond will sell at a premium.

EXAMPLE. Consider a twenty-year bond with a coupon rate of 11%, and a face value of $1000. The market value of the bond can be calculated by finding the present value of the dollar payments due to the bondholder using the market interest rate.

(a) Find the market value of the bond if the market interest rate is 13%.
(b) Find the market value of the bond if the market interest rate is 8%.

SOLUTION:

If the market interest rate was equal to the coupon rate of 11%, the market value of the bond would be:

$$V = \frac{\$110}{1.11} + \frac{\$110}{1.11^2} + \cdots + \frac{\$110}{1.11^{20}} + \frac{\$1000}{1.11^{20}}$$

$$a = \frac{110}{1.11}, \qquad r = \frac{1}{1.11}, \qquad n = 20$$

$$S_n = \frac{a(1 - r^n)}{1 - r}$$

$$S_{20} = \frac{\frac{110}{1.11}\left(1 - \frac{1}{1.11^{20}}\right)}{1 - \frac{1}{1.11}}$$

$$V = 875.97 \qquad + \qquad 124.03$$
$$= 1000.00$$

The market value is equal to the face value.

E=mc²

(a) If the market interest rate increases to 13%, then the market value of the bond would be:

$$V = \frac{\$110}{1.13} + \frac{\$110}{1.13^2} + \cdots + \frac{\$110}{1.13^{20}} + \frac{\$1000}{1.13^{20}}$$

$$a = \frac{110}{1.13}, \quad r = \frac{1}{1.13}, \quad n = 20$$

$$S_n = \frac{a(1 - r^n)}{1 - r}$$

$$S_{20} = \frac{\frac{110}{1.13}\left(1 - \frac{1}{1.13^{20}}\right)}{1 - \frac{1}{1.13}}$$

$$V = 772.72 \qquad + \qquad 86.78$$
$$= 859.50$$

The market value of the bond is $859.59, and the bond would be sold at a discount.

(b) If the market interest rate decreases to 8%, then the market value of the bond would be:

$$V = \frac{\$110}{1.08} + \frac{\$110}{1.08^2} + \cdots + \frac{\$110}{1.08^{20}} + \frac{\$1000}{1.08^{20}}$$

$$a = \frac{110}{1.08}, \quad r = \frac{1}{1.08}, \quad n = 20$$

$$S_n = \frac{a(1 - r^n)}{1 - r}$$

$$S_{20} = \frac{\frac{110}{1.08}\left(1 - \frac{1}{1.08^{20}}\right)}{1 - \frac{1}{1.08}}$$

$$V = 1080 \qquad + \qquad 214.55$$
$$= 1294.55$$

The market value of the bond is $1294.55, and the bond would be sold at a premium.

EXERCISE 11.1

1. Evaluate, giving your answers to the nearest cent.

(a) $V = \dfrac{\$100}{1.10} + \dfrac{\$100}{1.10^2} + \cdots + \dfrac{\$100}{1.10^{20}} + \dfrac{\$1000}{1.10^{20}}$

(b) $V = \dfrac{\$70}{1.07} + \dfrac{\$70}{1.07^2} + \cdots + \dfrac{\$70}{1.07^{20}} + \dfrac{\$1000}{1.07^{20}}$

(c) $V = \dfrac{\$90}{1.10} + \dfrac{\$90}{1.10^2} + \cdots + \dfrac{\$90}{1.10^{20}} + \dfrac{\$1000}{1.10^{20}}$

(d) $V = \dfrac{\$90}{1.08} + \dfrac{\$90}{1.08^2} + \dots + \dfrac{\$90}{1.08^{10}} + \dfrac{\$1000}{1.08^{10}}$

(e) $V = \dfrac{\$70}{1.09} + \dfrac{\$70}{1.09^2} + \dots + \dfrac{\$70}{1.09^{10}} + \dfrac{\$1000}{1.09^{10}}$

(f) $V = \dfrac{\$120}{1.10} + \dfrac{\$120}{1.10^2} + \dots + \dfrac{\$120}{1.10^{10}} + \dfrac{\$1000}{1.10^{10}}$

(g) $V = \dfrac{\$80}{1.09} + \dfrac{\$80}{1.09^2} + \dots + \dfrac{\$80}{1.09^{10}} + \dfrac{\$1000}{1.09^{10}}$

2. A twenty-year bond with a face value of $1000 has a coupon rate of 10%. The market interest rate is 9%.

(a) Will the bond sell at a premium or discount?
(b) What is the annual interest paid by the bond?
(c) Calculate the market value of the bond.
(d) What is the difference between the face value of the bond and the market value?

3. A twenty-year bond with a face value of $1000 has a coupon rate of 11%. Find the market value of the bond if the market interest rate drops to 9%.

4. A ten-year bond with a face value of $1000 has a coupon rate of 9%. Find the discounted value of the bond if the market interest rate rises to 12%.

5. A twenty-year bond with a face value of $1000 has a coupon rate of 10%. Find the premium value of the bond if the market interest rate drops to 7%.

6. A twenty-year bond with a face value of $1000 has a coupon rate of 8%.

(a) Calculate the discounted value of the bond if the market interest rate rises to 10%.
(b) Calculate the discounted value of the bond if the market interest rate rises to 12%.
(c) Calculate the market value of the bond, selling at a premium, if the market interest rate drops to 6%.
(d) Calculate the market value of the bond, selling at a premium, if the market interest rate drops to 4%.

(e) Show the results of parts (a) to (d) above by completing the following graph in your notebook and drawing a smooth curve.

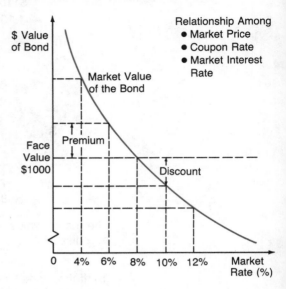

7. Research the following terms.
(a) convertible bonds
(b) protective covenants
(c) call provision
(d) call price
(e) call premium
(f) unsecured bond
(g) debenture

1. Write your age.
2. Multiply by 4.
3. Add 10.
4. Multiply by 25.
5. Add your "change" less than one dollar.
6. Subtract the number of days in a regular year.
7. Add 115.

Explain your answer.

11.2 YIELD RATE OF BONDS

The interest rate earned from buying a bond and holding it to maturity is called the yield to maturity of the bond.

If the price of a bond today is $1000, and it pays $80 per annum for 10 a, and then pays its face value of $1000, the yield to maturity (interest rate) on the bond would be 8% per year. However, if the bond is discounted to $850, the yield to maturity would be more than 8% per year. From the previous section, we know that we can find the market value of the bond using the equation:

$$\begin{array}{ccc} \text{Market Value} & = & \text{Present Value of} \\ \text{of the Bond} & & \text{Interest Payments} \end{array} \quad + \quad \begin{array}{c} \text{Present Value of} \\ \text{Principal Payment} \end{array}$$

$$\$850 = \underbrace{\frac{\$80}{1 + i} + \frac{\$80}{(1 + i)^2} + \cdots + \frac{\$80}{(1 + i)^{10}}} + \frac{\$1000}{(1 + i)^{10}}$$

EXAMPLE. Find the interest rate i, that will make the right side of the equation equal $850. Since the bond has been discounted, the yield rate will be higher than 8%. Proceed by taking trial rates and using interpolation.

SOLUTION:

Let the trial rate be 10%. Substitute i = 0.1 in the right side of the equation.

$$V = \underbrace{\frac{\$80}{1.1} + \frac{\$80}{1.1^2} + \cdots + \frac{\$80}{1.1^{10}}} + \frac{\$1000}{1.1^{10}}$$

$$a = \frac{80}{1.1}, \quad r = \frac{1}{1.1}, \quad n = 10$$

$$S_{10} = \frac{\dfrac{80}{1.1}\left(1 - \dfrac{1}{1.1^{10}}\right)}{1 - \dfrac{1}{1.1}}$$

$$V = 491.57 \quad + \quad 385.54$$

$$= 877.11$$

Since 877.11 > 850, the trial rate of 10% is too low.

Let the trial rate be 11%. Substitute i = 0.11 in the right side of the equation.

$$V = \underbrace{\frac{\$80}{1.11} + \frac{\$80}{1.11^2} + \cdots + \frac{\$80}{1.11^{10}}} + \frac{\$1000}{1.11^{10}}$$

$$a = \frac{80}{1.11}, \quad r = \frac{1}{1.11}, \quad n = 10$$

$$S_{10} = \frac{\dfrac{80}{1.11}\left(1 - \dfrac{1}{1.11^{10}}\right)}{1 - \dfrac{1}{1.11}}$$

$$V = 471.14 \quad + \quad 352.18$$

$$= 823.32$$

Since 823.32 < 850, the trial rate of 11% is too high.

The interest rate lies between 10% and 11%. Use interpolation to find the interest rate.

$$\begin{bmatrix} 877.11 \\ 850.00 \\ 823.32 \end{bmatrix} \quad \begin{bmatrix} 10\% \\ (10 + x)\% \\ 11\% \end{bmatrix}$$

$$\frac{27.11}{53.79} = \frac{x}{1\%}$$

$$x \doteq 0.503\,997\,025$$
$$\doteq 0.5$$

The estimated rate is, therefore, 10% + 0.5% = 10.5%.

Check.

$$V = \frac{\$80}{1.105} + \frac{\$80}{1.105^2} + \cdots + \frac{\$80}{1.105^{10}} + \frac{\$1000}{1.105^{10}}$$

$$a = \frac{80}{1.105}, \quad r = \frac{1}{1.105}, \quad n = 10$$

$$S_{10} = \frac{\dfrac{80}{1.105}\left(1 - \dfrac{1}{1.105^{10}}\right)}{1 - \dfrac{1}{1.105}}$$

$$V = 481.18 + 368.45$$
$$= 849.63$$

849.63 ≐ 850

EXERCISE 11.2

B 1. Solve for i using a calculator and interpolation, if necessary.

(a) $\$935.82 = \dfrac{\$80}{1 + i} + \dfrac{\$80}{(1 + i)^2} + \cdots +$
   $\dfrac{\$80}{(1 + i)^{10}} + \dfrac{\$1000}{(1 + i)^{10}}$

(b) $\$721.28 = \dfrac{\$75}{1 + i} + \dfrac{\$75}{(1 + i)^2} + \cdots +$
   $\dfrac{\$75}{(1 + i)^{20}} + \dfrac{\$1000}{(1 + i)^{20}}$

(c) $\$1132.19 = \dfrac{\$110}{(1 + i)} + \dfrac{\$110}{(1 + i)^2} + \cdots +$
   $\dfrac{\$110}{(1 + i)^{20}} + \dfrac{\$1000}{(1 + i)^{20}}$

(d) $\$1189.27 = \dfrac{\$105}{1 + i} + \dfrac{\$105}{(1 + i)^2} + \cdots +$
   $\dfrac{\$105}{(1 + i)^{20}} + \dfrac{\$1000}{(1 + i)^{20}}$

(e) $\$1000 = \dfrac{\$100}{1 + i} + \dfrac{\$100}{(1 + i)^2} + \cdots +$
   $\dfrac{\$100}{(1 + i)^{10}} + \dfrac{\$1000}{(1 + i)^{10}}$

(f) $\$926.06 = \dfrac{\$95}{1 + i} + \dfrac{\$95}{(1 + i)^2} + \cdots +$
   $\dfrac{\$95}{(1 + i)^{15}} + \dfrac{\$1000}{(1 + i)^{15}}$

2. A bond with twelve years remaining to maturity, a face value of $1000, and a coupon rate of 9%, is sold at a discount for $799.06. Find the yield rate of the bond.

3. A bond with eight years remaining to maturity, a face value of $5000, and a coupon rate of 12%, is sold at a premium for $5860. Find the yield rate of the bond.

4. A twenty-year bond with a face value of $5000 and a coupon rate of 6.5% is sold at a discount for $4450. Find the yield rate of the bond.

5. A ten-year bond with a face value of $1000 and a coupon rate of 9.5% is sold for $1090. Find the yield rate of the bond.

6. A twenty-year bond with a face value of $1000 and a coupon rate of 9% is sold at a premium for $1078. Find the yield rate of the bond.

7. A corporate bond with eleven years remaining to maturity, has a face value of $1000, and a coupon rate of 10%. The bond is offered for sale for $925. Find the yield rate of the bond.

8. A twenty-five-year bond with a face value of $10 000 and a coupon rate of 10% is sold for $9500. Find the yield rate of the bond.

9. A five-year bond with a face value of $2000 and a coupon rate of 8.5% is sold at a discount for $1900. Find the yield rate of the bond.

10. When Archie Lang's insurance policy matured, he bought a five-year bond for $9275. The face value of the bond is $10 000 and the coupon rate is 10%. Find the yield rate of the bond.

The market value of a bond is the present value of the interest payments added to the present value of the principal payment, or face value.

Present Value of Interest Payments:

$$\frac{R}{1 + I} + \frac{R}{(1 + I)^2} + \cdots + \frac{R}{(1 + I)^n}$$

Present Value of Principal Payment:

$$\frac{A}{(1 + I)^n}$$

Market Value

$$\frac{R}{1 + I} + \frac{R}{(1 + I)^2} + \cdots$$
$$+ \frac{R}{(1 + I)^n} + \frac{A}{(1 + I)}$$

The following program can be used to find the market value of a bond. In order to use the program, you will need to calculate the following:

R = Annual Interest Paid (in dollars)
I = Yield Rate as a Decimal
A = Face Value
N = Years to Maturity

```
NEW
10 PRINT "MARKET VALUE OF A BOND"
20 INPUT "ANNUAL INTEREST PAID";R
30 INPUT "YIELD RATE AS A DECIMAL";I
40 INPUT "FACE VALUE";A
50 INPUT "YEARS TO MATURITY";N
60 M=R*(1-(1+I)↑-N)/I+A*(1+I)↑-N
70 PRINT "THE MARKET VALUE IS $";M
76 PRINT " "
80 END
RUN
```

EXERCISE

Use the above program to solve the following questions.

1. What is the market value of a fifteen-year bond with a face value of $1000 and a coupon rate of 11.75%? The current market interest rate is 9.5%.

2. What is the market value of a twenty-year bond with a face value of $5000 and a coupon rate of 9.5%, when the market interest rate is 11.25%?

11.3 STOCKS

Money is invested in stocks by individuals and organizations such as pension funds with the expectation of making a profit. With stocks, this profit may come in the form of a dividend, or a capital gain. A dividend is a portion of the company's profits earned by shares. A capital gain is the profit made by buying at one price and selling at a higher price.

When stocks are purchased, part ownership of a company is bought in the form of shares. The two basic types of shares are common shares and preferred shares.

Stockholders are the holders of common stock and represent the ownership of the company. The amount of ownership in a corporation that an individual has is determined by the amount of common stock owned. For example, if a company has 1000 shares of common stock outstanding, then an individual who owns 100 shares of common stock owns 10% of the company. When a company makes a profit, the owners share in this surplus income. It is one of the functions of the board of directors to decide how much of the profit is to be used to expand the company, often called "ploughing back in", and how much is to be distributed to the shareholders as dividends.

Preferred stock has a par value, usually $100 per share, and a fixed rate of return, shown as a percentage of the par value. If the company has a profit, dividends are paid to preferred shareholders before shareholders of common stock.

There is a greater risk in investing in stocks than there is in buying bonds. Whereas the face value of a bond does not change, in buying stock, there is always the risk that the price of the shares can go down, creating a loss. However, when a person invests in a stock, the expected return outweighs the risk.

EXAMPLE 1. Helen Jonas owned 1500 shares of Toronto Dominion Bank on September 26. How much will she receive in dividends if a dividend of 33 cents is payable on November 1 to shareholders on record as of September 26?

SOLUTION:
Dividend declared: $0.33 per share
Dividends received: 1500 × $0.33
                        = $495.00

$$\frac{5\frac{3}{4}\% \times \$100 \text{ par value}}{4} = \$1.43\frac{3}{4}$$

Why must we divide by 4?

EXAMPLE 2. What are the total dividends receivable as of the date of record for 500 shares of Hawker Siddely Canada at $5\frac{3}{4}\%$ preferred and 2500 shares of Hawker Siddely common?

The following appeared on the financial pages of the newspaper:

DIVIDENDS
Corporation dividends are quarterly unless stated otherwise. Hawker Siddely Canada Ltd., eight cents, Oct. 16, record Sept. 27; $5\frac{3}{4}$ percent pfd. $1.43, Oct. 2, record Sept. 20

SOLUTION:
Common shares: $0.08 per share
Dividend receivable: 2500 × $0.08
                        = $200.00

$5\frac{3}{4}\%$ preferred shares = $1.4375 per share
Dividend receivable: 500 × $1.4375
                        = $718.75
Total dividend receivable:

          $200.00 + $718.75 = $918.75

EXERCISE 11.3

1. What dividend should be paid in each of the following shareholdings?

(a) quarterly dividend of $0.62 on 1000 shares of Lake Shore

(b) semi-annual dividend of $0.34 on 500 shares of Canada Paper

(c) quarterly dividend of $0.50 on 1000 shares of Bellcan

(d) quarterly dividend of $0.72 on 4000 shares of Canoil

2. What dividend per share should be paid for each of the following?

(a) quarterly dividend on an 8% preferred share with a $100 par value

(b) semi-annual dividend on a 6% preferred share with a $100 par value

(c) quarterly dividend on a 10% preferred share with a $100 par value

(d) quarterly dividend on a $6\frac{1}{4}$% preferred share with a $100 par value

3. When company directors declare a dividend, the amount to be paid to shareholders is divided by the number of shares to calculate the dividend. Calculate the dividend per share for each of the following.

(a) $50 000 to be paid in dividends on 100 000 outstanding common shares

(b) $250 000 to be paid in dividends on 1 000 000 outstanding common shares

(c) $400 000 to be paid in dividends on 2 000 000 outstanding common shares

(d) $100 000 to be paid in dividends on 150 000 outstanding common shares

4. What annual return should be expected on each of the following investments?

	Number of Shares	Stock	Indicated Dividend Rate	Par Value $
(a)	500	Maple Leaf	$5\frac{1}{2}$% pfd	100
(b)	700	Power Corp.	$4\frac{1}{4}$% pfd	100
(c)	100	Rolland	$4\frac{3}{4}$% pfd	100
(d)	300	United Ltd.	5% pfd	30
(e)	200	Ford	$6\frac{3}{4}$% pfd	100

To find the value of the holdings in a stock, we multiply:

$$\left(\begin{array}{c} \text{Value} \\ \text{of Shares} \end{array} \right) \times \left(\begin{array}{c} \text{Number} \\ \text{of Shares} \end{array} \right)$$

Commission on sales and purchases of stock are discussed in a later section.

5. Find the value of the following holdings.

(a) 100 shares of Argus preferred at $18.00

(b) 1000 shares of BC Rail common at $27.00

(c) 5000 shares of Bow Valley at $4.35

(d) 800 shares of Marconi at $42.50

(e) 2000 shares of Husky Oil at $9\frac{1}{4}$

(f) 2500 shares of Seagram common at $54\frac{1}{2}$

(g) 50 shares of IBM common at $178.00

(h) 1200 shares of Woodward A at $21\frac{3}{4}$

(i) 500 shares of Nor Tel common at $17\frac{1}{4}$

6. Find the costs of the following purchases, excluding commission.

(a) 200 shares of Inco at $20\frac{3}{4}$

(b) 175 shares of Xerox common at $27\frac{1}{4}$

(c) 300 shares of Sears common at $8\frac{1}{2}$

7. Calculate the proceeds from the following sales, excluding commission.

(a) 150 shares of Rio Algom at $21\frac{1}{2}$

(b) 300 shares of BC Phone at $24\frac{1}{4}$

(c) 400 shares of Canadian Tire common at $12\frac{3}{4}$

$$\left(\begin{array}{c} \text{Capital Gain} \\ \text{or} \\ \text{Loss} \end{array} \right) = \left(\begin{array}{c} \text{Proceeds} \\ \text{from} \\ \text{Sale} \end{array} \right) - \left(\begin{array}{c} \text{Cost} \\ \text{of} \\ \text{Purchase} \end{array} \right)$$

8. Find the capital gain, or loss, excluding commission from each of the following transactions.

(a) 1000 shares of Royal Bank purchased at $29\frac{1}{4}$, and sold at $31.00

(b) 200 shares of IBM purchased at $162.00, and sold at $177\frac{1}{2}$

(c) 500 shares of CP Ltd. purchased at $17\frac{1}{4}$, and sold at $15\frac{3}{4}$

11.4 STOCK MARKET QUOTATIONS

Who sets the price of a stock? The price is set by the people who are willing to trade it. Since the price depends on many individual decisions, it is impossible to predict with certainty what the price of a stock will be over a specified period of time. Unpredictable things such as world crises or the discoveries of new deposits of natural resources can have a great effect on stock prices for short periods. In order to buy and sell a stock, a broker is needed who will make these transactions at the stock market. When we mention the stock market, we usually think of one of the stock exchanges: Montreal, Toronto, Winnipeg, Calgary, Vancouver, or the New York Exchange. Over 1500 companies with a market value of over $1000 billion are listed on the New York Stock Exchange. The stock exchanges are a way to bring buyers and sellers together through brokers so that business can take place. Many smaller firms who do not meet the strict criteria for listing their stock on one of the exchanges will still have their stock traded through investment dealers as unlisted stocks.

Stock	Div	Bid or High	Ask or Low	Last Price	Chge	Vol	Last 52 wks High	Lo
Cancorn		$6\frac{1}{4}$	$6\frac{1}{4}$	$6\frac{1}{4}$		13600	$7\frac{3}{8}$	320
Cdn Tire	.20	$12\frac{1}{2}$	$12\frac{1}{2}$	$12\frac{1}{2}$	$-\frac{1}{4}$	100	$16\frac{1}{2}$	11
CTire A f	.20	$9\frac{3}{8}$	$9\frac{1}{8}$	$9\frac{1}{4}$		313508	11	$8\frac{1}{8}$
C Util A f	1.20	$17\frac{1}{2}$	$17\frac{1}{2}$	$17\frac{1}{2}$	$+\frac{1}{8}$	11465	$18\frac{3}{4}$	14
C Util B	1.20	$17\frac{1}{2}$	$17\frac{1}{4}$	$17\frac{1}{2}$		700	19	14
C Util $4\frac{1}{4}$ p	4.25	$49	51	49		nt	50	42
C Util 5 pr	5.00	$56	60	56		nt	56	$50\frac{1}{2}$

The transactions which take place in the stock market are published each day in the newspaper. A partial list of these stock quotations is given at left. The given information in abbreviated form from left to right is: the name of the stock, the dividend paid in the current year, the highest price paid per share that day, the lowest price per share paid that day, the last (or closing) price paid that day, the net change from the previous day's closing price, the total number of shares traded that day, and the highest and lowest prices paid in the current year. Some weekly summaries contain additional information.

To get an overall picture of what the market is doing we can consult the market indexes. An index reflects the average performance of a number of established stocks and indicates the general level of the market. If the selected index stocks are high, then the index average will also be high. Some of the major indexes are the TSE 300, the Dow Industrial, and the NYSE Industrial. These indexes are published daily in the financial pages of the newspaper.

Stock price trends

	Wedn. Close	Net Change	1985 High	1985 Low
TSE 300	2662.1	+ 10.5	2819.98	2348.55
Dow indust.	1300.40	+ 2.24	1359.54	1184.96
NYSE indust.	120.57	+ 0.14	128.89	108.38

An index taken alone is not sufficient information on which to make investment decisions. It is best to follow these indexes from day to day, and month to month in order to see the trend of prices. These trends are followed through charts and graphs. The charts and graphs of individual stocks can also be consulted so that more intelligent decisions can be made.

In charting an index or stock, axes and scales are used as before, but extended to provide the high, low, and the last sale, or "close" for each day.

In the chart at the right, the top of the vertical bar indicates the high entry, while the bottom of the bar indicates the low entry. The horizontal bar indicates the closing or last sale.

Daily — High / Close / Low

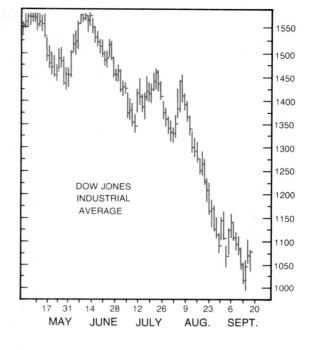

DOW JONES
INDUSTRIAL
AVERAGE

17 31 MAY   14 28 JUNE   12 26 JULY   9 23 AUG.   6 20 SEPT.

1. What was the high, low, and close for September 17, the last entry on the chart?

2. What periods had rising prices as indicated by the Dow Jones index?

3. What periods had falling prices as indicated by the Dow Jones index?

EXERCISE 11.4

1. Prepare a chart from the following information.

| Date | IBM | | |
	High	Low	Close
08 02	184	182	182
06	182	181	182
07	184	183	184
08	185	184	185
09	186	186	186
12	186	184	185
13	186	184	184
14	184	183	183
15	183	180	180
16	179	179	179
19	179	178	178
20	178	176	176
21	177	176	176
23	177	175	175
26	175	172	172
27	172	170	170
28	170	168	170
29	169	167	168
30	168	167	167

2. Prepare a chart from the following information.

| Date | TSE 300 | | |
	High	Low	Close
08 02	2742.25	2712.08	2719.2
06	2736.71	2694.56	2702.4
07	2786.35	2714.63	2732.5
08	2812.11	2766.24	2770.6
09	2819.98	2684.35	2741.5
12	2807.84	2645.30	2673.8
13	2798.24	2625.32	2651.6
14	2784.12	2632.41	2662.1
15	2780.56	2615.21	2651.2
16	2735.56	2584.25	2626.4
19	2732.25	2556.48	2590.8
20	2719.63	2518.78	2583.7
21	2710.05	2435.61	2512.6
22	2689.74	2348.55	2465.2
23	2578.35	2381.26	2436.2
26	2685.23	2486.23	2588.3
27	2672.86	2472.56	2602.3
28	2697.24	2504.21	2621.4
29	2713.42	2563.12	2685.5
30	2732.28	2614.60	2706.1

11.5 BUYING AND SELLING STOCK

Barbara Wilton decided to invest money in stocks. She wished to speculate on the probability that the price would increase as well as that she would receive a dividend, so she purchased common shares. In order to select an investment, she consulted the financial pages of a newspaper, read the comments of a market analyst, and decided to invest in Bank of Nova Scotia. Barbara went to see her stockbroker, who obtained a quote on Bank of Nova Scotia stock from the teletype linking his office computer to the computer at the stock exchange where the stock was listed. The call letters for Bank of Nova Scotia are BNS.

Barbara put in a purchase order for 200 shares of Bank of Nova Scotia (BNS) at $13.

Most trading is done in "board lots." Odd lots are sold, but are more difficult to buy and often harder to sell. Regular trading units of board lots are based on prices as follows:

under $0.10	1000 shares
$0.10 to $1.00	500 shares
$1.00 to $100.00	100 shares
over $100.00	10 shares

The stockbroker handles the rest of the transaction with the stock exchange. Barbara watched the transactions of the exchange displayed on a board at the broker's office where ticker tape information is shown. The board showed the following:

IBM	BNS	GMC	CIL
$185\frac{3}{8}$	2.13	$92\frac{1}{2}$ B/$93\frac{1}{4}$	$10.29\frac{5}{8}$

The information summarized on this portion of the tape is

10 shares of IBM sold at $185\frac{3}{8}$ per share

200 shares (2 board lots) of Bank of Nova Scotia at $13

a bid to buy 100 shares of General Motors at $92\frac{1}{2}$ and an offer to sell 100 shares of the same stock at $93\frac{1}{4}$

1000 shares of CIL sold at $29\frac{5}{8}$

Soon after, the broker received confirmation that the purchase for Barbara Wilton had been made.

For performing this service, the broker will charge Barbara a commission. Stockbrokers have published the following commission scale for buying and selling stock orders.

C = Commission ($) x = Value of Order ($) n = Number of Shares		
Selling Price	Rate of Commission	Formula
under $14.00	2.5% of order	C = 0.025x
$14.00 to $30.00	0.8$\frac{3}{4}$% of order + 22.575¢/share	C = 0.00875x + 0.22575n
over $30.00	1.64% of order	C = 0.0164x

Purchasers are also permitted to negotiate a lesser rate of commission with the broker.

Barbara negotiated a rate of 2% of the order. The cost of the order including commission is calculated as follows:

200 shares BNS @ $13.00:	$2600.00
Commission @ 2%: 0.02 × $2600	52.00
Total cost:	$2652.00

Barbara paid the broker $2652.00 and now owns 200 shares of Bank of Nova Scotia stock.

As well as following the stock in the newspaper quotations, or the ticker tape and computer display at the stockbrokers', it is also possible to get information from the stock market channel on television.

EXAMPLE 1. Find the cost to the purchaser of 200 shares of Cygnus A at $4.30 per share, and $2\frac{1}{2}$% commission.

SOLUTION:

Cost of shares:	200 × $4.30
	= $860.00
Commission:	0.025 × $860
	= $21.50
Cost to purchaser:	$860.00 + $21.50
	= $881.50

Commission is added to purchases.

The total cost is $881.50.

EXAMPLE 2. Find the net proceeds to the seller of 500 shares of Pacific Petroleum at $16.50 per share, and commission of $2\frac{1}{4}$%.

SOLUTION:

Proceeds:	500 × $16.50
	= $8250.00
Commission:	0.0225 × $8250.00
	= $185.63
Net proceeds:	$8250.00 − $185.63
	= $8064.37

Commission is deducted from proceeds.

EXAMPLE 3. An investor buys 600 shares of INCO at $24\frac{3}{8}$, and sells at $25\frac{1}{4}$. Calculate the person's profit or loss after paying 2% commission for each transaction.

SOLUTION:

Cost of shares:	$600 \times \$24.375$
	$= \$14\ 625.00$
Commission:	$0.02 \times \$14\ 625.00$
	$= \$292.50$
Total cost:	$\$14\ 625.00 + \$292.50$
	$= \$14\ 917.50$
Proceeds:	$600 \times \$25.25$
	$= \$15\ 150.00$
Commission:	$0.02 \times \$15\ 150.00$
	$= \$303.00$
Net proceeds:	$\$15\ 150.00 - \$303.00$
	$= \$14\ 847$
Profit or loss:	Net proceeds − Total cost
	$= \$14\ 847.00 - \$14\ 917.50$
	$= -\$70.50$

Although the stock did have a small increase, the investor had a loss of $70.50.

EXERCISE 11.5

B 1. Calculate the cost of each of the following purchases as shown on the ticker tape.

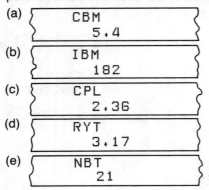

(a) CBM 5.4

(b) IBM 182

(c) CPL 2.36

(d) RYT 3.17

(e) NBT 21

2. Find the cost of the following purchases, excluding commission.

(a) 100 shares of Ashland preferred at $18.00
(b) 1000 shares of Ashland common at $6.25
(c) 300 shares of Baton B at $19.00
(d) 100 shares of Royal Bank at $34\frac{1}{2}$
(e) 100 shares of Jannock at $13\frac{1}{4}$
(f) 100 shares of Keeprite at $9\frac{3}{4}$
(g) 2000 shares of Meridian at $3.90

3. Calculate the commission on each of the following sales at 2%.

(a) 2000 shares of Dome Pete at $4.90
(b) 500 shares of Emco at $20\frac{1}{4}$
(c) 100 shares of Norcan at $16\frac{3}{8}$
(d) 100 shares of Labatt at $32\frac{5}{8}$
(e) 200 shares of Sony Corp. at $22\frac{1}{2}$

4. Calculate the commission on each of the following purchases at $2\frac{1}{2}$%.

(a) 200 shares of Stelco at $27\frac{1}{2}$
(b) 300 shares of Bow Valley at $14\frac{1}{4}$
(c) 100 shares of Torstar at $31\frac{3}{8}$
(d) 8000 shares of Petro Sun at $3.75

5. Calculate the total cost of the following purchases, including commission of $2\frac{1}{4}$%.

(a) 100 shares of Agnico Eagle at $17\frac{5}{8}$
(b) 200 shares of Bank of Montreal at $29\frac{1}{2}$
(c) 500 shares of Loblaw Co at $20\frac{1}{4}$
(d) 200 shares of Celanese at $9\frac{3}{8}$

6. Calculate the net proceeds from each of the following sales, including commission of 2%.

(a) 100 shares of Molson B at $18.00

(b) 100 shares of Northern Telephone at $46\frac{5}{8}$

(c) 300 shares of Scot Paper at $19\frac{1}{4}$

(d) 100 shares of Imperial Oil at $49\frac{1}{2}$

(e) 1000 shares of Echo Bay at $16\frac{3}{8}$

7. Calculate the profit or loss resulting from each of the following transactions.

(a) 300 shares of BC Rail were purchased at $25\frac{1}{4}$, and sold at $27.00. Commission was charged at $2\frac{1}{4}$% of the order on purchases, and 2% on sales.

(b) 100 shares of Genstar were purchased at $23\frac{3}{8}$, and sold at $30\frac{1}{2}$. Commission was charged at 2% on purchases, and 2% on sales.

(c) 50 shares of MacMillan were purchased at $24\frac{5}{8}$, and sold at $17.00. Commission was charged at $2\frac{1}{4}$% to buy, and $2\frac{1}{2}$% to sell.

(d) 1000 shares of Lanpar were purchased at $5.25, and sold at $4.90. Commission was charged at 2% to buy, and $2\frac{1}{4}$% to sell.

(e) 200 shares of Cominco A were purchased at $11\frac{3}{8}$, and sold at $17.00. Commission was charged at $2\frac{1}{4}$% on the purchase, and $2\frac{1}{2}$% on the sale.

8. Use the rates of commission stated earlier in this section to calculate the following.

(a) the total cost of 1000 shares of Noma A purchased at $14.00

(b) the total cost of 300 shares of Domtar purchased at $18\frac{1}{2}$

(c) the total cost of 200 shares of Lac Minerals purchased at $34\frac{1}{2}$

(d) the net proceeds from the sale of 1000 shares of Dome Canada which were sold at $7\frac{5}{8}$

(e) the net proceeds from the sale of 100 shares of Nova C which were sold at $83\frac{3}{4}$

(f) the net proceeds from the sale of 1000 shares of Mitel Corporation which were sold at $12\frac{5}{8}$

MICRO MATH

We can write a program to compute the profit or loss from the buying and selling of a stock using the work of this section. Let the number of shares be N, the purchase price be $P, and the rate of commission be B%.
The total cost is

$$X = N \times P + N \times P \times 0.01 \times B$$
$$= N \times P \times (1 + 0.01 \times B)$$

Let the selling price be $S, and the commission on selling be T%.
The net proceeds are

$$Y = N \times S - N \times S \times 0.01 \times T$$
$$= N \times S \times (1 - 0.01 \times T)$$

The profit or loss is found by subtracting.

$$(\text{Net proceeds}) - (\text{Total cost}) = Y - X$$

These statements form the basis of the following computer program which calculates the profit or loss from buying and selling a stock.

```
NEW
10 PRINT "PROFIT OR LOSS"
20 INPUT "NUMBER OF SHARES ";N
30 INPUT "PURCHASE PRICE $";P
40 INPUT "% COMMISSION/BUY";B
50 INPUT "SELLING PRICE $";S
60 INPUT "% COMMISSION/SELL";T
70 X=N*P*(1+0.01*B)
80 Y=N*S*(1-0.01*T)
90 PRINT "TOTAL COST = $";X
100 PRINT "NET PROCEEDS = $";Y
110 PRINT "PROFIT OR LOSS = ";Y-X
120 END
RUN
```

The program can be made reiterative by inserting the following lines in the program:

```
112 INPUT "ANOTHER QUESTION?
    Y OR N";Z$
113 IF Z$ = "Y" THEN 10
```

11.6 YIELD RATE FOR STOCKS

In order to compare investments to determine which are most profitable at the current prices and dividend rates, we look at the yields. The yield is the ratio of the dividend rate to the price per share.

$$\text{Yield} = \frac{\text{dividend per share}}{\text{price per share}} \times 100\%$$

The current value of an investment depends both on the dividend rate and the price per share. If two reliable stocks both paid $1.50 per share as an annual dividend, and one was priced at $35 per share while the other was priced at $70 per share, which gives the greatest percentage return?

Consider the returns on the two stocks mentioned above.

I. For the $35 stock,

$$\text{Yield} = \frac{\$1.50}{\$35} \times 100\%$$
$$= 4.29\%$$

II. For the $70 stock,

$$\text{Yield} = \frac{\$1.50}{\$70} \times 100\%$$
$$= 2.14\%$$

Although both stocks paid dividends of $1.50, we find that the lower priced stock produced the greater yield.

EXAMPLE 1.   Find the yield from shares of Ford of Canada paying a dividend rate of $6.65 per year with a current price of $136 per share.

SOLUTION:

$$\text{Yield} = \frac{\$6.65}{\$136} \times 100\%$$
$$= 4.89\%$$

The yield is 4.89% per year.

EXAMPLE 2.   BC Sugar paid a quarterly dividend of $0.30 per share. If the stock is currently trading at $23, find the yield.

SOLUTION:

Annual dividend rate: $4 \times \$0.30$
$$= \$1.20$$
$$\text{Yield} = \frac{\$1.20}{\$23} \times 100\%$$
$$= 5.22\%$$

The yield is 5.22% per year.

The ability of a company to make money can be determined by calculating the price/earnings ratio. The price/earnings ratio compares the price of the stock to the earnings per share of the company.

EXAMPLE 3. Computer Software Inc. with 100 000 shares outstanding, earned $473 000 last year. The current price is $64 per share. Find the price/earnings ratio.

SOLUTION:

$$\text{Earnings per share:} \quad \frac{\$473\ 000}{100\ 000}$$

$$= \$4.73$$

If the current price of a share is $64, then the price/earnings ratio is

$$\frac{64}{4.73} = 13.5$$

EXERCISE 11.6

1. Calculate the annual return and yield from the following investments.

(a) 200 shares of Municipal Finance at a current price of $9.00, paying a quarterly dividend of $0.25

(b) 500 shares of Steinberg at a current price of $29.00, paying a quarterly dividend of $0.73

(c) 300 shares of Dennison A at a current price of $12.00, paying a quarterly dividend of $0.25

(d) 400 shares of Dofasco at a current price of $27.00, paying a quarterly dividend of $0.28

(e) 500 shares of Imperial Life at a current price of $27.00, paying an annual dividend of $3.06

2. Calculate the price/earnings ratio (P/E) for each of the following.

	Stock	Closing Price ($)	Earnings ($) per Share
(a)	Nova G	$25\frac{7}{8}$	2.44
(b)	Daon 9 p	$6\frac{1}{2}$	0.95
(c)	Siltronic	2.45	0.38
(d)	Noranda	$15\frac{1}{4}$	1.10
(e)	Stuart Oil	$19\frac{3}{8}$	2.08

3. ABW Construction Ltd. pays a quarterly dividend of $1.35 per share. Find the yield if the market price of the stock is $129.00 per share.

4. Consolidated Vegetable Oil Company Limited has annual earnings for the current year of $1 495 000, with 250 000 outstanding shares. The current market price for the shares is $49.00. Find the price/earnings ratio.

5. Parkdale Manufacturing Limited has declared a semi-annual dividend of $3.40 per share. The shares have a market value of $84.50. What is the yield?

6. Johnston Explorations has current annual earnings of $10 500 000, with 625 000 shares outstanding. The market value is $53.00 per share. What is the price/earnings ratio?

A man has to take a wolf, a goat, and some lettuce across a river. The rowboat has room for the man plus either the wolf, or the goat, or the lettuce. If he takes the lettuce with him, the wolf will eat the goat. If he takes the wolf, the goat will eat the lettuce. Only when the man is present are the goat and lettuce safe from their enemies. How does the man get the wolf, goat, and lettuce across the river?

11.7 INFLATION

Inflation is an increase in the general level of prices for goods and services. Deflation, the opposite of inflation, is a reduction in the general level of prices. Many things happen during periods of inflation. At this time, the prices of goods are increasing, so that it is often necessary for consumers to borrow larger sums of money. This increase in total credit arises from the increasing prices and not necessarily from the wider use of credit.

During periods of inflation, borrowers tend to benefit at the expense of lenders. The lender receives a fixed payment, while the goods that the borrower purchased have increased in value. An example of this is in the purchase of a house, while the value of the house increases, the amount owing is constant. The increase in value becomes part of the owner's equity and the loan which was made in constant dollars is repaid in current dollars.

On the other hand, inflation has been the cause of a significant loss in the value of money saved and not invested in ways that keep up with inflation. In order to calculate the purchasing power in year A of a sum of money from year B, we use the formula:

$$\frac{\text{A dollars}}{\text{B dollars}} = \frac{\text{A Consumer Price Index}}{\text{B Consumer Price Index}}$$

A constant dollar is taken at 100 in the year 1971. In order to determine the effect of inflation, we monitor the consumer price index. The consumer price index is a comparison of the value of a dollar taken at 100 in 1971. When we say that the consumer price index in 1948 was 50, we mean that what cost $1.00 in 1971 cost $0.50 in 1948. When we say the consumer price index in 1978 was 150, we mean that what cost $1.00 in 1971 will cost $1.50 in 1978. The graph at the right shows the Canadian Consumer Price Index from 1913 to 1978.

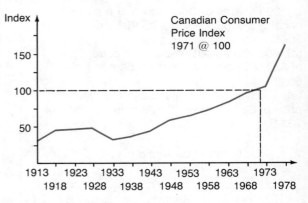

EXAMPLE.   From the graph, the consumer price index based on the value of a dollar in 1977 is 138. What is the value in 1971 dollars of a painting that was sold for $1200 in 1977?

SOLUTION:

$$\frac{\text{1971 dollars}}{\text{1977 dollars}} = \frac{\text{1971 Consumer Price Index}}{\text{1977 Consumer Price Index}}$$

$$\frac{\text{1971 dollars}}{1200} = \frac{100}{138}$$

$$\text{1971 dollars} = \frac{1200 \times 100}{138}$$

$$= \$869.57$$

The value of the painting in 1971 dollars is about $870.00.

In order for an investor to protect against inflation, a return on the amount invested must be equal to the rate of inflation. A return of 8% on an investment in a period of 8% inflation will just protect the purchasing power without giving any net return for taking the risk in an investment.

EXERCISE 11.7

1. The following table shows the percentage increase in the Canadian Consumer Price Index for the years 1977 to 1984.

Year	Change from Preceding Year	Consumer Price Index
1976	——	127
1977	8.7%	138
1978	9.0%	150
1979	9.2%	
1980	10.1%	
1981	12.5%	
1982	10.8%	
1983	5.8%	
1984	3.4%	
1985		
1986		
1987		
1988		

(a) Locate the changes from preceding years in the index and update the table.
(b) Calculate the Canadian Consumer Price Index for the remaining years.
(c) Draw a graph of the consumer price index using axes similar to the following.

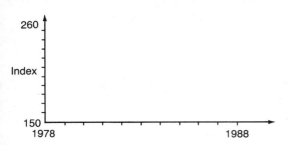

2. A bronze statue was purchased in 1948 for $700. The value of the statue increased annually according to the rate of inflation. What is the value of the statue in 1971 constant dollars?

3. A bond with a face value of $1000 was purchased for $1000 in 1958. Due to inflation, the value of the bond in constant dollars is reduced. What is the value of the bond in 1971 constant dollars?

4. Jack put a $100 bill in his safety deposit box at the bank in 1938. What was the purchasing power of the $100 when Jack tried to spend it in 1971?

5. A cottage lot cost $12 000 in 1978, and has increased in apparent value over the years due to inflation. Assuming the only increase in value was due to inflation, calculate the price of the lot in 1963.

Use the table or graph prepared in Question 1 to answer the following questions.

6. What is the purchasing power in 1984 of $1000 as compared to the purchasing power of the same amount in 1980?

7. What is the purchasing power of $5000 in 1982 as compared to the purchasing power of the same amount in 1958?

8. What is the value in 1984 dollars of a twenty-year bond with a face value of $2000 purchased in 1964?

9. In 1956 Toni bought 1000 shares of Fraser Valley Resources at $2.65 per share. The stock was sold in 1980 for $4.20. The stock never paid any dividends.
(a) What were the proceeds from the sale of the stock, excluding commission?
(b) What was the original cost of the stock?
(c) What was the original cost of the stock in 1980 dollars?
(d) What was Toni's profit or loss after considering inflation?

10. Heather bought 2000 shares of Consolidated Alberta Tar Sands for $6.50 per share in 1979. She sold the shares in 1984 for $9.20. What was Heather's profit after allowing for inflation?

11.8 MORTGAGES

A mortgage is a loan secured by real estate—land or buildings—and repaid in regular instalments. Mortgages to businesses and individuals are quite similar and differ mainly in the length of the term or the method of repayment. Commercial mortgages normally extend for ten to twenty years and often have quarterly payments. Residential mortgages require regular monthly or sometimes bimonthly payments, and generally extend for over twenty years. The interest rate on residential mortgages is usually higher than the rate on commercial mortgages because the latter are considered a lower risk. Mortgages are arranged through a mortgage loan company as well as the chartered banks, trust companies, and credit unions.

An open-end mortgage permits the borrower to make additional loans on the same property with the same privileges as the existing mortgage. A closed-end mortgage forbids the company from making further loans on the property with the same status. Thus, we have first mortgages and second mortgages. On default by the borrower, the first mortgage holder's claims are settled first. Second mortgages have a slightly higher rate of interest due to the increased risk of not being first for the settlement of claims.

Some mortgages have fixed rates, while amortized for longer periods, and are renegotiated every one to five years. Many mortgages now being made by banks have variable rates, which are adjusted every quarter. A variable rate may also be stated as a percent "over prime." For example, if the stated interest rate on a mortgage was "1% over prime," and the current prime rate was 8%, then the current rate of interest would be 9%.

When an individual or institution lends money as an investment, there is an expectation that the money will stay invested for the duration of the term to produce the expected return on the investment. Some mortgages have a provision for prepayment of principal on the anniversary dates once per year, while others permit prepayment by paying a bonus (or penalty) of several months' interest.

Although the house you live in does not usually yield money income, it does provide a service in the form of shelter. This is an example of non-money income. Savings that are invested in home ownership are invested tax-free. Very few people are able to make a home purchase with cash and usually need a loan in the form of a mortgage since they are unable to repay the loan in a lump sum. The payments are amortized over a long period of time, with equal monthly payments. These equal monthly payments comprise varying proportions of principal and interest. In this way, a mortgage is an annuity and the method for repayment is identical to that for any instalment loan. In practice, calculations are made using a table of monthly factors based on a principal of $1000.

Rate %	Factor	Rate %	Factor	Rate %	Factor	Rate %	Factor
10	.008 164 8461	13	.010 551 0740	16	.012 909 4570	19	.015 240 7000
$10\frac{1}{8}$	.008 264 8377	$13\frac{1}{8}$	.010 649 8909	$16\frac{1}{8}$	.013 007 1292	$19\frac{1}{8}$	.015 337 2563
$10\frac{1}{4}$	.008 364 7797	$13\frac{1}{4}$	.010 748 6596	$16\frac{1}{4}$	.013 104 7543	$19\frac{1}{4}$	.015 433 7666
$10\frac{3}{8}$	.008 464 6722	$13\frac{3}{8}$	.010 847 3799	$16\frac{3}{8}$	.013 202 3325	$19\frac{3}{8}$	.015 530 2312
$10\frac{1}{2}$	.008 564 5152	$13\frac{1}{2}$	.010 946 0522	$16\frac{1}{2}$	.013 299 8636	$19\frac{1}{2}$	.015 626 6499
$10\frac{5}{8}$	.008 664 3089	$13\frac{5}{8}$	.011 044 6762	$16\frac{5}{8}$	.013 397 3478	$19\frac{5}{8}$	.015 723 0229
$10\frac{3}{4}$	.008 764 0532	$13\frac{3}{4}$	.011 143 2522	$16\frac{3}{4}$	.013 494 7852	$19\frac{3}{4}$	.015 819 3502
$10\frac{7}{8}$	.008 863 7482	$13\frac{7}{8}$	.011 241 7802	$16\frac{7}{8}$	.013 592 1758	$19\frac{7}{8}$	.015 915 6318
11	.008 963 3940	14	.011 340 2602	17	.013 689 5196	20	.016 011 8678
$11\frac{1}{8}$	.009 062 9906	$14\frac{1}{8}$	.011 438 6923	$17\frac{1}{8}$	.013 786 8166	$20\frac{1}{8}$	.016 108 0583
$11\frac{1}{4}$	.009 162 5381	$14\frac{1}{4}$	.011 537 0764	$17\frac{1}{4}$	.013 884 0670	$20\frac{1}{4}$	.016 204 2033
$11\frac{3}{8}$	.009 262 0365	$14\frac{3}{8}$	.011 635 4128	$17\frac{3}{8}$	.013 981 2708	$20\frac{3}{8}$	.016 300 3028
$11\frac{1}{2}$	.009 361 4858	$14\frac{1}{2}$	.011 733 7014	$17\frac{1}{2}$	.014 078 4280	$20\frac{1}{2}$	.016 396 3569
$11\frac{5}{8}$	.009 460 8863	$14\frac{5}{8}$	.011 831 9423	$17\frac{5}{8}$	.014 175 5387	$20\frac{5}{8}$	.016 492 3656
$11\frac{3}{4}$	.009 560 2378	$14\frac{3}{4}$	.011 930 1355	$17\frac{3}{4}$	.014 272 6030	$20\frac{3}{4}$	.016 588 3290
$11\frac{7}{8}$	.009 659 5404	$14\frac{7}{8}$	.012 028 2811	$17\frac{7}{8}$	.014 369 6208	$20\frac{7}{8}$	.016 684 2471
12	.009 758 7942	15	.012 126 3791	18	.014 466 5922	21	.016 780 1200
$12\frac{1}{8}$	.009 857 9993	$15\frac{1}{8}$	.012 224 4297	$18\frac{1}{8}$	.014 563 5173	$21\frac{1}{8}$	.016 875 9478
$12\frac{1}{4}$	.009 957 1557	$15\frac{1}{4}$	.012 322 4327	$18\frac{1}{4}$	.014 660 3961	$21\frac{1}{4}$	.016 971 7304
$12\frac{3}{8}$	.010 056 2634	$15\frac{3}{8}$	.012 420 3883	$18\frac{3}{8}$	.014 757 2287	$21\frac{3}{8}$	.017 067 4679
$12\frac{1}{2}$	.010 155 3225	$15\frac{1}{2}$	.012 518 2966	$18\frac{1}{2}$	.014 854 0152	$21\frac{1}{2}$	.017 163 1604
$12\frac{5}{8}$	.010 254 3331	$15\frac{5}{8}$	.012 616 1575	$18\frac{5}{8}$	.014 950 7554	$21\frac{5}{8}$	.017 258 8079
$12\frac{3}{4}$	.010 353 2952	$15\frac{3}{4}$	.012 713 9712	$18\frac{3}{4}$	.015 047 4497	$21\frac{3}{4}$	.017 354 4104
$12\frac{7}{8}$	.010 452 2088	$15\frac{7}{8}$	.012 811 7377	$18\frac{7}{8}$	.015 144 0978	$21\frac{7}{8}$	.017 449 9680

For example, $1000 at 11% compounded semi-annually has an interest factor of 0.008 963 3940. To find the interest for one month on $20 000, we multiply

$$\$20\ 000 \times 0.008\ 963\ 3940 = \$179.27$$

EXAMPLE. The following is the information on a mortgage.

Principal: $60 000

Interest rate: 11%, compounded semi-annually

Amortization period: 25 a

Calculate:
(a) the amount of the monthly payment
(b) the interest at the end of the first month
(c) the proportion of interest and principal in the first payment
(d) the outstanding principal after the first payment
(e) the interest to be paid the second month
(f) the mortgage schedule for the first 6 months

SOLUTION:

From the table, the monthly interest factor is 0.008 963 3940

(a) Monthly payment:
    $1000 at 11% for 25 a is $9.63 per month
    For a loan of $60 000:
    60 × $9.63 = $577.80

> We get $9.63 from the table on the next page.

(b) Interest for the first month.

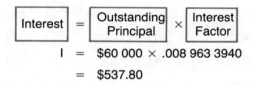

$$I = \$60\ 000 \times .008\ 963\ 3940$$
$$= \$537.80$$

(c) Proportion of principal and interest in the first payment.

$577.80 − $537.80 = $40.00

(d) Outstanding principal after the first payment.

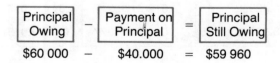

$60 000 − $40.000 = $59 960

(e) Interest for the second month.

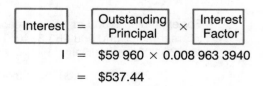

$$I = \$59\ 960 \times 0.008\ 963\ 3940$$
$$= \$537.44$$

(f) Mortgage schedule for the first 6 months.

Payment Number	Date of Payment	Total $ Payment	Interest Portion $	Principal Portion $	Outstanding Principal $
1	March 1	577.80	537.80	40.00	59 960.00
2	April 1	577.80	537.44	40.36	59 919.64
3	May 1	577.80	537.08	40.72	59 878.92
4	June 1	577.80	536.71	41.09	59 837.83
5	July 1	577.80	563.34	41.46	59 796.37
6	August 1	577.80	535.97	41.83	59 754.54

The calculations for the table of the above mortgage schedule is simplified using the memory key on the calculator as follows:

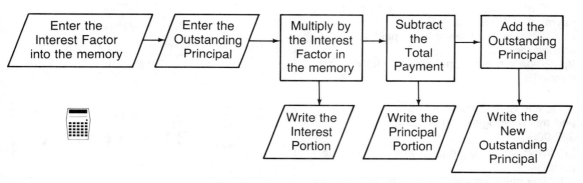

Monthly Payments Required To Amortize A $1000 Loan,
Interest Compounded Semi-annually

Amortization Period	Nominal Rate									
Years	8.0	8.5	9.0	9.5	10.0	10.5	11.0	11.5	12.0	12.5
5	20.22	20.45	20.68	20.91	21.15	21.38	21.62	21.86	22.10	22.34
10	12.07	12.33	12.58	12.84	13.10	13.37	13.64	13.91	14.18	14.46
15	9.49	9.77	10.05	10.33	10.62	10.92	11.21	11.51	11.82	12.12
20	8.29	8.59	8.90	9.20	9.52	9.83	10.16	10.48	10.81	11.14
25	7.64	7.96	8.28	8.61	8.94	9.28	(9.63)	9.97	10.32	10.67
30	7.25	7.59	7.93	8.28	8.63	8.98	9.34	9.70	10.06	10.43

Amortization Period	Nominal Rate								
Years	13.0	13.5	14.0	14.5	15.0	15.5	16.0	16.5	17.0
5	22.58	22.82	23.07	23.31	23.56	23.80	24.05	24.30	24.55
10	14.73	15.01	15.29	15.58	15.86	16.15	16.44	16.73	17.02
15	12.43	12.74	13.06	13.37	13.69	14.01	14.33	14.66	14.99
20	11.48	11.81	12.15	12.49	12.84	13.18	13.53	13.88	14.23
25	11.02	11.38	11.74	12.10	12.46	12.83	13.19	13.56	13.93
30	10.80	11.17	11.54	11.91	12.29	12.66	13.04	13.42	13.79

Research the following:
• first mortgage
• second mortgage
• mortgage covenant
• conventional mortgage
• insured mortgage
• discharge of mortgage
• equity

EXERCISE 11.8

B 1. Barb and Gerry Cormier purchase a house for $124 700, and have a down payment of $57 700. They have a mortgage on the balance at 10.5% compounded semi-annually, and amortized over 25 a. The term of the mortgage is 5 a.

(a) What is the principal of the mortgage?

(b) What is the amount of the monthly payment?

(c) Find the interest payable at the end of the first month.

(d) How much of the first payment is applied to the principal?

(e) What principal is outstanding after the first payment has been made?

(f) Calculate the interest to be paid in the second month.

(g) Prepare a mortgage schedule for the first 6 months.

2. Jessie Stern bought a house for $126 000, and has a mortgage on the property for $62 000.

(a) What is Jessie's equity on the day of the purchase?

(b) How much equity does Jessie have if he sells the house for $135 000?

(c) Is all of the monthly payment on the mortgage going into equity?

(d) Does Jessie's down payment contribute to the amount of equity?

(e) How does a change in the value of a house affect the equity?

3. The mortgage on the Halifax Inn is amortized over 25 a at 11% compounded semi-annually. The principal is $240 000.

(a) What is the principal of the mortgage?

(b) What is the amount of the monthly payment?

(c) Find the interest payable at the end of the first month.

(d) How much of the first payment is applied to the principal?

(e) What principal is outstanding after the first payment has been made?

(f) Calculate the interest to be paid in the second month.

(g) Prepare a mortgage schedule for the first 6 months.

MICRO MATH

The following program prints an amortization schedule. In order to print the table you will have to enter the principal, annual rate of interest, the number of times per year that the interest is compounded, the number of payments in a year, and the amount of money in each payment. The program will then print a table for the complete repayment schedule.

NEW

```
110 INPUT "PRINCIPAL";PRIN
120 INPUT "RATE PER ANNUM";RA
130 INPUT "NUMBER OF TIMES COMP./YEAR "
    ;NC
140 INPUT "PAYMENTS PER YEAR";NP
150 INPUT "PAYMENT";PAY
160 ROTCAF=(1+RA/(100*NC))↑(NC/NP)-1
170 PRINT "PAYMENT","INTEREST","EQUITY",
    "BALANCE"
180 PRINT
190 J=J+1
200 ITR=INT((PRIN* ROTCAF)*100+0.5)/100
210 PPAY=PAY-ITR
220 EQ=INT(PPAY*100+0.5)/100
230 PRIN=INT((PRIN-PPAY)*100+0.5)/100
240 PRINT J,ITR,EQ,PRIN
250 IF PRIN>0 THEN GOTO 190
260 END
```

RUN

Statements 130 and 170 in the above program should be entered on one line.

Print an amortization schedule for each of the following loans.

1. Principal: $25 000.00
   Rate: 10.5%
   Compounded semi-annually
   12 payments per year
   The monthly payment is
   (a) $500.00
   (b) $450.00
   (c) $650.00

2. Principal: $75 000.00
   Rate: 9.875%
   Compounded quarterly
   12 payments per year
   The monthly payment is
   (a) $1000.00
   (b) $1250.00
   (c) $1500.00

11.9 CALL PRICE OF A BOND

Some corporate bonds have a call provision. This means that the firm that first issued, or sold, the bond can repurchase, or "call" the bond at a stipulated call price. The call price is usually higher than the face value of the bond. The difference between the call price and the face value is called the call premium.

$$\boxed{\text{Premium} = \text{Call Price} - \text{Face Value}}$$

For example, if a ten-year bond with a $1000 face value and a coupon rate of 8% can be repurchased by the issuing firm for $1080, then the call premium is $80. The premium is often equal to the payment of one year's interest at the coupon rate. In some cases, the premium decreases over time so that it may be $80 in the first year, $77.50 in the second and so on to maturity. In order to compensate the bondholders for the inconvenience of having their investment called in, these bonds usually pay a higher coupon rate.

Assume that the $1000, 8%, ten-year bond mentioned above has a call provision that allows the corporation to repurchase the bond at any time for $1080. If the market interest rate drops to 6% after 2 a, due to changes in economic conditions, then the market value of the bond would rise as follows:

$$\frac{\$80}{1.06} + \frac{\$80}{1.06^2} + \cdots + \frac{\$80}{1.06^8} + \frac{\$1000}{1.06^8} = \$1124.20$$

The corporation would probably exercise its "call" since the bond could be repurchased for $1080 and the corporation can now borrow at 6% instead of 8%.

In this case, the bondholder is at a disadvantage since the market value of the bond after 2 a is $1124.20 — the present value of $80 per year for 8 a, plus the present value of the $1000 face value for 8 a, both at 6%. However, the bondholder receives $1080 from the corporation. A similar bond without the call provision would have had a coupon rate of about 7% instead of 8%.

We can calculate the yield rate of the bond for the two-year period in which it was held as follows:

$$\$1000 = \frac{\$80}{1 + i} + \frac{\$80}{(1 + i)^2} + \frac{\$1080}{(1 + i)^2}$$

The present value of $1000 is used on the left side since this is the face value — the value of the bond on the day it was purchased. The principal payment of $1080 is used since this is the amount that is received when the call is exercised after two years.

Solving the above equation for i, we have a rate of 11.8%. The resulting higher rate of interest is to compensate for reducing the term of the investment in a favourable money market.

11.10 REVIEW EXERCISE

1. Solve for i.

(a) $961.97 $= \dfrac{\$95}{(1 + i)} + \dfrac{\$95}{(1 + i)^2} + \cdots +$
$\dfrac{\$95}{(1 + i)^{10}} + \dfrac{\$1000}{(1 + i)^{10}}$

(b) $730.16 $= \dfrac{\$80}{(1 + i)} + \dfrac{\$80}{(1 + i)^2} + \cdots +$
$\dfrac{\$80}{(1 + i)^{20}} + \dfrac{\$1000}{(1 + i)^{20}}$

(c) $1234.85 $= \dfrac{\$115}{(1 + i)} + \dfrac{\$115}{(1 + i)^2} + \cdots +$
$\dfrac{\$115}{(1 + i)^{10}} + \dfrac{\$1000}{(1 + i)^{10}}$

(d) $1241.82 $= \dfrac{\$120}{(1 + i)} + \dfrac{\$120}{(1 + i)^2} + \cdots +$
$\dfrac{\$120}{(1 + i)^{15}} + \dfrac{\$1000}{(1 + i)^{15}}$

(e) $1115.19 $= \dfrac{\$125}{(1 + i)} + \dfrac{\$125}{(1 + i)^2} + \cdots +$
$\dfrac{\$125}{(1 + i)^5} + \dfrac{\$1000}{(1 + i)^5}$

(f) $974.27 $= \dfrac{\$105}{(1 + i)} + \dfrac{\$105}{(1 + i)^2} + \cdots +$
$\dfrac{\$105}{(1 + i)^8} + \dfrac{\$1000}{(1 + i)^8}$

2. A company has a net worth of $7 500 000 and there are 100 000 shares outstanding. What is the value of 500 shares in this company?

3. A company with 750 000 shares outstanding has declared a dividend of $1.47 per share.
(a) How much money will the company pay out in dividends?
(b) How much would you receive in dividends if you hold 500 shares?

4. Calculate the value of the yearly coupons for each of the following bonds.
(a) $5000 Government of Canada, 11%
(b) $2000 Industrial, 10.5%
(c) $8000 Canada Trust, 9.25%

5. Find the cost, including 2% commission of these stock purchases.
(a) 500 shares at $19.50
(b) 2000 shares at $6.35
(c) 300 shares at $15.00
(d) 1500 shares at $0.75
(e) 350 shares at $45.50

6. Find the proceeds, less $1\frac{3}{4}$% commission, from the following sales of stock.
(a) 1000 shares at $0.90
(b) 50 shares at $62.00
(c) 25 shares at $32.00
(d) 2000 shares at $14.50
(e) 1500 shares at $1.45

7. A bond with a face value of $15 000 and a rate of 9% is sold at 92% of the face value. Find the net proceeds from the sale of this bond if the commission is $0.50 per $100 face value.

8. Find the cost of the following bonds. The standard commission is $0.50 per $100 face value.
(a) a bond with a face value of $5000 and a rate of 8% is purchased at 90% of the face value
(b) a bond with a face value of $8000 and a rate of 9.5% is purchased at 88% of the face value
(c) a bond with a face value of $4000 and a rate of 10.5% is purchased at 102% of the face value

9. An investor buys a $20 000 bond at 85% of the face value, and sells it for 89% of the original price.
(a) What is the profit in dollars?
(b) What is the profit as a percentage of the purchase price?

10. Find the approximate yield for each of the following $1000 bonds.
(a) 9% coupon rate, selling at $840.00, maturing in 5 a
(b) 10.5% coupon rate, selling at $950.00, maturing in 10 a

11. Find the market value of the following $1000 bonds.

(a) a coupon rate of 9%, maturing in 9 a, to yield 12%

(b) a coupon rate of 12%, maturing in 10 a, to yield 13%

(c) a coupon rate of 6%, maturing in 11 a, to yield 10%

12. Find the market value of a bond with a face value of $1000, and a coupon rate of 9%, maturing in 8 a. The bond is to yield 12% compounded semi-annually.

13. Find the cost of a $20 000 bond, bearing 7.5% annual coupons, maturing in 5 a. The bond is to yield 12% compounded annually. The commission is $0.50 per $100 face value.

14. Find the market value of a bond with a face value of $2000, and a coupon rate of 10%, if the bond is to yield 12%. The bond is to mature in 6 a.

15. Find the annual return and yield for each of these investments.

(a) 200 shares, 8% preferred stock, par value $100 with a market value of $56.00

(b) 500 shares, $10\frac{1}{2}$% preferred stock, par value $50 with a market value of $30

(c) 1000 shares, 12% preferred stock, par value $100 with a market value of $72.50

16. 500 shares were purchaed at $16.50 and sold 30 d later at $14.25. Commission was charged at 3% on purchases and 2% on sales. What was the loss after paying commission?

17. 5000 shares of ETC were purchased at $3.50 on June 1. Another 3000 shares of ETC were purchased on June 15 at $2.75. All 8000 shares of ETC were sold on October 15 at $4.25. Commission was paid at $2\frac{3}{4}$% on the purchase, and $2\frac{1}{2}$% on the sale.

(a) What was the average price paid for the 8000 shares?

(b) What were the proceeds from the sale?

(c) What was the profit after commission?

18. A mortgage of $75 000 at 12% compounded semi-annually is amortized over 25 a. Calculate:

(a) the amount of the monthly payment

(b) the interest at the end of the first month

(c) the amount of principal in the first payment

(d) the mortgage schedule for the first 6 months

19. Edna Durant opened a store at a cost of $250 000. This is paid for with $50 000 from Edna's savings, and the balance from a mortgage with the Permanent Trust Company at 13% compounded semi-annually and amortized over 25 a.

(a) What is the principal of the mortgage?

(b) What is the amount of the monthly payment?

(c) How much of the first payment is applied to the principal?

(d) What principal is outstanding after the first payment has been made?

(e) Calculate the interest to be paid in the second month?

(f) Prepare a mortgage schedule for the first 6 months.

20. The consumer price index based on the value of a dollar three years ago is 1.24. What is the value in constant dollars of a statue that has just been purchased for $800?

21. The consumer price index based on the value of a dollar five years ago is 1.52. What is the value in constant dollars of a violin that has just been sold for $3500?

22. Antique farm equipment was purchased to decorate a restaurant for $12 000 in 1971. What is the price of the equipment in 1980 dollars?

11.11 CHAPTER 11 TEST

1. Solve for i.

(a) $\$1250.63 = \dfrac{\$125}{(1 + i)} + \dfrac{\$125}{(1 + i)^2} + \cdots + \dfrac{\$125}{(1 + i)^{12}} + \dfrac{\$1000}{(1 + i)^{12}}$

(b) $\$849.60 = \dfrac{\$80}{(1 + i)} + \dfrac{\$80}{(1 + i)^2} + \cdots + \dfrac{\$80}{(1 + i)^{10}} + \dfrac{\$1000}{(1 + i)^{10}}$

2. A twenty-year bond has a face value of $1000 and a coupon rate of 12%. Find the market value of the bond if the market interest rate is 10%.

3. A bond with 7 a remaining to maturity has a face value of $1000 and a coupon rate of 14%. The bond is sold at a premium for $1251.65. Find the yield rate of the bond.

4. What are the total dividends receivable for 200 shares of Reichold $10\frac{1}{2}$% preferred and 400 shares of Reichold common?

> **DIVIDENDS**
> Quarterly unless stated otherwise.
> Reichold Canada,
> $0.18, June 15, record May 31;
> $10\frac{1}{2}$ percent pfd. $0.26,
> June 10, record May 26

5. 200 shares of INCO were purchased at $18.75 and sold for $14.50.

(a) Calculate the total cost of the purchase if commission was charged at $2\frac{1}{4}$%.

(b) Calculate the net proceeds on the sale of these shares if the commission charged on the sale was $1\frac{3}{4}$%.

(c) Calculate the capital gain, or loss, from the purchase and sale of these shares excluding commission.

(d) Calculate the profit or loss after paying all commissions.

6.

Year	1977	1978	1979	1980
Change from preceding year	8.0%	9.0%	9.2%	10.1%
Consumer Price Index	138	150		

(a) Calculate the consumer price index for 1979 and 1980.
(b) An antique chair was sold in 1971 for $750.00. The value of the chair increased at the rate of inflation. What is the value of the chair in 1977?
(c) What is the purchasing power in 1978 of $1000 in 1977?

7. A house is purchased for $99 500 with a down payment of $50 000. The balance is secured by a mortgage at 11%, compounded semi-annually, and amortized over 25 a, for a term of 5 a. The interest factor for 11% semi-annually is 0.008 963 394. Find the monthly payment.

PROBABILITY AND STATISTICS

CHAPTER

12

Probability is the very guide of life.
Cicero

REVIEW AND PREVIEW TO CHAPTER 12

CALCULATOR MATH

EXERCISE

1. Simplify to the nearest hundredth.

(a) $\dfrac{421^2}{417} - 532$

(b) $\dfrac{3^4 + 8^2 + 7^3}{45}$

(c) $\dfrac{7.62 \times 4.98}{8.34} + 11.7$

(d) $\dfrac{7^5 + 8^3}{42.3} - 407.5$

(e) $\dfrac{81.8 \times 7.93}{5.24 + 8.6}$

(f) $\dfrac{0.082 - 0.076}{5.76 + 8.94} - 6$

(g) $\dfrac{(43.7 \times 8.14)^2}{56.2 \div 4.9}$

(h) $\left(\dfrac{6^4 - 3.7}{8.1 \div 5.2} - 4.6 \right) \times 3$

(i) $\dfrac{21.6 \times 3.9 - 11.7}{6.8 \times 0.25}$

(j) $\dfrac{(4.62 \times 0.53)^3}{8.1 - 5.8} - 6.6$

2. Find the value of each fraction in decimal form.

(a) $\dfrac{1}{3 + \dfrac{1}{2 + \dfrac{1}{5}}}$

(b) $12 + \dfrac{1}{6 + \dfrac{1}{5 + \dfrac{1}{6}}}$

(c) $322 + \dfrac{1}{4 + \dfrac{1}{13 + \dfrac{1}{7 + \dfrac{1}{6}}}}$

(d) $63 + \dfrac{1}{5 + \dfrac{1}{7 + \dfrac{1}{13 + \dfrac{1}{26}}}}$

3. Find the area of the shaded regions (use $\pi = 3.14$).

(a)

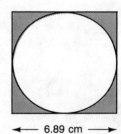

← 6.89 cm →

(b)

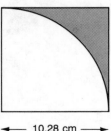

← 10.28 cm →

(c)

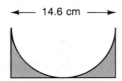

← 14.6 cm →

4. The formula to calculate the surface area of a sphere is

$$A = 4\pi r^2$$

(a) Find the surface area of a sphere whose radius is 35.7 cm (use $\pi = 3.1416$).

(b) If the radius of the sphere is doubled, by what factor is the area increased?

(c) If the radius of the earth is 6379 km, find the surface area of the earth.

(d) If 70.6% of the surface area of the earth is water, how much (km²) of the surface is land?

5. The formula to calculate the volume of a sphere is

$$V = \frac{4}{3}\pi r^3$$

(a) Find the volume of a sphere whose radius is 47.4 cm (use $\pi = 3.1416$).

(b) If the radius of the sphere is doubled, by what factor is the volume increased?

(c) If the radius of the earth is 6379 km, find the volume.

6. If the radius of the earth is 6379 km, how far will a satellite 295 km above the surface of the earth travel in one rotation? (Assume a circular orbit; use $\pi = 3.14$.)

7. The object of the puzzle is to move all of the disks to another peg one at a time. More than one disk may be placed on any peg, but they must always be arranged in order with the largest on the bottom. The idea is to complete the task in the least number of moves.

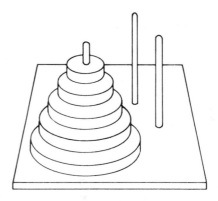

Formula for least number of moves:

$$M = 2^n - 1$$

Compute the value of M for each value of n.

Number of disks n	Number of moves M
3	
4	
5	
6	
7	
8	
9	
10	

12.1 MATHEMATICAL PROBABILITY

Tossing a coin is an example of an experiment involving probability. Other examples of experiments are "rolling a die," "dealing 5 cards from a deck," and "tossing a coin, then rolling a die." We call the set of all possible outcomes the sample space of the experiment and we can list these possible outcomes with the aid of a figure known as a tree diagram.

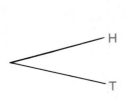

There are 2 possible outcomes when a coin is tossed.

There are 6 possible outcomes when a die is rolled.

In the coin-tossing experiment the sample space is {H, T}. Of the 2 equally likely outcomes only 1 is heads, and so we say the probability of tossing heads is $\frac{1}{2}$. In symbols, we write

$$P(\text{heads}) = \tfrac{1}{2}.$$

In general, if the sample space of an experiment consists of N equally likely outcomes and if S of those outcomes are considered successful (or favourable) for an event E, then we define probability as follows:

$$\frac{\text{probability of}}{\text{an event E}} = \frac{\text{number of successful outcomes}}{\text{total number of possible outcomes}}$$
$$\text{or} \qquad P(E) = \frac{S}{N}$$

When we say that the probability of obtaining a head in a coin toss is $\frac{1}{2}$, we do not mean that heads will occur exactly half the time.

However, after a large number of tosses we expect to obtain heads about half the time. Probability theory tells us what to expect in the long run, but not in a specific event. Even if heads have come up 20 times in a row, it would not be wise to bet a fortune that the next toss will be tails, because the probability of heads on the next toss is still $\frac{1}{2}$.

EXAMPLE 1. Two coins are tossed. What is the probability of obtaining

(a) at least one head?

(b) no head?

SOLUTION:

There are 4 equally likely outcomes which we list in a tree diagram.

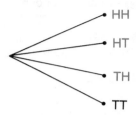

Three of the four outcomes contain a head and so,

(a) P (at least one head) = $\frac{3}{4}$.

(b) P (no head) = $\frac{1}{4}$.

Notice that the sum of the probabilities in Example 1 is 1. This always happens when we calculate the probabilities of the occurrence and non-occurrence of an event.

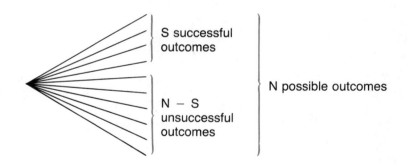

Let the probability of a successful outcome be $p = \dfrac{S}{N}$ and let the probability of an unsuccessful outcome be $q = \dfrac{N - S}{N}$. Adding we get

$$p + q = \frac{S}{N} + \frac{N - S}{N}$$

$$= \frac{S + (N - S)}{N}$$

$$= \frac{N}{N}$$

$$= 1$$

Thus,

$$p + q = 1$$

> If p is the probability that an event will occur, then $0 \leq p \leq 1$.
> If $p = 1$, the event is a certainty. If $p = 0$, it is impossible.
> The closer p is to 1, the more likely the event. The closer p is
> to 0, the less likely the event. If q is the probability that the
> event will not occur, then $q = 1 - p$.

EXAMPLE 2. (a) What is the probability of rolling a 7 with two dice?
(b) What is the probability of not rolling a 7?

SOLUTION:

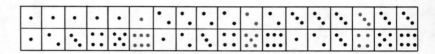

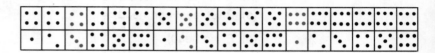

(a) Total number of possible outcomes: N = 36
Number of successful outcomes: S = 6

$$\therefore P(7) = \frac{6}{36}$$
$$= \frac{1}{6}$$

(b) P(not a 7) = 1 − P(7)
$$= 1 - \frac{1}{6}$$
$$= \frac{5}{6}$$

EXERCISE 12.1

A 1. A single die is rolled.
(a) How many possible outcomes are there?
(b) If a success is rolling a number greater than 2, how many successes are there?
(c) What is the probability of rolling a number greater than 2?
(d) What is the probability of rolling a number less than 3?

2. A letter is selected at random from the letters of the word BRAVE. What is the probability that the letter selected is
(a) a vowel?
(b) a consonant?

3. A student, trying to find the probability of rolling a 7 with two dice argued as follows: "There are 11 possible outcomes when 2 dice are rolled: 2, 3, 4, 5, 6, 7, 8, 9, 10, 11, and 12. Since only one of these is 7, the probability of rolling 7 is

$$P(7) = \frac{\text{number of successful outcomes}}{\text{total number of possible outcomes}}$$

$$= \frac{1}{11}."$$

What is wrong with the student's argument?

B 4. A digit is chosen at random from the number 6 893 512.

(a) List the sample space and draw a tree diagram.

(b) List the event that the digit is an odd number, that is, list the set of successful outcomes.

(c) What is the probability that the chosen digit is an odd number?

(d) What is the probability that the chosen digit is an even number?

5. A letter is chosen at random from the word STATISTICS.

(a) List the sample space.

(b) Find the probability that the letter is a vowel.

(c) Find the probability that the letter is a consonant.

6. A box contains 8 green marbles, 12 white marbles, and 4 blue marbles. You are to pick one marble at random. Find the probability of picking a marble which is

(a) green.

(b) white.

(c) blue.

(d) not green.

(e) not white.

(f) not blue.

7. A coin is tossed 3 times.

(a) List the sample space and draw a tree diagram.

(b) Find the probability of obtaining

(i) 3 heads.

(ii) exactly 2 tails.

(iii) at least 2 tails.

8. Find the probability of rolling a 5 with two dice.

9. A deck of playing cards is shuffled and a card is drawn. Find the probability that the card is

(a) a black card.

(b) a club.

(c) not a club.

(d) a king.

(e) the king of clubs.

(f) a red jack.

(g) not a jack.

(h) a face card.

10. An integer from 1 to 40, inclusive, is chosen at random. What is the probability that the integer

(a) is odd?

(b) is even?

(c) is less than 9?

(d) is divisible by 4?

(e) ends in a 6?

(f) is prime?

11. 25 000 lottery tickets have been sold and Jean has bought one of them. There are 10 prizes. What is the probability that Jean will win

(a) a prize?

(b) first prize?

(c) first or second prize?

Place a pack of 52 playing cards on a table so that the bottom card is flush with the edge of the table. By offsetting successive cards from the bottom to the top, get the topmost card to overhang the table's edge by more than one card length.

12.2 EMPIRICAL PROBABILITIES

In the preceding section we were able to calculate probabilities by determining theoretically the number of successful outcomes of an experiment. But in many useful applications of probability this is not possible. In empirical probability we make predictions based on past performances. When the weather forecast says that the probability of rain is 70%, what is meant is that, in the past, rain has occurred on 70% of the days with similar climatic conditions. When an actuary says that the probability of a five year old boy living to age 70 is $\frac{3}{5}$, he means that in the recent past 60% of the Canadian boys who have reached the age of 5 have lived to age 70.

Ever since the eighteenth century, when the first life insurance companies were formed, actuaries have calculated life insurance premiums by using mortality tables to find the probability of a person living to a certain age. For example, the table below is a modern Canadian mortality table which gives the number living at a given age per 100 000 live births.

CANADIAN LIFE TABLE

Age	Male			Female		
	Number living at each age	Number dying between each age and the next	Expectation of life (a)	Number living at each age	Number dying between each age and the next	Expectation of life (a)
At birth	100 000	2 002	69.34	100 000	1 544	76.36
1 a	97 998	126	69.76	98 456	113	76.56
2 a	97 872	92	68.85	98 343	72	75.64
3 a	97 780	83	67.91	98 271	60	74.70
4 a	97 697	69	66.97	98 211	56	73.74
5 a	97 628	232	66.02	98 155	179	72.79
10 a	97 396	267	61.17	97 976	157	67.91
15 a	97 129	682	56.33	97 819	262	63.02
20 a	96 447	872	51.71	97 557	279	58.18
25 a	95 575	730	47.16	97 278	315	53.34
30 a	94 845	773	42.50	96 963	433	48.51
35 a	94 072	1 037	37.83	96 530	644	43.71
40 a	93 035	1 645	33.22	95 886	988	38.99
45 a	91 390	2 569	28.77	94 898	1 465	34.37
50 a	88 821	4 060	24.52	93 433	2 236	29.86
55 a	84 761	6 042	20.57	91 197	3 301	25.53
60 a	78 719	8 675	16.95	87 896	4 804	21.39
65 a	70 044	11 469	13.72	83 092	7 097	17.47
70 a	58 575	13 787	10.90	75 995	10 371	13.85
75 a	44 788	14 812	8.47	65 624	14 387	10.63
80 a	29 976	13 644	6.41	51 237	17 609	7.88
85 a	16 332	9 841	4.74	33 628	17 008	5.67
90 a	6 491	4 891	3.43	16 620	11 358	3.99
95 a	1 600	1 409	2.45	5 262	4 427	2.76
100 a	191		1.71	835		1.89

Adapted from: *Canadian Life Table*, 1971, *Statistics Canada*

EXAMPLE. Use the Canadian Life Table to calculate the probability that a boy of age 5 will live to
(a) age 40.
(b) age 70.

SOLUTION:

(a) The table shows that out of the original sample of males, 97 628 were alive at age 5 and 93 035 of those boys were alive at age 40.

$$N = 97\ 628 \qquad S = 93\ 035$$

The probability of a five year old boy living to age 40 is

$$p = \frac{93\ 035}{97\ 628}$$
$$\doteq 0.953$$

(b) Similarly, the probability that a five year old boy will live to age 70 is

$$p = \frac{58\ 575}{97\ 628}$$
$$\doteq 0.600$$

The answer to part (b) of the Example does not mean that $\frac{3}{5}$ of every group of five year old boys will live to age 70. We cannot predict the life span of a single person or members of a small group. Empirical probabilities only apply to large populations and can depend on time and local conditions.

EXERCISE 12.2

In questions 1-3 use the Canadian Life Table and calculate the answers to 3 decimal places.

1. Find the probability that a twenty year old woman will live to

(a) age 40.
(b) age 70.

2. Find the probability that a man of the given age will live another 10 a.

(a) 30 a   (b) 40 a
(c) 50 a   (d) 60 a

3. Find the probability that a fifteen year old person will live to be the following ages if the person is

(a) a boy
　(i) 50   (ii) 70   (iii) 90
(b) a girl
　(i) 50   (ii) 70   (iii) 90

4. For this question you need a toothpick, needle, or matchstick. On a blank piece of paper construct several parallel lines so that the distance between the lines is equal to the length of the object that you are using. If the object is dropped at random on the paper within the set of parallel lines, determine the empirical probability that it will land touching one of the lines. Do this by throwing the object 100 times and calculating the cumulative value of $\frac{S}{N}$ after every 5 throws.

(In 1777 a French mathematician, Buffon, found the probability to be $\frac{2}{\pi}$. Does your value agree with his?)

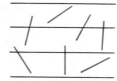

12.3 MUTUALLY EXCLUSIVE EVENTS

Two events are mutually exclusive if they cannot both occur at once. If an event E can occur as either of two mutually exclusive outcomes, then we can add their probabilities to obtain the probability of E.

EXAMPLE 1. A game consists of rolling a single die. You win if the die shows a 3 or a 5. What is the probability of winning?

SOLUTION:
Total number of possible outcomes: $N = 6$
Number of successful outcomes: $S = 2$

$$P(\text{winning}) = \frac{2}{6}$$
$$= \frac{1}{3}$$

Rolling a 3 and rolling a 5 are mutually exclusive events (the die cannot turn up 2 and 5 simultaneously), and so we add the probabilities.

$$
\begin{aligned}
P(\text{winning}) &= P(\text{rolling 2 or 5}) \\
&= P(2) + P(5) \\
&= \frac{1}{6} + \frac{1}{6} \\
&= \frac{2}{6} \\
&= \frac{1}{3}
\end{aligned}
$$

If an event can occur in two mutually exclusive ways which have probabilities p_1 and p_2, then the probability that the event will occur is the sum:
$$p = p_1 + p_2$$

When applying this rule, we must be certain that the two events are mutually exclusive. The following example shows that the addition rule cannot be used when events are not mutually exclusive.

EXAMPLE 2. An integer from 1 to 10 is chosen at random. What is the probability of choosing an even number or a number less than 5?

SOLUTION:
The sample space is $\{1, 2, 3, 4, 5, 6, 7, 8, 9, 10\}$. $\therefore N = 10$
The event is $\{1, 2, 3, 4, 6, 8, 10\}$. $\therefore S = 7$
The probability is $p = \frac{7}{10}$.

ANOTHER SOLUTION:
Let A be the event of choosing an even number.
Let B be the event of choosing a number less than 5.

$$p_1 = P(A) = \frac{5}{10} = \frac{1}{2}$$

$$p_2 = P(B) = \frac{4}{10} = \frac{2}{5}$$

$$p_1 + p_2 = \frac{1}{2} + \frac{2}{5} = \frac{9}{10}$$

Note that $p \neq p_1 + p_2$ because A and B are not mutually exclusive events. It is possible to choose a number which is simultaneously even and less than 5 (namely 2 or 4).

Since A and B are not mutually exclusive we must use

$$
\begin{aligned}
P(E) &= P(A \cup B) \\
&= P(A) + P(B) - P(A \cap B) \\
&= \frac{5}{10} + \frac{4}{10} - \frac{2}{10} \\
&= \frac{7}{10}
\end{aligned}
$$

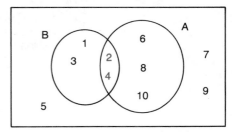

We subtract $P(A \cap B)$ since the numbers in $A \cap B = \{2, 4\}$ have been counted twice.

EXERCISE 12.3

A 1. Which of the following pairs of events are mutually exclusive?

(a) picking a green marble and picking a white marble
(b) having blue eyes and black hair
(c) dealing a spade and dealing a jack
(d) dealing a spade and dealing a club
(e) living in Ottawa and living in Nova Scotia
(f) living in Vancouver and living in British Columbia
(g) choosing an even number and choosing a prime number

B 2. A box contains 6 green marbles, 4 white marbles, 9 red marbles, and 5 blue marbles. You are to pick one marble at random. Find the probability of picking

(a) a green marble.
(b) a green or a white marble.
(c) a red or a blue marble.
(d) a red or a green marble.
(e) a white, or a red, or a blue marble.

3. A deck of playing cards is shuffled and a card is drawn. Find the probability that the card is

(a) a red ace or a black face card.
(b) a diamond or a black queen.

4. In the game of craps, the player wins on the first roll if he rolls a 7 or 11 with a pair of dice. He loses if he rolls a 2, 3, or 12 on the first roll.

(a) What is the probability of winning on the first roll?
(b) What is the probability of losing on the first roll?

5. Two dice are rolled. What is the probability that the result is

(a) greater than 8?
(b) less than 8?
(c) a prime number?

6. Three dice are rolled. Find the probability that the total is

(a) 2
(b) 3
(c) 4
(d) 5
(e) less than 6.

7. Four coins are tossed. What is the probability of getting

(a) 2 tails and 2 heads?
(b) 3 tails and a head?
(c) at least 2 tails?

12.4 INDEPENDENT EVENTS

Two events are called independent if neither has an influence on the other. For instance, if a die is rolled and a coin is tossed, the events of the die coming up 3 or 4 and the coin coming up heads are independent.

EXAMPLE 1. A game is played by rolling a die and then tossing a coin. You win if the die shows a 3 or a 4 and the coin shows heads. What is your probability of winning?

SOLUTION:
The tree diagram involves two events: first the roll of a die and then the toss of a coin.

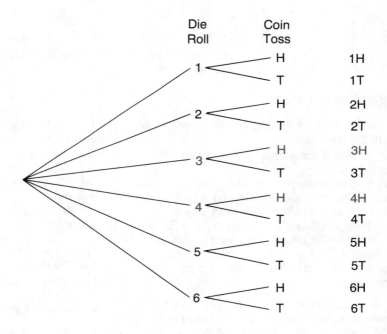

The sample space {1H, 1T, 2H, 2T, 3H, 3T, 4H, 4T, 5H, 5T, 6H, 6T} contains 12 equally likely outcomes. (For example, 5T means the die shows 5 and the coin shows tails.) Only 2 of the outcomes are favourable: 3H and 4H. Thus, N = 12 and S = 2.

$$P(\text{winning}) = \frac{2}{12}$$
$$= \frac{1}{6}$$

Note that this probability is the product of the individual probabilities of the two events.

$$P(3 \text{ or } 4) = \frac{2}{6}$$
$$= \frac{1}{3}$$

$$P(\text{heads}) = \frac{1}{2}$$

The probability of both events happening is

$$P(\text{winning}) = \frac{1}{3} \times \frac{1}{2}$$
$$= \frac{1}{6}$$

The reason for multiplying the probabilities can be seen in the tree diagram. There are 2 small branches on each of 6 main branches and so there are 12 possible outcomes which in turn means a probability of $\frac{1}{12}$ for each outcome.

If the probabilities of 2 independent events are p_1 and p_2, then the probability that both events will occur is the product

$$p = p_1 p_2.$$

$P(A \cap B) = P(A) \times P(B)$ for independent events.

EXAMPLE 2. (a) Suppose a marble is drawn at random from a bag containing 4 red and 3 blue marbles. Then the marble is replaced in the bag and another is drawn. What is the probability that both marbles are red?
(b) If the first marble is not replaced, what is the probability that both marbles are red?

SOLUTION:
(a) On the first draw $N = 7$ and $S = 4$.

$$P(\text{red}) = p_1$$
$$= \frac{4}{7}$$

Since the marble is replaced and the second draw is again random, it is independent of the first draw. Again the probability of getting a red marble is

$$p_2 = \frac{4}{7}$$

Thus, the probability that both marbles are red is

$$p_1 p_2 = \frac{4}{7} \times \frac{4}{7}$$

$$= \frac{16}{49}$$

(b) As in part (a)

$$P(\text{red}) = p_1$$

$$= \frac{4}{7}$$

If the first marble is red, there are 3 red marbles left in the bag.
$\therefore N = 6, S = 3$.

The probability that the second marble is red is

$$p_2 = \frac{3}{6}$$

$$= \frac{1}{2}$$

The probability that both marbles are red is

$$p_1 p_2 = \frac{4}{7} \times \frac{1}{2}$$

$$= \frac{2}{7}$$

In Example 2(b) note that if the first event has occurred before the second and we know its outcome, then we can use the outcome of the first event as one of the conditions in calculating the probability of the second event.

EXERCISE 12.4

A 1. What is the probability of tossing 3 heads in a row with a coin?

B 2. A game is played by first tossing a coin and then rolling a die. You win if the coin shows tails and the die shows an even number.

(a) Draw a tree diagram.
(b) Find your probability of winning.

3. A player rolls a pair of dice, one red and the other green.

(a) Find the probability of rolling a 2 on the red die and 3 on the green die.
(b) Find the probability of rolling a 2 on one die and a 3 on the other die.

4. A player rolls a pair of dice and picks a card from a deck of playing cards. What is the probability of rolling a 9 and picking a spade?

5. Two letters are chosen at random from the English alphabet. If y is considered to be a consonant, find the probability that
(a) both are vowels.
(b) both are consonants.

6. A pair of dice is rolled 4 times in a row. Find the probability of rolling less than 5 on every roll.

7. A bag contains 4 white, 3 blue, and 6 red marbles. A marble is drawn from the bag, replaced, and another marble is drawn. Find the probability that
(a) both marbles are red.
(b) both marbles are blue.
(c) the first marble is red and the second is blue.
(d) one marble is red and the other is blue.
(e) neither is red.

8. Do question 7 assuming that the first marble is not replaced.

9. A card is dealt from a deck of playing cards and is replaced. The deck is shuffled and another card is dealt. What are the probabilities of the following events?
(a) Both cards are diamonds.
(b) The first card is a diamond and the second is a club.
(c) One card is a diamond and the other is a club.
(d) Both cards are aces.
(e) A king and a jack are dealt.
(f) The ace of spades is dealt twice.
(g) The same card is dealt twice.

10. Do question 9 assuming that the first card is not replaced.

11. 5000 raffle tickets have been sold and Tom has bought 4 of them. There are 10 prizes. What is the probability that Tom will win
(a) first prize?
(b) first and second prizes?
(c) first, second, and third prizes?

12. A pair of dice is rolled. What is the probability that
(a) two 1s appear?
(b) the same number appears on each die?
(c) the sum will be an even number?
(d) the sum will be a number larger than 9?

13. A dime and a quarter are tossed, and a die is rolled. What is the probability of getting
(a) two heads and a 6?
(b) a head on the dime, a tail on the quarter, and a 2?
(c) a head on the quarter, a tail on the dime, and a number greater than 2?

14. A coin is tossed, a die is rolled, and a card is drawn from a pack of playing cards. What is the probability of getting
(a) a head, a 6 on the die, and a club?
(b) a tail, a 3 on the die, and a king?
(c) a tail, a 2 on the die, and the four of diamonds?

C 15. Three identical bags A, B, and C contain black and white marbles as follows: A, 4 black and 1 white; B, 2 black and 3 white; and C, 3 black and 7 white. A bag is chosen at random and a marble is chosen from that bag. Find the probability that the marble is white.

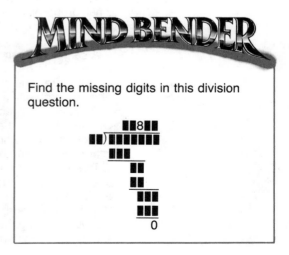

Find the missing digits in this division question.

12.5 STATISTICS

Statistics is concerned with the collection of numerical facts, called data, the organization and presentation of these facts, and finally the interpretation of the data.

The subject of statistics has been applied to many different areas, including insurance, medicine, advertising, electronics, television audiences, population growth, energy consumption, student enrollment, etc.

The following are some examples of the familiar use of statistics.

I. The Gallup Poll uses statistics to predict the outcome of elections.

II. Based on statistics, drivers under 25 a pay more for car insurance.

III. The Nielsen television ratings help determine which TV programs should be discontinued.

IV. Statistics show what brand of toothpaste is most effective in reducing tooth decay.

V. Schools are built or closed according to statistics.

VI. Betting odds on sports events are calculated using statistics.

You rarely live through a day without having statistics forced on you. Everyone is bombarded with statistics by advertisers, commentators, politicians, pollsters, and salesmen. Knowledge of statistics is fast becoming an important tool for everyone.

SAMPLING

The Nielsen television ratings are very important to producers of television programs. In order to determine the popularity of a show, the pollsters cannot ask every individual who watches television for a reaction to the show. What they do is take a sample. A relatively small group of TV viewers is asked for its reactions. From their comments, the polling company will then generalize for all television viewers.

A sample is a small group of individuals (or objects) selected to stand for a larger group called the population.

There are a number of problems that can arise in taking a sample. The first problem is that of sample size. If the sample is too small the individuals (or objects) may not be representative of the population. On the other hand a large sample is usually very costly.

A good sample should be large enough to contain sufficient information about the population.

The most important requirement of a sample is that it be a random one. This means that each individual in the population should have an equally likely chance (or probability) of being selected. Also the "target" population must be kept in mind. For example, it is unlikely that you would get an unbiased reaction on whether Sunday sports should be permitted if you sampled a crowd at a ball game.

EXERCISE 12.5

A 1. Selecting names for a survey from a telephone book does not produce a random sample. Why not?

2. (a) Conducting an election survey at a football game would omit a large selection of the population. Who?
(b) What type of survey would you conduct at a football game?

3. Surveying every 10th person that passes through a turnstile at a subway station does not give a random sample of a city's population. Why not?

4. In order to predict the results of an election a national magazine sent a questionnaire to all of its subscribers. What is wrong with this method of surveying?

5. The Ace Fireworks Company tests (samples) one out of every 500 firecrackers it makes.

(a) Why is it not feasible to sample the total population?
(b) What other industries would test a small percentage of the product?
(c) What industries test 100% of their products?

GIROLAMO CARDANO (1501-1576)

Cardano, the foremost pioneer of the theory of probability, was a man of many talents, active in a dozen arts and sciences. Aside from being a mathematician, he was one of the two most famous doctors of his day, the most widely read popularizer of science in the sixteenth century, a gambler, an astrologer, a dabbler in spiritualism, and an alchemist. While he was a student at the University of Padua, gambling was both his favourite activity and his main source of income. He applied his mathematical genius to gambling and discovered some rules of probability which were published in his *Book on Games of Chance*. Another of his books, *The Great Art*, established him as the major expert of his day in the field of algebra.

Cardano led a very stormy life. He was constantly involved in disputes of all sorts and was finally put in jail for heresy. Some men praised him for his genius, others believed him to be an evil spirit. The contradictions in his character are summed up by the following single sentence which Cardano wrote about himself: "Nature has made me capable in all manual work, it has given me the spirit of a philosopher and ability in the sciences, taste and good manners, voluptuousness, gaiety, it has made me pious, faithful, fond of wisdom, meditative, inventive, courageous, fond of learning and teaching, eager to equal the best, to discover new things and make independent progress, of modest character, a student of medicine, interested in curiosities and discoveries, cunning, crafty, sarcastic, an initiate in the mysterious lore, industrious, diligent, ingenious, living only from day to day, impertinent, contemptuous of religion, grudging, envious, sad, treacherous, magician and sorcerer, miserable, hateful, lascivious, solitary, disagreeable, rude, divinator, obscene, lying, obsequious, fond of the prattle of old men, changeable, irresolute, indecent, fond of women, quarrelsome, and because of the conflicts between my nature and soul I am not understood even by those with whom I associate most frequently."

12.6 FREQUENCY DISTRIBUTIONS

Following is the "raw data" resulting from fifty rolls of a single die.

14231   64536   11336   33265   46613
55522   42544   61243   54345   63253

We can tabulate the data in a frequency distribution table where we determine the number of times each value occurs.

Value	Tally mark	Frequency
1	⊪⊩ ⏐	6
2	⊪⊩ ⏐⏐	7
3	⊪⊩ ⊪⊩ ⏐	11
4	⊪⊩ ⏐⏐⏐⏐	9
5	⊪⊩ ⏐⏐⏐⏐	9
6	⊪⊩ ⏐⏐⏐	8

Formulating the data in a frequency table hides the order in which the values occurred. From the table, we see the frequency of each value. This information can be presented in a graph called a histogram, where the heights of the rectangles represent the frequencies. In this histogram, the width of each bar occupies a space 0.5 below the observed value to 0.5 above.

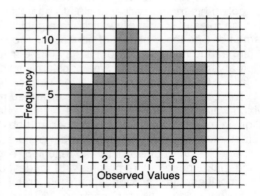

We can now read the frequencies from the vertical axis. When displaying a frequency distribution in a histogram, we are not really interested in the order, or in the individual values, but rather in general characteristics such as the shape of the distribution, most frequent scores, and range.

If we join the midpoints of the tops of the rectangles of the histogram, the graph formed is called the frequency polygon which is shown below. The distribution is extended one unit beyond the highest and lowest observed values so that the polygon is completed by drawing lines to these values on the baseline. Examination of the graph shows that the histogram and the frequency polygon showing the same data and drawn with the same scale are equal in area. This is significant because the area is proportional to the frequency of the values. Either a histogram or a frequency polygon can be used when we wish to compare the distributions of two or more groups of data.

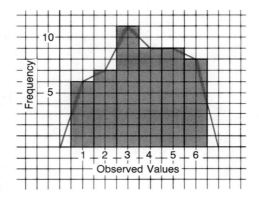

EXAMPLE. In a survey of cross sections of trees in a forest, data were recorded and the following frequency distribution was prepared. Prepare a histogram and frequency polygon.

Diameter in cm	Frequency
20.0 – 21.5	2
21.5 – 23.0	6
23.0 – 24.5	12
24.5 – 26.0	19
26.0 – 27.5	28
27.5 – 29.0	46
29.0 – 30.5	50
30.5 – 32.0	48
32.0 – 33.5	38
33.5 – 35.0	25
35.0 – 36.5	17
36.5 – 38.0	7
38.0 – 39.5	5
39.5 – 41.0	1

SOLUTION:

Histogram and frequency polygon of cross sections of trees.

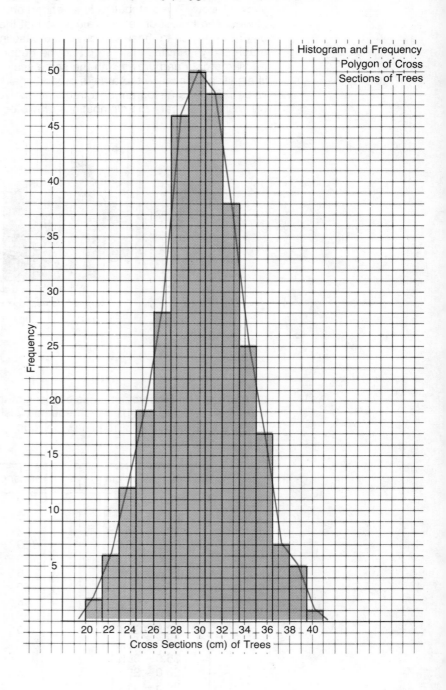

Histogram and Frequency Polygon of Cross Sections of Trees

Frequency

Cross Sections (cm) of Trees

EXERCISE 12.6

B 1. Tabulate the following data to find the frequency, then construct a corresponding histogram and frequency polygon.

4, 2, 4, 6, 1, 4, 2, 2, 3, 6, 1, 4, 4, 5, 2, 3, 4, 4, 5, 5, 1, 1, 6, 3, 4, 6, 3, 2, 3, 2, 4, 3, 3, 5, 2, 5, 6, 5, 2, 4, 6, 4, 6, 1, 2, 2, 5, 1, 3, 3.

2. Fifty applicants for employment at Bolder Graphics were required to answer an aptitude test consisting of 50 questions. The raw scores were

25, 25, 26, 27, 28, 28, 29, 30, 30, 30, 30, 31, 31, 31, 31, 32, 32, 32, 33, 33, 33, 34, 34, 34, 34, 34, 35, 35, 35, 35, 35, 35, 35, 35, 36, 36, 36, 36, 37, 37, 37, 37, 37, 38, 38, 38, 41, 41, 42, 43.

(a) Prepare a frequency distribution table and construct the corresponding histogram and frequency polygon.

(b) Prepare a frequency distribution table using class intervals of 2. (25 − 26, 27 − 28, 29 − 30, ..., 43 − 44).

3. Given the data for masses of 50 students in kilograms:

55, 56, 56, 57, 58, 59, 59, 60, 60, 62, 62, 63, 64, 64, 65, 66, 66, 67, 68, 69, 70, 70, 70, 71, 72, 72, 72, 72, 72, 73, 73, 74, 75, 75, 77, 77, 78, 78, 79, 79, 80, 80, 81, 83, 84, 86, 88, 89, 91, 98.

(a) Tabulate the data using two class intervals 50 − 75 and 76 − 100, then construct the corresponding histogram and frequency polygon.

(b) Tabulate the data using five equal class intervals, then construct the corresponding histogram and frequency polygon.

4. In order to determine whether or not to request a radar trap in front of a school where the speed limit was 45 km/h, a class of students recorded the following car speeds in km/h during 1 h.

39 57 42 40 52 43 48 51 41 42
53 24 45 49 41 53 55 48 51 51
44 50 57 48 50 47 51 48 57 43
36 54 50 45 29 49 57 47 42 49
43 51 50 53 51 48 47 51 50 45
54 42 44 52 42 58 45 53 51 38
61 45 51 50 57 49 47 47 42 48
44 57 41 43 35 59 54 51 50 50

(a) Prepare a histogram and frequency polygon.

(b) Should a speed trap be installed?

WORD LADDER

Start with the word "wheat" and change one letter at a time to form a new word until you reach "bread." The best solution has the fewest steps.

w h e a t
- - - - -
- - - - -
- - - - -
- - - - -
- - - - -
- - - - -
b r e a d

12.7 MEASURES OF CENTRAL TENDENCY

In the previous section we used graphs to provide a "picture" of the data in order to help analyze the data. Statisticians also describe data by indicating a centre of the distribution called a measure of central tendency. Three common measures of central tendency are the mean, median, and mode.

I. The MEAN:

The mean or average of a set of numbers is found by adding them together and dividing the total by the number of numbers added. For a set of values $x_1, x_2, x_3, ..., x_n$ the mean is

$$\bar{x} = \frac{x_1 + x_2 + x_3 + ... + x_n}{n}$$

II. The MEDIAN:

When a set of numbers is arranged in order (smallest to largest), then the middle number is the median. If there is an even number of numbers, then the median is the average of the middle two numbers. Knowing the median enables one to tell whether any number is in the top half or bottom half of the group.

III. The MODE:

The mode of a set of numbers is the number that occurs most often. If every number occurs only once, then we say there is no mode. It is also possible, however, for a set of numbers to have several modes.

EXAMPLE 1. A baseball team pays the following annual salaries to its players.

Number of players	Salary ($)
1	250 000
1	200 000
2	190 000
3	170 000
5	100 000
4	90 000
3	80 000
3	70 000
2	60 000
1	50 000

Determine the (a) mean, (b) median, and (c) mode of these salaries.

SOLUTION:

(a) For data of this type we can use the following method to determine the mean.

x_i	f	$f \times x_i$
250 000	1	250 000
200 000	1	200 000
190 000	2	380 000
170 000	3	510 000
100 000	5	500 000
90 000	4	360 000
80 000	3	240 000
70 000	3	210 000
60 000	2	120 000
50 000	1	50 000
TOTAL	25	2 820 000

$$\bar{x} = \frac{2\ 820\ 000}{25}$$
$$= 112\ 800$$

The mean salary is $112 800.

(b) There are 25 people on the payroll so the median is in the 13th position. The median salary is $90 000.

(c) The mode is $100 000.

EXAMPLE 2. A purchasing agent for a calculator manufacturer is considering three different brands of batteries to be used in calculators. The following data were received from a testing laboratory.

	Brand of Battery		
	Alpha	Beta	Delta
Mean lifetime (h)	12	12	12
Median lifetime (h)	12	8	15

How should the agent interpret these data?

SOLUTION:

The agent should see that the graph of the lifetime of the Alpha battery is symmetric.

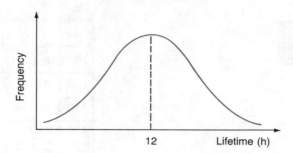

The data on the Beta battery indicate that most of these batteries wear out very quickly but a few have a high life expectancy. When the median is less than the mean the numbers probably "tail off" to the right.

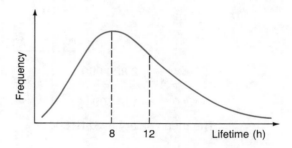

The data on the Delta brand suggest that a few batteries wear out very quickly but a lot of them last a long time. When the median is higher than the mean the numbers probably "tail off" to the left.

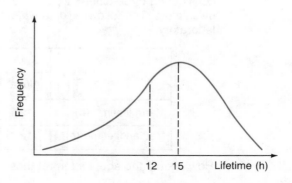

The purchasing agent may then assess the position as follows:

(a) If Alpha batteries are used, there will be a balance between the batteries that last a short time and those that last a long time.

(b) If Beta batteries are used, there will be many customers whose batteries won't last very long.

(c) If Delta batteries are used, there will be a few customers whose batteries will fail almost instantly. On the other hand, however, many customers will have batteries that last a long time.
Which one should he choose?

EXERCISE 12.7

1. State the measure of central tendency that best describes the following.
(a) the number of L/100 km you can expect from a new car
(b) the hat sizes to be kept in stock
(c) a set of class marks
(d) the amount of rainfall for a country
(e) the helmet size of a football team

2. Determine the mean, median, and mode(s) of the following sets of data.
(a) 5, 10, 16, 17, 7, 12, 15, 9, 12
(b) 15, 12, 10, 13, 10, 11, 13, 14
(c) 15, 18, 9, 13, 17, 10, 22, 16
(d) 5, 2, 3, 4, 2, 6, 9, 8, 7, 5, 2

3. The lifetime of a certain brand of tires has a mean of 50 000 km and a median of 40 000 km. Describe the performance of these tires.

4. The lifetime of a certain brand of TV picture tube has a mean of 18 000 h and a median of 2 700 h. Describe the performance of these picture tubes.

5. A firm pays the following annual salaries:
   1 President at $130 000
   2 Vice-presidents at $70 000
   1 Treasurer at $60 000
   3 Accountants at $50 000
   8 Sales staff at $35 000
   9 Consultants at $30 000
(a) Determine the mean, median, and mode.
(b) Which statistic best indicates the level of salary paid by the company? Why?

6. The sales staff for an automobile dealership have the following new car sales for the month of June.

Carol	15	Pete	12
Al	10	Sharon	14
Eric	9	Curt	9
Marty	7	Warren	11

(a) Find the mean, median, and mode of the sales.
(b) Suppose that the salesmanager is hiring a new salesman. What would be the most realistic estimate of expected sales he could give the new salesman?

7. A company vice-president gathered the following data on the length of time spent on coffee breaks by employees: mean 13 min; median 10 min. Describe the coffee break habits of the employees.

Put the numbers from 1 to 9 in the spaces to make the statements true.

$$\blacksquare - \blacksquare + \blacksquare = 10$$
$$\blacksquare + \blacksquare - \blacksquare = 10$$
$$\blacksquare \div \blacksquare + \blacksquare = 10$$

12.8 MEASURES OF DISPERSION

The measures of central tendency discussed in the previous section may not tell everything we want to know about a set of data. Two sets of data could have the same mean or median and yet differ greatly in their dispersion or spread.

One measure of dispersion is the range.

> The range of a set of numbers is found by subtracting the smallest number from the largest.

The range is very sensitive to extreme values, but does not tell us anything about how the numbers vary. For this reason we need another measure of the dispersion called the standard deviation (σ).

The standard deviation of a set of numbers is determined by finding (in order):

I. The mean of the numbers
II. The difference between each number and the mean
III. The squares of each of these differences
IV. The mean of the squares
V. The square root of this mean

The five steps may be abbreviated in the formula

$$\sigma = \sqrt{\frac{(x_1 - \bar{x})^2 + (x_2 - \bar{x})^2 + \ldots + (x_n - \bar{x})^2}{n}}$$

EXAMPLE 1. A basketball coach measured the heights of his players and found them to be 173 cm, 177 cm, 177 cm, 179 cm, 181 cm, 181 cm, 182 cm, 182 cm, 183 cm, and 185 cm. Find the standard deviation of this set of data.

SOLUTION:

Heights	Differences	Squares
x_i	$(x_1 - \bar{x})$	$(x_i - \bar{x})^2$
173	-7	49
177	-3	9
177	-3	9
179	-1	1
181	1	1
181	1	1
182	2	4
182	2	4
183	3	9
185	5	25

Heights	Squares
sum = 1800	sum = 112
$\bar{x} = \dfrac{1800}{10}$	$\text{mean}\atop\text{of squares}$ $= \dfrac{112}{10}$
= 180	= 11.2

$$\sigma = \sqrt{11.2}$$
$$\doteq 3.3$$

The standard deviation $\sigma \doteq 3.3$.

NORMAL DISTRIBUTION

When data are distributed normally, 68% of the population will lie within one standard deviation of the mean and 95% of the population will lie within two standard deviations of the mean. Almost all (99.7%) of the population will lie within three standard deviations of the mean.

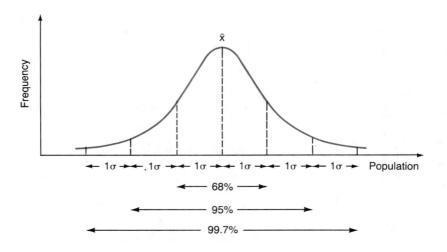

In order to facilitate calculations the distribution is simplifed as follows.

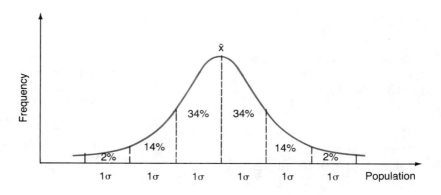

EXAMPLE 2. The quality controller at a candy factory determined that the mean mass of Krispy Nut chocolate bars is 70 g with a standard deviation of 3 g. Assuming a normal distribution,
(a) what percent of the bars have masses between 70 g and 76 g?
(b) if the company makes 8000 bars per day, how many of them will have masses less than 67 g?

SOLUTION:
Draw a normal curve.

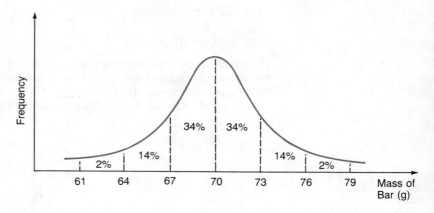

The standard deviation is 3 g, so we can mark the horizontal axis accordingly.

(a) 48% of the bars have masses between 70 g and 76 g.

(b) 16% of the bars have masses less than 67 g.

$$0.16 \times 8000 = 1280$$

1280 bars will have masses less than 67g.

EXERCISE 12.8

A 1. State the range of the following.
 (a) 3, 7, 15, 19, 24
 (b) 13, 5, 36, 48, 9
 (c) 54, 13, 25, 61, 47, 8

B 2. Determine the standard deviation for the following sets of data.
 (a) 4, 5, 8, 9, 9, 9, 10, 10, 12, 14
 (b) 20, 24, 17, 15, 28, 25, 22, 20, 17, 12, 20

3. A hospital determines that the average stay in the hospital for pneumonia cases is 7 d with a standard deviation of 2 d. Assuming a normal distribution,
(a) what percent of pneumonia cases stay in longer than 9 d?
(b) what percent of cases stay in between 3 d and 7 d?

4. The amount of detergent used daily by a hospital is normally distributed with a mean of 150 kg and a standard deviation of 10 kg.

(a) What percent of the days will they use less than 130 kg?

(b) What percent of the days will they use more than 160 kg?

5. The Long Life Light Company advertises that its light bulbs have a mean life of 900 h with a standard deviation of 50 h. Assuming normal distribution,

(a) what percent of the bulbs will last between 900 h and 1000 h?

(b) what percent of the bulbs will last longer than 1000 h?

(c) if a business purchases 3000 bulbs, how many can be expected to last less than 850 h?

6. A tire manufacturer advertises that its Mac III truck tire has a mean life of 60 000 km with a standard deviation of 4000 km. The local streets and sanitation department buys 600 tires. Assuming a normal distribution,

(a) what percent of the tires will last longer than 56 000 km?

(b) what percent of the tires will last less than 52 000 km?

(c) how many tires should last longer than 68 000 km?

(d) how many tires will last between 56 000 km and 64 000 km?

7. The following data show the times (in minutes) that it takes firemen to reach fires in their area.

Station 1	Station 2
5.5	7
5	8
6.5	7.5
5.5	7
4.5	6.5
7	6
3	6.5
6	7
4	7.5
3	7

(a) Find the mean and the standard deviation of the response time for Station 1.

(b) Find the mean and the standard deviation of the reponse time for Station 2.

(c) What could account for the difference in the means?

(d) What could account for the difference in the standard deviation?

(e) Can we use this data to determine which station area had the better fire protection? Explain.

8. The mean number of fielding errors committed by major league second basemen in one season is 7 with a standard deviation of 1. For third basemen, the mean is 9 with a standard deviation of 2.

(a) What would account for the differences in the mean?

(b) If the second baseman for the Yankees commits 9 errors in a season and the third baseman for the Blue Jays commits 11 errors, who is the better fielder?

9. The graph below shows the masses of pennies that were put into circulation at the same time. Curve A shows the mass distribution when the coins were new and the other two curves show the mass distribution after four years (B) and eight years (C) of circulation.

(a) As circulation time increases, what happens to the mean mass of the coins? Why?

(b) As circulation time increases, what happens to the standard deviation? Why?

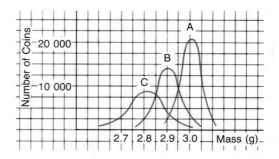

12.9 REVIEW EXERCISE

1. A word is selected at random from the sentence *The sky is blue*. Find the probability that the number of letters in the selected word is
(a) 2
(b) 3
(c) 5
(d) less than 5.

2. A ball is drawn at random from a box containing 12 red, 15 green, 5 black, and 8 white balls. Find the probability that the ball is
(a) red.
(b) green.
(c) black.
(d) white.
(e) red or green.
(f) not white.
(g) neither red nor white.
(h) blue.
(i) green, or white, or black.

3. A number is formed by arranging the digits 1, 2, 3 in random order.
(a) List the sample space.
(b) What is the probability that the number is
    (i) greater than 100?
    (ii) greater than 200?
    (iii) greater than 220?
    (iv) greater than 330?

4. Find the probability of the following events.
(a) Drawing a club or a diamond from a deck of cards.
(b) Rolling a number less than 10 with a pair of dice.
(c) Obtaining at least one head when a coin is tossed 3 times.
(d) Not drawing an ace from a deck of cards.
(e) Arriving at a traffic light when it is green if it stays green for forty seconds, red for thirty seconds, and yellow for two seconds.
(f) Drawing a black card with an even number on it from a deck of cards.

5. Lynn's laundry contains 2 pairs of green socks, 3 pairs of blue socks, and a pair of black socks. When the dryer stops a green sock is pulled out. If a second sock is then pulled out, what is the probability that it will match the first sock?

6. A test consists of multiple choice questions. Each question has five choices, one of which is correct. An unprepared student makes random guesses for each answer. What is the probability that the student answers
(a) the first question correctly?
(b) the first 3 questions correctly?
(c) the first 3 questions incorrectly?

7. Find the probability of tossing a coin 10 times and getting tails every time.

8. Each letter of the alphabet is printed on a separate slip of paper. The 26 slips are then mixed in a bowl. Find the probability of each of the following, assuming that each slip is replaced before the next slip is drawn.
(a) Choosing 2 slips and getting NO
(b) Choosing 3 slips and getting YES
(c) Choosing 4 slips and getting MATH

9. Do question 8 assuming that the slips are not replaced.

10. There are three children in the Evans family. What is the probability that
(a) all of the children are boys?
(b) at least one child is a girl?
(c) at least two of the children are girls?
(d) there are no boys in the family?

11. Two people get into an elevator. They can get off at any floor from 2 to 7. What is the probability that they will both get off at the same floor?

12. Terry took a true-false test. He did not know the answers to the last three questions, so he guessed.
(a) What is the probability that all three answers were wrong?
(b) What is the probability that at least one of the answers was correct?
(c) What is the probability that all three answers were correct?

13. A pair of dice is rolled. What is the probability that

(a) the sum is 9?
(b) the sum is an odd number less than 6?
(c) the sum is less than 2?

14. Determine the mean, median, and mode(s) of the following sets of data.

(a) 10, 7, 5, 9, 13, 12, 17, 5, 15
(b) 11, 12, 15, 20, 22, 25, 24, 21, 12, 11, 10, 12, 15
(c) 10, 16, 13, 9, 10, 12, 11, 14, 9, 10, 15, 9

15. Determine the range and standard deviation for the following sets of data.

(a) 13, 9, 15, 12, 11
(b) 31, 28, 21, 25, 24, 30, 33, 27, 36, 35

16. The lifetime of a certain brand of calculator battery has a mean of 60 h and a median of 50 h. Describe the performance of these batteries.

17. After one season the manager of a baseball team determined that the team had a mean batting average of 0.242. The median batting average was 0.263. Describe the batting performance of the team.

18. The members of a police department have a mean height of 180 cm with a standard deviation of 4 cm.

(a) What percent of the department are taller than 172 cm?
(b) What percent of the department are shorter than 184 cm?
(c) If the department has 450 employees, how many are taller than 176 cm?
(d) How many have heights between 172 cm and 180 cm?

19. A company uses an automated packaging device to produce 50 g bags of Karmel Korn. The machine needs frequent checking to see if it is actually putting 50 g in each bag. The following are masses of thirty bags of Karmel Korn.

```
54 50 47 50 51 50
53 50 47 51 50 51
52 49 46 52 50 49
52 48 48 53 49 49
51 48 49 52 49 50
```

(a) Find the mean and standard deviation of this data.
(b) What problems will be encountered if the standard deviation gets too high?
(c) Can any problems result if the median goes below 50 g?

12.10 CHAPTER 12 TEST

1. A digit is chosen at random from the number 386 157.
(a) What is the probability that the chosen digit is an even number?
(b) What is the probability that the chosen digit is an odd number?

2. Two dice are rolled. Find the probability that the result is
(a) 6
(b) not 6
(c) less than 6

3. (a) A bag contains 2 green, 5 blue, and 3 red balls. A ball is selected at random from the bag, it is replaced, and another ball is selected. Find the probability that both balls are green.
(b) If the first ball is not replaced, find the probability that both balls are green.

4. Find the mean, median, and mode(s) of the following data.

8, 10, 4, 8, 3, 9, 6, 8, 7

5. Find the range and standard deviation of the following data.

12, 13, 9, 7, 9

6. The quality controller at a company making light bulbs determines that the mean life of its bulbs is 1000 h with a standard deviation of 100 h. Assuming normal distribution,
(a) what percent of the bulbs will last between 900 h and 1100 h?
(b) what percent of the bulbs will last less than 900 h?

TRIGONOMETRY

CHAPTER

13

During the second century B.C., the first trigonometric table
apparently was compiled by the astronomer Hipparchus of Nicaea,
who thus earned the right to be known as 'the father of trigonometry'.

Carl Boyer

SIMILAR TRIANGLES

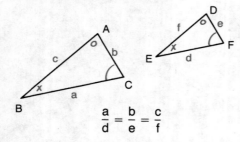

$$\frac{a}{d} = \frac{b}{e} = \frac{c}{f}$$

EXERCISE

1. Find the length of the indicated side.

(a)

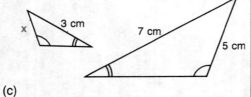

(b)

(c)

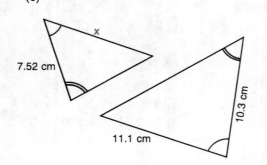

(d)

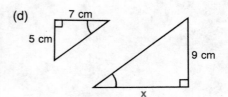

(e)

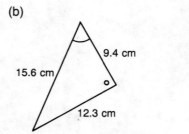

2. Find the lengths of the indicated sides.

(a)

(b)

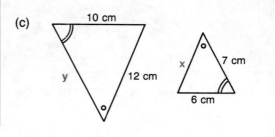

(c)

THE PYTHAGOREAN THEOREM

EXERCISE

1. Find the length of the indicated side in the following.

(a)

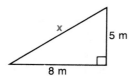

(b)

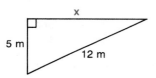

(c)

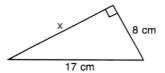

(d)

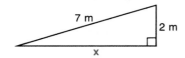

(e)

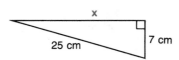

(f)

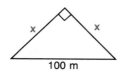

(g)

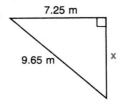

(h)

(i)

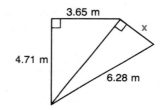

EQUATIONS

EXERCISE

1. Solve the following equations.

(a) $\dfrac{x}{7} = \dfrac{11}{35}$

(b) $\dfrac{3x}{11} = \dfrac{25}{18}$

(c) $\dfrac{5x}{38} = \dfrac{21}{18}$

(d) $\dfrac{35}{11} = \dfrac{x}{5}$

(e) $\dfrac{2}{3} = \dfrac{7}{x}$

(f) $\dfrac{51}{5} = \dfrac{17}{3x}$

(g) $\dfrac{x}{4.25} = \dfrac{16.8}{7.35}$

(h) $\dfrac{1.2x}{5.37} = \dfrac{21.6}{18.5}$

(i) $\dfrac{x}{1.25 \times 1.74} = \dfrac{5.28}{6.75}$

13.1 TRIGONOMETRIC RATIOS OF AN ACUTE ANGLE

The following relationships for any right triangle are true.

In $\triangle ABC$, where $\angle A = 90°$

I. $a^2 = b^2 + c^2$

II. $\angle A = \angle B + \angle C$

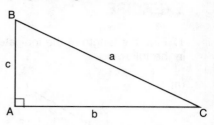

In trigonometry (from the Greek "trigonon" meaning triangle, and "metrikos" meaning measure), we relate the angles of a triangle to the sides using a new set of formulas.

If we have a triangle, with $\angle P = 90°$, and we designate $\angle QRP = \theta$, then with reference to θ, the sides of $\triangle PQR$ are described as follows:

RQ is the hypotenuse of the triangle
QP is the opposite side to θ
RP is the adjacent side to θ

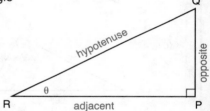

Naming the sides of the right triangle in this manner, we define the primary trigonometric ratios:

$$\text{sine of } \theta = \frac{\text{opposite side}}{\text{hypotenuse}}$$

$$\text{cosine of } \theta = \frac{\text{adjacent side}}{\text{hypotenuse}}$$

$$\text{tangent of } \theta = \frac{\text{opposite side}}{\text{adjacent side}}$$

These definitions are abbreviated to:

$$\sin \theta = \frac{\text{opp}}{\text{hyp}}$$

$$\cos \theta = \frac{\text{adj}}{\text{hyp}}$$

$$\tan \theta = \frac{\text{opp}}{\text{adj}}$$

We also define the reciprocal trigonometric ratios as follows:

$$\text{cosecant of } \theta = \frac{\text{hypotenuse}}{\text{opposite side}}$$

$$\text{secant of } \theta = \frac{\text{hypotenuse}}{\text{adjacent side}}$$

$$\text{cotangent of } \theta = \frac{\text{adjacent side}}{\text{opposite side}}$$

We abbreviate these to:

$$\csc \theta = \frac{\text{hyp}}{\text{opp}} \qquad \sec \theta = \frac{\text{hyp}}{\text{adj}} \qquad \cot \theta = \frac{\text{adj}}{\text{opp}}$$

EXAMPLE 1. Find the six trigonometric ratios of θ in the given triangle.

SOLUTION:

$$\sin \theta = \frac{3}{5} \qquad \cos \theta = \frac{4}{5} \qquad \tan \theta = \frac{3}{4}$$

$$\csc \theta = \frac{5}{3} \qquad \sec \theta = \frac{5}{4} \qquad \cot \theta = \frac{4}{3}$$

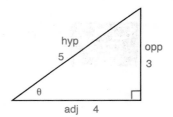

EXAMPLE 2. In △ABC, ∠A = 90°, AB = 12 cm, and AC = 5 cm. Find the six trigonometric ratios of the acute angles.

SOLUTION:

Using the Pythagorean theorem,

$$BC^2 = AB^2 + AC^2$$
$$= 12^2 + 5^2$$
$$= 144 + 25$$
$$= 169$$
$$BC = \sqrt{169}$$

∴ BC = 13 cm

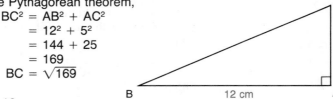

Taking ∠B as the designated angle,
AC is the opposite side,
AB is the adjacent side.

$$\sin B = \frac{5}{13} \qquad\qquad \csc B = \frac{13}{5}$$

$$\cos B = \frac{12}{13} \qquad\qquad \sec B = \frac{13}{12}$$

$$\tan B = \frac{5}{12} \qquad\qquad \cot B = \frac{12}{5}$$

Taking $\angle C$ as the designated angle,
AB is the opposite side,
AC is the adjacent side.

$$\sin C = \frac{12}{13} \qquad\qquad \csc C = \frac{13}{12}$$

$$\cos C = \frac{5}{13} \qquad\qquad \sec C = \frac{13}{5}$$

$$\tan C = \frac{12}{5} \qquad\qquad \cot C = \frac{5}{12}$$

EXAMPLE 3. If $\cos \theta = \frac{2}{5}$, find the other five trigonometric ratios of the given angle.

SOLUTION:

From the given ratio, $\cos \theta = \frac{2}{5}$, we let the hypotenuse be 5 units in length, the adjacent side 2 units, and the opposite side x units. Using the Pythagorean theorem,

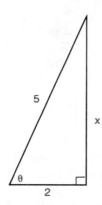

$$x^2 + 2^2 = 5^2$$
$$x^2 = 25 - 4$$
$$x^2 = 21$$
$$x = \sqrt{21}$$

$$\sin \theta = \frac{\sqrt{21}}{5} \qquad\qquad \csc \theta = \frac{5}{\sqrt{21}}$$

$$\tan \theta = \frac{\sqrt{21}}{2} \qquad\qquad \sec \theta = \frac{5}{2}$$

$$\cot \theta = \frac{2}{\sqrt{21}}$$

The trigonometric ratios of an angle are independent of the lengths of the sides of the triangle.

Given $\triangle ABC$ with $\angle B = 90°$, and $B'C' \parallel BC$ from $\triangle ABC$, we have defined

$$\sin A = \frac{BC}{AC}$$

and from $\triangle AB'C'$

$$\sin A = \frac{B'C'}{AC'}$$

Let us check to see if our definition is consistent.

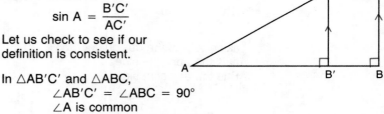

In $\triangle AB'C'$ and $\triangle ABC$,

$$\angle AB'C' = \angle ABC = 90°$$
$$\angle A \text{ is common}$$
$$\therefore \angle AC'B' = \angle ABC \text{ (SATT)}$$
$$\therefore \triangle AB'C' \sim \triangle ABC$$

and $\dfrac{B'C'}{AC'} = \dfrac{BC}{AC}$

Hence, sin A is independent of the lengths of the sides of the triangles. In a similar manner we can prove that the other five ratios are independent of the sides of the triangle.

EXERCISE 13.1

A 1. State the six trigonometric ratios for each of the indicated angles.

(a)

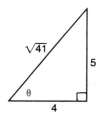

(b)

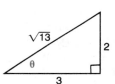

(c)

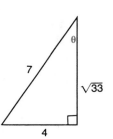

(d)

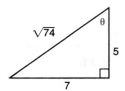

(e)

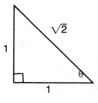

(f)

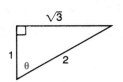

(g)

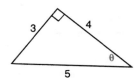

(h)

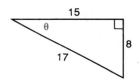

B 2. Find the length of the unknown side and then state the six trigonometric ratios of the indicated angle.

(a)

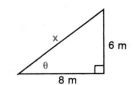

(b)

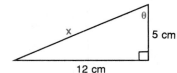

(c)

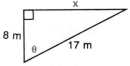

(d)

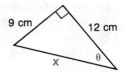
9 cm
12 cm
x
θ

3. Find the length of the unknown side and state the six trigonometric ratios of each of the acute angles.

(a)

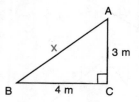
A
x
3 m
B
4 m
C

(b)

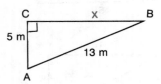
C
x
B
5 m
13 m
A

(c)

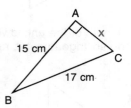
A
x
15 cm
C
17 cm
B

(d)

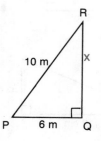
R
x
10 m
P
6 m
Q

4. If $\sin \theta = \frac{1}{2}$, and $\angle\theta$ is acute, make a diagram, and find the other five ratios.

5. If $\sec \theta = \frac{13}{5}$, and $\angle\theta$ is acute, make a diagram, and find the other five ratios.

6. If $\tan \theta = 2$, and $\angle\theta$ is acute, make a diagram, and find the other five ratios.

7. If $\cos \theta = \frac{\sqrt{3}}{2}$, and $\angle\theta$ is acute, make a diagram, and find the other five ratios.

C 8.

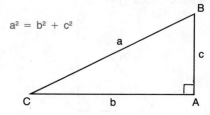
$a^2 = b^2 + c^2$
B
a
c
C
b
A

(a) Given △ABC with sides having lengths a, b, c as shown, find the six trigonometric ratios of $\angle B$ and $\angle C$ in terms of a, b, and c.
(b) Noting that $\angle B + \angle C = 90°$, find
  (i) a ratio of $\angle B$ equal to sin C.
  (ii) a ratio of $\angle B$ equal to cos C.
  (iii) a ratio of $\angle B$ equal to tan C.

9. In △ABC, $\angle C = 90°$, a = 5, c = $\sqrt{26}$.

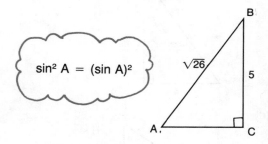
$\sin^2 A = (\sin A)^2$
B
√26
5
A
C

(a) Find the six trigonometric ratios of each of the acute angles A and B.
(b) Prove the following.
  (i) $\sin^2 A + \cos^2 A = 1$
  (ii) $\sin^2 B + \cos^2 B = 1$
  (iii) $\dfrac{\sin A}{\cos A} = \tan A$

10. The trigonometric ratios of an acute angle of a right triangle are independent of the lengths of the sides of the triangle. Prove this statement for the following.
(a) $\cos \theta$
(b) $\tan \theta$

13.2 TRIGONOMETRIC RATIOS OF SPECIAL ANGLES

If we consider the following special triangles, we can find the trigonometric ratios of 30°, 45°, and 60°. Since the trigonometric ratios of the acute angles of a right triangle are independent of the lengths of the sides, we choose convenient values as follows.

In an isosceles right triangle, △ABC, let AC = 1, BC = 1, and ∠C = 90°. From our work in geometry we can find

$$AB = \sqrt{2}, \qquad \angle A = 45°, \qquad \text{and} \qquad \angle B = 45°.$$

Then

$$
\begin{array}{ll}
\sin 45° = \dfrac{1}{\sqrt{2}} & \csc 45° = \sqrt{2} \\[2ex]
\cos 45° = \dfrac{1}{\sqrt{2}} & \sec 45° = \sqrt{2} \\[2ex]
\tan 45° = 1 & \cot 45° = 1
\end{array}
$$

In △ABD, AB = BD = DA = 2 and ∠B = ∠D = ∠A = 60°.

AC ⊥ BD, so that C bisects BD, making BC = 1, and ∠BAC = 30°.

We find AC = √3 using the Pythagorean Theorem.

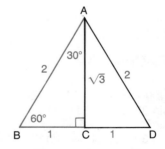

Using △ABC, and ∠ABC:

Then

$$
\begin{array}{ll}
\sin 60° = \dfrac{\sqrt{3}}{2} & \csc 60° = \dfrac{2}{\sqrt{3}} \\[2ex]
\cos 60° = \dfrac{1}{2} & \sec 60° = \dfrac{2}{1} \\[2ex]
\tan 60° = \dfrac{\sqrt{3}}{1} & \cot 60° = \dfrac{1}{\sqrt{3}}
\end{array}
$$

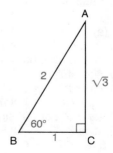

For ∠BAC:

$$\sin 30° = \frac{1}{2} \qquad \csc 30° = \frac{2}{1}$$

$$\cos 30° = \frac{\sqrt{3}}{2} \qquad \sec 30° = \frac{2}{\sqrt{3}}$$

$$\tan 30° = \frac{1}{\sqrt{3}} \qquad \cot 30° = \frac{\sqrt{3}}{1}$$

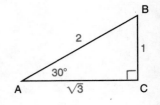

EXAMPLE. Find the value of sin² 45° + 2 sin 30° sec 60°

SOLUTION:

$$\sin^2 45° + 2 \sin 30° \sec 60° = \left(\frac{1}{\sqrt{2}}\right)^2 + 2\left(\frac{1}{2}\right)\left(\frac{2}{\sqrt{3}}\right)$$

$$= \frac{1}{2} + \frac{2}{\sqrt{3}}$$

$$= \frac{1}{2} + \frac{2\sqrt{3}}{3}$$

$$= \frac{3 + 4\sqrt{3}}{6}$$

EXERCISE 13.2

B 1. Find the value of the following.
(a) sin 30° sin 45° sin 60°
(b) sin² 30° + cos² 30°
(c) sin 30° cos 30° + sin 60° cos 60°
(d) sin 60° cos 30° + sin 30° cos 60°
(e) 2 sin 30° cos 30°
(f) 5 sec 30° tan 60°

2. (a) cos² 60° + 3 sec² 30°
(b) 3 sin² 45° + 4 cos² 45°
(c) sin 60° cos 60° tan 60°
(d) sec² 45° csc² 45° − 1
(e) sin 30° + cos 60° + sec 60°
(f) 8 cos² 45° − cot 45°
(g) tan 60° cot 60° sin 60° csc 60°
(h) 2 csc² 30° − tan² 30°
(i) 5 tan 45° cot 45° − sin² 60°

C 3. Prove.
(a) sin² 30° + cos² 30° = sin² 60° + cos² 60°
(b) 1 + tan² 45° = sec² 45°
(c) csc² 60° = 1 + cot² 60°
(d) cos 60° sec 30° = tan 30°

Put the numbers from 1 to 9 in the
spaces to make the statements true.

■ × ■ − ■ = 2
■ + ■ ÷ ■ = 2
■ − ■ + ■ = 2

13.3 TRIGONOMETRIC RATIOS OF ANY ACUTE ANGLE

In the previous section we found ratios of 30°, 45°, and 60° using special triangles. We can find ratios of other acute angles from mathematical tables as found on pages 464-465 or by using a calculator. Tables of trigonometric ratios have been computed using formulas from more advanced mathematics. These same formulas are used in calculators equipped with the trigonometric functions.

A calculator with the trigonometric functions is often limited to the primary trigonometric ratios. In order to use the calculator for all ratios we must first establish the following relationships.

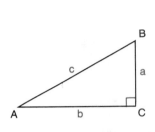

$$\frac{1}{\sin A} = \frac{1}{\frac{a}{c}} = \frac{c}{a} = \csc A$$

$$\frac{1}{\cos A} = \frac{1}{\frac{b}{c}} = \frac{c}{b} = \sec A$$

$$\frac{1}{\tan A} = \frac{1}{\frac{a}{b}} = \frac{b}{a} = \cot A$$

Hence we can find the reciprocal ratios using the following relationships

$$\csc \theta = \frac{1}{\sin \theta} \qquad \sec \theta = \frac{1}{\cos \theta} \qquad \cot \theta = \frac{1}{\tan \theta}$$

EXAMPLE. Find the ratio cot 54°.

SOLUTION:

(a) From tables cot 54° ≐ 0.7265

(b) Using a calculator.

Press 5 4 tan ¹/ₓ

and get

Display 0.7265

(nearest ten thousandth)

EXERCISE 13.3

A 1. Find each ratio using a calculator or the tables of trigonometric ratios.

(a) sin 25°   (b) cot 15°
(c) tan 45°   (d) sec 62°
(e) csc 21°   (f) cos 42°
(g) csc 81°   (h) cos 54°
(i) sec 28°   (j) sin 15°
(k) tan 72°   (l) cos 18°
(m) cot 53°   (n) csc 54°
(o) sin 37°   (p) tan 58°

2. Find the measure of the angle.

(a) sin θ = 0.9063   (b) csc θ = 1.1547
(c) tan θ = 2.246   (d) sin θ = 0.4226
(e) cot θ = 1.2349   (f) sec θ = 1.0403
(g) cos θ = 0.4226   (h) sec θ = 5.2408
(i) cos θ = 0.8192   (j) tan θ = 0.7002
(k) csc θ = 2.3662   (l) cot θ = 0.8391
(m) sin θ = 0.9272   (n) tan θ = 0.0175
(o) cot θ = 1.2349   (p) csc θ = 2.3662

13.4 SOLUTION OF RIGHT TRIANGLES

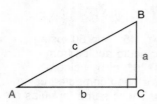

We say that a triangle is solved when we are able to state the measure of the three angles and the three sides. We facilitate this work by agreeing to label triangles so that the small letter a represents the length of the side opposite ∠A.

EXAMPLE 1. Given △ABC, where ∠A = 90°, b = 5.2 cm, and c = 4.7 cm, find the measure of the angles.

SOLUTION:
From the diagram,

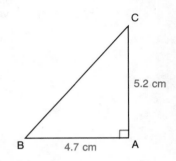

$$\tan B = \frac{5.2}{4.7}$$

$$\doteq 1.1064$$

$$\angle B \doteq 48$$

$$\angle C = 90 - \angle B$$

$$\doteq 90 - 48$$

$$\doteq 42$$

∴ ∠B ≐ 48° and ∠C ≐ 42°.

Calculator
tan B ≐ 1.1064
Press

1 . 1 0 6 4
INV **tan**

EXAMPLE 2. Given △ABC, where ∠B = 90°, a = 27 m, and ∠C = 52°, find b.

SOLUTION:
From the diagram,

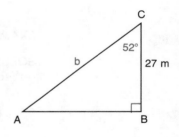

$$\frac{b}{27} = \sec 52°$$

$$b = 27 \sec 52°$$

$$\doteq 27(1.6243)$$

$$\doteq 43.9$$

∴ b ≐ 43.9 m

5 2 cos ¹/ₓ
× 2 7 =

The length of a side is found using the relationship

$$\frac{\text{unknown}}{\text{known}} = \text{trigonometric ratio}$$

EXAMPLE 3. Solve △ABC if ∠C = 90°, a = 1.3 m, and ∠A = 64°.

SOLUTION:
From the diagram,

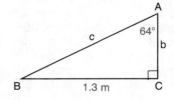

$$\angle B = 90 - \angle A$$

$$= 90 - 64$$

$$= 26$$

∴ ∠B = 26°

$$\frac{b}{1.3} = \cot 64°$$
$$b = 1.3 \cot 64°$$
$$\doteq 1.3(0.4877)$$
$$\doteq 0.634$$

$$\frac{c}{1.3} = \csc 64°$$
$$c = 1.3 \csc 64°$$
$$\doteq 1.3(1.1126)$$
$$\doteq 1.446$$

$$\therefore \angle B = 26°, \quad b \doteq 0.634 \text{ m}, \quad \text{and} \quad c \doteq 1.446 \text{ m}.$$

EXERCISE 13.4

1. Find the measure of the indicated angle.

(a)

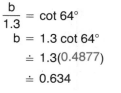

(b)

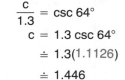

(c)

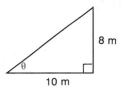

(d)

(e)

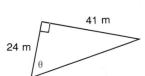

(f)

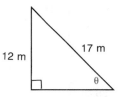

(g)

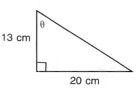

(h)

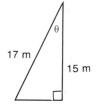

2. Find the length of the indicated side.

(a)

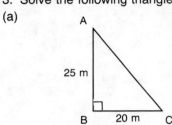

(b)

(c)

(d)

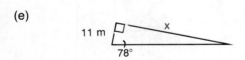

(e)

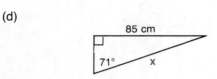

(f)

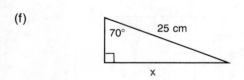

(g)

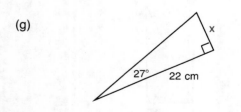

(h)

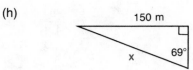

3. Solve the following triangles.

(a)

(b)

(c)

(d)

(e)

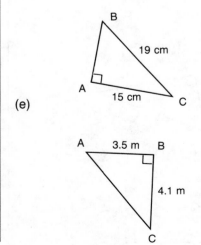

(f)

(f)

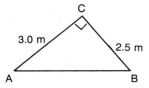

(g)

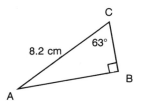

(h)

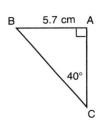

4. Solve the following triangles.
(a) △ABC, ∠A = 90°, a = 15 m, b = 11 m
(b) △DEF, ∠E = 90°, ∠D = 35°, f = 100 cm
(c) △GHI, ∠I = 90°, ∠H = 80°, i = 120 cm
(d) △JKL, ∠K = 90°, j = 50 m, k = 60 m

5. Solve the following triangles.
(a) △ABC, ∠A = 90°, a = 3.75 m,
    b = 2.15 m
(b) △ABC, ∠B = 90°, a = 21.6 cm,
    c = 32.7 cm
(c) △ABC, ∠A = 90°, ∠B = 38°, c = 4.35 m
(d) △ABC, ∠C = 90°, ∠A = 67°, c = 10.7 cm

6. Find the length of BC.
(a)

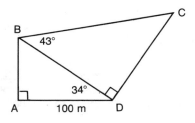

(b)

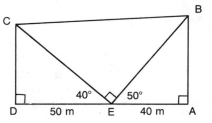

(c)

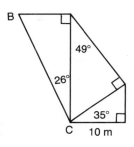

(d)

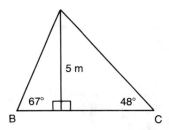

13.5 PROBLEM SOLVING WITH RIGHT TRIANGLES

In this section we will apply some of the computational skills developed earlier to real world situations. It is important to make a reasonably accurate diagram, identify the proper angles and use the correct ratios. The following diagrams help to describe some of the terminology that will be used.

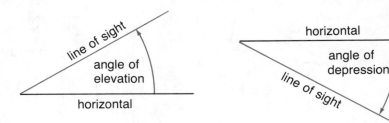

EXAMPLE 1. Find the height of a tree that casts a shadow 25.7 m long when the angle of elevation of the sun is 51°.

SOLUTION:

READ

Shadow = 25.7 m
Angle of Elevation = 51°
Height = ?

PLAN

Draw a diagram. Express the height in terms of the length of the shadow and the angle of elevation.

$$\frac{h}{25.7} = \tan 51°$$

SOLVE

$$\frac{h}{25.7} = \tan 51°$$
$$h = 25.7 \tan 51°$$
$$\doteq 25.7(1.2349)$$
$$\doteq 31.7$$

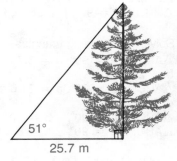

ANSWER

The height of the tree is approximately 31.7 m.

EXAMPLE 2. From opposite ends of the Simplon Tunnel (started in 1898) between Italy and Switzerland, the angles of elevation of the peak are 14° and 28°. If the peak is 2130 m above the tunnel, find the length of the tunnel.

SOLUTION:

READ

The angles of elevation of the peak are 14° and 28°. The peak is 2130 m above the tunnel.
Length of tunnel = ?

PLAN

Draw a diagram.

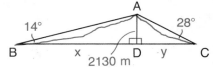

Write the length as (x + y) m. Express x and y in terms of the height and the angles of elevation.

In △ABD,

$$\frac{x}{2130} = \cot 14°$$
$$x = 2130 \cot 14°$$
$$\doteq 2130(4.0108)$$
$$\doteq 8543$$

SOLVE

In △ACD,

$$\frac{y}{2130} = \cot 28°$$
$$y = 2130 \cot 28°$$
$$\doteq 2130(1.8807)$$
$$\doteq 4007$$

$$x + y \doteq 8543 + 4007$$
$$\doteq 12\,550 \text{ m}$$

ANSWER

The length of the tunnel is approximately 12 550 m.

EXAMPLE 3. From two tracking stations, A and B, 450 km apart a satellite is sighted at C, directly above the line joining A to B. At the moment of sighting, ∠CAB = 39° and ∠CBA = 45°. Find the height of the satellite.

SOLUTION:

Let the height of the satellite be h km.

In △CAD,
$$\frac{AD}{h} = \cot 39°$$
$$AD = h \cot 39°$$

In △CBD,
$$\frac{BD}{h} = \cot 45°$$
$$BD = h \cot 45°$$

Add.

$$AD + BD = h \cot 39° + h \cot 45°$$
$$450 = h(\cot 39° + \cot 45°)$$
$$\doteq h(1.2349 + 1.0000)$$
$$\doteq h(2.2349)$$
$$\frac{450}{2.2349} \doteq h$$
$$201 \doteq h$$

∴ the height of the satellite is approximately 201 km.

EXAMPLE 4. From a point on level ground the angle of elevation of the top of the CN Tower is measured and found to be 31°. From a point 130 m closer the angle of elevation is 35°. Find the height of the tower.

SOLUTION:
Let the height of the tower be h metres.
Label the diagram as shown.

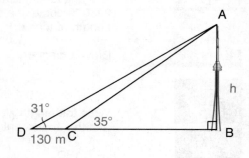

In △ABD,

$$\frac{BD}{h} = \cot 31°$$

$$BD = h \cot 31°$$

In △ABC,

$$\frac{BC}{h} = \cot 35°$$

$$BC = h \cot 35°$$

$$CD = BD - BC$$
$$130 = h \cot 31° - h \cot 35°$$
$$= h(\cot 31° - \cot 35°)$$
$$\doteq h(1.6643 - 1.4281)$$
$$\doteq h(0.2362)$$
$$\frac{130}{0.2362} \doteq h$$
$$550 \doteq h$$

∴ the CN Tower is approximately 550 m high.

EXERCISE 13.5

B 1. The angle of elevation of the top of a building is 72° from a point 54 m from the foot of the building. Find the height of the building.

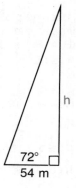

h

72°
54 m

2. A tower 115 m high casts a shadow 24 m long. Find the angle of elevation of the sun.

3.

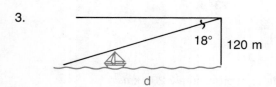

18° 120 m

d

From the top of a cliff 120 m above the water, the angle of depression of a boat on the water is 18°. How far is the boat from the cliff?

4. A 150 m tower is to be secured using four guy wires each making an angle of 55° with the ground. Find the length of wire required for each line, allowing 4 m for fastening.

5. Kelly lets out 175 m of kite string, then estimates that the angle of elevation of the kite is 30°. What is the height of the kite, if Kelly holds the string 1.2 m above the ground?

6.

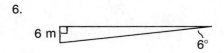

6 m

6°

A railroad underpass drops 6 m and a safe angle of descent is 6°. Find the length of the approach.

7. What angle will a 72 m guy wire make with the ground if it secures a tower from a point 45 m up the tower?

8. A tree casts a 23 m shadow when the angle of elevation of the sun is 52°.
(a) Find the height of the tree.
(b) Find the length of the shadow when the angle of elevation of the sun is 38°.

9. A surveyor wishes to find the height of an inaccessible cliff BC. To do this he sets up his transit at A, measures ∠CAB, lays off a baseline AD perpendicular to AB, then measures ∠ADB. He records the following data: ∠CAB = 62°, ∠ADB = 51°, AD = 84.5 m. Find the height of the cliff.

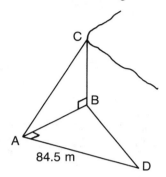

84.5 m

10. From a window of one building, the angles of elevation of the top and depression of the bottom are respectively 38° and 51°. Find the height of the second building if the buildings are 42 m apart.

11.

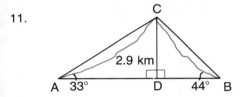

2.9 km

A 33°   D   44° B

The peak of Mont Blanc, the highest of the Alps, is 4.8 km above sea level. If the angles of elevation from points A and B are 33° and 44° as in the diagram, calculate the length of AB, the distance through the mountain at an elevation of 1.9 km.

12.

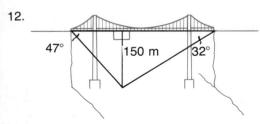

47°   150 m   32°

A bridge is 150 m above the water. From the ends of the bridge, the angles of depression of a buoy moored in the water directly below the bridge are 32° and 47°. Find the length of the bridge.

13. From a point 120 m from the foot of the building, the angles of elevation of the top and bottom of the flagpole are 49° and 46° respectively. Find the height of
(a) the building.
(b) the flagpole.

14. From two tracking stations A and B, 350 km apart, a UFO is sighted at C above AB, making ∠CAB = 32°, and ∠CBA = 54°. Find the height of the UFO.

15. From a boat on the water the angle of elevation of the top of a cliff is 31°. From a point 300 m closer to the cliff, the angle of elevation is 33°. Find the height of the cliff.

16. From the top of a cliff 185 m high, the angles of depression of two channel buoys in the same line of sight on the water are 13° and 15°. How far apart are the buoys?

17. From the roof of a building, the angles of depression of the top and bottom of a 10 m utility pole are 33° and 52°. Find the height of the building.

18.

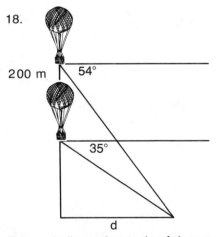

200 m   54°

35°

d

From a balloon, the angle of depression of a ground marker is 35°. From a point 200 m higher, the angle of depression is 54°. Find the distance from the marker to the point on the ground directly beneath the balloon.

19. A person in a balloon finds the angle of depression of a bonfire is 29°, when the balloon is 160 m above the level ground. After further vertical ascension, the angle of depression is 34°. How much higher is the second position than the first?

13.6 REVIEW EXERCISE

1. Find the length of the unknown side and then state the six trigonometric ratios of the indicated angle.

(a)

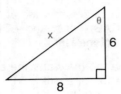

(b)

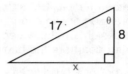

(c)

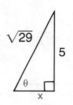

2. Find the length of the unknown side and state the six trigonometric ratios of each acute angle.

(a)

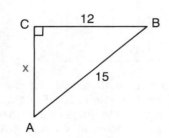

(b)

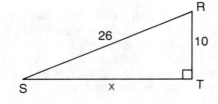

(c)

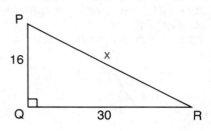

3. If csc $\theta = \dfrac{13}{12}$, and $\angle\theta$ is acute, make a diagram and find the other five trigonometric ratios.

4. If tan $\theta = 3$, and $\angle\theta$ is acute, draw a diagram and find the other five trigonometric ratios.

5. Without using tables or a calculator, find the six trigonometric ratios of
(a) 45°
(b) 30°
(c) 60°
then evaluate:
(i) sin² 45° + cos² 45°
(ii) sin² 30° + sin² 60°
(iii) $\dfrac{\sin 30° + \sin 60°}{\sin 45°}$
(iv) 2 tan 30° cot 60°
(v) sin 30° cos 60° tan 45°

6. Find each ratio.
(a) sin 36°
(b) cos 72°
(c) tan 51°
(d) csc 11°
(e) sec 44°
(f) cot 81°

7. Find the measure of the angle.
(a) sin $\theta = 0.8746$
(b) cos $\theta = 0.7986$
(c) tan $\theta = 1.6643$
(d) csc $\theta = 5.7588$
(e) sec $\theta = 4.1336$
(f) cot $\theta = 1.0355$

8. Solve the following triangles.

(a)

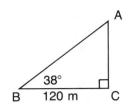

(b)

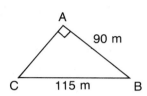

(c)

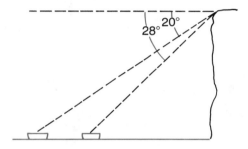

(d)

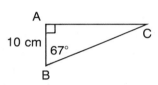

(e)

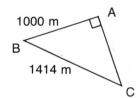

(f)

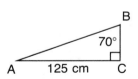

9. A tree casts a shadow 120 m long when the angle of elevation of the sun is 34°. Find the height of the tree.

10. The angle of elevation of the top of a 125 m cliff from a boat on the water is 18°. How far is the boat from the foot of the cliff.

11. From opposite ends of a tunnel through a mountain, the angles of elevation of the peak of the mountain are 37° and 49°. Find the length of the tunnel if the peak is 2750 m above the tunnel.

12. From the top of a cliff 200 m high, the angles of depression of two boats on the water are 20° and 28°.

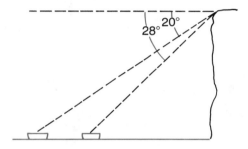

How far apart are the boats?

13. From a point on the water, the angle of elevation of the top of a cliff is measured and found to be 30°. From a point 350 m further out, the angle of elevation is 27°. Find the height of the cliff.

14. In △ABC, ∠ACB = 90° and ∠ABC = 37°. In △BCD, ∠CBD = 90°, BD = 120 m, and ∠BDC = 78°. Find the length of AC.

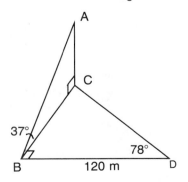

13.7 CHAPTER 13 TEST

1. State the six trigonometric ratios of each acute angle.

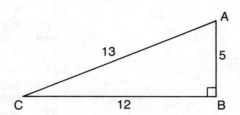

2. Without using tables or a calculator, evaluate.
(a) $\sin^2 45° + \cos^2 60°$
(b) $4 \sec 30° \tan 60°$

3. Solve the triangle.

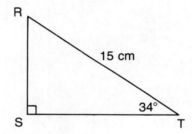

4. Solve the triangle.

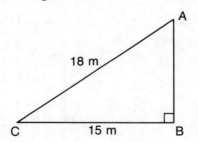

5. From a point 100 m from the foot of a building, the angle of elevation of the top of the building is 52°. Find the height of the building.

6. From the top of a cliff 80 m above the water, the angle of depression of a boat on the water is 23°. How far is the boat from the cliff?

13.8 CUMULATIVE REVIEW FOR CHAPTERS 9 TO 13

1. State the first five terms of the sequence whose nth term is given.
(a) $t_n = n^2 - 5$ (b) $t_n = 3n + 2$

2. List the first five terms of the sequences defined by the following functions.
(a) $f(n) = n^2$ (b) $f(n) = 2^n$

3. Find the first five terms of the sequences defined by the following recursion formulas.
(a) $t_1 = 3$, $t_n = 5t_{n-1}$, $n > 1$
(b) $t_1 = -2$, $t_{n+1} = t_n + 2$
(c) $t_1 = 1$, $t_2 = 2$, $t_{n+2} = t_{n+1} - t_n$
(d) $t_1 = 3$, $t_2 = 3$, $t_{n+2} = t_{n+1} + t_n$

4. Write the first three terms and a general term for each arithmetic sequence given the following values for a and d.
(a) $a = 5$, $d = 6$
(b) $a = -3$, $d = 4$
(c) $a = -6$, $d = -2$
(d) $a = x + y$, $d = x - y$

5. Write the first three terms and a general term for each of the following geometric sequences given the following values for a and r.
(a) $a = 1$, $r = 2$ (b) $a = 2$, $r = -3$
(c) $a = -3$, $r = 3$ (d) $a = 1$, $r = x^{-1}$

6. Find a, d, and t_n for each of the following arithmetic sequences.
(a) $t_8 = 16$ and $t_{24} = 40$
(b) $t_{10} = 12$ and $t_{13} = 28$

7. Find a, r, and t_n for each of the following geometric sequences.
(a) $t_3 = 8$ and $t_6 = 64$
(b) $t_2 = 162$ and $t_5 = -6$

8. Write each series in expanded form and find the sum.
(a) $\sum_{n=1}^{5} = (2n - 3)$

(b) $\sum_{n=1}^{5} = \dfrac{n - 1}{n}$

(c) $\sum_{i=1}^{6} = (i - 1)^2$

9. Find the sum of the following arithmetic series.
(a) $S_{100} = 2 + 5 + 8 + \cdots$
(b) $3 + 6 + 9 + \cdots + 999$
(c) $1000 + 999 + 998 + \cdots + 1$

10. Find the indicated sums of the following geometric series.
(a) $S_6 = 4 + 8 + 16 + \cdots$
(b) $S_8 = \dfrac{1}{4} + \dfrac{1}{2} + 1 + \cdots$

11. For a given sequence, $t_1 = 2$ and $t_6 = -64$.
(a) Find the first four terms of the sequence if it is arithmetic.
(b) Find the first four terms of the sequence if it is geometric.

12. Find the compound amount of the following.
(a) $5000 at 9.5% for 3 a compounded semi-annually
(b) $12 000 at 12% for 3 a compounded monthly
(c) $7000 at 16% for 5 a compounded quarterly

13. Find the present value of each of the following.
(a) $3000 due in 2 a at 10% compounded semi-annually
(b) $1000 due in 1 a at 12% compounded monthly
(c) $2000 due in 2 a at 12% compounded quarterly

14. How long will it take $1000 to accumulate to $2500 if it is invested at 10% compounded semi-annually?

15. Find the amount of each of the following annuities.
(a) 24 monthly payments of $100 each into an account that pays 1% per month
(b) 12 semi-annual payments of $500 each into an account that pays 10% compounded semi-annually

16. Find the present value of each of the following annuities.

(a) 12 monthly payments of $258 each at 12% compounded monthly
(b) 4 payments of $1500 each at 11% compounded annually

17. Find the periodic rent for each of the following annuities.

(a) 24 monthly payments to amount to $2500 at 1.5% per month
(b) an amount of $18 000 in 5 a at 10% compounded annually
(c) 20 semi-annual payments at 12% semi-annually with a present value of $20 000

18. Solve for i.

(a) $1048.68

$$= \frac{\$110}{1 + i} + \frac{\$110}{(1 + i)^2} + \cdots + \frac{\$110}{(1 + i)^7} + \frac{\$1000}{(1 + i)^7}$$

(b) $1308.87

$$= \frac{\$90}{1 + i} + \frac{\$90}{(1 + i)^2} + \cdots + \frac{\$90}{(1 + i)^{10}} + \frac{\$1000}{(1 + i)^{10}}$$

19. A company with 350 000 shares has declared a dividend of $1.65 per share.

(a) What is the total paid out in dividends?
(b) What dividends will be received by a shareholder who has 200 shares?

20. What is the cost, including a commission of 2%, for 2000 shares of Amalgamated Peat selling at $4.65 per share?

21. Find the proceeds, less a commission of 1.5%, from the sale of 2000 shares of Canadian Eureka at $10.45 per share.

22. A five-year bond with a face value of $5000 has a coupon rate of 8%. Find the market value of the bond if the market interest rate rises to 12%.

23. 500 shares of Western Petroleum were purchased at $21.50 per share and sold 45 d later at $28.00 per share. Commission was charged at a rate of 3% on the purchase, and 2% on the sale. What was the profit or loss after paying the commission?

24. Find the mean, median, and mode(s), and calculate the standard deviation of the following set of data.

```
25 34 45 87 67 32 21 80 75 63
36 56 48 57 95 12 32 56 55 27
44 58 87 52 24 65 95 85 42 26
29 37 85 64 15 29 86 88 29 63
```

25. A box contains 5 red balls, 2 blue balls, and 8 white balls. One ball is drawn at random from the box. What is the probability that the ball is

(a) white?   (b) red?
(c) not blue?   (d) green?
(e) red or blue?   (f) not white?

26. A coin is tossed and if it comes up heads, a die is rolled. What is the probability of rolling an even number?

27. Solve each of the following triangles.
(a)

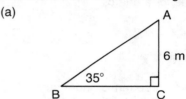

(b)

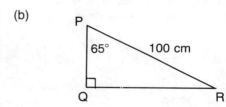

(c)

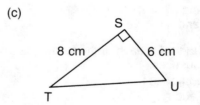

(d)

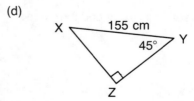

28. If cos θ = 0.5, and θ is acute, find the other five trigonometric ratios.

29. Without using tables or a calculator, make a diagram and find the six trigonometric ratios of each of the following angles.

(a) 45°   (b) 30°   (c) 60°

30. Without using tables or a calculator, find the value of each of the following.

(a) sin 30° cos 30°
(b) sin² 60° + cos² 60°
(c) sin² 45° + cos² 45°

31. From the top of a cliff 250 m high, the angles of depression of two boats on the water are 18° and 28°. How far apart are the boats?

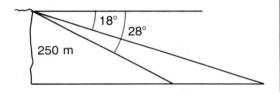

32. A flagpole casts a shadow 105 m long when the angle of elevation of the sun is 25°. How long is the shadow when the angle of elevation is 50°?

33. In the following diagram, △ABC lies in a vertical plane, while △BCD lies in a horizontal plane. Find the length of AB.

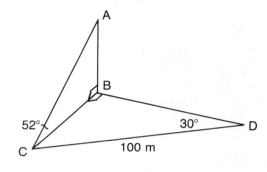

ANSWERS

REVIEW AND PREVIEW TO CHAPTER 1
ROUNDING AND APPROXIMATING
1. (a) 36 500   (b) 36 100   (c) 4100   (d) 255 600   (e) 4520   (f) 67 500
2. (a) 0.26   (b) 0.06   (c) 2.62   (d) 1.21   (e) 32.65   (f) 0.38
3. (a) 1.265   (b) 32.566   (c) 0.564   (d) 0.568   (e) 0.568   (f) 1.006
4. (a) 25 060   (b) 370   (c) 28   (d) 0   (e) 12 260   (f) 60
5. (a) 120   (b) 180   (c) 400   (d) 72   (e) 30
6. (a) 5600   (b) 0.375   (c) 150 000   (d) 600   (e) 540 000

SET NOTATION
1. (a) $\{-1, 0, 1, ...\}$ (b) $\{8, 9, 10, ...\}$
(c) $\{0, 1, 2, 3, 4, 5\}$ (d) $\{3\}$
(e) $\{-2, -1, 0, 1, ...\}$ (f) $\{0, 1, 2, 3\}$
(g) $\{-3, -2, -1, 0, 1, 2, 3\}$ (h) $\{-3, -2, -1, 0, 1, 2\}$
(i) $\{..., -6, -5, -4, 3, 4, 5, ...\}$ (j) $\{-2, -1, 0, 1, 2, 3\}$

INTERSECTION AND UNION
1. (a) $\{2, 4, 6, 7, 8, 9, 10\}$ (b) $\{6, 8\}$
2. (a) $\{10, 20\}$ (b) $\{5, 10, 15, 20, 25, 30, 40, 50, 60\}$
3. (a) $\{2, 4, 6\}$ (b) $\{1, 2, 3, 4, 5, 6, 8, 10\}$
(c) $\{2, 4, 6\}$ (d) $\{1, 2, 3, 4, 5, 6, 8, 10\}$
(e) $\{4, 6\}$ (f) $\{1, 2, 3, 4, 5, 6, 7, 8, 9, 10\}$
(g) $\{2, 4, 6\}$ (h) $\{2, 4, 6, 7, 9\}$
(i) $\{2, 4, 6\}$ (j) $\{2, 4, 6, 7, 9\}$
4. (a) $\{-1, 0, 1, 2\}$ (b) $\{..., -2, -1, 0, 1, 2, ...\}$
(c) $\{..., -6, -5, -4, 4, 5, 6, ...\}$ (d) $\{..., -2, -1, 0, 1, 2\}$
(e) $\{-2, -1, 0, 1, 2, 3, 4\}$ (f) $\{..., -4, -3, 0, 1, 2, ...\}$

EXERCISE 1.1
1. (a) 0.125   (b) 0.25   (c) 0.375   (d) 0.625   (e) -0.875
(f) 0.42   (g) $0.\overline{1}$ (h) $0.\overline{45}$ (i) $-0.\overline{538\,461}$ (j) $1.\overline{36}$
(k) $1.\overline{6}$ (l) 0.6
2. (a) $\frac{3}{8}$ (b) $\frac{43}{200}$ (c) $\frac{309}{500}$ (d) $\frac{7}{200}$ (e) $\frac{4}{11}$ (f) $\frac{5}{9}$
(g) $\frac{235}{999}$ (h) $\frac{523}{110}$ (i) $\frac{1}{2}$ (j) $\frac{1}{7}$ (k) $\frac{4769}{999}$ (l) $\frac{1}{55}$
3. (a) $\frac{1}{3}$ (b) $\frac{1}{12}$ (c) $\frac{1}{11}$ (d) $\frac{1}{27}$ (e) $\frac{2}{9}$ (f) $\frac{4}{55}$
4. (a) $\frac{1}{2}$ (b) $\frac{2}{5}$ (c) $\frac{7}{10}$ (d) 1
5. (a) $\frac{1}{9}$ (b) $\frac{1}{3}$ (c) $\frac{4}{11}$ (d) $\frac{52}{333}$

EXERCISE 1.2
1. c, e, h

EXERCISE 1.3
1. (a) $\sqrt{125}$ (b) $\sqrt{28}$ (c) $-\sqrt{99}$ (d) $\sqrt{32}$ (e) $\sqrt{63}$ (f) $\sqrt[3]{54}$
(g) $\sqrt[3]{-40}$ (h) $\sqrt[4]{48}$ (i) $\sqrt{150}$
2. (a) $3\sqrt{3}$ (b) $7\sqrt{2}$ (c) $12\sqrt{12}$ (d) $5\sqrt{3}$ (e) $10\sqrt{2}$ (f) -3
(g) $-4\sqrt[3]{2}$ (h) $2\sqrt[4]{3}$ (i) $3\sqrt[3]{2}$
3. (a) $7\sqrt{3}$ (b) $\sqrt{5}$ (c) $4\sqrt{3}$ (d) $9\sqrt{6} - 3\sqrt{5}$ (e) $-\frac{1}{2}\sqrt{2}$
(f) $8\sqrt{2} - 7\sqrt{3}$
4. (a) $2\sqrt{3} - \sqrt{6}$ (b) $2\sqrt{15} - 4\sqrt{3}$ (c) $6\sqrt{5} - 3\sqrt{10}$ (d) $6 + 3\sqrt{2}$ (e) $\sqrt{ab} - 2\sqrt{a}$
(f) $a + \sqrt{ab}$ (g) $4 - 2\sqrt{3}$ (h) $11 - 6\sqrt{2}$ (i) $2\sqrt{6} - 2\sqrt{15}$
5. (a) $2\sqrt{3}$ (b) $5\sqrt{5}$ (c) $5\sqrt{2}$ (d) $2\sqrt{5}$ (e) $12\sqrt{2}$ (f) $7\sqrt{2}$
(g) $2\sqrt{2}$ (h) $2\sqrt[3]{2}$ (i) $4\sqrt[3]{2}$ (j) $3\sqrt{3}$ (k) $5\sqrt[3]{2}$ (l) $2\sqrt[3]{3}$
6. (a) $\sqrt{45}$ (b) $\sqrt{28}$ (c) $\sqrt{96}$ (d) $\sqrt{20}$ (e) $\sqrt[3]{16}$ (f) $\sqrt[3]{81}$
(g) $-\sqrt{75}$ (h) $-\sqrt[3]{24}$ (i) $\sqrt{54}$ (j) $\sqrt[4]{48}$ (k) $\sqrt[3]{135}$ (l) $-\sqrt{44}$
7. (a) $15\sqrt{2}$ (b) $11\sqrt{5}$ (c) $8\sqrt{5} + \sqrt{3}$ (d) $26\sqrt{2} - 7\sqrt{3}$

8. (a) $6\sqrt{2} - 6\sqrt{3}$ (b) $10\sqrt{3} - 15\sqrt{2}$ (c) $10\sqrt{2} - 4\sqrt{5}$ (d) $19 - 6\sqrt{2}$ (e) $5 - 2\sqrt{6}$
   (f) $62 - 20\sqrt{6}$ (g) 17   (h) 11   (i) 62   (j) $53 + 36\sqrt{2}$

EXERCISE 1.4

1. (a) $\{x \in R \mid -2 < x \leq 3\}$ (b) $\{x \in R \mid -1 \leq x \leq 1\}$
   (c) $\{x \in R \mid -3 < x < 1\}$ (d) $\{x \in R \mid x > -1\}$
2. (a) $\{x \in R \mid -3 < x < 3, x \neq 0\}$ (b) $\{x \in R \mid -3 \leq x \leq 3, x \neq 0\}$
   (c) $\{x \in R \mid x \leq 2\}$ (d) $\{x \in R \mid x \neq -1\}$

EXERCISE 1.5

1. (a) closure $(+)$ (b) neutral element $(\times)$ (c) commutative $(\times)$ (d) associative $(+)$
   (e) commutative $(+)$ (f) inverse element $(-)$ (g) closure $(\times)$ (h) distributive
2. (a) trichotomy   (b) symmetric
   (c) completeness   (d) transitive property of order
   (e) transitive property of order   (f) transitive — axiom of equality
   (g) symmetric — axiom of equality or commutative   (h) symmetric — axiom of equality or commutative
3. (a) $a < b$ (b) $a = a$ (c) $a > b$ (d) $3a + 3b = 3(a + b)$
   (e) $x + 3 = 2x + 1$ (f) $a + b > a - b$ (g) $-1 < 3c$ (h) $0.25 = \frac{1}{4}$
4. (a) $y = 0$ (b) $y < 0$ (c) $y > 0$ (d) $y = 0$
   (e) $y > 0$ (f) $y = 0$ and $y = 1$
5. Answers vary

EXERCISE 1.6

1. (a) 2^4 (b) 2   (c) 2^5 (d) 2^{15} (e) 2^9 (f) 2^{10}
   (g) 2^7 (h) 2^5 (i) 2^{12}
2. (a) 3^3 (b) 3^5 (c) 3^6 (d) 3^9 (e) 3^4 (f) 3^9
   (g) 3^7 (h) 3^7 (i) 3^{12}
3. (a) 5^2 (b) 5^3 (c) 5^4 (d) 5^4 (e) 5^4 (f) 5^6
   (g) 5^5 (h) 5^2 (i) 5^{12}
4. (a) 72   (b) 54   (c) 48   (d) 36   (e) 216   (f) 36
   (g) 10   (h) 6   (i) 27
5. (a) 9   (b) -27 (c) 81   (d) -81 (e) -729 (f) 10
   (g) 4   (h) 18   (i) -9
6. (a) -24 (b) 576   (c) 13   (d) 20   (e) 36   (f) 16
   (g) 1   (h) 35   (i) 36
7. (a) 1.1025   (b) 1.4641   (c) 4.913   (d) 1.5625   (e) $1.\overline{003}$
   (f) 3.8416   (g) 2.488 32   (h) 1.061 363 55   (i) 3.051 757 81

EXERCISE 1.7

1. (a) $2123.04   (b) $3184.56   (c) $6700.48   2. (a) $545   (b) $1638.04   (c) $10 305.16

EXERCISE 1.8

1. (a) 2^9 (b) 2^3 (c) $2^{(3+a)}$ (d) $2^{(b-4)}$ (e) 2^{15} (f) 2^{3a}
2. (a) 3^6 (b) 3^3 (c) $3^{(a+2)}$ (d) $3^{(m-5)}$ (e) 3^9 (f) 3^{2m}
   (g) 4^{mn} (h) $3^{(a-1)}$
3. (a) $9^4 \times 13^4$ (b) $9^4 x^4$ (c) $\dfrac{5^{18}}{6^{18}}$ (d) $\dfrac{x^5}{6^5}$ (e) 63^8 (f) $6^5 \times 72^5$

   (g) —   (h) $x^3 y^3$ (i) π^5 (j) π^6 (k) $\dfrac{2^{10}}{3^{10}}$ (l) $\dfrac{x^6}{y^6}$

   (m) $(2.78)^8$ (n) $(-2)^7$ (o) $\dfrac{9}{a^2}$ (p) $\left(\dfrac{a}{b}\right)^8$ (q) —   (r) $8x^6$

4. (a) $-27x^{18}$ (b) $-12x^9$ (c) 3^{18} (d) a^{m+n+p} (e) 2^{24} (f) a^{mnp}
   (g) 4^4 (h) x^{3m}
5. (a) 7   (b) $\frac{1}{9}$ (c) 256   (d) 16   (e) 5   (f) 1
6. (a) $-a^{3p+q}$ (b) x^{6n+8} (c) 3^{10n} (d) 2^{4n-4} (e) 2^{3n}
7. (a) $x^7 y^9$ (b) —   (c) $\frac{1}{4}x^{38}y^8$ (d) $-x^{39}y^{12}$ (e) $\dfrac{5y^3}{6x^4}$ (f) $\dfrac{a^{11}}{b^4}$

   (g) $4y^4$ (h) $\dfrac{x}{4y^2}$

EXERCISE 1.9

1. (a) 1   (b) $\frac{1}{9}$ (c) $\frac{1}{8}$ (d) $\frac{1}{9}$ (e) -1 (f) 1
   (g) $\frac{1}{10\ 000}$ (h) $\frac{1}{837}$

2. (a) $\frac{1}{x^8}$ (b) $\frac{x^2}{y^2}$ (c) $\frac{1}{a^3b^4}$ (d) $\frac{1}{a^2}$ (e) a^{10} (f) $\left(\frac{y}{x}\right)^3$

3. (a) x^{-3} (b) $2ab^{-4}$ (c) $\pi x^2 y$

4. (a) $\frac{8}{25}$ (b) 1024   (c) $\frac{27}{25}$ (d) 4   (e) 9   (f) 2
   (g) $\frac{5}{4}$ (h) $\frac{10}{21}$ (i) 2   (j) 625   (k) $\frac{45}{19}$ (l) $\frac{7}{4}$

5. (a) $\frac{343x^6}{y^9}$ (b) $\frac{a^5b^{10}}{243}$ (c) $\frac{64a^3c^9}{b^6d^{12}}$ (d) $\frac{a^2}{b^3}$ (e) $a^2b^2x^2y^4$
   (f) $\frac{a^4}{b^2} - 2\frac{a^2}{b} + 1$ (g) $a^6 + a^5 - 5a^2$ (h) $2x^2 - \frac{1}{x^2} - 1$ (i) $\frac{1}{6^{2n}}$ (j) $\frac{1}{x^{2n}} - \frac{1}{y^{2m}}$

EXERCISE 1.10

1. (a) $9658.92   (b) $18 655.38   (c) $7875.66   (d) $3655.96   (e) $3867.65
   (f) $457.17   (g) $6143.17
2. $5499.49   3. $4319.78   4. $5131.58

EXERCISE 1.11

1. (a) $\sqrt[9]{2}$ (b) $\sqrt{37}$ (c) $\sqrt[3]{x}$ (d) $\sqrt[3]{4}$ (e) $\sqrt{8} = 2\sqrt{2}$
   (f) $\sqrt[4]{27}$ (g) $\sqrt[5]{a^2}$ or $(\sqrt[5]{a})^2$ (h) $\sqrt[7]{x^4}$ or $(\sqrt[7]{x})^4$ (i) $\frac{1}{\sqrt{2}}$ (j) $\frac{1}{\sqrt[5]{7}}$
   (k) $\frac{1}{\sqrt{a^3}}$ (l) $\sqrt[11]{81}$

2. (a) $3^{\frac{1}{2}}$ (b) $19^{\frac{1}{2}}$ (c) $23^{\frac{1}{7}}$ (d) $x^{\frac{1}{4}}$ (e) $7^{\frac{2}{3}}$ (f) $7^{\frac{2}{3}}$
   (g) $6^{\frac{4}{5}}$ (h) $13^{\frac{5}{3}}$ (i) $a^{\frac{2}{5}}$ (j) $a^{\frac{5}{6}}$ (k) $5^{-\frac{1}{2}}$ (l) $7^{-\frac{3}{4}}$

3. (a) 5   (b) 4   (c) 27   (d) 1   (e) $\frac{1}{6}$ (f) 4
   (g) 3   (h) $\frac{1}{2}$ (i) 4   (j) $\frac{1}{8}$ (k) -2 (l) 4

4. (a) 16   (b) 128   (c) 100   (d) $\frac{1}{4}$ (e) 27   (f) $\frac{1}{125}$
   (g) 256   (h) 3   (i) 36   (j) 343   (k) 2   (l) 162
   (m) $\frac{2}{3}$ (n) $\frac{12}{7}$ (o) $\frac{125}{512}$ (p) 96   (q) 0.08   (r) 3
   (s) 8   (t) 0   (u) 7   (v) $\frac{5}{6}$ (w) 3   (x) $\frac{5}{108}$

5. (a) $2^{\frac{5}{6}}$ (b) $3^{\frac{8}{9}}$ (c) $x^2y^{\frac{1}{2}}$ (d) a^3b^4 (e) ab^2c^3
   (f) $x + 3x^{\frac{2}{3}}$ (g) $2x^2y^{\frac{1}{2}}$ (h) $\frac{16x^6}{y^2}$ (i) $\frac{40\sqrt{5}\,x^3y^{\frac{9}{2}}}{z^{\frac{3}{2}}}$ (j) $\frac{a^{\frac{9}{2}}x}{b^9y^{\frac{24}{5}}}$
   (k) $\frac{y^{\frac{1}{4}}}{x^{\frac{1}{24}}}$ (l) $a^{\frac{n}{2}}$

6. (a) 1.319 507 91   (b) 0.172 427 286   (c) 90.597 458   (d) 2.080 083 82
   (e) 0.341 278 752   (f) 5.179 474 68

7. (a) $3^{\frac{1}{2}}$ (b) $7^{\frac{1}{4}}$ (c) $3^{\frac{3}{2}}$ (d) $14^{-\frac{1}{3}}$

EXERCISE 1.12

1. (a) 5   (b) 3   (c) 6   (d) 2   (e) 4   (f) 2
   (g) 3   (h) 3   (i) 4   (j) 2

2. (a) 6   (b) 3   (c) 3   (d) 16   (e) 2   (f) 4
   (g) 6   (h) 2   (i) $x \in N$ (j) $x \in N$, x even

3. (a) 2   (b) -7 (c) 5
   (d) $x = 0, \pm 2, \pm 4, \pm 6, ...$ (e) 2   (f) 1
   (g) 1   (h) $x \in I$ (i) 4
   (j) $-\frac{5}{2}$

4. (a) $\frac{3}{2}$ (b) $-\frac{1}{3}$ (c) $\frac{2}{3}$ (d) $\frac{1}{4}$ (e) $\frac{1}{4}$ (f) -2
   (g) $\frac{16}{15}$ (h) -1 (i) 4   (j) -2 (k) 0   (l) 0

(m) $\frac{1}{2}$ (n) $\frac{1}{2}$

EXERCISE 1.13

1. 120   2. 512   3. 7   4. 3   5. 792   6. 16 horses, 19 cowboys
7. 11 plain, 14 with frame   8. $(1 \times 25¢) + (1 \times 10¢) + (6 \times 5¢) + (2 \times 1¢)$ 9. 1053
10. 3   11. 4:22   12. 2401   13. $16 - 4\pi$ 14. \$16 965.10

1.14 REVIEW EXERCISE

1. (a) 0.6 (b) 0.75   (c) 0.375   (d) $0.\overline{285\ 714}$ (e) $0.\overline{18}$
   (f) $0.1\overline{3}$ (g) 0.14   (h) $0.\overline{857\ 142}$ (i) $0.\overline{72}$
2. (a) $\frac{53}{200}$ (b) $\frac{13}{4}$ (c) $\frac{13}{200}$ (d) $\frac{7}{20}$ (e) $\frac{256}{999}$ (f) $\frac{127}{495}$
   (g) $\frac{7}{9}$ (h) $\frac{128}{165}$ (i) $\frac{3}{11}$ (j) $\frac{3}{10}$ (k) $\frac{2}{11}$ (l) $\frac{2}{9}$
   (m) $\frac{5}{11}$
3. (a) $\sqrt{45}$ (b) $\sqrt{x^2y}$ (c) $\sqrt{9x^4yz}$ (d) $\sqrt[3]{-x^6y^3z}$ (e) $\sqrt[4]{16x^5}$
   (f) $\sqrt[4]{16x^8y^4z}$
4. (a) $2\sqrt{10}$ (b) $xy\sqrt{y}$ (c) $4x^2y\sqrt{y}$ (d) $5yz\sqrt{5xz}$ (e) $-3x\sqrt[3]{y^2}$
   (f) $2x\sqrt[5]{x}$
5. (a) 2   (b) $2\sqrt{2}$ (c) $2\sqrt{x}$ (d) 3   (e) 3   (f) $\frac{1}{9}$
6. (a) $8\sqrt{2} - 6\sqrt{3}$ (b) $-2\sqrt{x} - 2\sqrt{y}$ (c) $3\sqrt{2}$ (d) $9\sqrt{3}$ (e) $6\sqrt{5}$
   (f) $-\sqrt{6} - \sqrt{3}$
7. (a) $6\sqrt{5} + 15\sqrt{2}$ (b) $6\sqrt{2} - 2\sqrt{6}$ (c) $6 - 2\sqrt{15}$ (d) $6 - 6\sqrt{2}$ (e) $x\sqrt{y} + y\sqrt{x}$
   (f) $6x - 3\sqrt{xy}$ (g) $9x$ (h) $12x + 2\sqrt{6x}$
9. (a) closure $(+)$ (b) commutative $(\times)$ (c) closure $(\times)$ (d) commutative $(\times)$
   (e) inverse element $(+)$ (f) commutative $(+)$ (g) neutral element $(\times)$ (h) distributive
10. (a) $x = 0$ (b) $x < 0$ (c) $x = 0$ (d) $x = 0$ (e) $x < 0$ (f) $x > 0$
11. (a) 2^5 (b) 5^2 (c) 7^2 (d) 5^3 (e) 3^5 (f) 2^7
   (g) 3^5 (h) 5^5 (i) 3^4
12. (a) 36   (b) 24   (c) -54 (d) -16 (e) 216   (f) 54
   (g) -1 (h) 1   (i) 9
13. (a) $4x^5$ (b) $-2x^9$ (c) x^8 (d) $3x^6$ (e) a^{x+y+z} (f) a^{xyz}
   (g) $-8x^6$ (h) $-12x^7$ (i) $3x^2$ (j) $3x^3$
14. (a) 125   (b) x^{3a} (c) 3^{2n} (d) 3   (e) x^{n+3} (f) x^{2n+2}
   (g) 2^{14n+6} (h) 3^{-2n-10}
15. (a) 1   (b) undefined   (c) $\frac{1}{5}$ (d) -1 (e) $\frac{1}{100}$ (f) -1
16. (a) 5   (b) $\dfrac{1 + 2^3}{2^3 - 1}$ (c) $\frac{1}{6}$ (d) $\frac{1}{7}$ (e) $\frac{1}{6}$ (f) $\frac{1}{ab}$
17. (a) \$3184.56   (b) \$2924.65   (c) \$5634.13   (d) \$12 155.06   (e) \$1349.23
   (f) \$7092.60
18. (a) \$1404.93   (b) \$1425.76   (c) \$1418.52   (d) \$1430.77
19. (a) \$793.83   (b) \$920.87   (c) \$2181.08   (d) \$2948.32   (e) \$4582.24
20. \$11 193.23   21. \$1400.57
22. (a) $\sqrt{5}$ (b) $\sqrt[3]{34}$ (c) $\sqrt[5]{6}$
23. (a) $7^{\frac{1}{2}}$ (b) $a^{\frac{3}{2}}$ (c) $a^{\frac{2}{5}}$
24. (a) 5   (b) 9   (c) 5   (d) 125   (e) $\frac{2}{5}$ (f) $-\frac{1}{2}$
25. (a) -2 (b) $\frac{1}{4}$ (c) 0   (d) 3
26. 7, 6   27. 7, 5   28. 2, 4
29. (a) 4950   (b) 2550   (c) 2500   (d) 10 000
30. (a) 0.000 000 000 037 cm/r   (b) 0.000 019 166 cm/km

1.15 CHAPTER 1 TEST

1. (a) $0.\overline{27}$ (b) $\frac{7}{11}$
2. (a) $12\sqrt{2}$ (b) $\sqrt{245}$ (c) $8\sqrt{3} - 7\sqrt{5}$
4. (a) 75   (b) -8 (c) $14x^4$ (d) 2^{3n-1} (e) $-12x^7y^8$
   (f) $\frac{17}{60}$ (g) $2^{\frac{5}{8}}$
5. $\frac{13}{8}$ 6. 228 m   7. \$7049.61   8. \$6655

REVIEW AND PREVIEW TO CHAPTER 2

DISTRIBUTIVE PROPERTY

1. (a) $2x + 12$ (b) $3x + 12$ (c) $4x - 20$ (d) $35 - 7m$
 (e) $16 - 12x$ (f) $5x - 5y$ (g) $-4x + 14$ (h) $-21m + 28n$
 (i) $-3m + 7$ (j) $6x^2 - 12x + 15$ (k) $-6t^2 + 15t - 12$ (l) $-3x^2 + 4xy - y^2$
 (m) $2x^2 - 14x$ (n) $5x^3 - 15x^2 - 20x$ (o) $m^3 + 3m^2 - m$ (p) $3m^2 - 3mn$

2. (a) $6x^2y - 8xy^2$ (b) $-2x^3y + 2x^2y^2 - 2xy^3$ (c) $3t^5 - 6t^4 - 12t^3$
 (d) $2\pi rw - 2\pi rx$ (e) $-3\pi m + 2\pi n$ (f) $1.2x - 2y$
 (g) $-0.6x^2 + 0.8x$ (h) $-14x + 1.4y$ (i) $0.3mx - 0.7nx$
 (j) $3t^2 - 7t$ (k) $-48x^3 + 73x^2y$ (l) $0.004m - 0.005$
 (m) $-0.03m^3 + 0.06m^2$ (n) $30x^4y^2 - 7.5x^3y^3$ (o) $12m^3nt - 16mn^3t + 20mnt^3$
 (p) $-4x^2 + 3.5xy$

3. (a) $5x + 32$ (b) $26x + 53$ (c) $46w - 33$ (d) $20m - 22$
 (e) $2x - 3$ (f) $12t - 28$ (g) $7x^2 - 3x + 9$ (h) $x^2 + 9x - 2$
 (i) $-2x - 10y$ (j) $8x^2 - 2xy - 4y^2$ (k) $5x^2 - 46x + 1$ (l) $10t^2 - 8t - 7$
 (m) $8m + 3n - 6$

4. (a) -6 (b) 9 (c) -11 (d) -12 (e) 11 (f) 24
 (g) 3 (h) 1 (i) 19

FUNCTION NOTATION

1. (a) 5 (b) 6 (c) 12 (d) 24 (e) 4 (f) 0
2. (a) 4 (b) 19 (c) -2 (d) -17 (e) 40 (f) -47
3. (a) -8 (b) -33 (c) 16 (d) -24 (e) -24 (f) 136
4. (a) 0 (b) -4 (c) 6 (d) 0 (e) 0 (f) -24

MONOMIALS

1. (a) 81 (b) 15 (c) 9 (d) -18 (e) -18 (f) -81
 (g) 9 (h) -9 (i) 8
2. (a) 12 (b) -32 (c) 4 (d) -4 (e) 4 (f) -8
 (g) -10 (h) 2 (i) $\frac{1}{9}$
3. (a) $4x^2$ (b) $15x^2$ (c) $8x^2$ (d) $-30x^2$ (e) $8x^3$ (f) $6x^7$
 (g) $-8x^4$ (h) $10x^7$ (i) $6x^7$ (j) $-5x^5$ (k) $6x^5$ (l) $-2x^7$
4. (a) $2x$ (b) 5 (c) x^3 (d) 1 (e) $4x$ (f) $-4x^3$
 (g) $-3x^2$ (h) $-7x^4$ (i) $6x^3$ (j) $-6x^2$ (k) $4x^5$ (l) 9
5. (a) $15x^6$ (b) $4x$ (c) $4x^2$ (d) $32x^9$ (e) $-6x^{11}$ (f) $81x^6$
 (g) $-3x^3$ (h) $-7x$ (i) $64x^{10}$ (j) $-4x^2$ (k) 16 (l) $25x^6$
6. (a) -30 (b) 38 (c) 16 (d) -1350 (e) -8640 (f) -240
 (g) -1875 (h) 10 (i) 108 (j) 8 (k) 1 (l) $\frac{15}{2}$

EXERCISE 2.1

1. (a) monomial, 3rd degree   (b) monomial, 6th degree   (c) monomial, 2nd degree
 (d) binomial, 2nd degree   (e) trinomial, 1st degree   (f) trinomial, 2nd degree
 (g) trinomial, 7th degree   (h) monomial, 5th degree   (i) binomial, 4th degree
 (j) trinomial, 2nd degree

2. (a) $-3x + 4x^2 + 16x^3 - 5x^5$ (b) $5x + 17x^2 - 3x^3 + 15x^4$
 (c) $5x^3y + 5x^2y^3 - 4x^3y^3$ (d) $2x^2y - 4x^2y^2 + 8x^3y - 3x^5y^2$
 (e) $4x^3y - x^5 - 5x^6 + 11x^5y^2$

3. (a) rectangle, 1st degree   (b) triangle, 1st degree   (c) square, 1st degree
 (d) circle, 1st degree   (e) quadrilateral, 1st degree   (f) line, 1st degree

4. (a) rectangle, 2nd degree   (b) circle, 2nd degree   (c) parallelogram, 2nd degree
 (d) trapezoid, 1st degree   (e) ring, 2nd degree

5. (a) cuboid or rectangular prism, 3rd degree   (b) cylinder, 3rd degree
 (c) sphere, 3rd degree

6. (a) 9 (b) -36 (c) 12 (d) 5 (e) 0 (f) 14
 (g) -23 (h) 16 (i) -23 (j) -6 (k) 0 (l) -16
7. (a) 21 (b) 61 (c) 213
8. (a) -14 (b) 261 (c) -120
9. (a) 12 (b) 0 (c) 6
10. (a) 68 (b) 8 (c) 2
11. (a) -247 (b) -52 (c) -52
12. (a) 0 (b) 1 (c) 76 (d) -2 (e) 10 (f) 145
13. (a) 3 (b) 6 (c) 33 (d) 0 (e) -27 (f) 108
14. (a) 1 (b) 13 (c) 85 (d) 157 (e) No   (f) No
15. (a) 4 (b) 6 (c) No   (d) No   (e) No   (f) >

16. (a) 0 (b) 0 (c) Yes (d) <

EXERCISE 2.2

1. (a) $6x - 2y - 4x$ (b) $-2x^2 + 2x - 1$ (c) $-2xy + 2xz - 2yz$ (d) $-4x^3 - x^2 + 2x$
   (e) $x - 4y - 3z - 8$ (f) $10x^2 - 5x - 1$
2. $-6x + 8$
3. $4x + 7y - 5$
4. (a) $2x - 8y + 5$ (b) $2x^2 - 7x + 7$ (c) $7x + 5y - 5z$ (d) $4x^2 + 6x - 14$
   (e) $10x - 5y + 7$ (f) $2x^2 + 7x - 3$
5. $3x^2 + x + 10$
6. $-2x^2 + 7x$
7. (a) $-7x + 15y + 7w$ (b) $8xy - 2xz + 9yz$ (c) $11r - 11s - 3t$
   (d) $x^2 + 3x + 13$ (e) $7m^2 + 4m - 7$ (f) $11x + 22xy - 9yz$
   (g) $x^2 + x + 4$ (h) $6.6u - 11.2v + 7.1w$ (i) $5.4x^2 - 1.2x + 3.1$
8. (a) $5x^2 + x - 1$ (b) $-x^2 - 4x + 1$ (c) $9xy + 3x + 4$ (d) $7x - 2y - 2xy$
   (e) $5x^2 - 9x + 9$
9. (a) $2x + 3y + 4z$ (b) $-5x^2 + 2x + 10$ (c) $2x + 8y + 4z$
   (d) $2u - 2v - w$ (e) $3m + 4n - 4$ (f) $x + 3y - 4z$
   (g) $6xy - 4y$ (h) $-0.5x + 0.4y$ (i) $-4.8x - 2.2y - 0.4z$
   (j) $1.8x^2 - 10.2x + 5$
10. (a) -1 (b) 95 (c) -49 (d) 9 (e) 65
11. (a) 7 (b) -51 (c) 24 (d) -32 (e) 80
12. (a) 14 (b) -3 (c) 5 (d) 27
13. (a) $-2x^2 + 9x - 8$ (b) $7x - 5y + 8$ (c) 6 (d) $2x^2 + 8x - 6$ (e) $-3x - 2y - 2w$
14. $-3u + 6v + w$ 15. $x^2 - 6x + 8$ 16. $x^2 + 9xy - 4y^2$ 17. $xy - xz + 7yz$
18. $-2x^2 + 8x - 6$ 19. $-3m - 4mn + 5n$ 20. (a) $10x^2 - 10x - 8$ (b) $x^2 + 7x + 14$

EXERCISE 2.4

1. (a) $3x + 3y$ (b) $x^2 + xy$ (c) $3x - 6$ (d) $x^2 + 7x$ (e) $4x^2 - 20x$
   (f) $6x - 4x^2$
2. (a) $x^2 + 6x + 9$ (b) $x^2 - 4$ (c) $m^2 - 2mx + x^2$
   (d) $r^2 - 49$ (e) $x^2 - y^2$ (f) $4m^2 + 4m + 1$
   (g) $4x^2 - 9y^2$ (h) $1 - 2x + x^2$ (i) $9x^2 - 24xy + 16y^2$
   (j) $25s^2 + 30st + 9t^2$ (k) $4 - 12st + 9s^2t^2$ (l) $9x^4 - 4y^2$
3. (a) $x^2 + 5x + 6$ (b) $y^2 + 9y + 20$ (c) $y^2 - 10y + 21$ (d) $t^2 - t - 12$
   (e) $x^2 - 4x - 21$ (f) $m^2 - 12m + 35$ (g) $t^2 + 16t + 55$ (h) $x^2 + 21x + 110$
   (i) $x^4 + 3x^2 - 18$ (j) $x^4 - x^2 - 2$ (k) $x^6 - 2x^3 - 48$ (l) $5 - 4x - x^2$
   (m) $80 - 18x + x^2$ (n) $56 - t - t^2$
4. (a) $-x - 14$ (b) $2x^2 - 17x + 22$ (c) $13x - 15y + 7$
   (d) $-9r - 12s + 15t$ (e) $4x - 11$ (f) $15x - 20y - 19$
   (g) $4x^2 - 18x - 18$ (h) $7x^2 - 18x + 28$ (i) $-x^2 - 2x + 1$
   (j) $-8m^2 + 6m$ (k) $3x_1 - 8x_2 + 11x_3$ (l) $6x^2 - 12xy + 22y^2$
5. (a) $3x^2 + 19x + 20$ (b) $6t^2 + 17t + 7$ (c) $6x^2 - 11x + 4$
   (d) $6m^2 - 25m + 24$ (e) $20x^2 - x - 12$ (f) $6r^2 + 19r - 7$
   (g) $3 - 23y + 30y^2$ (h) $-6m^2 - 13m + 5$ (i) $6x^2 - 7xy - 3y^2$
   (j) $12x^2 - 55xy + 50y^2$ (k) $6w^2 + 7wx - 33x^2$ (l) $56x^2 - 33xy - 14y^2$
   (m) $15x^4 - 2x^3 - 8x^2$ (n) $-3m^4 - 4m^3 + 4m^2$
6. (a) $x^2 + y^2 + z^2 + 2xy + 2xz + 2yz$ (b) $w^2 + x^2 + y^2 - 2wx - 2wy + 2xy$
   (c) $4x^2 + y^2 + z^2 + 4xy + 4xz + 2yz$ (d) $4w^2 + 9x^2 + y^2 - 12wx + 4wy - 6xy$
   (e) $16x^4 + 24x^3 + x^2 - 6x + 1$ (f) $25m^2 + 40m - 30mn - 24n + 9n^2 + 16$
7. (a) $2x^3 + 7x^2 + 8x + 3$ (b) $6w^3 - 11w^2 - 2w + 3$
   (c) $8m^4 + 8m^3 - 4m^2 + 11m - 3$ (d) $8w^2 + 3x^2 + 8y^2 - 14wx + 16wy - 14xy$
   (e) $5x^4 + 11x^3 - 19x^2 - 2x + 2$ (f) $3x^2 - 12y^2 - 2z^2 + 5xy - xz + 10yz$
   (g) $x^5 - 2x^4 - x^3 + x^2 - 2x + 3$ (h) $x^6 + x^5 - x^4 - 9x^3 - 13x^2 - 11x - 3$
   (i) $2m^4 - 9m^3 + 4m^2 + 13m + 5$ (j) $3x^4 - 10x^3 + 23x^2 - 32x + 16$
8. (a) $12x^2 + 53x - 54$ (b) $-9t^2 - 90t + 96$ (c) $-9m^2 - 44m - 49$ (d) $9m^2 + 60m - 18$
   (e) $12x^2 - 8x - 102$ (f) $-21x^2 - 27$ (g) $x^2 + 27x - 49$ (h) $10x^2 - 8x + 9$
   (i) $-11x^2 + 17x - 34$ (j) $-2x^2 + 40xy - 35y^2$ (k) $3w^2 + 17wx - 24x^2$ (l) $x^2 - 14xy + y^2 + 5$
   (m) $-2x^3 - 8x^2 + 27x + 8$ (n) $4r^2 + 9s^2 - 16t^2 - 17rs - 12rt + 3st - r + s + 3t$
9. (a) $6x^3 + 11x^2 - 47x + 20$ (b) $2x^3 - 3x^2y - 17xy^2 + 30y^3$
   (c) $w^2 + x^2 + y^2 + z^2 + 2wx + 2wy + 2wz + 2xy + 2xz + 2yz$

   (d) $x^2 - \dfrac{1}{x^2}$ (e) $m^2 + 1 - \dfrac{6}{m^2}$ (f) $-x^2 - x + 4 + \dfrac{1}{x} - \dfrac{1}{x^2}$

EXERCISE 2.5

1. (a) $5(x + 2)$ (b) $2(x - 3)$ (c) $2(x - 2y)$ (d) —
   (e) $x(3y - 4)$ (f) $5(2m - 1)$ (g) $x(3y + 4w)$ (h) $2(t^2 + 2t - 3)$
   (i) $mn(5 - 6t)$ (j) $4x(2x - 1)$ (k) $7r(1 - 2r)$ (l) $2s(w + 2x + 3y)$
   (m) $(x - 1)(r + t)$ (n) $(w - z)(7 - 5x)$
2. (a) $5x^2(1 + 5x - 6x^2)$ (b) $x^2(x^3 - x^2 + x - 1)$
   (c) $4x^2(7x + 6)$ (d) —
   (e) $2m(3m^2 - 5mt + 3t^2)$ (f) $7rst(2 - r + 3s - 4t)$
   (g) $xy(x^2y^2 - x + y)$ (h) $13m^2n(1 - 2n)$
   (i) $4p^3q^3(4p - 2q - 1)$ (j) $3xy(6xy + 3 - x^2y^3)$
   (k) —   (l) $x^2(x^4 - x^3 + x^2 - x - 1)$
   (m) $4p^2x^4(9 - 4px + 6p^3)$ (n) $m^9n^{11}(m^6 - 3mn - 4n^4)$
   (o) $12x^9y^7(2 - 4x^2y^6 + 3xy^3)$ (p) $2x^8y^4(1 - 3x^9y^{14} + 2y^5 - 5x^6y^{17})$
3. (a) $(p + q)(2x + 3)$ (b) $(x - 1)(2m + 5)$ (c) $(x - y)(9m - 14n)$ (d) $(m + 4)(8 - 9x)$
   (e) $(m + n)(5ty - 6x)$ (f) $(x - 3)(1 + y)$ (g) $(m + 7)(x - 1)$ (h) $(x^2 - 2x - 1)(2 - y$
   (i) $7m(x - y)(2m - 1)$ (j) $(x + y)(x - y + 3)$ (k) $(x - 4)(2x - 3)$ (l) $(2m + 1)(5m - 4)$
   (m) $(x - 5)(x - 5 - m)$ (n) $(x - y)(x - y - 1)$ (o) $(y + 2)(x - 7)$ (p) $(m - n)^2(m - n - 2$
4. (a) $(x + y)(m + n)$ (b) $(x + 3)(t + w)$ (c) $(x + y)(m - 3)$ (d) $(x + 1)(x^2 + 1)$
   (e) $(x^2 + y^2)(x - y)$ (f) $(a + 3)(m + n)$ (g) $(1 - x)(1 + x^2)$ (h) $(x - y)(5 - x)$
   (i) $(x + y)(3x + 1)$ (j) $(m - n)(4x - y)$ (k) $(x - y)(2m - 3n)$ (l) $(x - y)(2m + 3n)$
5. $2\pi r(r + h)$
6. (a) $(x + y)(2m - 3)$ (b) $2(m + n)(2x - 1)$ (c) $(x - y)(6n - 5)$ (d) $(x - 4)(7m + 3)$
   (e) $(3x - y)(5m + 2)$ (f) $(m - n)(3x + 1)$

EXERCISE 2.6

1. (a) 44   (b) 40   (c) 5   (d) -38 (e) 120   (f) -77
2. (a) 1   (b) -44 (c) 18   (d) -51 (e) -4 (f) -36
3. (a) $5]x - 7]x + 3$ (b) $3]x + 4]x - 5]x + 2$ (c) $3]x + 0]x - 2]x + 5$
   (d) $4]x + 3]x + 0]x - 7$ (e) $3]x - 7]x + 5]x - 3$ (f) $-2]x + 0]x + 7]x - 11$
4. (a) $3]x + 2]x - 5]x + 3$ (b) $-2]x + 3]x - 4]x + 7$ (c) $5]x - 2]x + 3]x + 6$
   (d) $1]x - 3]x + 2]x - 1$ (e) $6]x + 0]x - 3]x + 5$ (f) $-2]x - 3]x + 0]x + 6$
5. (a) $3]x + 5]x + 3; 25$ (b) $2]x - 3]x + 1; 3$ (c) $-1]x + 5]x - 1; 5$
   (d) $3]x - 5]x + 1; 3$ (e) $6]x + 1]x - 3; 23$ (f) $5]x - 3]x + 4; 18$
6. (a) $5]x - 2]x + 3]x - 1; -163$ (b) $3]x - 5]x + 2]x + 7; -125$ (c) $-1]x + 5]x - 3]x + 2; 83$
   (d) $2]x - 6]x + 0]x - 9; -117$ (e) $4]x - 3]x + 5]x; -150$ (f) $3]x + 0]x - 2]x - 6; -81$
7. (a) $2]x + 0]x - 5]x + 7; 13$ (b) $3]x + 7]x - 5]x + 3; 17$
   (c) $2]x - 5]x + 0]x + 6; -1$ (d) $6]x - 3]x + 0]x + 5; -55$
   (e) $-2]x + 5]x - 2]x + 3; -12$ (f) $3]x + 0]x - 2]x + 3]x - 1; 233$
8. (a) 35   (b) -5 (c) -1 (d) -8 (e) 27
9. (a) $-4.9]t + 24.4]t + 1.4$ (b) 20.9 m; 30.6 m; 30.5 m; 20.6 m; 0.9 m;
   (c) 31.77 m

EXERCISE 2.7

1. (a) $(x + 2)$ (b) $(x - 3)$ (c) $(t - 11)$ (d) $(y + 8)$ (e) $(m^2 - 10)$ (f) $(x - 8)$
   (g) $(x + 1)$ (h) $(x + 3)$ (i) $(7 - m)$
2. (a) 3, 4   (b) 18, 2   (c) $-6, 3$ (d) —   (e) $8, -5$ (f) $-4, -4$
   (g) 15, 5   (h) $-9, 3$ (i) $-4, -1$
3. (a) $(x + 5)(x + 2)$ (b) $(y - 3)(y - 5)$ (c) $(m + 7)(m + 3)$ (d) $(w - 8)(w + 7)$
   (e) —   (f) $(s - 12)(s + 8)$ (g) $(t + 16)(t + 20)$ (h) —
   (i) $(x + 15)(x + 11)$ (j) $(w - 6)(w - 7)$ (k) $(z + 8)(z - 5)$ (l) $(m - 9)(m + 2)$
   (m) $(x - 4)(x - 4)$ (n) —   (o) $(n - 4)(n - 9)$ (p) $(x - 5)(x - 4)$
   (q) $(7 + x)(6 - x)$ (r) $(11 - y)(8 + y)$
4. (a) $2(x + 1)(x + 1)$ (b) $3(q - 5)(q + 3)$ (c) $(m + 7)(m + 4)$ (d) $5(p - 7)(p - 2)$
   (e) $2(x^2 - 2x - 10)$ (f) $(x - 13)(x - 9)$ (g) $(12 - m)(5 + m)$ (h) $10(y - 8)(y + 5)$
   (i) —   (j) $(x + 6)(x + 8)$ (k) $2(12 - x)(5 + x)$ (l) $(s - 3)(s - 3)$
   (m) $3(x^2 - 12x + 28)$ (n) $(m - 6)(m - 6)$ (o) $(w + 11)(w - 7)$ (p) $(x - 16)(x - 2)$
5. (a) $(x + 16y)(x - 10y)$ (b) $(s + 2t)(s + 7t)$ (c) $(x + 4y)(x - 3y)$
   (d) $(t - 7)(t - 8)$ (e) $(x - 11y)(x + 5y)$ (f) $(m - 17)(m - 24)$
   (g) —   (h) $(x + 13y)(x - 7y)$ (i) $(x - 7)(x - 6)$
   (j) $(x^2 - 10)(x^2 + 9)$ (k) $(a^3 + 7)(a^3 - 5)$ (l) $(s - 5t)(s - 4t)$
   (m) $2(y - 12)(y + 2)$ (n) $(z + 14)(z + 8)$ (o) $(m - 31)(m + 24)$
   (p) $(m + 12n)(m - 20n)$ (q) $(x + 22y)(x - 30y)$ (r) $(x^2 - 12)(x^2 + 4)$

EXERCISE 2.8

1. (a) $(x + 2)$ (b) $(m - 6)$ (c) $(n - 2)$ (d) $(x - 3)$ (e) $(x^3 - 18)$ (f) $(x + 4)$
   (g) $(1 + s)$ (h) $(8 - t)$ (i) $(9 - x)$
2. (a) $-3, -5$ (b) $4, 1$ (c) $9, -2$ (d) $-4, 3$ (e) —   (f) $7, -4$
   (g) $-9, 2$ (h) $-9, -3$ (i) —
3. (a) $(3x + 4)(2x - 5)$ (b) $(3x + 5)(4x + 1)$ (c) $(2t + 1)(11t + 1)$ (d) —
   (e) $(2m - 3)(3m + 1)$ (f) $(2y + 7)(y - 4)$ (g) $(4q + 3)(3q + 5)$ (h) $(4x - 3)(5x + 3)$
   (i) $(6x + 1)(2x - 7)$ (j) —   (k) $(5t - 3)(2t - 5)$ (l) —
   (m) $(3q - 1)(2q - 7)$ (n) $(m + 7)(4m - 5)$ (o) $(y + 8)(3y - 2)$ (p) $(3x + 7)(x + 6)$
   (q) $(s + 9)(4s - 5)$ (r) $(5x + 4)(4x - 1)$ (s) $(2x - 5)(10x - 7)$ (t) $(4m - 3)(9m + 5)$
   (u) $(8t + 3)(2t - 3)$ (v) —   (w) $(s + 3)(4s + 9)$ (x) $(4x - 3)(2x + 1)$
4. (a) $2(4x + 1)(3x - 1)$ (b) $(2x + 7)(5x - 3)$ (c) $3(2x + 1)(x - 5)$
   (d) $(3x + y)(2x + 5y)$ (e) $(4x - 5y)(3x + 7y)$ (f) $10(6m + n)(m + 6n)$
   (g) $(8x - 5y)(3x - 4y)$ (h) $(5t + 4s)(3t + 2s)$ (i) $2(1 + 7x)(1 - 6x)$
   (j) $y(x + 4)(5x - 2)$ (k) $(4x^2 + 7)(x^2 + 7)$ (l) $2(x^2 + xy + y^2)$
   (m) —   (n) $(3 - 10x)(5 + 2x)$
5. (a) $(2x + 5)(4x + 9)$ (b) $(2y + 5)(10y - 3)$ (c) $(5t - 4)(8t - 3)$
   (d) $(6x + 7)(8x + 3)$ (e) $(7m - 5)(6m - 3)$ (f) $(10x + 7)(4x - 1)$
   (g) $(2m + 7)(4m + 9)$ (h) $(8y - 3)(6y - 1)$ (i) $(5s - 11)(4s + 3)$
   (j) $(3 - 7x)(5 - 6x)$ (k) $(6 + 7t)(7 - 8t)$ (l) $(4m + 9)(7m + 11)$

EXERCISE 2.9

1. (a) $(x - 3)(x + 3)$ (b) $(x - 3)^2$ (c) $(x + 4)^2$ (d) $(x + 6)(x - 6)$
   (e) $(x - 6)^2$ (f) $(x + 8)(x - 8)$
2. (a) $(2x + 3)(2x - 3)$ (b) $(2x - 3)^2$ (c) $(3x + 5)(3x - 5)$ (d) $(4x + 7)(4x - 7)$
   (e) $(5x + 6)(5x - 6)$ (f) $(5x + 1)(5x - 1)$
3. (a) $(x + 11)(x - 11)$ (b) $(x - 10)^2$ (c) $(5x + 2y)^2$
   (d) $(9x - 1)^2$ (e) $(10x + 3y)^2$ (f) $(5x + 12)(5x - 12)$
4. (a) $(x^2 + 1)(x + 1)(x - 1)$ (b) $(9x^2 + 1)(3x + 1)(3x - 1)$
   (c) $(25x + 9)(25x - 9)$ (d) $[(x + 1)(x - 1)]^2$
   (e) $(4x^2 + 25)(2x + 5)(2x - 5)$ (f) $(x^2 - 3y^2)^2$
5. (a) 9996 (b) 9997 (c) 9951 (d) 9775 (e) $999\,999$ (f) $999\,975$
6. (a) $5(x + 2)(x - 2)$ (b) $3(x - 2)^2$ (c) $7(x + 7)(x - 7)$ (d) $2(9x^2 + 6x + 4)$
   (e) $5(3x - 2)^2$ (f) $2(2x + 5)(2x - 5)$

EXERCISE 2.10

1. (a) $(a + b)(x - y)$ (b) $(a + b)(x + y)$ (c) $(2a - 3b)(x + y)$ (d) $(2x + 3)(y - 5)$
   (e) $(3x - 2)(2y + 1)$ (f) $(6x + 1)(2y - 1)$
2. (a) $(3 + a + b)(3 - a - b)$ (b) $(a - b + 1)(a - b - 1)$ (c) $(5 - x - y)(5 + x + y)$
   (d) $(2x - 1 + y)(2x - 1 - y)$ (e) $(x - y + z)(x + y - z)$ (f) $(a + b + c)(a + b - c)$
3. (a) $(a - b + 5)(a - b - 5)$ (b) $(x - 3 + y)(x - 3 - y)$ (c) $(x + 5 + y)(x + 5 - y)$
   (d) $(2x - 1 + 3y)(2x - 1 - 3y)$ (e) $(4 + 2x - y)(4 - 2x + y)$ (f) $(10 - a - 3)(10 + a + 3)$
4. (a) $a(x + y + b)(x + y - b)$ (b) $(b + x - y)(b - x + y)$ (c) $3(a + b)(a - b + 4)$
   (d) $(6 + x + y)(6 - x - y)$

EXERCISE 2.11

1. (a) $8x; x \neq 0$ (b) $2t^3; t \neq 0$ (c) $5; x, y \neq 0$ (d) $-12rs; r, s \neq 0$
   (e) $-9mn^2; m, n \neq 0$ (f) $5mn; m, n \neq 0$ (g) $-8xyz^5; x, y, z \neq 0$ (h) $-10m^3n^7; m, n \neq 0$
   (i) $3; m, n \neq 0$ (j) $5r^5s^6t; r, s \neq 0$ (k) $-6x^5y^7; x, y, m \neq 0$ (l) $-5z^2; x, y, z \neq 0$
2. (a) $\dfrac{1}{x}; x \neq 0$ (b) $\dfrac{1}{y}; x, y \neq 0$ (c) $\dfrac{1}{2t^2}; t \neq 0$ (d) $\dfrac{5}{y^5}; y \neq 0$
   (e) $\dfrac{1}{4xy}; x, y \neq 0$ (f) $\dfrac{n}{2}; m, n \neq 0$
3. (a) $m + n$ (b) $2x - 5y$ (c) $3x - 2; x \neq 0$
   (d) $5s + 1$ (e) $-x + 2; x \neq 0$ (f) $m^2 - 2m - 4; m \neq 0$
4. (a) $x + 2y; x, y \neq 0$ (b) $4m^3n - 3mn^2 - 1; m, n \neq 0$
   (c) $x^2m^4 - 3xm^2 - 6m; x, m \neq 0$ (d) $3xy^2 - y^4 + 2x^2y^2; x \neq 0$
5. (a) $-x + 2y$ (b) $-t + 2t^2; t \neq 0$
   (c) $-7s + 5r + 2t; r, s, t \neq 0$ (d) $-4mn^3 + 1 + 3n^4; m, n \neq 0$
   (e) $-15x^2yz^2 + 3x^3; x, y, z \neq 0$ (f) $-3xy^2 + 2z + 1; x, y, z \neq 0$
6. (a) $4x + 2 + \dfrac{1}{x}; x \neq 0$ (b) $6x^2 - 4x + 2 + \dfrac{1}{x^2}; x \neq 0$

(c) $\frac{m^3}{3} - 2m + 3; m \neq 0$

(d) $3t^2 - 2t - 1 - \frac{1}{7t}; t \neq 0$

(e) $xy^2 - \frac{y^3}{2} + \frac{3}{2xy}; x, y \neq 0$

(f) $3w^2 - 2w - \frac{1}{w} + \frac{3}{17w^2}; w \neq 0$

(g) $-10m^2n + 4mn^2 + \frac{2}{mn} - \frac{5}{2m^2n^2}; m, n \neq 0$

(h) $-21rs^3 + 27r^2s^4 - 6rs + \frac{3}{s}; r, s \neq 0$

(i) $\frac{-x^2}{8y^3} - 3xy^2 + y^3 + \frac{1}{8x^3}; x, y \neq 0$

EXERCISE 2.12

1. (a) $3x^5 + x^3 - 4x^2 + 7$
   (b) $-x^4 + 2x^2 - 5x + 1$
   (c) $2m^5 - 4m^3 + 3m^2 + 5m$
   (d) $-s^4 + rs^3 + 2r^2s^2 + r^4$
   (e) $2x^5 - 5x^4 + 4x^2 + 3x$
   (f) $4x^4 - 3x^3y + 2x^2y^2 + xy^3$
   (g) $-t^5 + t^4 + 4t^3 - 5t^2 + 1$
   (h) $5m^5 - 3m^3 + 2m^2 + m$

2. (a) $(6 \times 3) + 1 = 19$
   (b) $(3 \times 2) + 1 = 7$
   (c) $(1 \times 7) + 4 = 11$
   (d) $(2 \times 20) + 6 = 46$
   (e) $(2 \times 9) + 3 = 21$
   (f) $(10 \times 7) + 2 = 72$
   (g) $(5 \times 12) + 5 = 65$
   (h) $(10 \times 9) + 8 = 98$
   (i) $(7 \times 4) + 2 = 30$

3. (a) $2x + 5, x \neq -1$
   (b) $x^2 + 2x - 3; R = 1, x \neq 4$
   (c) $2y^2 - 6y + 7; R = -2, y \neq 4$
   (d) $3w - 4; R = -10, w \neq \frac{7}{2}$
   (e) $3m - 5, m \neq -\frac{4}{3}$
   (f) $t^2 - 3t + 5; R = -5, t \neq -\frac{5}{2}$
   (g) $2m^2 - 3m - 4; R = 9, m \neq -\frac{1}{2}$
   (h) $3x^2 - 2x + 7, x \neq \frac{5}{4}$
   (i) $4y^2 - 5; R = 16, y \neq -\frac{3}{2}$

4. (a) $8x^2 - 1; R = -6$
   (b) $x + 4$
   (c) $m^2 - 2m + 1; R = 4$
   (d) $2x + 7; R = -4$
   (e) $t^2 + 2t - 1; R = -6$
   (f) $4r^2 + 2r + 1; R = 8$
   (g) $5s^3 - 2$
   (h) $x^3 + 2x^2 + 4x + 8; R = -3$
   (i) $x^4 + x^3 + x^2 + x + 1; R = -1$

5. (a) $x + y; R = y^2$
   (b) $x - y; R = 3y^2$
   (c) $x - 3y; R = -2y^2$
   (d) $x^2 + xy + y^2$
   (e) $x^2 + xy + y^2$
   (f) $x^2 - xy + 2y^2$
   (g) $x^2 - 2xy - y^2; R = 2y^4$
   (h) $2m^2 - 3mn + n^2$

6. $3x + 4$

7. $x^6 + 2x^5 + 4x^4 + 8x^3 + 16x^2 + 32x + 64$

EXERCISE 2.13

1. (a) $2, 11, 37, 118$
   (b) $1, 2, -6, 15$
   (c) $8, 12, 4, 4$

2. (a) $x^2 + 7x + 24; R = 67$
   (b) $x^2 - 4x + 7; R = -16$
   (c) $2x^2 - 6x + 13; R = -32$
   (d) $5x^2 - 2x + 1; R = -1$
   (e) $-2x^2 + 9x - 27; R = 73$

3. (a) $x^2 - 6x + 25; R = -70$
   (b) $x^2 + 7x + 15; R = 33$
   (c) $x^2 + 6x + 6; R = 12$
   (d) $x^2 - 2x + 4; R = 22$
   (e) $4x^2 + 6x + 9$
   (f) $x^2 - x + 4; R = -9$

4. (a) $x^3 + x^2 - 4x + 9; R = -15$
   (b) $2x^3 + 15x^2 + 57x + 233; R = 934$
   (c) $3x^3 - 15x^2 + 73x - 358; R = 1795$
   (d) $x^3 + 4x^2 + 12x + 29; R = 91$
   (e) $x^2 + 5x + 25; R = 5$
   (f) $x^3 - 4x^2 + 9x - 6; R = 4$
   (g) $x^3 + 3x^2 + 11x + 40; R = 125$
   (h) $x^4 - x^3 + 4x^2 + 2; R = 5$
   (i) $x^4 + 7x^3 + 35x^2 + 168x + 830; R = 4153$

5. (a) $x^2 + 3x - 5; R = -10$
   (b) $-4x^2 + 7x - 1; R = -3$
   (c) $x^2 + 3; R = 1$
   (d) $x^3 - 5x + 7; R = -3$
   (e) $x^3 + 3x^2 - 5; R = -30$
   (f) $2x^3 + 5x^2 - 7x + 3; R = 3$
   (g) $3x^3 + 2x^2 - x - 6$

EXERCISE 2.14

1. (a) m
   (b) wy
   (c) 1
   (d) 1
   (e) $3x + 3y$
   (f) $7m - 7n$
   (g) $x - 1$
   (h) $y - x$
   (i) 2
   (j) 2

2. (a) $x \neq 0$
   (b) $x \neq 4$
   (c) $t \neq -3$
   (d) $x \neq 1, -3$
   (e) $r \neq -5$
   (f) $m \neq 0, -8$
   (g) $x \neq m$
   (h) $x \neq -\frac{1}{2}, \frac{2}{3}$

3. (a) $\frac{2}{5}$
   (b) $\frac{12x}{5}$
   (c) $-2t$
   (d) $\frac{4m}{3n}$
   (e) $\frac{mx}{ny}$
   (f) $\frac{5xy}{w}$
   (g) $\frac{-7y^2z}{x^2}$
   (h) $\frac{1}{x^5}$
   (i) $\frac{1}{4x^3y^2}$
   (j) $\frac{-m}{3n^6}$
   (k) $\frac{x + 2}{x - 2}$
   (l) $\frac{m - 3}{m + 3}$

4. (a) $\frac{x + 4}{x + 3}, x \neq 7, -3$
   (b) $\frac{x}{x + 7}, x \neq -7$
   (c) $\frac{2}{t - 1}, t \neq 1, 2$
   (d) $\frac{2(m + 3)}{(m - 2)(m + 2)}, m \neq 2, -2$
   (e) $\frac{2(x + y)}{x - y}, x \neq y$
   (f) $\frac{2}{x + 7}, x \neq 5, -7$

(g) $\dfrac{x - 7}{x - 1}$, $x \neq 1, -7$ (h) $\dfrac{(x - 3)(x - 2)}{(x - 6)(x + 1)}$, $x \neq 6, -1$ (i) $-x - 2$, $x \neq 2$

(j) -1, $x \neq 4$

5. (a) $\dfrac{2x - 1}{x - 2}$, $x \neq 2, -2$ (b) $\dfrac{2x + 3}{2x - 3}$, $x \neq \frac{3}{2}, -\frac{1}{3}$

(c) $-x - y$, $x \neq y$ (d) $\dfrac{x}{x + 4}$, $x \neq 5, -4$

(e) $\dfrac{(2m + 1)(m + 3)}{(2m + 3)(m + 1)}$, $m \neq -1, -\frac{3}{2}$ (f) $\dfrac{-m - 4}{m - 2}$, $m \neq 2, \frac{3}{2}$

(g) $\dfrac{6(m - 3)}{2m + 1}$, $m \neq -3, -\frac{1}{2}$ (h) $\dfrac{x^2 + 2x + 4}{3x + 5}$, $x \neq 2, -\frac{5}{3}$

(i) $\dfrac{-m}{x + y}$, $x \neq y$ (j) $\dfrac{3}{(x - 2)(x^2 + 4)}$, $x \neq 2, -2$

(k) $\dfrac{4(x + y)}{(x - y)}$, $x \neq y$ (l) $\dfrac{(2x - 1)(x + 4)}{(3x + 5)(x - 2)}$, $x \neq 2, -\frac{5}{3}$

6. (a) $\dfrac{1}{x + 4}$ (b) $x + 2$ (c) $\dfrac{7x - 8}{9x - 10}$ (d) $\dfrac{x + 4}{(x + 3)(x - 4)}$

(e) $\dfrac{x - 3}{2x + 3}$ (f) $\dfrac{x + 2}{3}$ (g) $\dfrac{x - y}{x^2 - xy + y^2}$ (h) $\dfrac{-1}{2x - 1}$

(i) $\dfrac{-1}{2(x + 3)}$ (j) $\dfrac{5x - 2y}{2x - y}$

EXERCISE 2.15

1. (a) $\dfrac{x}{w}$ (b) $\dfrac{y}{x}$ (c) $\frac{5}{12}$ (d) $\dfrac{2x}{y}$ (e) $\dfrac{7}{3t}$ (f) xy

(g) $\frac{3}{5}$ (h) $\dfrac{y}{2x}$ (i) x^2 (j) $\dfrac{1}{x}$ (k) $\dfrac{m^2}{n^2}$ (l) $\frac{1}{9}$

(m) $\frac{1}{2}$ (n) 3 (o) $\frac{1}{7}$ (p) $\dfrac{y^2}{x^2}$ (q) $\dfrac{1}{(x + y)^2}$ (r) $\dfrac{x + 3}{x + 4}$

(s) 1 (t) $\dfrac{x + 2}{x - 3}$ (u) $\dfrac{x + y}{y}$

2. (a) $\dfrac{10y^2}{7x}$ (b) $\dfrac{3x^3}{y}$ (c) $\dfrac{6x}{y^2}$ (d) $\dfrac{4y^4}{z}$ (e) $\dfrac{r^2s^2}{x}$ (f) $\dfrac{3mn}{10xy^3}$

3. (a) 2 (b) $3x$ (c) $14t^3$ (d) $-\frac{2}{3}$ (e) $\dfrac{-2x}{x - 2}$ (f) $\dfrac{x + y}{4}$

4. (a) $\dfrac{x + 4}{x + 1}$ (b) $\dfrac{x + 4}{x - 6}$ (c) $\dfrac{1}{(x - 9)(x + 6)}$ (d) $\dfrac{t + 7}{t - 6}$

(e) 1 (f) 1 (g) -5

5. (a) 1 (b) $-\dfrac{3x + 1}{3x - 1}$ (c) 1 (d) $\dfrac{4x - 5}{x - 3}$ (e) $\dfrac{m + 8}{m + 2}$

6. (a) 2 (b) $\dfrac{(x + 2)(x + 5)}{(x + 6)}$ (c) $\dfrac{(x - 3y)}{2(2x - y)}$

7. (a) $\dfrac{2(x - 2)}{x}$ (b) $m(m + 1)$ (c) $\dfrac{-1}{x(x + 4)}$

EXERCISE 2.16

1. (a) 42 (b) 24 (c) 360 (d) 300 (e) 1008 (f) 252
2. (a) x^2y^2 (b) $4x^2y^2$ (c) $6m^2n^2$ (d) $24m^3n^4$ (e) $60x^3y$ (f) $270tx^5y^4$
3. (a) $(x - 3)(x + 3)(x - 4)$ (b) $mn(x - y)$
   (c) $2(x + 3)^2$ (d) $x(x - y)(x + y)$
   (e) $30(x - 2)(x + 2)$ (f) $5(x + 3)(x + 5)(x - 1)$
   (g) $(t + 1)(t - 1)(t + 6)(t + 7)$ (h) $(m + 5)(m - 4)(m - 2)$
4. (a) $(x + y)(x - y)^2$ (b) $(m - 3n)(m + n)^2$ (c) $(2x - 1)(x + 3)(3x + 2)$
   (d) $(5m - 3)(2m + 3)(3m - 4)$ (e) $5(3r - 4)(3r + 4)(5r + 1)$ (f) $(3x + 2)(2x - 3)(2x - 5)$

EXERCISE 2.17

1. (a) 42 (b) 12 (c) 24
   (d) xy (e) $6x^3$ (f) $3x^3y^2$
   (g) $2(m + 2)$ (h) $4(x - 1)(x + 2)$ (i) $x(x - 3)$

(j) $(m - 1)(m + 2)(m + 3)$ (k) $(x + 1)(x + 2)(x - 4)$ (l) $t^2(t + 5)(t + 7)$

2. (a) $\dfrac{8}{x}$ (b) $\dfrac{4}{y}$ (c) $\dfrac{8}{m}$ (d) $\dfrac{5}{x + 1}$ (e) $\dfrac{3x - 5}{x - 4}$

(f) $\dfrac{5x + 5y}{8}$ (g) $\dfrac{2x^2}{(x + 5)(x + 3)}$ (h) $\dfrac{2m^2 - 3m + 4}{m - 5}$

3. (a) $\frac{5}{4}$ (b) $\frac{7}{24}$ (c) $2x$ (d) $\dfrac{8m + 7n}{6}$ (e) $\dfrac{3x - 3}{20}$

(f) $\dfrac{8x^2 + 5}{30}$ (g) $\dfrac{70 + 2m + 7n}{14}$ (h) $\dfrac{-18t - 49}{24}$

4. (a) $\dfrac{17}{6x}$ (b) $\dfrac{49}{30x}$ (c) $\dfrac{2x^2 - 3x + 4}{x^3}$ (d) $\dfrac{-9y^2 + 12y + 10}{18y^3}$

(e) $\dfrac{2t^2 + 3st - 2s^2}{s^2t^2}$ (f) $\dfrac{-x + 4y - 1}{xy}$ (g) $\dfrac{my - x + y^2}{y}$ (h) $\dfrac{w^2 - x + wx}{w}$

(i) $\dfrac{3x^2 + 1}{x}$ (j) $\dfrac{4x - 1 + x^2}{x}$

5. (a) $\dfrac{7x - 5}{(x + 1)(x - 3)}$ (b) $\dfrac{11w + 20}{w(w + 4)}$ (c) $\dfrac{3m}{(m - 4)(m - 2)}$ (d) $\dfrac{7}{t - 2}$

(e) $\dfrac{23x^2 + 115x - 24}{12(x + 5)}$ (f) $\dfrac{x + 3y}{(x - y)(x + y)}$ (g) $\dfrac{10}{2r - 1}$ (h) $\dfrac{-t^2 + 22t}{(t - 4)(t + 5)}$

(i) $\dfrac{2x^2 + 10x + 13}{(x + 2)(x + 3)}$ (j) $\dfrac{-6m + 14}{(m - 1)(m - 3)}$ (k) $\dfrac{9x^2 - 4x - 56}{(x - 4)(x + 2)}$ (l) $\dfrac{3x^2 - xy + 2y^2}{(x - y)(x + y)}$

6. (a) $\dfrac{7x - 10}{(x - 4)(x + 3)(x + 5)}$ (b) $\dfrac{2x + 8}{x(x + 1)(x + 2)}$ (c) $\dfrac{-w^2 - 9w}{(w - 1)(w - 5)(w - 6)}$ (d) $\dfrac{x + 6}{(x + 4)(x + 3)}$

(e) $\dfrac{-t + 3}{(t + 4)(t + 1)(t + 5)}$ (f) $\dfrac{3x + 5}{x(x - 5)(x + 1)}$ (g) $\dfrac{3x + 16}{(x - 4)(x + 4)}$ (h) $\dfrac{-2m + 18}{(m - 3)(m + 3)}$

(i) $\dfrac{4x + 3y}{xy(x - y)}$ (j) $\dfrac{w^2 - wx + x^2}{(w - x)(w + x)}$

7. (a) $\dfrac{2x + 4}{(x + 3)(x + 1)}$ (b) $\dfrac{-4}{(x - 2)(x - 1)}$ (c) $\dfrac{x^2 + 6x - 15}{(x - 3)(x + 3)(x - 1)}$

(d) $\dfrac{3w + x - 4y}{(w - x)(w - y)(x - y)}$ (e) $\dfrac{x^2 - 8x + 6}{(x + 3)(x - 3)}$ (f) $\dfrac{x^3 + 3x^2 - 9x - 5}{2(x + 1)(x - 1)}$

8. (a) $\dfrac{-1}{(2x - 1)(x + 2)(3x - 2)}$ (b) $\dfrac{-t + 3}{(2t - 3)(t + 5)}$ (c) $\dfrac{3x - 4}{(x + 1)(x - 2)}$

(d) $\dfrac{x^2 + 6x + 1}{(x + 5)(x - 5)}$ (e) $\dfrac{x^2 - 2x + 3}{2x^2 - x - 1}$ (f) $\dfrac{rst(r^2 + s^2 + t^2)}{t^2 - r^2 - s^3}$

(g) $\dfrac{3x^2 - 2x}{3x + 1}$ (h) $-\dfrac{x^2 + 1}{2x}$

2.18 REVIEW EXERCISE

1. (a) $x^2 + 18 + 77$ (b) $4x^2 + 20x + 25$ (c) $9x^2 - 1$ (d) $t^2 + 4t - 21$
(e) $m^2 + m - 72$ (f) $m^4 - 6m^2 - 16$

2. (a) $5(x + 6)$ (b) $4m(2m - 1)$ (c) $(x - y)(3m + 2n)$ (d) $(x + 6)(x + 3)$
(e) $(m - 7)^2$ (f) $(n - 6)(n + 4)$ (g) $(x - 4)(x + 4)$ (h) $(x - 7)(x - 4)$
(i) $(y - 5)(y + 4)$ (j) $(2x - 5)(2x + 5)$ (k) $(t + 4)^2$ (l) $(1 - 5w)(1 + 5w)$

3. (a) $m \neq 0$ (b) $x \neq 5$ (c) $n \neq 7$ (d) $x \neq 5, -2$

4. (a) $\dfrac{15}{x^2}$ (b) $\dfrac{8}{x}$ (c) $\frac{5}{3}$ (d) $\dfrac{2}{x}$ (e) $\dfrac{-1}{x + 2}$ (f) $\frac{3}{4}$

(g) $\dfrac{12}{(x + 2)^2}$ (h) x^2 (i) $\dfrac{1}{x^2}$ (j) 1

5. (a) $5x + 1$ (b) $7m - 23n$
(c) $5x^2 + x$ (d) $12x^2 + 23x - 24$
(e) $-21t^2 + 31t - 4$ (f) $12x^4 - 41x^3 + 35x^2$
(g) $x^4 - 6x^3 + 23x^2 - 42x + 49$ (h) $6m^3 + m^2 - 14m - 8$
(i) $-6s^3 + 11s^2 - 14s + 5$ (j) $3w^4 - 7w^3 - 10w^2 + 19w - 5$
(k) $-22x^2 - 48x - 10$ (l) $18x^2 - 17xy + 32y^2$
(m) $2x^3 - 36x^2 + 13x - 9$

6. (a) $(m + 5)(m + 3)$ (b) $2(x - 4)(x + 3)$ (c) $(2x - 3)(x - 7)$
(d) $(3x + 7)(2x + 9)$ (e) $(5x - 4)(3x - 1)$ (f) $(3m + 2)(4m - 5)$
(g) $(4t - 5)(3t - 7)$ (h) $2(4n + 5)(n - 6)$ (i) $(6r - 1)(2r + 5)$

(j) $(x + y)(m + n)$　　　　(k) $(x - y)(m - 4)$　　　　(l) $(s + t)(w + 5)$
(m) $(4m - 3n)(2m + 5n)$　　(n) $(5p - q)(4p + 3q)$　　(o) $(2x + 5y)(3x + 4y)$
(p) $3(3s - 5t)(2s + 5t)$
7. (a) $(5x - 6y)(5x + 6y)$　　　　　　(b) $(1 - 10m)(1 + 10m)$
(c) $(t^2 - 6xy)(t^2 + 6xy)$　　　　　　(d) $(3x - 5)^2$
(e) $(4m - 3n)^2$　　　　　　　　　　(f) $(2r + 5s)^2$
(g) $(x + y - 6)(x + y + 6)$　　　　　(h) $(5m - x + y)(5m + x - y)$
(i) $(m + 2n - 2r - 2s)(m + 2n + 2r + 2s)$　(j) $(x + 2y - m - 3n)(x + 2y + m + 3n)$
(k) $(x - 4y - 6m)(x - 4y + 6m)$　　　(l) $(3t - x + 3y)(3t + x - 3y)$
8. (a) $6x + 1$　　　　　　　　(b) $3x^2 - 2x + 1$　　　　　(c) $-5x^2y^4 + 6xy^6 - 7x^2y^5$
(d) $1 - 3mn + 4m^3n^2$　　　(e) $-4 + 3rt - 5t^3$　　　　(f) $n^2 + 2m^2 - 3m^3n^5$
9. (a) $x^2 - x - 1$　　　　(b) $x^2 - 3x - 5$　　　　(c) $m^3 - 3m + 1$　　　(d) $t^2 - 3t + 1$
(e) $3x^2 - 2xy + y^2$
10. (a) $15x^2y$　　　　(b) $\dfrac{2y^4}{x^4}$　　　(c) $\dfrac{x + 4}{x + 5}$　　　(d) 1　　　(e) $\dfrac{x}{x + 2}$　　　(f) $\dfrac{x - 6}{x - 7}$

(g) $\dfrac{x + 7}{x}$

11. (a) $\dfrac{2x + 7}{(x + 4)(x + 3)}$　　　　(b) $\dfrac{-t + 1}{(t - 3)(t - 4)}$　　　(c) $\dfrac{-2x - 33}{(2x + 3)(2x - 3)}$

(d) $\dfrac{6x + 10}{(x + 1)(x + 2)(x + 3)}$　(e) $\dfrac{-m^2 - m}{(m - 1)(m - 2)(m + 3)}$　(f) $\dfrac{-x}{(x + 2)(x - 4)(x - 3)}$

(g) $\dfrac{10x + 4}{(x + 2)(x + 2)(x - 2)}$　(h) $\dfrac{11x + 8}{(2x + 1)(3x + 2)(x - 4)}$　(i) $\dfrac{7x + 11}{(x + 2)(x + 1)}$

(j) $\dfrac{3m - 24}{(m + 2)(m - 4)}$　　　(k) $\dfrac{2x^2 - 7x + 8}{(x - 4)(x - 2)}$

12. (a) $\dfrac{4x + 7}{4x - 1}$　　　(b) 1　　　(c) $\dfrac{x - 5}{x + 6}$

2.19 CHAPTER 2 TEST

1. (a) 52　　　　　　　(b) -33
2. (a) $7x^2 - 2x + 1$　　(b) $-x^3 + 8x - 10$
3. $5x^3 - 5x^2 - 7x + 13$
4. (a) $9x^2 - 4$　　　　　(b) $15x^2 + 29x - 14$　　(c) $5x^2 - 10x - 19$
5. (a) $3x(x^2 - 4x + 5)$　(b) $(x + 4y)(x - 4y)$　　(c) $(x + 4)(x - 6)$　　　(d) $(2x + 3)(x - 6)$
6. $3x^2 - 5x + 2$
7. (a) $4x^2 - 5x + 7; x \neq 0$　　　(b) $x - 5; x \neq -4$
8. (a) $\dfrac{x + 3}{x^2 - 4}; x \neq -3, -2, 2, 3$　(b) $\dfrac{x^2 + 11x + 5}{x^2 + 3x^2 - 16x - 48}$

REVIEW AND PREVIEW TO CHAPTER 3

SLOPE

1. (a) $\frac{12}{7}$　　(b) $\frac{4}{3}$　　(c) $\frac{1}{8}$　　(d) -2　　(e) $\frac{3}{5}$　　(f) $-\frac{3}{4}$

(g) $\frac{3}{5}$　　(h) $\frac{2}{9}$　　(i) $-\frac{6}{5}$　　(j) no slope　(k) 0　　(l) $\frac{10}{11}$

(m) no slope
3. (a) 6　　　(b) 3　　　(c) -14　　(d) 4　　(e) 4　　(f) -3
4. (a) $(3, 4), (1, 6), (-3, 10)$　　　　　(b) $(5, -7), (-5, 3), (-3, 1)$
(c) $(4, 1), (9, 6), (-4, -7)$　　　　　(d) $(6, 10), (1, 5), (-3, -1)$
(e) $(3, 7), (-1, 3), (-2, 2)$　　　　　(f) $(4, 11), (1, 5), (-2, -1)$
(g) $(5, -7), (-12, 10), (-8, 6)$　　　(h) $(3, 2), (0, 4), (-3, 6)$

GEOMETRY

1. (a) $56°$　　(b) $38°$　　(c) $64°$　　(d) $63°$　　(e) $30°$　　(f) $30°$
2. 1 and 11; 3 and 4; 5 and 7; 6 and 12; 10 and 13

PARALLEL LINES, TRANSVERSALS, AND ANGLES

1. (a) $a = 110°, b = c = 70°$　　(b) $a = c = 65°, b = 115°$　　(c) $a = 30°, b = 70°, c = 80°$
(d) $a = 70°, b = 30°, c = 80°$　(e) $a = 75°, b = 55°, c = 50°$　(f) $a = 60°, b = 80°, c = 40°$
(g) $a = 60°, b = 80°, c = 40°$　(h) $a = b = 60°, c = 120°$
2. (a) $a = 40°, b = 80°, c = 60°, d = 60°$　　(b) $a = 85°, b = 60°, c = 35°, d = 35°$
(c) $a = 40°, b = 50°, c = 90°, d = 90°$　　(d) $a = 70°, b = 45°, c = 65°, d = 65°$
(e) $a = 85°, b = 75°, c = 20°, d = 20°$　　(f) $a = 30°, b = 60°, c = 60°, d = 90°$

EXERCISE 3.1

1. Answers vary
2. (a) 3, 5   (b) 6, 2   (c) 4, -8 (d) 7, -4 (e) 3, -4 (f) —, -6
   (g) -2, —   (h) 5, -2 (i) 2, -1 (j) $\frac{1}{2}$, $-\frac{1}{3}$
3. parallel to y-axis
4. parallel to x-axis
5. $y = 6$
6. $x = 4$
7. (a) $3x - y - 4 = 0$ (b) $3x - 2y - 7 = 0$ (c) $2x - y - 1 = 0$ (d) $4x + y + 7 = 0$
   (e) $2x + 6y + 7 = 0$ (f) $4x - 3y + 72 = 0$
8. (a) $y = -2x + 7$ (b) $y = -\frac{1}{2}x + \frac{3}{2}$ (c) $y = -3x + 4$ (d) $y = 2x - 6$
   (e) $y = 5x - 1$ (f) $y = \frac{3}{2}x$ (g) $y = -\frac{5}{2}x + \frac{9}{2}$ (h) $y = \frac{1}{5}x - \frac{6}{5}$
   (i) $y = \frac{2}{3}x - \frac{4}{3}$ (j) $y = -\frac{7}{3}$
10. (a) 3   (b) 4   (c) 2   (d) 4   (e) 4   (f) ± 1

EXERCISE 3.2

1. (a) $y - 2 = 4(x - 3)$ (b) $y - 4 = 6(x - 1)$ (c) $y - 5 = 1(x + 2)$
   (d) $y + 2 = 5(x + 7)$ (e) $y + 4 = -2(x + 3)$ (f) $y = -4(x - 6)$
   (g) $y - 5 = 2x$ (h) $y + 2 = 0$
2. (a) $y = 2x + 3$ (b) $y = -4x + 5$ (c) $y = 5x - 3$ (d) $y = \frac{1}{2}x - 7$
3. (a) 3, -7 (b) $-\frac{1}{2}$, 5   (c) -5, -4 (d) 7, 0   (e) $\frac{2}{3}$, $-\frac{1}{3}$ (f) 2, -7
   (g) -3, 4   (h) 2, -3 (i) -5, 2   (j) 7, -3 (k) 4, -5 (l) $-\frac{3}{2}$, $\frac{5}{2}$
   (m) $-\frac{1}{3}$, $-\frac{2}{3}$ (n) 6, -4
4. (a) $3x - y - 5 = 0$ (b) $5x - y + 19 = 0$ (c) $x + y + 1 = 0$ (d) $7x - y + 5 = 0$
   (e) $6x + y - 30 = 0$ (f) $x - 2y - 10 = 0$ (g) $y - 6 = 0$ (h) $x - 4 = 0$
   (i) $x + 2y - 2 = 0$ (j) $x - 3y + 4 = 0$
5. (a) $2x - y + 13 = 0$ (b) $4x + y + 11 = 0$ (c) $3x - 2y + 22 = 0$ (d) $x + 2y - 6 = 0$
6. (a) $4x - 3y = 0$ (b) $4x - 3y + 18 = 0$ (c) $4x - 3y - 29 = 0$ (d) $4x - 3y - 5 = 0$
7. (a) $y = 3x - 1$ (b) $y = -2x - 3$ (c) $y = 2x - 6$ (d) $y = -0.5x$
   (e) $y = \frac{1}{2}x - 3$ (f) $y = -6$ (g) $y = \frac{3}{4}x + \frac{25}{4}$ (h) $y = 0.2x + 18$
8. (a) $2x - y = 0$ (b) $x - y + 1 = 0$ (c) $x - y + 1 = 0$ (d) $x + 2y - 7 = 0$
   (e) $x + 6y + 8 = 0$ (f) $6x + 5y + 16 = 0$ (g) $7x - 13y + 34 = 0$ (h) $9x + 5y - 43 = 0$
   (i) $4x - 9y - 21 = 0$ (j) $10x + 7y + 11 = 0$ (k) $12x - 22y - 19 = 0$ (l) $60x - 44y - 21 =$
   (m) $15x - 50y + 117 = 0$ (n) $10x + 3y + 27 = 0$ (o) $2x + y + 8 = 0$ (p) $x - 4 = 0$
9. (a) -4, 7   (b) $-\frac{3}{2}$, $\frac{5}{2}$ (c) $\frac{2}{3}$, $-\frac{1}{3}$ (d) $\frac{1}{2}$, -2 (e) $\frac{1}{4}$, $\frac{1}{2}$ (f) $-\frac{2}{3}$, 0
10. (a) $3x - y - 12 = 0$ (b) $2x + y - 10 = 0$ (c) $x + y + 3 = 0$
    (d) $5x - y + 10 = 0$ (e) $x - 4y = 0$ (f) $x - 5y + 1 = 0$
    (g) $5x + 10y - 7 = 0$ (h) $65x + 25y + 338 = 0$
11. (a) $2x + y - 6 = 0$ (b) $2x - y + 4 = 0$ (c) $5x - 2y - 10 = 0$ (d) $6x + y + 6 = 0$
    (e) $x - 2y + 8 = 0$ (f) $7x - 3y - 21 = 0$
12. $36x + 77y - 1000 = 0$

EXERCISE 3.3

1. (i) 2, $-\frac{1}{2}$ (ii) $\frac{1}{4}$, -4 (iii) $-\frac{3}{4}$, $\frac{4}{3}$ (iv) -4, $\frac{1}{4}$ (v) $-\frac{3}{2}$, $\frac{2}{3}$
   (vi) $\frac{5}{6}$, $-\frac{6}{5}$ (vii) -1, 1   (viii) 0, no slope   (ix) 1, -1
2. (a) neither   (b) parallel   (c) perpendicular   (d) neither
   (e) perpendicular   (f) neither   (g) perpendicular   (h) parallel
3. (a) parallel   (b) perpendicular   (c) neither   (d) perpendicular
   (e) parallel   (f) parallel   (g) perpendicular   (h) perpendicular
4. (a) parallel   (b) neither   (c) perpendicular   (d) neither
5. (a) $3x - y - 6 = 0$ (b) $2x + y + 7 = 0$ (c) $x - 3y + 16 = 0$ (d) $x + 2y + 7 = 0$
   (e) $3x - 5y - 5 = 0$ (f) $3x - 2y + 27 = 0$ (g) $5x + 2y - 25 = 0$ (h) $x + 4y - 8 = 0$
6. (a) $3x + 2y - 12 = 0$ (b) $x + 1 = 0$ (c) $x - 5y - 10 = 0$ (d) $y + 2 = 0$
10. (a) $\frac{2}{3}$ (b) $-\frac{3}{2}$ (c) $-\frac{8}{3}$
11. (a) $-\frac{3}{2}$ (b) 15   (c) 10

12. $4x - 5y = 0$ and $x + y - 9 = 0$
13. $Ax + By - As - Bt = 0$
14. $ax + by + (-a^2 - b^2) = 0$
15. $ax + by + (-a^2 - 2b) = 0$

EXERCISE 3.4

1. (b) 82.5 cm   (c) $y = \frac{3}{4}x + 30$ (d) spring constant   (e) length of coil

2. (b) 430.7 cm³   (c) 64.5°C   (d) $v = \frac{4}{3}t + 364$
   (e) 448 cm³   (f) coefficient of expansion   (g) volume at 10°C
3. (a) $t = 0.025\,p$ (b) 0.025   (c) $2000   (d) $112 000   (e) 25
4. (b) $105   (c) $f = 3s - 105$ (d) $75   (e) rate of fine   (f) 35 km/h
5. 3.36 g   6. $280   7. 45.5 months   8. 61.7 mL
9. (b) 70 L   (c) $L = -\frac{1}{6}d + 100$ (d) 60 L
   (e) gas consumption (L/km)   (f) 100 L   (g) 600 km
   (h) 16.7 L/100 km
11. (a) 45   (b) 3125   (c) 16 000   (d) 400
12. (a) $V = \dfrac{22\,400}{P}$
13. (a) $P = kRQ$ $k = 7$ (b) 6825
14. (a) $D = \dfrac{kE}{F}$ $k = 21$ (b) 105
15. (a) $X = \dfrac{kY^2}{Z}$ $k = 18$ (b) 242
16. (a) $C = kpn$ $k = 60$ (b) $15 750 000
17. (a) $R = \dfrac{k\ell}{d^2}$ $k = 0.3$ (b) 15 000 Ω
18. (a) $L = \dfrac{kwd^2}{\ell}$ $k = 150$ (b) 450 kg

EXERCISE 3.6

1. (a) (4, 5) consistent, independent   (b) (6, −14) consistent, independent
   (c) inconsistent   (d) consistent, dependent
   (e) (−1, −3) consistent, independent   (f) (0, 2) consistent, independent
2. Answers vary
3. $2x - y - 1 = 0$
4. $s = 4, t = 5$

EXERCISE 3.7

1. (a) (7, 5)   (b) (5, 4)   (c) (8, 2)   (d) (4, 2)   (e) (2, 5)   (f) (9, 2)
   (g) (3, 1)   (h) (5, 5)   (i) (3, 1)   (j) (1, 4)   (k) (2, 1)   (l) (4, 2)
2. (a) (2, 4)   (b) (4, 11)   (c) (−5, −7)   (d) (−3, −6)   (e) (2, 7)   (f) (6, 3)
   (g) (3, 7)   (h) (1, −6)   (i) (−1, −4)   (j) (−4, −5)
3. (a) (−4, 3)   (b) ∅   (c) (−7, 0)
   (d) (−3, 8)   (e) (−11, −19)   (f) $(-\frac{1}{2}, 5)$
   (g) $(-3, \frac{1}{3})$ (h) infinitely many solutions   (i) $(\frac{1}{2}, -\frac{1}{5})$
   (j) (9, −2)
4. (a) (2, 1)   (b) (5, 2)   (c) (−2, −3)   (d) (2, −1)   (e) (−2, −4)   (f) (−2, 7)
5. (a) (4, 8)   (b) (2, −3)   (c) (10, 6)   (d) (0.2, 0.3)   (e) (−0.6, 0.2)
   (f) (−0.5, −0.7)   (g) (3, −4)   (h) (4, 9)   (i) (−2, −3)   (j) (3, −1)
   (k) (3, 5)   (l) (5, −2)
6. (a) consistent, dependent   (b) inconsistent   (c) consistent, independent
   (d) consistent, independent   (e) consistent, dependent   (f) inconsistent
7. (4, 3), (−4, −5), (−2, 5)
8. (a) $(\frac{1}{2}, \frac{1}{3})$ (b) $(\frac{1}{4}, 3)$ (c) $(-1, \frac{1}{2})$ (d) (6, −2)

EXERCISE 3.8

1. 471, 292   2. 562, 385   3. 97, 84   4. 35, 24   5. 81, 72
6. $3700 at 8%, $8300 at 9%   7. $300 000 at 7%, $35 000 at 10%   8. 16
9. 45 km/h, 585 km/h   10. 30 km/h, 270 km/h   11. $x = $3, y = -$2$

12. $m = \$4, n = -\$3$
13. 8 L of 40%, 12 L of 30%
14. 70 L of 50%, 30 L of 40%
15. 1200 L of 40%, 300 L of 30%
16. 4 km/h, 20 km/h
17. 5 km/h
18. 6 kg of raisins, 4 kg of fruit
19. 34 one bedroom, 26 two bedroom
20. $0.15
21. 86
22. 28
23. $A = 4, B = 3$
24. $A = 4, B = -2$
25. $m = 3; b = -4$
26. $\dfrac{m + n}{2}, \dfrac{m - n}{2}$
27. 1 km/h

EXERCISE 3.9

1. Answers vary
2. (a) 150   (b) 650
3. (a) 4000   (b) 24 000
4. (a) loss of $100 000   (b) 20   (c) 30
5. (a) profit of $270   (b) 50   (c) 250
6. (a) surplus of corn   (b) shortage of corn

EXERCISE 3.10

1. (a) ① × 3, ② × 2 (b) ① × 3, ② × 2 (c) ① × e, ② × b (d) ① × m, ② × n

2. (a) $(a, -b)$ (b) $\left(\dfrac{a + b}{2}, \dfrac{a - b}{2}\right)$

   (c) $(2a, 3a)$ (d) $\left(\dfrac{1}{a}, \dfrac{1}{b}\right)$

   (e) $(a, -b)$ (f) $(-c, -d)$
   (g) $(m - n, m + n)$ (h) $(a - b, -a - b)$

   (i) $\left(\dfrac{st - nw}{ms - nr}, \dfrac{rt - mw}{nr - ms}\right)$ (j) $\left(\dfrac{b_2c_1 - b_1c_2}{a_1b_2 - a_2b_1}, \dfrac{a_2c_1 - a_1c_2}{a_2b_1 - a_1b_2}\right)$

3. (a) $\left(\dfrac{d}{ad - bc}, \dfrac{-c}{ad - bc}\right)$ (b) $\left(\dfrac{-b}{ad - bc}, \dfrac{a}{ad - bc}\right)$

EXERCISE 3.12

1. (a) yes   (b) yes   (c) no   (d) no   (e) yes   (f) no
   (g) no   (h) yes
2. (a) $(1, 2, 3)$ (b) $(3, 2, 1)$ (c) $(-3, 2, 1)$ (d) $(5, 6, -2)$ (e) $(-3, 0, 7)$
   (f) $(3, -5, -2)$ (g) $(6, -3, 0)$ (h) $(-4, -3, -2)$ (i) $(3, 5, -2)$ (j) $(-2, -2, -$
   (k) $(-3, -1, 0)$ (l) $(3, 0, -4)$
3. (a) $(5, -3, -4)$ (b) $(\frac{1}{2}, 4, -3)$ (c) $(\frac{1}{2}, -\frac{1}{3}, \frac{1}{4})$
   (d) $(3, 1, -2)$ (e) $(6, -4, -2)$ (f) $(3, 2, 4)$
   (g) $(0.1, -0.3, -0.2)$ (h) $(2, 1, 3, 2)$ (i) $(6, 8, 12)$
   (j) $(12, 6, -6)$

EXERCISE 3.13

1. (a) $\dfrac{1}{a + b} \neq \dfrac{1}{a} + \dfrac{1}{b}$ (b) $\dfrac{8 + c}{8} = 1 + \dfrac{c}{8}$ (c) $\dfrac{x}{x + y} \neq \dfrac{1}{1 + y}$

   (d) $\sqrt{x^2} = x$ (e) $\dfrac{1 + \sqrt{x}}{1 - x} \neq \dfrac{1}{1 - \sqrt{x}}$ (f) $(x + y)^3 \neq x^3 + y^3$

   (g) $\sqrt{4 + m^2} \neq 2 + m$ (h) $\dfrac{1}{x - 1} = \dfrac{x^2 + x + 1}{x^3 - 1}$ (i) $\sqrt[3]{a + b} \neq \sqrt[3]{a} + \sqrt[3]{b}$

2. $a = 1, b = 2, c = 2, d = 3, e = 5$
3. 648
4. (a) $x = 14, y = 229$ (b) infinitely many

3.14 REVIEW EXERCISE

1. (a) $2x - 3y + 3 = 0$ (b) $4x - y - 7 = 0$ (c) $x + 6y + 8 = 0$ (d) $6x - y - 10 = 0$
   (e) $x + 2y + 5 = 0$ (f) $3x - 2y - 30 = 0$
2. (a) $y = -3x + 4$ (b) $y = 4x - 11$ (c) $y = -\frac{1}{2}x + \frac{7}{4}$ (d) $y = \frac{8}{3}x - \frac{1}{3}$
4. (a) $2x - y - 5 = 0$ (b) $6x - y + 3 = 0$ (c) $3x + y + 1 = 0$ (d) $y = 0$
   (e) $x - 2y + 2 = 0$ (f) $x = -3$
5. (a) $3x - y - 3 = 0$ (b) $2x + y - 2 = 0$ (c) $5x + y + 8 = 0$ (d) $6x - y + 2 = 0$
   (e) $y = -8$ (f) $x = 4$ (g) $x + 2y + 4 = 0$ (h) $2x - y - 19 = 0$
6. (a) $m = 2, b = -7$ (b) $m = -3, b = -2$ (c) $m = 3, b = 2$ (d) $m = -\frac{3}{4}, b = 3$

(e) $m = \frac{2}{3}$, $b = -\frac{5}{3}$ (f) $m = \frac{1}{6}$, $b = -\frac{2}{3}$

7. (a) $3x - y + 1 = 0$ (b) $x + 4y - 19 = 0$ (c) $3x + 2y + 15 = 0$ (d) $4x + y - 12 = 0$

8. $y = 6$ 9. $x = -4$ 13. $k = -\frac{3}{2}$ 14. $k = 4$

15. (b) $c = 20n + 500$ (c) $2320   (d) price/meal
     (e) Mississippi Queen rent

16. $997.50

17. (a) $t = kp$ (b) 0.037   (c) $1961   (d) $42 000

18. (a) $(2, 3)$ (b) $(2, 1)$ (c) $(-3, 2)$ (d) $(-4, -5)$ (e) $(\frac{1}{2}, -3)$ (f) $(-3, -\frac{1}{3})$

19. (a) $(8, -6)$ (b) $(-4, -5)$ (c) $(0.2, -0.3)$ (d) $(-1, -3)$

20. 243, 175   21. 11, 9   22. 140 L of 40%, 60 L of 30%

23. (a) 20 h   (b) 1110 h   24. (b) $m = 280$ (c) $I = \dfrac{280}{R}$

25. (a) $U = \dfrac{RM_1M_2}{r}$ (b) $k = 6.7 \times 10^{-11}$ Nm² kg⁻²   (c) $U \to \infty$, $U \to 0$
     (d) 4.7×10^9 J   Energy required to place satellite in orbit.

3.15 CHAPTER 3 TEST

2. (a) $4x + y - 16 = 0$ (b) $x + y - 6 = 0$ (c) $3x + 5y + 16 = 0$ (d) $2x + y - 9 = 0$
     (e) $y = -5$ (f) $x = 5$

4. $1833

5. (b) $120   (c) $y = 4x - 200$ (d) $60.00   (e) rate of fine   (f) 50 km/h

6. (a) $(4, -1)$ (b) $(4, 3)$ (c) $(6, 7)$ (d) $(5, -6)$

7. 9 and 11

8. 600 L of 20%, 1400 L of 30%

REVIEW AND PREVIEW TO CHAPTER 4

DISTANCE BETWEEN TWO POINTS

1. (a) $2\sqrt{5}$ (b) $7\sqrt{2}$ (c) $10\sqrt{2}$ (d) $2\sqrt{17}$ (e) $\sqrt{277}$ (f) $2\sqrt{109}$
     (g) $5\sqrt{5}$ (h) 17   (i) 10   (j) 8   (k) $5\sqrt{5}$ (l) $\sqrt{4.5}$
     (m) $\sqrt{130}$ (n) $\sqrt{6.29}$ (o) $\sqrt{34}$ (p) $\sqrt{a^2 + b^2}$

4. (a), (d)   5. Answers vary   6. $(9,0)$, $(-3,0)$ 7. $(11,0)$, $(-13,0)$ 8. $(0, -3)$, $(0, -11)$

10. $(0, -4)$ 12. $\sqrt{277}$ 13. $15 + \sqrt{65}$ 14. 22

15. $\sqrt{34} + \sqrt{37} + 5\sqrt{2} + \sqrt{13}$ 16. yes   17. no

18. $a + b + \sqrt{a^2 + b^2}$ 19. 48

EXERCISE 4.1

1. (a) $\frac{4}{7}$ (b) 12   (c) 16   (d) $\frac{4}{7}$ (e) $\frac{11}{9}$ (f) $\dfrac{k - 10}{10}$

     (g) $\dfrac{1}{c}$ (h) $4(x + 1)$ (i) $\dfrac{6}{r}$ (j) $\frac{2}{7}$ (k) ± 3 (l) 4, 3

2. (a) 15   (b) $\frac{27}{10}$ (c) 9   (d) 13   (e) $\frac{27}{4}$ (f) $\frac{28}{15}$

     (g) $\frac{19}{3}$ (h) 4   (i) 4   (j) $-\frac{17}{4}$ (k) $\dfrac{as}{r}$ (l) $\frac{1}{4}$

     (m) ± 5 (n) -7

3. (a) $\frac{10}{3}$, 63   (b) 90, 126   (c) $\frac{1}{5}, \frac{2}{15}$ (d) ± 16, ± 2

4. (a) $\frac{2}{5}$ (b) $\frac{9}{25}$ (c) $\frac{8}{11}$

5. (a) $\frac{5}{3}$ (b) $\frac{2}{7}$ (c) $\pm \frac{5}{4}$ (d) 7

6. (a) ± 12 (b) ± 10 (c) ± 2 (d) $\pm 3\sqrt{7}$ (e) $\pm x^2$ (f) $\pm 6x$

7. (a) $\frac{28}{3}$ (b) 20   (c) 1   (d) $\dfrac{bc}{a}$

8. 15, 35   9. 72, 45   10. 40   11. 45   12. $\frac{25}{168}$

EXERCISE 4.2

1. (a) 6, 9   (b) 3, 5   (c) 14, 2   (d) 4, 12   (e) 15, 6   (f) 1, $\frac{3}{4}$

(g) 1, 15   (h) 4, 2

2. (a) $\frac{2}{3}$, 54   (b) 6, $\frac{63}{2}$ (c) 14, $\frac{7}{9}$ (d) $-\frac{1}{3}$, $\frac{10}{3}$ (e) $\frac{2}{3}$, $\frac{9}{2}$, 9   (f) $\frac{35}{2}$, 25, $\frac{65}{2}$

     (g) 4, 11   (h) 4, -2

3. $\frac{2}{3}$

4. 2.5 kg nitrogen, 1.5 kg phosphorus

5. (a) 416$\frac{2}{3}$ g peanuts, 166$\frac{2}{3}$ g almonds   (b) 250 g peanuts, 150 g cashews, 100 g almonds

6. 42, 18, 12   7. 45, 60, 90, 135   8. 20°, 60°, 100°

EXERCISE 4.3

1. 9927   2. 1154   3. 342 km   4. 166.25 N   5. 40 g   6. 610 g

7. 160 g   8. 63 kg   9. 1800 km   10. 90.1 kg   11. 800 m³   12. 8, 12, 16

13. 270 kg : 180 kg : 45 kg   14. 218.5 g : 9.5 g : 114 g : 152 g

EXERCISE 4.4

1. (a) (3, 5)   (b) (5, 11)   (c) $(-4, -3)$ (d) $(-4, 6\frac{1}{2})$ (e) $(-4\frac{1}{2}, -\frac{1}{2})$

     (f) $(5, -6\frac{1}{2})$ (g) $(-4, 5)$ (h) $(3, -2\frac{1}{2})$ (i) $(1.3, -0.8)$ (j) $(-1.3, -4.8)$

2. $(-2, 9)$ 3. $(-11, -4)$ 4. $(-5, 3)$, $(-2, -2)$, $(1, -7)$ 5. $(2\sqrt{2}, 3\sqrt{3})$

6. 6, 1   7. $3x + 2y - 7 = 0$ 9. Answers vary   10. (4, 6), (6, 9)

EXERCISE 4.5

2. 12, 9   3. 10, 26   4. (a) $2\frac{11}{12}$ cm, $2\frac{1}{3}$ cm   (b) $4\frac{4}{7}$ cm, $5\frac{1}{7}$ cm   (c) $9\frac{1}{6}$ cm, $8\frac{2}{5}$ cm

EXERCISE 4.6

1. (a) internally, 2:1   (b) externally, 3:2   (c) internally, 3:1

     (d) externally, 2:5   (e) internally, 4:5   (f) internally, 6:5

2. (a) internally, 5:6   (b) internally, 1:2   (c) externally, 3:5

     (d) externally, 6:5   (e) internally, 5:3   (f) internally, 1:1

3. (a) 1:3   (b) 2:9   (c) 3:1

4. 4 cm, 14 cm

5. 40 cm, 48 cm

6. 24 cm, 42 cm

7. (a) $(3, 2\frac{1}{3})$ (b) $(-1\frac{3}{5}, -4)$ (c) $(2\frac{3}{4}, -\frac{1}{4})$ (d) $(2\frac{5}{9}, \frac{5}{9})$ (e) $(1\frac{2}{5}, \frac{3}{5})$ (f) $(\frac{6}{7}, -2\frac{5}{7})$

8. (a) (16,7)   (b) $(6\frac{1}{2}, 4)$ (c) $(-11, 4)$ (d) $(24, -10)$ (e) $(-8\frac{1}{3}, -\frac{2}{3})$ (f) $(-4, 8\frac{2}{5})$

9. $(-1, 0)$, $(-4, -2)$

10. (20, 7)

11. $\left(\dfrac{ax_2 + bx_1}{a + b}, \dfrac{ay_2 + by_1}{a + b}\right)$

EXERCISE 4.7

1. 26, 14, 8   2. Answers vary

4.8 REVIEW EXERCISE

1. (a) $\frac{36}{7}$ (b) $\frac{30}{7}$ (c) $\frac{48}{5}$ (d) $\frac{5}{2}$ (e) $\frac{5}{7}$ (f) 14

2. (a) $\frac{7}{2}$, 2   (b) $\frac{45}{4}$, $\frac{27}{4}$ (c) $\frac{40}{3}$, $\frac{56}{3}$ (d) $\frac{4}{15}$, $\frac{6}{5}$

3. 72, 96, 192

4. 38, 76, 114, 171

5. (a) 896 g   (b) 4.5 g

6. 340 km

7. (a) 55 g   (b) 820 g

8. $120 000, $240 000, $480 000

9. (a) (6, 6)   (b) $(-4, 6)$ (c) $(-\frac{13}{2}, -\frac{13}{2})$ (d) (7, 4)

10. (2, 0)

11. $(-\frac{1}{2}, -\frac{3}{2})$, (3,3), $(\frac{13}{2}, \frac{15}{2})$

14. 39, 36   15. 12, 13.5   16. $\frac{24}{5}$, $\frac{28}{5}$ 17. 12 cm, 8 cm

4.9 CHAPTER 4 TEST

1. (a) $\frac{21}{2}$ (b) $\frac{17}{4}$ 2. (a) $\frac{6}{5}, \frac{9}{5}$ (b) $\frac{9}{2}, \frac{80}{3}$ 3. (a) (4, 11)   (b) (1, −9)

5. $\frac{40}{3}$, 15   6. 12.9 cm, 8.1 cm   7. $(\frac{14}{5}, \frac{16}{5})$ 8. (8, −5)

4.10 CUMULATIVE REVIEW FOR CHAPTERS 1 TO 4

1. (a) 0.8   (b) $0.\overline{27}$ (c) $0.\overline{571\,428}$
2. (a) $\frac{63}{200}$ (b) $\frac{4}{9}$ (c) $\frac{587}{990}$
3. (a) $\sqrt{6}$ (b) 2   (c) $\sqrt{5x}$
4. (a) $2\sqrt{2} + 12\sqrt{3}$ (b) $12\sqrt{3} - 14\sqrt{2}$ (c) $5\sqrt{6} + 25\sqrt{2}$
   (d) $6\sqrt{x} + 5\sqrt{y}$ (e) $2\sqrt{3} + 6$ (f) $2\sqrt{10} + 2\sqrt{30}$
6. (a) $9x^{10}$ (b) $-10x^7$ (c) $-8x^{12}$ (d) $4x^3$
7. $7092.60
8. $3153.97
9. 13 and 14
10. (a) $-4x - 68$ (b) $4x^2 - x$ (c) $20x^2 + 21x - 27$
    (d) $-2x^2 - 31x - 57$ (e) $2x^3 - 7x^2 - 10x + 24$ (f) $6x^4 - 2x^3 - 21x^2 - 9x - 1$
11. (a) $(m - 4)(m + 3)$ (b) $(2x - 3)(2x + 3)$ (c) $(3x + 5)(2x + 3)$
    (d) $(5x - 7)(3x - 2)$ (e) $(7x + 3)(4x - 5)$ (f) $(x - y - 4)(x - y + 4)$
    (g) $(a + b)(x + y)$ (h) $(x + 3)^2 - (m + 1)^2$
12. (a) $4x + 1$ (b) $2x + 2$ (c) $x^2 - 3x - 23$; R -54
    (d) $x^2 + 5x - 1$ (e) $x^2 - x - 3$
13. (a) $\dfrac{x + 3}{x - 1}$ (b) $\dfrac{t + 4}{t - 3}$
14. (a) $\dfrac{2x + 3}{x^2 + 3x + 2}$ (b) $\dfrac{t - 4}{t^2 - 3t + 2}$ (c) $\dfrac{4x + 10}{x^3 + 6x^2 + 11x + 6}$
    (d) $\dfrac{3x + 7}{x^3 - x^2 - 8x + 12}$
16. (a) $4x - y - 18 = 0$ (b) $x + y + 4 = 0$
17. (a) $2x - y - 2 = 0$ (b) $x - 2y + 3 = 0$
18. $4x - y - 5 = 0$
19. $3x + 2y + 13 = 0$
21. $550
22. (b) $C = 50n + 200$ (c) $850   (d) rate per person   (e) basic cost
23. (a) (3, 4)   (b) (−3, −1)   (c) (8, 9)   (d) (4, 2)
24. 361 and 243
25. (a) (1, −1, 2)   (b) (−2, −3, 4)
26. 400 L of 40%; 600 L of 30%
27. $(3, \frac{5}{3})$
28. (28, 36)
29. 160 000

REVIEW AND PREVIEW TO CHAPTER 5

ALGEBRAIC MANIPULATION

1. (a) $\dfrac{b - 9}{b^2 - 9}$ (b) $\dfrac{3x^2 + 20x - 32}{49 - x^2}$

2. $R = \dfrac{R_1 R_2}{R_1 + R_2}, R_1 = \dfrac{R_2 R}{R_2 - R}, R_2 = \dfrac{R_1 R}{R_1 - R}$

3. $P_1 = \dfrac{P_2 V_2 T_1}{V_1 T_2}, V_1 = \dfrac{P_2 V_2 T_1}{P_1 T_2}, T_1 = \dfrac{P_1 V_1 T_2}{P_2 V_2}; P_2 = \dfrac{P_1 V_1 T_2}{V_2 T_1}, V_2 = \dfrac{P_1 V_1 T_2}{P_2 T_1}, T_2 = \dfrac{P_2 V_2 T_1}{P_1 V_1}$

4. $x = \dfrac{y - 1}{y + 1}$

5. $y = \dfrac{z}{xz - 1}, z = \dfrac{y}{xy - 1}$

6. $m = \dfrac{2E}{v^2}, v = \sqrt{\dfrac{2E}{m}}$

7. $s = \dfrac{bc(m - 1)}{mc - b}, b = \dfrac{msc}{s + c(m - 1)}, c = \dfrac{sb}{b + m(s - b)}$

CONTINUED FRACTIONS

1. (a) -1
   (b) $\frac{43}{23}$
   (c) $\frac{2x + 3}{x + 2}$
   (d) $1 - x$
   (e) $\frac{3x - 1}{2x - 1}$
   (f) $\frac{x^2 + x + }{x}$

   (g) $\frac{2(x - 1)}{3(x + 1)}$
   (h) $\frac{x + 2}{x^2}$
   (i) $\frac{1}{x + 1}$

EXERCISE 5.1

1. (a) 2 (b) -1 (c) 8 (d) 2.6 (e) -13 (f) 3
   (g) 302 (h) -28 (i) $2 + 3a$ (j) $2 + 3b$
2. (a) -3 (b) 0 (c) 0 (d) -4 (e) 96 (f) 60
   (g) 60 (h) 1 (i) $a^2 - 4$ (j) $a^2 - 4$
3. (a) 1 (b) 28 (c) -13 (d) 27 (e) -3 (f) 10
   (g) 0 (h) 17 (i) 16
4. (a) A = {Frank, Ed, Suzanne, Gord}, B = {12, 14, 15, 17}
   (b) A = {Monique, Glen, Bob, Fred}, B = {3, 10, 16}
   (c) A = {-1, 0, 1, 2}, B = {-2, 0, 2, 3}
   (d) A = {-2, -1, 0, 1, 2}, B = {1}
5. (a) 13 (b) 5 (c) 260 (d) 5.69 (e) 40 (f) 488
   (g) 9.76 (h) 5.0201
6. (a) $-1\frac{1}{3}$ (b) $\frac{14}{17}$ (c) $\frac{6}{23}$ (d) $-\frac{7}{8}$ (e) $\frac{-16}{23}$ (f) $\frac{103}{195}$

   (g) $\frac{7}{36}$ (h) $\frac{a + 3}{2a - 5}$

7. (a) 3 (b) 13 (c) 273 (d) 8.3125
   (e) 2451 (f) 1333 (g) $3 + \sqrt{2}$ (h) 21.362 508 01
8. (a) 2 (b) 3 (c) 216 (d) 1024 (e) 2 (f) -4
9. (a) $3a^2 + a - 1$ (b) $3a^2 + 7a + 3$
   (c) $12a^2 + 2a - 1$ (d) $3(a^2 + 2ab + b^2) + a + b - 1$
   (e) $3a^2 - a - 1$ (f) $3x^2 - x - 1$
10. (a) 5 (b) 9 (c) 2.5
11. B = {791, 793, 796}
12. (a) B = {1, 2, 5} (b) B = {-4, -2, 0, 2, 4}
13. (a) $y \in R$ (b) $y \in R$
14. (a) {$x \in R \mid x \neq 3$} (b) {$x \in R \mid x \neq -2$} (c) {$x \in R \mid x \geq 2$} (d) {$x \in R \mid -3 \leq x \leq$
15. (a) 50 m (b) after 20 s
16. (a) 19 602.22 (b) 2.013 673 (c) 4312.842 (d) 2.695 262
17. (a) 1 (b) 95 (c) -8 (d) -4
   (e) $3x^2 - 11$ (f) $9x^2 + 24x + 11$

18. (a) $\frac{1}{8}$ (b) $1\frac{1}{7}$ (c) $\frac{3}{7}$ (d) $1\frac{3}{4}$ (e) $\frac{1}{x + 2}$ (f) $\frac{x + 2}{x + 1}$

19. (a) 3, -4 (b) 4, -5 (c) 6, -7
20. 1, 4
21. 2

EXERCISE 5.2

1. (i) (a) 0 (b) 2 (c) -2 (d) -1 (e) 1 (f) A = {-2, -1, 0, 1, 2}
      (g) B = {-2, 2, 0, -1, 1}
   (ii) (a) 2 (b) -1 (c) 0 (d) -1 (e) 0 (f) A = R
      (g) {$y \mid -1 \leq y \leq 2$}
   (iii) (a) -1 (b) -1 (c) -2 (d) 1 (e) 2 (f) $x \in R$
      (g) {$y \mid 1 \leq y \leq -2$} and {$y - 2 \leq y \leq -1$}

EXERCISE 5.3

1. (a) (i) up (ii) up (iii) down (iv) down (v) up (vi) up
      (vii) down (viii) down (ix) down
   (b) (i) (0, 0) (ii) (0, 3) (iii) (0, 0) (iv) (0, 5) (v) (0, -6) (vi) (0, 7)
      (vii) (0, -7) (viii) (0, 3.4) (ix) (0, 0)
   (c) (i) to (ix) x = 0
2. (a) $y = 4x^2$ (b) $y = -3x^2$ (c) $y = -\frac{1}{2}x^2$ (d) $y = x^2 + 4$ (e) $y = -2x^2 -$
   (f) $y = 6x^2 - 4$ (g) $y = -3x^2 + 6$ (h) $y = -\frac{1}{3}x^2$ (i) $y = 4x^2 - 7$
3. (a) {$y \mid y \geq 0$} (b) {$y \mid y \geq 0$} (c) $y \mid y \leq 0$} (d) {$y \mid y \leq 0$} (e) {$y \mid y \geq 5$}

(f) $\{y \mid y \geqslant -1\}$ (g) $\{y \mid y \leqslant 4\}$ (h) $\{y \mid y \geqslant 1\}$ (i) $\{y \mid y \geqslant -5\}$ (j) $\{y \mid y \geqslant -2\}$

(k) $\{y \mid y \leqslant -3\}$ (l) $\{y \mid y \leqslant 3\}$

4. (a) $y = 3x^2$ (b) $y = \frac{1}{4}x^2$ (c) $y = 3x^2 - 2$ (d) $y = 2x^2 + 3$

5. $y = 2x^2 - 5$

6. $y = \frac{1}{2}x^2 - 5$

EXERCISE 5.4

1. (a) (i) up   (ii) down   (iii) up   (iv) down   (v) down   (vi) up

     (vii) up   (viii) down   (ix) up   (x) up   (xi) down   (xii) up

   (b) (i) $(-1, 0)$ (ii) $(5, 0)$ (iii) $(4, 2)$ (iv) $(-6, 0)$ (v) $(1, 6)$ (vi) $(3, 0)$

     (viii) $(-6, -10)$ (viii) $(3, -3)$ (ix) $(\frac{1}{2}, 0)$ (x) $(3, -0.7)$ (xi) $(-5, -10)$ (xii) $(1, 8)$

   (c) (i) $x = -1$ (ii) $x = 5$ (iii) $x = 4$ (iv) $x = -6$ (v) $x = 1$ (vi) $x = 3$

     (vii) $x = -6$ (viii) $x = 3$ (ix) $x = \frac{1}{2}$ (x) $x = 3$ (xi) $x = -5$ (xii) $x = 1$

2. (a) $y = 2(x - 5)^2$ (b) $y = -3(x + 4)^2$ (c) $y = \frac{1}{2}(x + 6)^2$

   (d) $y = (x - 3)^2 + 2$ (e) $y = -2(x + 3)^2 + 5$ (f) $y = 4(x - 4)^2 - 3$

   (g) $y = -5(x + 6)^2 - 7$ (h) $y = \frac{1}{3}(x + 5)^2$ (i) $y = -\frac{1}{2}(x - 6)^2 + 6$

   (j) $y = -11x^2 + 2$ (k) $y = -5(x + 7)^2 + 2$ (l) $y = 7(x + 3)^2 - 4$

3. (a) $\{y \mid y \geqslant 0\}$ (b) $\{y \mid y \leqslant 0\}$ (c) $\{y \mid y \geqslant 0\}$ (d) $\{y \mid y \leqslant 0\}$ (e) $\{y \mid y \geqslant 2\}$

   (f) $\{y \mid y \leqslant -1\}$ (g) $\{y \mid y \geqslant -5\}$ (h) $\{y \mid y \leqslant 4\}$ (i) $\{y \mid y \leqslant 6\}$ (j) $\{y \mid y \leqslant 3\}$

   (k) $\{y \mid y \geqslant 1\}$ (l) $\{y \mid y \geqslant 7\}$

4. (a) $y = (x - 3)^2 + 2$ (b) $y = -2(x + 1)^2 - 3$ (c) $y = 4(x + 3)^2 + 6$

   (d) $y = -3(x - 2)^2 - 4$ (e) $y = \frac{1}{2}(x - 2)^2 - 6$ (f) $y = -(x + 4)^2 + 3$

5. (a) 5   (b) 3, -1 (c) -4 (d) 1, 3   (e) 2, 6   (f) -7

6. (a) $a = 2, q = 2$ (b) $a = -2, q = 5$ (c) $a = 3, q = 5$

EXERCISE 5.5

1. (a) $y = (x + 4)^2 - 16$ (b) $y = (x + 3)^2 - 9$ (c) $y = (x - 6)^2 - 36$

   (d) $y = (x - 1)^2 - 1$ (e) $y = (x - 5)^2 - 25$ (f) $y = (x - 7)^2 - 49$

   (g) $y = (x + \frac{1}{2})^2 - \frac{1}{4}$ (h) $y = (x - \frac{3}{2})^2 - \frac{9}{4}$ (i) $y = (x - \frac{5}{2})^2 - \frac{25}{4}$

   (j) $y = (x + \frac{1}{4})^2 - \frac{1}{16}$ (k) $y = (x - \frac{3}{5})^2 - \frac{9}{25}$ (l) $y = (x - 0.4x)^2 - 0.16$

   (m) $y = (x + \frac{1}{8})^2 - \frac{1}{64}$ (n) $y = (x + \frac{1}{3})^2 - \frac{1}{9}$ (o) $y = (x + 0.1)^2 - 0.01$

   (p) $y = (x - 0.6)^2 - 0.36$ (q) $y = (x - \frac{1}{6})^2 - \frac{1}{36}$ (r) $y = (x + 1.2)^2 - 1.44$

2. (a) $\{y \mid y \geqslant -2\}$ (b) $\{y \mid y \geqslant -5\}$ (c) $\{y \mid y \geqslant -16\}$ (d) $\{y \mid y \geqslant -16\}$

   (e) $\{y \mid y \geqslant -\frac{9}{4}\}$ (f) $\{y \mid y \geqslant \frac{3}{4}\}$ (g) $\{y \mid y \geqslant -\frac{17}{16}\}$ (h) $\{y \mid y \geqslant -\frac{19}{9}\}$

   (i) $\{y \mid y \geqslant -\frac{9}{100}\}$

3. (a) $\{y \mid y \geqslant -3\}$ (b) $\{y \mid y \geqslant 1\}$ (c) $\{y \mid y \leqslant 0\}$ (d) $\{y \mid y \leqslant 12\}$

   (e) $\{y \mid y \leqslant \frac{29}{2}\}$ (f) $\{y \mid y \leqslant 11\}$ (g) $\{y \mid y \geqslant \frac{49}{4}\}$ (h) $\{y \mid y \geqslant -\frac{13}{2}\}$

   (i) $\{y \mid y \leqslant 7\}$ (j) $\{y \mid y \leqslant \frac{9}{4}\}$ (k) $\{y \mid y \leqslant 12\}$ (l) $\{y \mid y \geqslant \frac{13}{8}\}$

   (m) $\{y \mid y \geqslant \frac{31}{32}\}$ (n) $\{y \mid y \leqslant \frac{7}{24}\}$ (o) $\{y \mid y \leqslant 1.92\}$

EXERCISE 5.6

1. (a) $(3, -25)$ (b) $(-3, 31)$ (c) $(1, -1)$ (d) $(1, -2)$ (e) $(6, 15)$ (f) $(-5, -4)$

   (g) $(3, -\frac{7}{2})$ (h) $(\frac{1}{4}, \frac{3}{4})$

2. $(16, 16)$ 3. $(-3, 3)$ 4. 500 m   5. 150 m × 150 m   6. 200 m × 400 m

7. $700   8. $30   9. 200 m × 300 m   10. $29   11. 30   12. 54 m

EXERCISE 5.8

1. (a) $a = 2, b = 3, c = -7$ (b) $a = 2, b = -1, c = -3$ (c) $a = -1, b = -4, c = 0$

   (d) $a = 3, b = 0, c = -4$ (e) $a = 1, b = 2, c = -3$ (f) $a = 2, b = 0, c = 7$

   (g) $a = 5, b = -3, c = 7$ (h) $a = -4, b = 3, c = -2$ (i) $a = 7, b = 5, c = -3$

2. (a) $(\frac{1}{2}, -\frac{49}{4})$ (b) $(-1, 2)$ (c) $(1, -3)$ (d) $(-1, 6)$ (e) $(\frac{1}{4}, -\frac{39}{8})$ (f) $(\frac{1}{6}, \frac{47}{12})$

(g) (2, −1)   (h) (6, 8)   (i) (0, −7)   (j) $(\frac{3}{2}, -\frac{9}{2})$

3. (a) $y = x^2 - x + 2$ (b) $y = x^2 - 5x + 4$ (c) $y = x^2 + 3x - 2$
   (d) $y = x^2 - 3x + 2$ (e) $y = x^2 + 5x - 3$ (f) $y = -2x^2 - 3x + 4$
   (g) $y = \frac{1}{2}x^2 + x - 1$ (h) $y = 3x^2 + x + 2$

4. (a), (c), (e), (f), (h)

5.9 REVIEW EXERCISE

1. (a) −3   (b) 7   (c) 11   (d) 84   (e) 0.4   (f) 0
   (g) $3\frac{1}{4}$ (h) 52   (i) $1 - 2a$ (j) $b^2 + 3$

2. (a) 15   (b) 17
   (c) −3   (d) 1
   (e) Domain = {Ron, Glen, Sue}   (f) Range = {15, 16, 17}
   (g) Domain = $\{x \in R \mid -3 \leqslant x \leqslant 2\}$ (h) Range = $\{y \in R \mid -3 \leqslant y \leqslant 2\}$

3. (a) 1   (b) −1
   (c) 2   (d) 1
   (e) 1   (f) −1
   (g) Domain = $\{x \in R \mid -3 \leqslant x \leqslant 2\}$ (h) Range = $\{y \in R \mid -1 \leqslant y \leqslant 2\}$

4. (a) (i) up   (ii) up   (iii) down   (iv) down   (v) up
       (vi) up   (vii) down   (viii) down   (ix) down
   (b) (i) (0, 0)   (ii) (0, −4)   (iii) (0, 3)   (iv) (1, 0)   (v) (−3, 4)
       (vi) (4, −3)   (viii) (0, −5)   (viii) (−1, 2)   (ix) (4, −7)
   (c) (i) $x = 0$ (ii) $x = 0$ (iii) $x = 0$ (iv) $x = 1$ (v) $x = -3$
       (vi) $x = 4$ (viii) $x = 0$ (viii) $x = -1$ (ix) $x = 4$

5. (a) $(x - 4)^2 - 16$ (b) $(x - 6)^2 - 36$ (c) $(x - 5)^2 - 25$
   (d) $(x - \frac{1}{2})^2 - \frac{1}{4}$ (e) $(x + \frac{3}{2})^2 - \frac{9}{4}$ (f) $(x - \frac{2}{3})^2 - \frac{4}{9}$
   (g) $(x - 0.2)^2 - 0.04$ (h) $(x + \frac{1}{6})^2 - \frac{1}{36}$ (i) $(x + \frac{5}{2})^2 - \frac{25}{4}$

6. (a) 2   (b) $\frac{3}{5}$ (c) $-\frac{144}{145}$ (d) −194   (e) 1472   (f) $\frac{99}{101}$
   (g) $\pi - 2\pi^2 + 3\pi - 4$ (h) $-\frac{3}{5}$

7. (a) $2a^2 - 3a + 1$ (b) $2a^2 - 7a + 6$ (c) $2(a^2 - 2ab + b^2) - 3a + 3b + 1$
   (d) $18a^2 - 9a + 1$ (e) $2a^2 + 3a + 1$ (f) $2x^2 + 3x + 1$

11. (a) (1, 4)   (b) (2, 6)   (c) (1, −2)   (d) $(-1, \frac{13}{2})$ (e) $(-\frac{1}{4}, -\frac{7}{8})$ (f) (−2, −6)

12. 300 m × 600 m
13. 300 m × 600 m

5.10 CHAPTER 5 TEST

1. (a) 6   (b) −9   (c) −0.552   (d) $a^6 + a^4 - 3a^2$
3. (a) $x \in R$ (b) $x \in R$
4. (a) (−5, −9); $x = -5$ 5. $y = 3(x - 2)^2 - 1$ 6. $(\frac{3}{2}, \frac{33}{4})$ 7. 50, 50

REVIEW AND PREVIEW TO CHAPTER 6

FACTORING QUADRATIC EXPRESSIONS

1. (a) $(x + 4)(x + 3)$ (b) $(x + 2)(x + 5)$ (c) $(y - 5)(y - 2)$ (d) $(w - 5)(w - 3)$
   (e) $(x - 4)(x + 2)$ (f) $(s - 7)(s + 3)$ (g) $(x + 5)(x - 2)$ (h) $(x - 4)(x + 4)$
   (i) $(x - 5)(x + 5)$ (j) $(x + 5)(x + 5)$ (k) $(x - 7)(x - 7)$ (l) $(w + 10)(w - 7)$
   (m) $(x + 5)(x - 3)$ (n) $(t - 4)(t + 3)$ (o) $(r + 6)(r - 4)$ (p) $(w - 9)(w + 5)$
   (q) $(t - 1)(t - 1)$ (r) $(x + 10)(x - 3)$ (s) $(x + 7)(x + 4)$ (t) $(w - 10)(w - 4)$
   (u) $(x + 9)(x - 3)$ (v) $(t - 5)(t + 4)$ (w) $(x + 11)(x - 8)$ (x) $(x - 10)(x + 10)$
2. (a) $(2x + 1)(x + 3)$ (b) $(2x - 5)(x - 1)$ (c) $(2w - 3)(3w + 1)$ (d) $(3w + 4)(w - 5)$
   (e) $(3y - 1)(2y + 1)$ (f) impossible   (g) $(2x + 3)(2x + 3)$ (h) $(2w - 1)(5w + 2)$
   (i) $(2w + 5)(w + 2)$ (j) $(2x - 3)(2x + 3)$ (k) $(6t + 5)(5t - 4)$ (l) $(7s + 3)(2s + 5)$
   (m) $(6x - 7)(4x - 3)$ (n) $(3w + 5)(4w + 3)$ (o) $(4t - 3)(3t - 4)$ (p) $(2x + 5)(2x + 5)$
   (q) $(5 - 3x)(2 - x)$ (r) impossible

INEQUALITIES

1. (a) −22   (b) −14   (c) −3   (d) $\frac{10}{7}$ (e) 1   (f) 4
   (g) −5   (h) $\frac{9}{5}$ (i) 6   (j) 17   (k) 0   (l) $\frac{14}{3}$

(m) $\frac{15}{14}$ (n) -2 (o) $\frac{19}{11}$ (p) $\frac{17}{13}$

2. (a) $x > 1$ (b) $w < 6$ (c) $t \leqslant 5$ (d) $x \geqslant -12$ (e) $t < \frac{9}{13}$ (f) $x < -6$

(g) $x \leqslant \frac{1}{2}$ (h) $x \leqslant 8$

RADICALS
1. (a) $4\sqrt{2} + 3\sqrt{3}$ (b) $-\sqrt{2}$ (c) $2\sqrt{3} - 18\sqrt{5}$
   (d) $16\sqrt{10} - 7\sqrt{3}$ (e) $19\sqrt{7} - 12\sqrt{6}$ (f) $-2\sqrt{17} - 13\sqrt{13}$
   (g) $53\sqrt{3}$ (h) $6\sqrt{5}$
2. (a) $15\sqrt{6} - 20\sqrt{3} - 8\sqrt{2} + 12$ (b) $14 - 4\sqrt{6}$ (c) 147
   (d) -14 (e) $2\sqrt{15} - 18\sqrt{3}$ (f) 67
3. (a) $x + 2\sqrt{x} - 3$ (b) $x - 8\sqrt{x} + 16$ (c) $2x - 7\sqrt{x} + 3$
   (d) $x + 7 + 5\sqrt{x + 1}$ (e) $x - 4 + 2\sqrt{x - 5}$ (f) $9x + 19 - 6\sqrt{x + 2}$
   (g) $x - 2 - 2\sqrt{x - 3}$ (h) $x + 7 - 4\sqrt{x + 3}$ (i) $4x - 3 + 4\sqrt{x - 1}$

EQUATIONS
1. (a) 9 (b) -1 (c) 3 (d) 5 (e) 6 (f) 15
   (g) 10 (h) 30 (i) 44 (j) 8 (k) -13 (l) $-6\frac{2}{3}$
   (m) $-2\frac{1}{3}$ (n) $-3\frac{1}{4}$ (o) 0 (p) -10 (q) 0 (r) 13
2. (a) 30 (b) 17 (c) -2 (d) $3\frac{2}{3}$ (e) -79 (f) -34
   (g) $-\frac{2}{9}$ (h) 10.5 (i) 9.25 (j) -27 (k) $4\frac{6}{7}$ (l) 7
   (m) -13 (n) $\frac{1}{5}$ (o) 3

EXERCISE 6.1

1. (a) $-4, 2$ (b) $-5, 1$ (c) $-7, 1$ (d) $2, -1$ (e) $-3, -1$ (f) $4, 2$
   (g) $-5, 4$ (h) $-5, -1$ (i) $2, -2$ (j) $-5, -3$
2. (a) $1, -3$ (b) $3, -3$ (c) $\frac{1}{2}, -2$ (d) 3 (e) $-\frac{2}{3}, -1$ (f) $1, 5$
3. $-4 \leqslant x \leqslant -3, 3 \leqslant x \leqslant 4$

EXERCISE 6.2

1. (a) $-3, 1$ (b) $1, 4$ (c) $-5, 4$ (d) $7, -9$ (e) $11, 7$ (f) $-7, -8$
   (g) $\frac{1}{3}, -\frac{5}{3}$ (h) $\frac{3}{4}, 5$ (i) $-\frac{5}{2}, -\frac{10}{3}$ (j) $-\frac{7}{4}, -\frac{1}{3}$ (k) $-\frac{3}{8}, \frac{5}{7}$ (l) $-\frac{11}{9}, -\frac{14}{3}$
2. (a) $4, -3$ (b) $-6, -3$ (c) $5, -4$ (d) $-5, -3$ (e) $11, -7$ (f) $-9, -13$
   (g) $9, 14$ (h) -4 (i) $5, -8$
3. (a) $\frac{1}{2}, -2$ (b) $-\frac{1}{3}, -2$ (c) $\frac{5}{2}, 1$ (d) $\frac{2}{3}, \frac{1}{2}$ (e) $-\frac{3}{4}, -\frac{5}{3}$ (f) $\frac{3}{2}, -7$
   (g) $-\frac{8}{3}, 4$ (h) $\frac{5}{2}, \frac{-10}{3}$ (i) $-1, -\frac{5}{6}$ (j) $-\frac{1}{4}, -\frac{7}{2}$ (k) $5, -\frac{9}{4}$ (l) $-\frac{1}{2}, 7$
   (m) $\frac{2}{5}, -\frac{5}{3}$ (n) $\frac{3}{2}, -\frac{5}{4}$ (o) $-\frac{8}{5}, -3$
4. (a) $-\frac{3}{2}, 4$ (b) $-\frac{4}{5}, \frac{3}{2}$ (c) $\frac{2}{3}, \frac{5}{2}$ (d) $-\frac{1}{2}, 5$ (e) $-\frac{1}{2}, -4$ (f) $-\frac{4}{5}, \frac{5}{2}$
   (g) $-\frac{1}{10}, -\frac{7}{3}$ (h) $\frac{7}{8}, \frac{2}{7}$ (i) $\frac{9}{5}, -6$
5. (a) $\frac{5}{2}, 1$ (b) $-3, 2$ (c) $\frac{1}{2}, -5$ (d) $-5, 3$ (e) $-\frac{4}{5}, 1$ (f) $-\frac{3}{2}, 1$
   (g) $\frac{1}{2}, -2$
6. (a) $-\frac{3}{2}, 2$ (b) $-\frac{3}{4}, 2$ (c) $-\frac{13}{7}, 2$ (d) $-30, 15$ (e) $\frac{4}{3}, 4$ (f) $-\frac{3}{7}, 2$
   (g) $3, -2$ (h) $\dfrac{-3 \pm \sqrt{5}}{4}$
7. (a) $x^2 - 7x + 12 = 0$ (b) $x^2 - 3x - 10 = 0$ (c) $x^2 + 11x + 28 = 0$
   (d) $6x^2 - 5x + 1 = 0$ (e) $20x^2 + 19x + 3 = 0$ (f) $x^2 - (r + s)x + (rs) = 0$

EXERCISE 6.3

1. (a) $0, -\frac{7}{2}$ (b) $-4, 4$ (c) 0 (d) ± 2 (e) $0, \frac{2}{3}$ (f) $\pm\frac{5}{2}$
   (g) 0 (h) $0, \frac{4}{5}$ (i) $\pm\frac{3}{10}$ (j) ± 9 (k) $0, \frac{6}{7}$ (l) $\pm\frac{1}{5}$
   (m) $0, \frac{2}{3}$ (n) 0 (o) $0, -\frac{2}{3}$

EXERCISE 6.4

1. (a) 9   (b) 16   (c) 25   (d) 1   (e) 4   (f) 36
   (g) 81   (h) 121   (i) $\frac{9}{4}$ (j) $\frac{1}{4}$ (k) $\frac{49}{4}$ (l) $\frac{1}{16}$
   (m) $\frac{1}{9}$ (n) $\frac{1}{100}$ (o) $\frac{9}{100}$

2. (a) ± 4 (b) ± 6 (c) $\pm\sqrt{7}$ (d) $\pm\frac{5}{2}$ (e) ± 7 (f) ± 3
   (g) $\pm\sqrt{10}$ (h) $\pm 4\sqrt{2}$

3. (a) $-7, -1$ (b) $2 \pm\sqrt{3}$ (c) $-1 \pm\sqrt{7}$ (d) $5 \pm 2\sqrt{2}$ (e) $-7 \pm 3\sqrt{3}$ (f) $6 \pm 2\sqrt{3}$
   (g) $-\frac{1}{2} \pm\sqrt{6}$ (h) $\frac{1 \pm\sqrt{3}}{2}$ (i) $\frac{-1 \pm\sqrt{5}}{3}$ (j) $\frac{3 \pm\sqrt{7}}{4}$ (k) $\frac{-10 \pm\sqrt{6}}{4}$ (l) $\frac{4 \pm\sqrt{2}}{3}$

4. (a) $4, -2$ (b) $2 \pm\sqrt{3}$ (c) $-3 \pm\sqrt{11}$ (d) $-1 \pm\sqrt{12}$ (e) $-4 \pm\sqrt{11}$
   (f) $-5 \pm\sqrt{17}$ (g) $4, -1$ (h) $\frac{5 \pm\sqrt{17}}{2}$

5. (a) $\frac{-4 \pm\sqrt{6}}{2}$ (b) $\frac{4 \pm\sqrt{10}}{2}$ (c) $\frac{3 \pm\sqrt{3}}{3}$ (d) $\frac{-5 \pm\sqrt{65}}{10}$ (e) $\frac{5}{2}, -1$ (f) $\frac{3 \pm\sqrt{13}}{2}$
   (g) $-1 \pm\sqrt{3}$ (h) $\frac{-2 \pm\sqrt{10}}{3}$

6. (a) $-1 \pm\sqrt{c + 1}$ (b) $\frac{k \pm\sqrt{4 + k^2}}{2}$ (c) $\frac{-b \pm\sqrt{32 - b^2}}{2}$ (d) $\frac{-10 \pm\sqrt{28a - 25}}{2a}$
   (e) $\frac{1 \pm\sqrt{k^3 + 1}}{k}$

EXERCISE 6.5

1. (a) $a = 2, b = 7, c = -1$ (b) $a = 5, b = 4, c = -7$ (c) $a = 3, b = 2, c = -7$
   (d) $a = 9, b = -4, c = -7$ (e) $a = 1, b = -7, c = 1$ (f) $a = -2, b = -9, c = 4$
   (g) $a = 2, b = 0, c = 7$ (h) $a = 5, b = -9, c = 0$ (i) $a = -1, b = \sqrt{2}, c = 4$

2. (a) $-2, -4$ (b) $5, -3$ (c) $\frac{1}{2}, 1$ (d) $\frac{3}{5}, \frac{3}{2}$ (e) $0, \frac{3}{7}$ (f) $\pm\frac{4\sqrt{5}}{5}$
   (g) $-\frac{1}{2}$ (h) $1 \pm\sqrt{5}$ (i) $\frac{1 \pm\sqrt{21}}{2}$ (j) $-1 \pm\sqrt{7}$ (k) $\frac{-4 \pm\sqrt{22}}{2}$ (l) $\frac{1 \pm\sqrt{15}}{7}$

3. (a) $\frac{3}{2}, -1$ (b) $-3 \pm\sqrt{11}$ (c) $\frac{1 \pm\sqrt{7}}{3}$ (d) $4 \pm\sqrt{2}$ (e) $\frac{-4 \pm\sqrt{22}}{3}$ (f) $\frac{1 \pm\sqrt{11}}{5}$
   (g) $1 \pm\sqrt{11}$ (h) $\frac{-7 \pm 3\sqrt{5}}{2}$ (i) $\frac{-4 \pm\sqrt{19}}{3}$ (j) $\frac{-7 \pm\sqrt{61}}{6}$ (k) $\pm\frac{\sqrt{5}}{5}$ (l) $\frac{-1 \pm\sqrt{1}}{2}$

4. (a) $0.3, -0.2$ (b) $2.3, 1.2$ (c) $2.6, -5.6$ (d) $3.3, -0.2$ (e) $4, 0.5$ (f) $-9.5, 3$

5. (a) $1 \pm\sqrt{6}$ (b) $\frac{1 \pm\sqrt{7}}{2}$ (c) $\frac{3 \pm\sqrt{29}}{2}$ (d) $\frac{3 \pm\sqrt{29}}{2}$ (e) $\frac{1 \pm\sqrt{33}}{4}$ (f) $\frac{3 \pm\sqrt{65}}{4}$
   (g) $\frac{3 \pm\sqrt{5}}{2}$ (h) $2 \pm\sqrt{7}$ (i) $\frac{7 \pm\sqrt{105}}{4}$ (j) $7 \pm\sqrt{58}$ (k) $\frac{3 \pm\sqrt{57}}{3}$ (l) $\frac{-5 \pm\sqrt{}}{2}$
   (m) $\frac{7}{2}, -1$ (n) $4 \pm\sqrt{21}$

6. (a) $3.4495, -1.4495$ (b) $2.1861, -0.6861$ (c) $1.6861, -1.1861$ (d) $0.8860, -3.3860$
   (e) $1.2808, -0.7808$ (f) $4.5414, -1.5414$

7. (a) $\frac{3\sqrt{2}}{2}, -\sqrt{2}$ (b) $\frac{\sqrt{3} \pm\sqrt{11}}{4}$ (c) $\frac{\sqrt{6} \pm\sqrt{22}}{4}$ (d) $(k - 1), 1$

EXERCISE 6.6

1. $14, 16; -14, -16$ 2. $17, 19; -17, -19$ 3. $16, -17$ 4. $15, 16$
5. $11, 12, 13; -11, -12, -13$ 6. $21, -20$ 7. $4, 16$ 8. $11, 19$
9. $22, 24; -22, -24$ 10. $13, 15$ 11. 12   12. 9
13. $4, 6, 8; -4, -6, -8$ 14. $21, -18$ 15. $10, 17; -10, -17$ 16. $9, 12$ 17. $10, 2$
18. $7\ m \times 13\ m$ 19. 3.1 cm   20. 11.0 cm   21. 2.3 cm   22. 20 m   23. 10 m
24. 20 m   25. $60\ m \times 40\ m$ 26. 500 km/h   27. 90 km/h   28. 50 km/h   29. 600 km/h
30. (a) 45 m   (b) 6.3 s   (c) 7.4 s   31. 6 km/h

EXERCISE 6.7

1. (a) $5 + 3i$ (b) $5 + 3i$ (c) $8 + 2i$ (d) $1 + 2i$ (e) $3 - 7i$ (f) 5
2. (a) $2i$ (b) $5i$ (c) $i\sqrt{2}$ (d) $10i$ (e) $i\sqrt{5}$ (f) $2i\sqrt{5}$
   (g) $2i\sqrt{3}$ (h) $3i\sqrt{2}$ (i) $4i$ (j) -1 (k) -2 (l) -3

| (m) -5 | (n) $+5$ | (o) $-i$ | (p) 1 | (q) -7 | (r) -10 |

(s) 21 (t) i

3. (a) -5 (b) 2 (c) 2i (d) $7 + i$ (e) 4 (f) $7 - 3i$

(g) 15 (h) 1 (i) 0

4. (a) $1 \pm i\sqrt{3}$ (b) $1 \pm i\sqrt{5}$ (c) $1 \pm i\sqrt{6}$ (d) $1 \pm \sqrt{7}$ (e) $\dfrac{-4 \pm i\sqrt{2}}{2}$

(f) $\pm 3i$ (g) $\pm \dfrac{i\sqrt{14}}{2}$ (h) $\dfrac{1 \pm 3i}{5}$ (i) $\dfrac{1 \pm i\sqrt{5}}{3}$ (j) $\dfrac{1 \pm i\sqrt{13}}{7}$

(k) $\dfrac{-1 \pm i\sqrt{31}}{4}$ (l) $5, -\frac{1}{3}$

5. (a) $-1 \pm i$ (b) $1 \pm i$ (c) $3 \pm \sqrt{7}$ (d) $\dfrac{-1 \pm i\sqrt{7}}{4}$

EXERCISE 6.8

1. (a) imaginary (b) real, distinct (c) real, distinct (d) imaginary

(e) real, distinct (f) real, distinct (g) real, distinct (h) real, equal

(i) imaginary (j) real, distinct (k) imaginary (l) real, distinct

2. (a) 2 (b) 0 (c) 1

3. (a) 0 (b) 21 (c) -31 (d) -24 (e) 0 (f) 64

(g) -47 (h) 41 (i) 49 (j) 289 (k) -11 (l) 25

4. (a) real, distinct (b) real, distinct (c) imaginary (d) real, distinct

(e) real, equal (f) real, distinct (g) real, distinct (h) real, distinct

(i) real, distinct

5. (a) 4 (b) $k < -\frac{9}{8}$ (c) $k < 1$ (d) $k > \frac{3}{4}$

(e) $k > 8$ or $k < -8$ (f) $\pm\frac{1}{2}$ (g) $k < -\frac{4}{3}$ (h) 2

(i) $-1, -5$ (j) $k < -1$ or $k > 3$

6. (a) 0 or 4 (b) $k < 0$ or $k > 4$ (c) $0 < k < 4$

7. $a = c$

8. $y = \frac{1}{3}x^2 - 3x + 6$

EXERCISE 6.9

1. (a) $-6, 8$ (b) $3, -4$ (c) $4, \frac{3}{2}$ (d) $-\frac{4}{3}, -1$ (e) $0, -\frac{5}{6}$ (f) $-\frac{7}{2}, 0$

(g) $\frac{2}{3}, \frac{5}{3}$ (h) $-\frac{1}{6}, -5$ (i) $2, -\frac{4}{3}$ (j) $\frac{6}{5}, \frac{7}{5}$ (k) $\frac{2}{3}, \frac{7}{9}$ (l) $\frac{4}{3}, -\frac{7}{3}$

2. (a) $x^2 - 4x + 3 = 0$ (b) $x^2 + 5x + 6 = 0$ (c) $x^2 + 9x - 7 = 0$ (d) $x^2 - 9 = 0$

(e) $x^2 + 5x = 0$ (f) $x^2 + 7x - 7 = 0$

3. (a) $\frac{2}{3}, 2$ (b) $1, -11$ (c) $-2, -3$ (d) $-\frac{10}{3}, -\frac{10}{3}$

4. (a) $x^2 - 11x + 28 = 0$ (b) $x^2 + 3x - 18 = 0$ (c) $2x^2 + 7x - 4 = 0$

(d) $12x^2 + 11x + 2 = 0$ (e) $5x^2 - 3x = 0$ (f) $x^2 - 36 = 0$

(g) $x^2 - 2x - 2 = 0$ (h) $x^2 - 6x + 1 = 0$ (i) $x^2 + 4 = 0$

(j) $x^2 - 2x + 10 = 0$ (k) $4x^2 - 8x - 1 = 0$ (l) $9x^2 - 18x + 11 = 0$

5. $\frac{1}{2}, -7$

6. $x^2 - x - 8 = 0$

7. $\frac{1}{5}, -2$

8. $x^2 + 4x + 9 = 0$

9. $2 + \sqrt{2}, -4$

10. $5x^2 + 3x + 2 = 0$

11. (a) -2 (b) $\frac{5}{14}$ (c) 6 (d) -10

12. $x^2 - 12x + 4 = 0$

EXERCISE 6.10

1. (a) $\pm 3, \pm 2$ (b) $\pm 1, \pm 2i$ (c) 1, 2, 4 (d) $1, \dfrac{3 \pm \sqrt{5}}{2}$ (e) 4

(f) $\pm 2, \pm 1$

2. (a) $-1, 1, 3$ (b) $2, 3, -1, 6$ (c) $4, -1, 2, 1$ (d) $\pm 2, \pm i$ (e) $\pm\sqrt{6}, \pm 2i$

(f) 1, 2

3. (a) $1 \pm \sqrt{7}, 1$ (b) $\pm 3, \pm 2$ (c) $-2, 3$ (d) $\pm\dfrac{\sqrt{2}}{2}, \pm 3i$ (e) $\pm 2, \pm 1$

(f) $\frac{3}{2}$, -2, $\frac{1}{2}$, -1

4. (a) $\pm \sqrt{2 \pm \sqrt{3}}$ (b) $\pm \sqrt{5 \pm 8\sqrt{2}}$ (c) $\dfrac{-3 \pm \sqrt{3}}{2}$

(d) no real roots   (e) $\dfrac{-(5 + \sqrt{13}) \pm \sqrt{22 + 10\sqrt{13}}}{4}$

EXERCISE 6.11

1. (a) 8   (b) 17   (c) $\emptyset$ (d) 6   (e) 6   (f) -3
   (g) 6   (h) 2   (i) $\frac{9}{4}$
2. (a) 26   (b) 9   (c) 21   (d) $\emptyset$ (e) 3   (f) 8
   (g) $\emptyset$ (h) 25
3. (a) 2   (b) 10, 15   (c) 12   (d) 1, 2   (e) 10   (f) 2
   (g) 5   (h) $\emptyset$
4. (a) 1   (b) 2, 6   (c) 0   (d) 1   (e) $\emptyset$ (f) -5
5. (a) 28   (b) 10   (c) 3   (d) 16   (e) 2   (f) 1, -3
   (g) 0   (h) $\frac{1}{4}$
6. (a) $1 + \sqrt{2}$ (b) $1 + \sqrt{3}$

EXERCISE 6.12

3. (a) $\{x \mid x < -5\} \cup \{x \mid x > -1\}$ (b) $\{x \mid 2 < x < 4\}$
   (c) $\{x \mid -5 < x < 3\}$ (d) $\{x \mid x < -2\} \cup \{x \mid x > 2\}$
   (e) $\{x \mid x < -2\} \cup \{x \mid x > 0\}$ (f) $\{\ \}$
   (g) $\{x \mid x \leq -4\} \cup \{x \mid x \geq 3\}$ (h) $\{x \mid 3 \leq x \leq 4\}$
   (i) $\{x \mid -5 < x < -4\}$ (j) $\{x \mid x \leq -7\} \cup \{x \mid x \geq 2\}$
4. (a) $\{x \mid x < -4\} \cup \{x \mid x > -3\}$ (b) $\{x \mid x < 2\} \cup \{x \mid x > 3\}$
   (c) $\{x \mid x \leq 1\} \cup \{x \mid x \geq 4\}$ (d) $\{x \mid -3 \leq x \leq 4\}$
   (e) $\{x \mid -3 \leq x \leq 3\}$ (f) $\{x \mid x \leq 0\} \cup \{x \mid x \geq 5\}$
   (g) $\{x \mid x < -3\} \cup \{x \mid x > \frac{1}{2}\}$ (h) $\{x \mid \frac{3}{4} \leq x \leq 5\}$
   (i) $\{x \mid -\frac{3}{2} < x < 4\}$ (j) $\{x \mid x < 1\} \cup \{x \mid x > \frac{5}{2}\}$
5. (a) $\{x \mid -3 < x \leq \frac{5}{2}\}$ (b) $\{x \mid x < 1 - \sqrt{3}\} \cup \{x \mid x > 1 + \sqrt{3}\}$
   (c) $\{x \mid x < 1 - \sqrt{2}\} \cup \{x \mid x > 1 + \sqrt{2}\}$ (d) $\{x \mid -1 < x < 4\}$
   (e) $\{x \mid -\frac{11}{3} \leq x \leq 4\}$ (f) $\{x \mid x < -3\} \cup \{x \mid x > 1\}$
   (g) $\{x \mid x < -2\} \cup \{x \mid x > 6\}$ (h) $\{x \mid 2 < x \leq \frac{10}{3}\}$

6.13 REVIEW EXERCISE

1. (a) 6, -7 (b) -5, -11 (c) 10, -3 (d) $-\frac{1}{2}$, 4   (e) $\frac{4}{3}$, $\frac{5}{2}$ (f) $-\frac{7}{5}$, $-\frac{11}{3}$
   (g) 0, $\frac{4}{3}$ (h) 0, $-\frac{12}{5}$ (i) $-\frac{5}{4}$
2. (a) $3i$ (b) $2i\sqrt{2}$ (c) -3 (d) $-i$ (e) -6 (f) -14
   (g) -1 (h) -10
3. (a) real, distinct   (b) real, equal   (c) imaginary   (d) real, distinct
   (e) real, distinct
4. (a) 7   (b) 17   (c) -1 (d) $\frac{25}{4}$
5. (a) 5, 5   (b) $-\frac{3}{2}$, $-\frac{7}{2}$ (c) $\frac{6}{7}$, 0   (d) 0, $\frac{8}{13}$
6. (a) $x^2 - 6x - 2 = 0$ (b) $x^2 + 7x + 6 = 0$ (c) $x^2 - 5 = 0$ (d) $x^2 + 2x = 0$
7. (a) 6, $-\frac{5}{2}$ (b) $\pm\frac{2}{3}$ (c) 4, $\frac{1}{6}$ (d) $-\frac{6}{5}$, $-\frac{3}{2}$ (e) $\frac{5}{2}$, 1   (f) -3, -5
   (g) $-\frac{5}{3}$, $\frac{1}{3}$ (h) $\frac{5}{7}$, $\frac{1}{2}$ (i) $\frac{3}{8}$, $-\frac{5}{2}$
8. (a) 0, $-\frac{14}{5}$ (b) $\dfrac{5 \pm \sqrt{13}}{2}$ (c) $-\frac{7}{6}$, -1 (d) $\pm\dfrac{i\sqrt{10}}{2}$ (e) $\dfrac{2 \pm \sqrt{2}}{2}$ (f) $\dfrac{3 \pm i\sqrt{39}}{4}$
   (g) -1, 2   (h) $-\frac{1}{2}$, -4
9. (a) 1, $-\frac{1}{6}$ (b) $\dfrac{-1 \pm i\sqrt{3}}{2}$ (c) $\frac{5}{2}$, -1 (d) $-1 \pm i\sqrt{5}$
   (e) $\dfrac{-1 \pm \sqrt{13}}{2}$ (f) $\dfrac{-1 \pm i\sqrt{23}}{4}$ (g) $\dfrac{2 \pm i\sqrt{5}}{3}$ (h) $\dfrac{-3 \pm \sqrt{57}}{4}$

10. (a) $1 \pm \sqrt{10}$　　　(b) $0.4, -0.5$　　　(c) $-3 \pm \sqrt{11}$　　　(d) $\dfrac{5 \pm i\sqrt{47}}{4}$　　　(e) $-5, 2$　　　(f) $\dfrac{3 \pm \sqrt{105}}{24}$

11. $-9, -8$ or $8, 9$
12. $9, 10, 11$ or $-9, -10, -11$
13. 8
14. 18.4 m
15. 100 km/h
16. (a) real, distinct　　　(b) real, equal　　　(c) imaginary　　　(d) real, distinct
　　(e) imaginary　　　(f) imaginary
17. (a) $\frac{9}{4}$　　　(b) $k < \frac{7}{16}$　　　(c) $k > \frac{-13}{24}$
18. (a) $x^2 + 13x + 40 = 0$　　　(b) $12x^2 - x - 1 = 0$　　　(c) $x^2 - 4x - 8 = 0$
　　(d) $x^2 - 4x + 13 = 0$
19. $-\frac{1}{3}, -4$
20. $x^2 - 2x - 5 = 0$
21. (a) $\pm 1, \pm 3$　　　(b) $-2, 4, -1, 3$　　　(c) $1, 2 \pm \sqrt{3}$　　　(d) $-1, 4, \dfrac{3 \pm i\sqrt{7}}{2}$
22. (a) 3　　　(b) $\emptyset$　　　(c) 3　　　(d) 7　　　(e) 10　　　(f) $\emptyset$
23. (a) $\{x \mid -3 \leqslant x \leqslant 4\}$　　　(b) $\{x \mid x < 0\} \cup \{x \mid x > 6\}$　　　(c) $\{x \mid -2 < x < 5\}$
　　(d) $\{x \mid x \leqslant -5\} \cup \{x \mid x \geqslant -3\}$　　　(e) $\{x \mid x < -\frac{1}{2}\} \cup \{x \mid x > 3\}$　　　(f) $\{x \mid -\frac{2}{3} \leqslant x \leqslant 4\}$

6.14 CHAPTER 6 TEST

1. (a) $3, 5$　　　(b) $4, -\frac{1}{2}$　　　2. (a) $\dfrac{-2 \pm i\sqrt{13}}{6}$　　　(b) $-1 \pm i$　　　3. $\dfrac{3 \pm \sqrt{33}}{6}$

4. $2 \pm \dfrac{1}{\sqrt{2}}$　　　5. $11 + 10i$　　　6. (a) imaginary　　　(b) real, distinct　　7. $\pm 2\sqrt{2}$

8. $x^2 + 8x + 8 = 0$　　　9. $8, 9, 10$

REVIEW AND PREVIEW TO CHAPTER 7

DISTANCE BETWEEN TWO POINTS
1. (a) $\sqrt{281}$　　　(b) $\sqrt{178}$　　　(c) $5\sqrt{5}$　　　(d) 17　　　(e) 25　　　(f) $5\sqrt{2}$
　　(g) $\sqrt{205}$

MIDPOINT OF A LINE SEGMENT
1. (a) $(5, 4)$　　　(b) $(-5, 5)$　　　(c) $(-9, -8)$　　　(d) $(\frac{15}{2}, \frac{3}{2})$　　　(e) $(-\frac{7}{2}, -\frac{7}{2})$　　　(f) $(\frac{19}{2}, -\frac{21}{2})$

　　(g) $(-\frac{13}{2}, 14)$

SLOPE
1. (a) 1　　　(b) 0　　　(c) $-\frac{2}{3}$　　　(d) undefined　　　(e) $\frac{4}{37}$　　　(f) 1

　　(g) $\frac{15}{11}$

EQUATIONS OF LINES
1. $2x - y = 0$　　　2. $4x + y + 23 = 0$　　　3. $x - y + 1 = 0$　　　4. $x + 2y - 11 = 0$
5. $3x - y - 9 = 0$　　　6. $x + 2y + 19 = 0$　　　7. $2x + 3y = 0$　　　8. $x + 2y = 0$

LINEAR SYSTEMS
1. (a) $3, 4$　　　(b) $-3, 7$　　　(c) $-6, \frac{1}{2}$　　　(d) $-3, 0$

THE PYTHAGOREAN THEOREM
1. 11.40 cm　　　2. 15.56 cm　　　3. 13.75 cm　　　4. 12.04 cm
5. 7.21 cm　　　6. 3.46 cm　　　7. 2 cm

EXERCISE 7.1

1. (a) $x^2 + y^2 = 25$　　　(b) $x^2 + y^2 = 81$　　　(c) $x^2 + y^2 = 121$　　　(d) $x^2 + y^2 = 3$　　　(e) $x^2 + y^2 = 20$
　　(f) $x^2 + y^2 = m^2$
2. (a) 6　　　(b) 8　　　(c) 12　　　(d) 4　　　(e) $2\sqrt{6}$　　　(f) 1
3. (a) $y = 5$　　　(b) $x = -2$　　　(c) $x^2 + y^2 = 25$
　　(d) $(x - 2)^2 + (y - 5)^2 = 49$　　　(e) $y = 2$　　　(f) $y = x - 3$
4. (a) $x^2 + y^2 = 25$　　　(b) $x^2 + y^2 = 20$　　　(c) $x^2 + y^2 = 36$　　　(d) $x^2 + y^2 = 4$
5. $x^2 + y^2 = 20$
6. (c) 3　　　(f) 10

EXERCISE 7.2

1. (a) $(0, 0)$, 1   (b) $(0, 2)$, 5   (c) $(-3, 0)$, 6   (d) $(5, -4)$, 10   (e) $(2, -1)$, 4
    (f) (a, b), c
2. (a) $(x - 2)^2 + (y - 5)^2 = 7$ (b) $(x + 2)^2 + (y - 3)^2 = 25$ (c) $(x + 3)^2 + (y + 4)^2 = 25$
    (d) $(x - 5)^2 + (y + 2)^2 = 9$ (e) $x^2 + (y - 3)^2 = 36$ (f) $(x + 4)^2 + y^2 = 32$
    (g) $(x + 3)^2 + (y + 8)^2 = 100$ (h) $(x - a)^2 + (y - b)^2 = c^2$ (i) $(x - a)^2 + (y + b)^2 = c^2$
3. (a) $x^2 + y^2 = 25$ (b) $(x - 2)^2 + (y - 5)^2 = 9$ (c) $(x + 3)^2 + (y + 2)^2 = 100$
    (d) $x^2 + (y - 5)^2 = 90$ (e) $(x + 3)^2 + y^2 = 25$
4. (a) $(-3, 0)$; 6   (b) $(0, 2)$; 3   (c) $(-5, 4)$; 5   (d) $(1, 1)$; $\sqrt{5}$ (e) $(2, 1)$; 3
    (f) $(5, 6)$; 8   (g) $(-\frac{5}{2}, \frac{3}{2})$; $2\sqrt{2}$
5. (a) $(x - 2)^2 + y^2 = 25$ (b) $(x - 5)^2 + (y + 1)^2 = 25$

EXERCISE 7.3

1. (b) $y = -7x$
2. (b) $(-2, 4)$ (c) -2 (d) $\frac{1}{2}$
3. (a) —   (b) 4, $\sqrt{106}$ (c) $\sqrt{90}$
4. (a) $5\sqrt{3}$ (b) $\sqrt{76}$ (c) $2\sqrt{21}$ (d) $\sqrt{10}$ (e) $\sqrt{114}$ (f) 8
    (g) 37   (h) $2\sqrt{33}$
5. (b) $-\frac{4}{3}$ (c) $\frac{3}{4}$ (d) $4y = 3x + 25$
6. (a) $3y = -2x + 13$ (b) $5y = -3x + 34$ (c) $7y = -5x - 74$ (d) $2y = x - 10$
    (e) $y = 6$ (f) $3y = 5x + 34$ (g) $5y = x - 39$ (h) $4y = 3x$
8. (a) $2x + y = 0$
9. (a) $y = \dfrac{a + b}{a - b}x$
10. (a) $\left(\dfrac{p + q}{2}, \dfrac{q - p}{2}\right)$ (b) $\dfrac{q - p}{p + q}$ (c) $\dfrac{p + q}{q - p}$
11. (a) r, $\sqrt{b^2 + a^2}$ (b) $\sqrt{b^2 + a^2 - r^2}$
12. (a) $\dfrac{b}{a}$ (b) $-\dfrac{a}{b}$ (c) $by = -ax + a^2 + b^2$

EXERCISE 7.4

1. $(1, 2)$, $(-1, 0)$ 2. $(-1, -2)$, $(2, 1)$ 3. $(-1, -8)$, $(-2, -9)$ 4. $(3, 13)$, $(-2, -7)$
5. $(1, -2)$ 6. $(0, 5)$, $(5, 0)$ 7. $(0, -5)$, $(3, 4)$ 8. $(0, 2)$, $(3, 0)$
9. $(-1, 1)$, $(3, 9)$ 10. $(3, -4)$ 11. $(9, 6)$, $(1, 2)$
12. $(10 + i, 1 + i)$, $(10 - i, 1 - i)$ 13. $(-3, -1)$ 14. $\pm 5\sqrt{17}$ 15. -1

EXERCISE 7.5

1. $(0, \pm 5)$, $(4, \pm 3)$ 2. $(\pm 1, 0)$, $(0, -1)$ 3. $(-4, -3)$, $(-3, -4)$, $(3, 4)$, $(4, 3)$
4. $(\frac{1}{2}, 4)$, $(-\frac{1}{2}, -4)$, $(2, 1)$, $(-2, -1)$ 5. $(1, \pm 3)$, $(-1, \pm 3)$ 6. $(\pm i\sqrt{3}, -3)$, $(\pm\sqrt{2}, 2)$

7. $(1, 1)$, $(-1, 1)$ 8. $(0, 5)$, $(\pm 1, -4)$ 9. $(\pm 3, 0)$ 10. $\pm\sqrt{\dfrac{a + b}{a^2 + b^2}}$, $\pm\sqrt{\dfrac{b - a}{b^2 + a^2}}$

EXERCISE 7.6

3. no

7.7 REVIEW EXERCISE

1. (a) $(0, 0)$; 7   (b) $(0, 0)$; 8   (c) $(2, 1)$; 4   (d) $(-3, -5)$; 3   (e) $(7, -1)$; 5
    (f) $(2, 0)$; 9   (g) $(0 -5)$; 10
2. (a) $x^2 + y^2 = 16$ (b) $x^2 + y^2 = 5$ (c) $(x + 3)^2 + (y - 5)^2 = 49$
    (d) $(x - 4)^2 + (y + 2)^2 = 10$ (e) $x^2 + y^2 = 53$ (f) $(x - 3)^2 + (y - 3)^2 = 18$
3. (a) $(x + 3)^2 + (y - 4)^2 = 68$ (b) $x^2 + (y - 3)^2 = 82$ (c) $(x + 4)^2 + y^2 = 17$
    (d) $(x + 3)^2 + (y + 3)^2 = 18$
4. (a) $(-4, 3)$; 6   (b) $(1, -4)$; 2   (c) $(3, 4)$; 5
5. (b) $x + 3y - 10 = 0$ (c) $3x - y = 0$
6. (a) $4\sqrt{2}$ (b) $\sqrt{33}$ (c) $2\sqrt{10}$
7. (a) $5y = -x + 26$ (b) $2y = x + 15$
8. (a) $(-2, 0)$, $(1, 3)$ (b) $(0, 4)$, $(4, 0)$ (c) $(3, 4)$ (d) $(3, 0)$, $(1, -2)$

9. (a) $(\pm i\sqrt{5}, -5), (\pm 2, 4)$ (b) $(\pm 2\sqrt{5}, 0)$
10. (a) $(x - 2)^2 + (y + 5)^2 = 25$ (b) $(x + 4)^2 + (y - 2)^2 = 16$ (c) $(x - 1)^2 + (y - 1)^2 = 73$
11. 9, 13

7.8 CHAPTER 7 TEST

1. (a) $(0, 0)$; 9   (b) $(-4, 5)$; 6   2. (a) $x^2 + y^2 = 9$ (b) $(x + 2)^2 + (y - 1)^2 = 36$
3. $(4, -2)$; $\sqrt{31}$ 4. 5   5. $5x - 12y + 169 = 0$
6. $(3, 4), (-3, -4)$
7. $(0, -5), (3, 4), (-3, 4)$

REVIEW AND PREVIEW TO CHAPTER 8

FUNCTIONS

1. (a) -4 (b) -2 (c) -8
   (d) $-3\frac{7}{8}$ (e) -3.919 (f) $a^3 - 2a^2 + a - 4$
   (g) $8x^3 - 8x^2 + 2x - 4$ (h) $-x^3 - 2x^2 - x - 4$
2. (a) 3   (b) -24 (c) -93
   (d) $1\frac{15}{16}$ (e) $4\sqrt{2} - 4$ (f) 0.7984
   (g) $-4x - x^4$ (h) $8x - 16x^4$
3. (a) (i) 1   (ii) -1 (b) (i) 1   (ii) -2.5 (c) (i) 2   (ii) -2
   (d) (i) 0   (ii) 2   (e) (i) $\{x \mid -3 \le x \le 3\}$ (ii) $\{x \mid -2 \le x \le 3\}$
   (f) (i) $\{y \mid -2 \le y \le 2\}$ (ii) $\{y \mid -3 \le y \le 2\}$

LINEAR FUNCTIONS

1. (a) $y = 2x + 5$ (b) $y = -\frac{1}{2}x - 6$ (c) $y = -3x - 5$ (d) $y = 4x - \frac{7}{2}$ (e) $y = \frac{3}{7}y - \frac{4}{7}$
   (f) $y = x - 3$

QUADRATIC FUNCTIONS

1. (a) $y = x^2 + 2$, $x \in R$, Range $= \{y \mid y \ge 2\}$
   (b) $y = -x^2 + 1$, $x \in R$, Range $= \{y \mid y \le 1\}$
   (c) $y = (x + 1)^2 - 1$, $x \in R$, Range, $\{y \mid y \ge -1\}$
   (d) $y = -(x - \frac{1}{2})^2 + \frac{1}{4}$, $x \in R$, Range $= \{y \mid y \le \frac{1}{4}\}$
   (e) $y = (x - 4)^2 - 12$, $x \in R$, Range $= \{y \mid y \ge -12\}$
   (f) $y = (x + \frac{5}{2})^2 + \frac{15}{4}$, $x \in R$, Range $= \{y \mid y \ge \frac{15}{4}\}$
   (g) $y = -2(x - \frac{3}{2})^2 + \frac{23}{2}$, $x \in R$, Range $= \{y \mid y \le \frac{23}{2}\}$
   (h) $y = 3(x + 1)^2 + 5$, $x \in R$, Range $= \{y \mid y \ge 5\}$

TRANSFORMATIONS

1. (a) $(5, 6)$ (b) $(2, 10)$ (c) $(2, 12)$ (d) $(2, \frac{3}{2})$ (e) $(2, -6)$ (f) $(-2, 6)$
   (g) $(6, 2)$
2. (a) $(-1, 3)$ (b) $(4, 5)$ (c) $(4, -9)$ (d) $(4, -3\frac{1}{2})$ (e) $(4, 3)$ (f) $(-4, -3)$
   (g) $(-3, 4)$
3. (a) (i) $(3, 12)$ (ii) $(3, 3)$ (iii) $(6, 8)$ (iv) $(-2, 6)$ (v) $(3, 18)$ (vi) $(3, 3)$ (vii) $(-3, 6)$
   (viii) $(6, 3)$ (ix) $(3, -6)$ (x) $(3, -12)$
   (b) (i) $(-5, 8)$ (ii) $(-5, -1)$ (iii) $(-2, 4)$ (iv) $(-10, 2)$ (v) $(-5, 6)$ (vi) $(-5, 1)$
   (vii) $(5, 2)$ (viii) $(2, -5)$ (ix) $(-5, -2)$ (x) $(-5, -4)$

EXERCISE 8.1

1. (a) quadratic   (b) constant   (c) root   (d) linear   (e) rational
   (f) polynomial   (g) linear   (h) root   (i) polynomial   (j) rational
2. (a) $x \in R$ (b) $x \in R$ (c) $\{x \mid x \ne 4\}$
   (d) $\{x \mid x \ne -2\}$ (e) $\{x \mid x \ne 0, x \ne 1\}$ (f) $\{x \mid x \ne 4, x \ne -3\}$
   (g) $\{x \mid x \ne -1, x \ne -5\}$ (h) $\{x \mid x \ne \pm 1\}$ (i) $x \in R$
   (j) $\{x \mid x \ne 1, x \ne 2\}$ (k) $\{x \mid x \ge 0\}$ (l) $\{x \mid x \ge 0\}$
   (m) $\{x \mid x \ge 1\}$ (n) $\{x \mid x \ge -1\}$
5. (b) $c = \begin{cases} 2.50 & 0 < t \le 3 \\ 0.3t + 1.6 & t > 3 \end{cases}$ (c) \$3.70   (d) 10 min
6. (b) $c = \begin{cases} 1.00 & 0 < d \le 0.4 \\ 1.25d + 1.5 & d > 0.4 \end{cases}$ (c) \$19.75

7. (a) $\{x \mid x \neq 0, \frac{3}{4}, -\frac{3}{2}\}$ (b) $x \in R$ (c) $\{x \mid 0 \geqslant x \geqslant 2\}$

(d) $\{x \mid -2 \leqslant x \leqslant 1\}$

10. (b) No

EXERCISE 8.2

1. (a) odd   (b) even   (c) even   (d) neither   (e) odd   (f) even

2. (i) $3x^2 + 1$, even   (ii) $-x^3 - 2x$, odd   (iii) $-x^3 + 2x^2$, neither   (iv) $1 - x$, neither   (v) $-\dfrac{1}{x}$, odd

(vi) $x^4 + \dfrac{1}{x^2}$, even   (vii) $x^2 - \dfrac{1}{x}$, neither   (viii) $\dfrac{-x}{1 + x^2}$, odd

3. (a) (i) even   (ii) odd   (iii) odd   (iv) even

EXERCISE 8.3

1. (a) shift 8 units upward   (b) shift 3 units downward   (c) shift 3 units to the right

(d) shift 6 units to the left   (e) shift $\frac{1}{2}$ unit upward   (f) shift $\frac{1}{2}$ unit to the left

(g) shift 2 units to the right   (h) shift 2 units downward   (i) shift 1 unit upward

(j) shift 5 units to the left

EXERCISE 8.4

1. (a) vertical stretch
   (b) reflect in x-axis
   (c) vertical shrink
   (d) stretch and reflect in the x-axis
   (e) vertical shrink and reflect in the x-axis
   (f) vertical stretch
   (g) vertical stretch, then shift 1 unit upward
   (h) reflect in x-axis, then shift 6 units upward
   (i) shift 1 unit to the left, stretch vertically
   (j) shift 2 units to the right, stretch vertically
   (k) vertical shrink, then shift 5 units downward
   (l) vertical shrink, reflect in the x-axis, then shift 4 units upward

EXERCISE 8.5

1. (a) Domain = {elephant, monkey}, Range = {peanut, banana}
   (b) Domain = {3, 4, 5, 6}, Range = {17, 0, 1}
   (c) Domain = {π}, Range = {1, 2, 3}
   (b) is a function

2. (a) Domain = {−2, −1, 0, 1, 2}, Range = {1, 2, 3}
   (b) Domain = {−2, −1, 0, 1, 2}, Range = {1, 2, 3}
   (c) Domain = $\{x \in R \mid -2 \leqslant x \leqslant 2\}$, Range = $\{y \in R \mid -2 \leqslant y \leqslant 3\}$
   (d) Domain = $\{x \in R \mid -3 \leqslant x \leqslant 4\}$, Range = $\{y \in R \mid -2 \leqslant y \leqslant 3\}$
   (e) Domain = $\{x \in R \mid -2 \leqslant x \leqslant 3\}$, Range = $\{y \in R \mid -3 \leqslant y \leqslant 2\}$
   (f) Domain = $\{x \in R \mid -2 \leqslant x < 3\}$, Range = {−2, −1, 0, 1, 2}
   (a), (d), and (f) are functions

3. (a) {(Jane, Joe), (Judy, Jane), (Judy, Joe), (Jim, Judy), (Jim, Jane), (Jim, Joe), (John, Jim), (John, Judy), (John, Jane), (John, Joe)}
   Domain = {Jane, Judy, Jim, John}, Range = {Joe, Jane, Judy, Jim}
   (b) {(Jane, Joe), (Jane, Judy), (Jane, Jim), (Jane, John), (Judy, Joe), (Judy, Jane), (Judy, Jim), (Judy, John)}
   Domain = {Jane, Judy}, Range = {Joe, Jane, Judy, Jim, John}

4. (a) {(2, 2), (2, 4), (2, 6), (2, 8), (3, 3), (3, 6), (3, 9), (4, 4), (4, 8), (5, 5), (6, 6), (7, 7), (8, 8), (9, 9)}

6. (b) and (e) are functions

EXERCISE 8.6

1. (a) (i) {(2, 0), (3, 0), (6, 4), (2, 8)}
       (ii) {(−2, 1), (−1, 2), (0, 3), (1, 1)}
       (iii) {(1, −1), (2, −2), (3, −3)}
       (iv) {($\frac{1}{3}$, 3), ($\frac{1}{4}$, 4), (3, 3), (4, 4)}

   (b) (i) Domain = {2, 3, 6}, Range = {0, 4, 8}
       (ii) Domain = {−2, −1, 0, 1}, Range = {1, 2, 3}
       (iii) Domain = {1, 2, 3}, Range = {−1, −2, −3}
       (iv) Domain = {$\frac{1}{3}$, $\frac{1}{4}$, 3, 4}, Range = {3, 4}

2. (a) yes
   (b) (i) Ann   (ii) Paul   (iii) Jack   (iv) Jean   (v) -2 (vi) -1 (vii) 3   (viii) -2
       (ix) 1   (x) 3   (xi) 3   (xii) 0
4. In (a) and (c), f^{-1} is a function.
5. In (a), (c), and (e), f^{-1} is a function.
6. (a) $f^{-1}(x) = \dfrac{x - 7}{5}$ (b) $f^{-1}(x) = \dfrac{2 - x}{3}$ (c) $f^{-1}(x) = \sqrt{x + 5}$ (d) $f^{-1}(x) = x^2$

   (e) $f^{-1}(x) = x^3$ (f) $f^{-1}(x) = x^{\frac{1}{3}} - 1$ (g) $f^{-1}(x) = \frac{1}{x} - 1$ (h) $f^{-1}(x) = (x - 1)^{\frac{1}{4}}$

EXERCISE 8.7
7. $N = 2^t$ 11. $D = K\sqrt[3]{N}$ (K a constant)   12. $m = C\ell^3$ (C a constant)

8.8 REVIEW EXERCISE
1. (a) polynomial   (b) polynomial   (c) root   (d) linear   (e) quadratic
   (f) rational   (g) constant
2. (a) shift 2 units to the left   (b) shift 2 units upward
   (c) vertical stretch   (d) reflect in x-axis
   (e) stretch and reflect in x-axis   (f) vertical shrink
   (g) shift 1 unit to the right   (h) vertical stretch, then shift 1 unit downward
3. (a) $\{x \mid -3 \leqslant x < 3\}$, $\{y \mid -3 \leqslant y < 3\}$ (b) $\{x \mid -2 \leqslant x < 2\}$, $\{y \mid -2 \leqslant y < 2\}$
   (c) $x \in R$, $\{y \mid -1 \leqslant y \leqslant 2\}$ (d) $\{x \mid -3 \leqslant x \leqslant 2\}$, $\{y \mid -2 \leqslant y \leqslant 3\}$
   (e) $\{x \mid -3 \leqslant x \leqslant 2\}$, $\{y \mid -2 \leqslant y \leqslant 3\}$ (f) $\{x \mid 1 \leqslant x \leqslant 2\}$, $\{y \mid 1 \leqslant y \leqslant 2\}$
   (c) and (e) represent functions
4. (a) $x \in R$ (b) $\{x \mid x \geqslant -2\}$ (c) $\{x \mid x \neq -\frac{1}{2}\}$ (d) $\{x \mid x \neq 0, 1\}$
8. (a) even   (b) neither   (c) odd
9. (a) $\{(2, 2), (4, 2), (4, 4), (6, 2), (6, 6), (8, 2), (8, 4), (8, 8), (10, 2), (10, 10), (12, 2), (12, 4), (12, 6), (12, 12)\}$
11. In (a) and (c) f^{-1} is a function.
12. (a) $f^{-1}(x) = 6x + 7$ (b) $f^{-1}(x) = \sqrt{\dfrac{x - 7}{3}}$ (c) $f^{-1}(x) = x^4 + 7$ (d) $f^{-1}(x) = \dfrac{3x + 1}{2 - x}$

8.9 CHAPTER 8 TEST
1. (a) $\{x \mid x \neq -6\}$ (b) $\{x \mid x \geqslant \frac{3}{2}\}$
3. (a) shift 5 units downward   (b) shift 5 units to the right
   (c) stretch and reflect in the x-axis   (d) vertical shrink, then shift 3 units upward
5. (a) $\{(4, 3), (6, 3), (6, 5), (8, 3), (8, 5), (8, 7), (10, 3), (10, 5), (10, 7), (10, 9)\}$
   (b) Domain $= \{4, 6, 8, 10\}$, Range $= \{3, 5, 7, 9\}$
6. (a) no
7. $f^{-1}(x) = \dfrac{x^2 - 1}{2}$

8.10 CUMULATIVE REVIEW FOR CHAPTERS 5 TO 8
1. (a) 115   (b) 91   (c) $-3\frac{1}{4}$ (d) $3x^2 - 2x - 5$

   (e) $3\pi^2 + 2\pi - 5$ (f) $3a^2 + 14a + 11$
3. (a) $-3, 2$
4. (a) $-0.6, 1.6$
5. (a) $(3, 1), x = 3$
6. (a) $-1 \pm \sqrt{7}$ (b) $\dfrac{-1 + i\sqrt{3}}{2}$ (c) $\dfrac{3 \pm i\sqrt{23}}{4}$ (d) $\dfrac{-4 \pm \sqrt{10}}{3}$

   (e) $1, \frac{1}{2}$ (f) $5 \pm i$

7. $2\frac{15}{16}$
8. No change
9. (a) $\dfrac{3 \pm \sqrt{5}}{2}$ (b) $\pm 2, \pm \sqrt{2}$ (c) 1.3819
10. 416.46 km/h
11. (a) real and equal   (b) imaginary   (c) real   (d) imaginary
    (e) real   (f) imaginary
12. $k < \frac{1}{4}$
13. $\{x \mid -3 < x < 4\}$

14. (a) $x^2 + y^2 = 36$ (b) $(x - 2)^2 + (y - 3)^2 = 1$ (c) $(x + 1)^2 + (y - 5)^2 = 2$
    (d) $(x - 2)^2 + (y + 4)^2 = 9$ (e) $(x + 6)^2 + y^2 = 100$ (f) $(x + 1)^2 + (y + 2)^2 = 5$
15. $(-2, 1)$, 4
16. centre at $(6, 3)$, radius is $\sqrt{5}$
17. $(1, -2)$
18. (a) $5x + 6y = 61$ (b) $3\sqrt{61}$
19. (b) $4y + x - 17 = 0$ (c) $y = 4x$
20. (a) $\{x \mid x \geqslant 2\}$ (b) $\{x \mid x \neq 0, -1\}$ (c) $\{x \mid x \neq -1, -3\}$ (d) $\{x \mid x \neq \pm 3\}$

REVIEW AND PREVIEW TO CHAPTER 9

EXPONENTS

1. (a) $6a^8$ (b) $4x^3$ (c) $8y^{12}$ (d) 1
2. (a) $15x^{-1}$ (b) $2xy^2$ (c) a^6b^9 (d) $\sqrt[3]{3b}$
3. (a) x^6 (b) 1 (c) x^2 (d) $2ab^2$
4. (a) 5 (b) 18 (c) 26 (d) 9 (e) $\frac{8}{9}$ (f) -5
   (g) $\frac{9}{4}$ (h) $\frac{1}{5}$ (i) 4
5. (a) 2 (b) -2 (c) -3 (d) 2 (e) 3 (f) 6
   (g) -3 (h) 4 (i) $\frac{7}{2}$ (j) 3 (k) 0 (l) 4

LINEAR SYSTEMS

1. (a) $-1, -2$ (b) 5, 1 (c) $-3, -4$ (d) $-3, 6$ (e) 8, 9 (f) 0.4, -1
   (g) 6, -10 (h) 1, -4

FUNCTIONS

1. (a) 14 (b) 32 (c) -13 (d) 2
2. (a) 17 (b) 121 (c) -35 (d) -51
3. (a) 8 (b) 32 (c) $\frac{1}{4}$ (d) 1
4. (a) 9 (b) 81 (c) $-\frac{1}{3}$ (d) $-\frac{1}{27}$
5. (a) 4 (b) 21 (c) -6 (d) -33
6. (a) 2 (b) $\frac{9}{7}$ (c) $\frac{1}{2}$ (d) $\frac{7}{9}$
7. (a) $\frac{26}{5}$ (b) $\frac{82}{9}$ (c) $-\frac{50}{7}$ (d) $-\frac{145}{12}$

EXERCISE 9.1

1. (a) 2, 4, 6, 8, 10 (b) $-1, -3, -5, -7, -9$ (c) 2, 4, 8, 16, 32
   (d) 1, 2, 4, 8, 16 (e) 1, 4, 9, 16, 25 (f) 6, 12, 24, 48, 96
   (g) 3, 5, 7, 9, 11 (h) 1, 3, 5, 7, 9 (i) $1, \frac{1}{2}, \frac{1}{3}, \frac{1}{4}, \frac{1}{5}$
   (j) $2, \frac{3}{2}, \frac{4}{3}, \frac{5}{4}, \frac{6}{5}$ (k) $\frac{1}{2}, \frac{1}{4}, \frac{1}{8}, \frac{1}{16}, \frac{1}{32}$ (l) $0, \frac{1}{4}, \frac{2}{8}, \frac{3}{16}, \frac{4}{32}$
2. (a) $t_n = 5n$; 25, 30, 35 (b) $t_n = 5^n$; 625, 3125, 15 625
   (c) $t_n = 2n$; 16, 32, 64 (d) $t_n = 1 + 2(n - 1)$; 9, 11, 13
   (e) $t_n = n^2$; 25, 36, 49 (f) $t_n = \frac{3^n}{3}$; 81, 243, 729
   (g) $t_n = \frac{1}{n}$; $\frac{1}{5}, \frac{1}{6}, \frac{1}{7}$ (h) –
   (i) $t_n = 2(-\frac{1}{2})^{n-1}$; $\frac{1}{8}, -\frac{1}{16}, \frac{1}{32}$ (j) $t_n = an$; 5a, 6a, 7a
   (k) $t_n = ar^{n-1}$; ar^4, ar^5, ar^6 (l) $t_n = a + (n - 1)d$; $1 + 3d, 1 + 4d, 1 + 5d$
3. (a) 3, 6, 9, 12, 15 (b) $-3, -1, 1, 3, 5$ (c) 0, 3, 8, 15, 24
   (d) $3, 3^2, 3^3, 3^4, 3^5$ (e) $1, 3, 3^2, 3^3, 3^4$ (f) 2, 8, 26, 80, 242
   (g) $-1, 1, -1, 1, -1$ (h) 3, $-6, 9, -12, 15$ (i) $-1, 1, 3, 5, 7$
   (j) $0, \frac{1}{3}, \frac{1}{2}, \frac{3}{5}, \frac{2}{3}$ (k) 1, 3, 6, 10, 15 (l) $0, \frac{2}{5}, \frac{6}{7}, \frac{4}{3}, \frac{20}{11}$
4. (a) 1, 3, 5, 7, 9 (b) 2, 4, 6, 8, 10 (c) $-2, 4, -8, 16, -32$
   (d) $2, \frac{4}{3}, \frac{6}{5}, \frac{8}{7}, \frac{10}{9}$ (e) $1, \frac{1}{2}, \frac{1}{3}, \frac{1}{4}, \frac{1}{5}$ (f) $\frac{1}{2}, \frac{1}{4}, \frac{1}{8}, \frac{1}{16}, \frac{1}{32}$
   (g) 2, 7, 12, 17, 22 (h) 1, 4, 9, 16, 25 (i) 2, 4, 8, 16, 32
   (j) $0, \frac{1}{3}, \frac{1}{2}, \frac{3}{5}, \frac{2}{3}$ (k) 0, 1, 3, 6, 10 (l) $0, \frac{3}{2}, \frac{8}{3}, \frac{15}{4}, \frac{24}{5}$
5. (a) n (b) $5 - n$ (c) $2(-\frac{1}{2})^{n-1}$ (d) 2^{n-4}
   (e) 2^{n+1} (f) $7 - 3n$ (g) $(-1)^{n+1}$ (h) $4n - 2$

(i) $(2)(3)^{n-1}$ (j) x^{n-1} (k) $(4 - n)a + nb$ (l) $a^{4-n}b^n$

6. 3, 1, -3, -11, -27
7. $t_n = 4n + 1$
8. (a) 2, 5, 8, 11, 14   (b) $t_n = 3n - 1$
   (c) (i) 74   (ii) 2999
9. (a) 3, -2, -7, -12, -17 (b) $8 - 5n$
   (c) (i) -242 (ii) -2492
10. (a) 3, 5, 7, 9, 11   (b) 1, 6, 16, 36, 76   (c) 3, 1, -1, -3, -5 (d) 0, 2, 4, 6, 8
    (e) 2, 6, 12, 20, 30   (f) 1, 1, 2, 3, 5

EXERCISE 9.2

1. (b), (c), (e), (h), (j), (l)
2. (a) 1, 5, 9, 13, 17   (b) 3, 8, 13, 18, 23
   (c) -8, -11, -14, -17, -20 (d) x, $x + y$, $x + 2y$, $x + 3y$, $x + 4y$
   (e) $x + 1, 2x + 3, 3x + 5, 4x + 7, 5x + 9$ (f) $5, 5 + x, 5 + 2x, 5 + 3x, 5 + 4x$
   (g) $5m, 8m, 11m, 14m, 17m$ (h) $3, 4 + x, 5 + 2x, 6 + 3x, 7 + 4x$
   (i) $5 + x, 8 + x, 11 + x, 14 + x, 17 + x$ (j) $2a, a, o, -a, -2a$
   (k) 0.5, 2.0, 3.5, 5.0, 6.5   (l) $3a, 2 + 2a, 4 + a, 6, 8 - a$
3. (a) 26, 94   (b) 70, 210   (c) 52, 703
   (d) -78, -414 (e) 4, 232   (f) $a + 12b, a + 184b$
   (g) $21x, 201x$ (h) $a - 6b, a - 103b$ (i) $67, 1 + 3n$
   (j) $289, -17 + 6n$ (k) $205, -5 + 7n$ (l) $a + 138, a + (n - 1)6$
4. (a) $a = 4, d = 3, t_n = 1 + 3n$ (b) $a = -3, d = 5, t_n = 8 + 5n$
   (c) $a = 42, d = 2, t_n = 40 + 2n$ (d) $a = -19, d = 7, t_n = -26 + 7n$
   (e) $a = 67, d = -5, t_n = 72 - 5n$ (f) $a = -\dfrac{128}{13}, d = -\dfrac{33}{13}, t_n = -\dfrac{95}{13} - \dfrac{33n}{13}$
   (g) $a = -277, d = -\dfrac{10}{3}, t_n = -\dfrac{821}{3} - \dfrac{10n}{3}$ (h) $a = 3 - 22x, d = \dfrac{9}{2}x, t_n = 3 + \dfrac{9nx}{2} - \dfrac{53}{2}x$
5. (a) 64   (b) 56   (c) 32   (d) 31   (e) 165   (f) 27
   (g) 129   (h) 28
6. 146
7. 89
8. 55
9. (a) 1090, 1180, 1270, 1360   (b) $4600
10. $\dfrac{1871}{13}$
11. 5
12. 0 or 4
13. 5
14. (a) 7, 9, 11, 13, 15   (b) $2k + 5$ (c) $2k + 3$ (d) 2
15. (a) 14   (b) 11, 17
   (c) 18, 21, 24, 27, 30, 33, 36   (d) $\dfrac{9}{7}, -\dfrac{45}{7}, -\dfrac{99}{7}, -\dfrac{153}{7}, -\dfrac{207}{7}, -\dfrac{261}{7}$
16. 6, 12
17. 7 m
18. 18 a
19. 5
20. $7.20, $1.50
21. $30 000
22. (a) 750   (b) 1350

EXERCISE 9.3

1. (b), (e), (f), (g), (i), (j), (k)
2. (a) $32, 2^n$ (b) $3125, 5^{n-1}$ (c) $96, (3)(2)^{k-1}$ (d) $1, 2^{6-k}$
   (e) $32, 2^n(-1)^{n-1}$ (f) $1, (-1)^{n-1} 2^{7-n}$ (g) $-\dfrac{1}{27}, (-3)^{5-n}$ (h) $\dfrac{1}{64}, (\dfrac{1}{2})^n$
   (i) $(2x)^{10}, (2x)^n$ (j) $\left(\dfrac{x}{2}\right)^7, \left(\dfrac{x}{2}\right)^{n-1}$ (k) x^{44}, x^{94} (l) $-3x^{-9}, -3x^{-49}$
3. (a) 6   (b) 9   (c) 9   (d) 10   (e) 7   (f) 8
   (g) 15   (h) 14   (i) 7   (j) 11   (k) 6   (l) 19
4. (a) $a = 4, r = 3, t_n = (4)(3)^{n-1}$
   (b) $a = 3, r = 2, t_n = (3)(2)^{n-1}$
   (c) $a = \pm 7, r = \pm 4, t_n = (7)(4)^{n-1}$ or $(-1)^n(7)(4)^{n-1}$
   (d) $a = 256, r = \dfrac{1}{2}, t_n = 2^{9-n}$

(e) $a = -243$, $r = \frac{1}{3}$, $t_n = (-1)^{n-1}(3)^{6-n}$

(f) $a = \pm 3$, $r = \pm 4$, $t_n = (3)(4)^{n-1}$ or $(-1)^n (3)(4)^{n-1}$

(g) $a = \pm \frac{4}{3}$, $r = \pm \sqrt{3}$, $t_n = \frac{4}{3}(-1)^{n-1}(3)^{n-1}$

(h) $a = \frac{3}{16}$, $r = 2$, $t_n = (3)(2)^{n-5}$

(i) $a = \dfrac{(2)(3)^{10}}{5^5}$, $r = \frac{5}{3}$, $t_n = (2)(3)^{11-n}(5)^{n-6}$

(j) $a = 5x^2$, $r = x^2$, $t_n = 5x^{2n}$

(k) $a = 1$, $r = 2x$, $t_n = (2x)^{n-1}$

(l) $a = \pm 128k^{10}$, $r = \pm \dfrac{1}{2k}$, $t_n = (-1)^{n-1}(2)^{8-n}k^{11-n}$

5. (a) 5   (b) 7   (c) 9   (d) 7   (e) 6   (f) 10
    (g) 8   (h) 13

6. -1, 25   7. $\frac{1}{5}$, 6   8. \$1512.63   9. \$327, \$356.43, \$388.50, $300(1 + 0.09)^n$

10. \$680.24   11. 1024   12. (b) exponential   13. 10 617   14. 50%   15. 10%

16. (a) 35, 245, 1715, or -35, 245, -1715 (b) 81, 243, 729, or -81, 243, -729
    (c) 24, 12, 6, 3   (d) 486, 162, 54, 18, 6

17. (a) ± 4 (b) ± 30 (c) $\pm\sqrt{mn}$

18. $a = x + y$, $r = 2$

19. 64

20. (a) 523.3   (b) 350   (c) 2

EXERCISE 9.4

1. (a) $[2(1) + 1] + [2(2) + 1] + [2(3) + 1] + [2(4) + 1]$
    $= 24$

(b) $2(1) + 2(2) + 2(3) + 2(4) + 2(5) + 2(6)$
    $= 42$

(c) $\frac{1}{1} + \frac{1}{2} + \frac{1}{3} + \frac{1}{4} + \frac{1}{5}$
    $= 2\frac{17}{60}$

(d) $(-1)^1 + (-1)^2 + (-1)^3 + (-1)^4 + (-1)^5$
    $= -1$

(e) $2^1 + 2^2 + 2^3 + 2^4$
    $= 30$

(f) $\dfrac{1}{2(1) - 1} + \dfrac{1}{2(2) - 1} + \dfrac{1}{2(3) - 1} + \dfrac{1}{2(4) - 1} + \dfrac{1}{2(5) - 1}$
    $= 1\frac{248}{315}$

(g) $[2(2) - 1] + [2(3) - 1] + [2(4) - 1] + (2(5) - 1]$
    $= 24$

(h) $1(1 + 1) + 2(2 + 1) + 3(3 + 1) + 4(4 + 1) + 5(5 + 1) + 6(6 + 1)$
    $= 112$

2. (a) $1 + 0 - 1 - 2 - 3 = -5$ (b) $-1 + 4 - 9 + 16 = 10$
(c) $0 + 3 - 8 + 15 - 24 = -14$ (d) $2 + 5 + 8 + 11 + 14 = 40$
(e) $0 - 1 - 2 - 3 - 4 = -10$ (f) $0 + 1 + 4 + 9 = 14$
(g) $0 - 1 - 2 - 3 - 4 = -10$ (h) $-9 + 16 - 25 + 36 - 49 = -31$
(i) $1 + 2 + 4 + 8 + 16 = 31$ (j) $\frac{1}{2} + \frac{1}{3} + \frac{1}{4} + \frac{1}{5} + \frac{1}{6} = \frac{87}{60}$
(k) $3 + \frac{3}{2} + 1 + \frac{3}{4} + \frac{3}{5} = 6\frac{17}{20}$ (l) $0 + 1 + \frac{4}{3} + \frac{3}{2} + \frac{8}{5} + \frac{5}{3} = \frac{71}{10}$

3. (a) $\displaystyle\sum_{n=1}^{6} 2n$ (b) $\displaystyle\sum_{i=1}^{n} i^2$ (c) $\displaystyle\sum_{n=0}^{4} 3 + 2n$ (d) $\displaystyle\sum_{k=1}^{n} 2^{k-1}$

(e) $\displaystyle\sum_{k=1}^{n} 3k - 2$ (f) $\displaystyle\sum_{k=1}^{n} 3(2)^{k-1}$ (g) $\displaystyle\sum_{i=0}^{3} 5x^i$ (h) $\displaystyle\sum_{k=2}^{5} (k + 1)x$

(i) $\displaystyle\sum_{k=1}^{5} (-1)^{k+1} x^k$ (j) $\displaystyle\sum_{k=1}^{4} (k)a^k$ (k) $\displaystyle\sum_{k=1}^{4} \frac{1}{k + 1}$ (l) $\displaystyle\sum_{k=1}^{4} \frac{k}{x^k}$

EXERCISE 9.5

1. (a) 39   (b) 153   (c) -90 (d) 52   (e) 512   (f) 120x

2. (a) 1 001 000   (b) 1 000 000   (c) 500 500   (d) $\frac{n}{2}[1 + n]$ (e) 33 975
    (f) 100 500

3. (a) 1 499 500   (b) −989 000   (c) 1 503 500   (d) −999 000
4. (a) 495   (b) 1350   (c) 350   (d) 1638   (e) 50   (f) −225
    (g) 1250   (h) 488
5. (a) 20 200   (b) 7155   (c) 0   (d) 42   (e) 399   (f) −3630
    (g) 130.5   (h) −62.5
6. (a) 10 000   (b) 4 002 000   (c) 9560
7. (a) 12   (b) 30   (c) 20   (d) 25   (e) 6, 10   (f) 15
8. 78
9. second, $7.50
10. 925
11. $132 000

12. (a) $\frac{n}{2}[n + 1]$ (b) (i) 5050   (ii) 500 500   (iii) 2 001 000
13. (a) k^2 (b) $k(k + 1)$
14. 93
15. (a) 120   (b) 5   (c) 11

16. $\frac{n}{2}[3n − 1]$ 17. $n[−2n − 1]$ 18. 2, 5, 8, 11, 14   19. (a) 45   (b) 360   20. 816

EXERCISE 9.6

1. (a) 2550   (b) 242   (c) 2343   (d) −3280   (e) 504   (f) 1698
2. (a) 511   (b) 3280   (c) 342   (d) 2735   (e) 364.5   (f) 300
3. 126   4. 341   5. (a) $3\frac{51}{64}$ (b) $89\frac{7}{32}$ 6. 175   7. $19 687.50

8. (a) 150   (b) −109   (c) $363\frac{31}{32}$ (d) $\frac{5(5^{1000} − 1)}{4}$ (e) $1 − (\frac{1}{2})^{96}$
    (f) 25

EXERCISE 9.7

1. (a) 2   (b) 8   (c) 10   2. $\frac{2}{9}$ 3. 2

EXERCISE 9.8

2. (a) yes   (b) yes   (c) no   (d) yes   (e) yes   (f) no
    (g) yes   (h) yes   (i) no
3. (a) no   (b) no   (c) yes   (d) yes

9.9 REVIEW EXERCISE

1. (a) 3, 6, 9, 12, 15   (b) 0, −2, −4, −6, −8   (c) 1, 4, 9, 16, 25
    (d) 2, 4, 8, 16, 32   (e) 2, 1, $\frac{2}{3}$, $\frac{1}{2}$, $\frac{2}{5}$ (f) −1, 1, 3, 5, 7
2. (a) a$t_n = 4n + 1$, 25, 29, 33   (b) $t_n = 3^n$; 243, 729, 2187
    (c) $t_n = 128(\frac{1}{2})^{n-1}$; 4, 2, 1   (d) $t_n = 51 + (n − 1)(−5)$; 26, 21, 16
    (e) $t_n = (−2)^{n-1}$; −32, 64, −128
3. (a) −1, 2, 5, 8, 11   (b) 2, −3, −8, −13, −18   (c) 3, 9, 27, 81, 243
    (d) 1, 8, 27, 64, 125   (e) −3, 2, 7, 12, 17   (f) $−\frac{1}{3}$, 0, $\frac{1}{5}$, $\frac{1}{3}$, $\frac{3}{7}$
4. (a) 4, 11, 18, 25, 32   (b) 3, 0, −3, −6, −9   (c) 2, 5, 11, 20, 32
5. (a) 4, 7, 10   (b) −2, 0, 2   (c) −7, −8, −9   (d) 3, 3 + x, 3 + 2x
6. (a) $t_n = 2n + 2$, 18   (b) $t_n = 23 − 3n$, −10   (c) $t_n = −2n − 5$, −35
    (d) $t_n = 6n − 3$, 123   (e) $t_n = 5 − 6n$, −109
7. (a) $a = −3$, $d = 1$, $t_n = n − 4$ (b) $a = 7$, $d = 4$, $t_n = 4n + 3$
    (c) $a = 3$, $d = −1$, $t_n = 4 − n$ (d) $a = 4$, $d = −3$, $t_n = 7 − 3n$
8. (a) 24   (b) 11   (c) 31   (d) 49
9. 93
10. 29
11. $8 − 3n$

12. (a) $t_n = 2^n$, 32   (b) $t_n = (\frac{1}{3})^{n-4}$, $\frac{1}{9}$ (c) $t_n = (−3)^{n-1}$, −27
13. (a) $a = 3$, $r = 2$, $t_n = (3)(2)^{n-1}$ (b) $a = −2$, $r = 3$, $t_n = −2(3)^{n-1}$
    (c) $a = ±3$, $r = ±4$, $t_n = ±3(4)^{n-1}$
14. (a) 10   (b) 8
15. (a) $2 + \frac{3}{2} + \frac{4}{3} + \frac{5}{4} + \frac{6}{5} + \frac{7}{6} = 8\frac{9}{20}$ (b) $0 + \frac{1}{3} + \frac{2}{4} + \frac{3}{5} + \frac{4}{6} = 2\frac{1}{10}$

(c) $1 + 2 + 4 + 8 + 16 = 31$
16. (a) 374 750   (b) $-219\,500$
17. (a) 4020   (b) -2400 (c) 7880
18. (a) 2706   (b) -1860
19. (a) 510   (b) -182 (c) 1008
20. (a) 2555   (b) 3280
21. (a) 12, 94.5   (b) 250, 7812
22. (a) $-144, -88, -32, 24$ (b) 3, 6, 12, 24
23. $S_{64} = 2^{64} - 1$ 24. \$15 741.06   25. 1824 seats   26. 420 m

9.10 CHAPTER 9 TEST

1. (a) 4, 11, 18   (b) $-4, -6, -8$ 2. (a) 2, -4, 8   (b) $-1, -3, -9$
3. (a) 62   (b) 98   (c) $3n - 1$ 4. (a) 81   (b) 729   (c) 3^{n-1}
5. $a = -2, d = 4, t_{22} = 82$ 6. $a = \pm 3, r = \pm 2, t_7 = \pm 192$
7. 91   8. 30 000   9. 63   10. -820 11. -1023

REVIEW AND PREVIEW TO CHAPTER 10

PERCENT

1. (a) 0.15   (b) 0.25   (c) 0.18   (d) 0.025
2. (a) 35%   (b) 37.5%   (c) 62.5%   (d) 112.5%   (e) 25%   (f) 62.5%
    (g) 87.5%   (h) 137.5%
3. (a) 3.5   (b) 9   (c) 11   (d) 8   (e) 3.5   (f) 36.05
    (g) 88   (h) 26.75
4. (a) 25%   (b) 100%   (c) 133%   (d) 75%   (e) 20%   (f) 4%
5. (a) 48   (b) 320   (c) 458.3   (d) 416.7   (e) 100   (f) 80
    (g) 300
6. (a) equal   (b) 15% of 20   (c) 60% of 10   (d) 25% of 50
7. (a) \$8.99   (b) 12.5%   (c) \$31.56   (d) \$34.44   (e) \$59.08   (f) \$50
    (g) \$70.07
8. (a) \$42.38   (b) 26.5%

MARK-UP AND DISCOUNT

1. \$37.05   2. \$13 840.25   3. 11.6%
4. (a) 1.856%   (b) 2.735%   (c) 3.076%   (d) 2.051%   (e) 2.168%
5. \$23.96
6. (a) \$2580   (b) \$18 920
7. \$19.95   8. \$89.94

COMMISSION

1. (a) \$7770   (b) \$1942.50   (c) \$121 730
2. \$760.86
3. 1.5%
4. (a) \$540.00   (b) \$475.00   (c) \$510.00   (d) \$575.00   (e) \$610.00
5. \$183.75
6. (a) 0.4%   (b) 0.5%   (c) 0.3%   (d) 1.2%

SIMPLE INTEREST

1. (a) $\frac{1}{2}$ (b) $\frac{2}{3}$ (c) $\frac{6}{73}$ (d) $\frac{18}{73}$ (e) $\frac{80}{73}$ (f) $\frac{3}{2}$
2. (a) \$165, \$665   (b) \$252, \$1452   (c) \$862.50, \$3362.50
    (d) \$280, \$5880   (e) \$748.12, \$11 248.12
3. (a) \$8.22; \$1008.22   (b) \$39.04; \$2539.04   (c) \$40.68; \$1540.68   (d) \$94.68; \$3694.68

EXERCISE 10.1

1. (a) $n = 14; i = 0.05$ (b) $n = 40; i = 0.02$ (c) 9% per annum, semi-annual
    (d) 8%
2. (a) \$8144.47   (b) \$9766.95   (c) \$2745.58
3. (a) \$7764.85   (b) \$2764.85
4. (a) \$1430.77   (b) \$1425.76   (c) \$1418.52
5. (a) 9 a   (b) 8 a   (c) 7 a
6. 6 a   7. \$1693.85   8. \$6381.54   9. 4.44%   10. \$53 053.80

EXERCISE 10.2

1. (a) \$6219.64   (b) \$1682.43   (c) \$393.79   (d) \$1104.67   (e) \$14 167.43

2. $4419.36
3. $14 213.60
4. (a) $698.92   (b) $701.38   (c) $704.96
5. $1385.22   6. $3280.48   7. $3193.84   8. $8151.36   9. $7651.41   10. $11 521.51

EXERCISE 10.3

1. (a) $14 889.04   (b) $5814.07   (c) $2153.84   (d) $12 577.89
2. (a) $165.56   (b) $232.48   (c) $118.64   (d) $394.45
3. $487.50   4. $3406.16   5. $68 666.25   6. $2148.92   7. $2822.74
8. (a) $7512.90   (b) $12 655.19   (c) $41 689.20   (d) $4727.60
9. $945.95
10. $28 750.90

EXERCISE 10.4

1. (a) $4077.10   (b) $16 935.54   (c) $2258.06   (d) $38 608.68
2. (a) $505.23   (b) $803.70   (c) $263.61   (d) $867.08
3. $3782.79   4. $166.07   5. $3069.58   6. $65 039.68
7. (a) $42 023.81   (b) $39 023.81
8. $358.94   9. $17 866.35
10. (a) $286 748.37   (b) $911 748.37   (c) $911 748.37 vs $1 125 000

EXERCISE 10.5

2. May 1, 1988   3. (a) decrease   (b) increase   4. $2921.24   5. $2531.25
6. $876.02   7. $128.58   8. $7238.04   9. $59 238.21
10. (a) $3230.77   (b) $3640.52   11. $153.26   12. $1294.28

EXERCISE 10.6

1. $36.05   2. $242.46   3. $800.91   4. $361.51   5. $133.27
6. $7.24, $82.40, $7.40   7. (a) $188.29   (b) $188.29   (c) $960.00
8. $978.68   9. (a) $4628.55   (b) $7906.23

EXERCISE 10.7

1. (a) 9.2%   (b) 12.55%   (c) 15.56%   (d) 10.381%   (e) 8.243%
2. 0.16%   3. 0.683%   4. $7087.50
5. (a) 10%   (b) 10.25%   (c) 10.381%   (d) 10.471%
6. 9.27%   7. 16.08%

EXERCISE 10.8

1. 1.943%   2. 5.83%   3. 0.949%   4. (a) 2.643%   (b) 2.713%
5. 2.96%   6. 2% per quarter 7. 2.59%   8. 0.98%   9. 0.94%

EXERCISE 10.9

1. $8090.61   2. $4878.05   3. $3212.85   4. $1525   5. $15 542.28

10.11 REVIEW EXERCISE

1. (a) $4557.91   (b) $110.46   (c) $1389.16   (d) $3125.42   (e) $12 942.05
2. (a) $790.31   (b) $2468.11   (c) $1025.76   (d) $1024.19   (e) $393.79   (f) $270.32
3. (a) $629.86   (b) $121.55   (c) $2040.73   (d) $521.53   (e) $880.71
4. 7 a   5. 6.2%   6. $426.47   7. 8 a, 9 months
8. (i) by $3.98   9. $4889.80   10. $1987.96   11. $451.97
12. (a) $307.60   (b) $18.92
13. (a) $18 392.80   (b) $7097.25   (c) $3879.92   (d) $25 750.71
14. (a) $413.90   (b) $697.44   (c) $278.06   (d) $821.77
15. (a) $3440.98   (b) $32 795.58   (c) $12 231.19
16. (a) $582.78   (b) $1970.97   (c) $299.54
17. $1788.04   18. $580.36   19. $166.07   20. $7657.74   21. $3660.17   22. $88 148.14
23. (a) $2697.35   (b) $239.65
24. $486.25

10.12 CHAPTER 10 TEST

1. $13 665.16   2. $698.92   3. $287.48   4. $15 075.25
5. $2165.23   6. $522.76   7. $9 292.83   8. $2067.26

PERCENTS AND MONEY

1. (a) $2500   (b) $8625   (c) $8160   (d) $6355
2. (a) 40%   (b) 11.1%   (c) 33.3%   (d) 33.3%
3. (a) $500 000   (b) $433 333.33   (c) $1 125 000   (d) $763 636.36
4. (a) $2425   (b) $3040   (c) $2650   (d) $3220
5. (a) $2400   (b) $29 900   (c) $452.05   (d) $3698.63
6. (a) $5154   (b) $8700   (c) $7500   (d) $15 000

INVESTING MONEY

1. (a) $670.05   (b) $3174.34   (c) $212 032.46
2. (a) $767.90   (b) $3368.12   (c) $1904.76
3. (a) $5394.69   (b) $3444.03
4. (a) $3010.78   (b) $2497.76   (c) $5426.25
5. $348.21

INTERPOLATION

1. (a) 17   (b) 73   2. $805.75   3. $1016.92   4. 7.99% or 8%

EXERCISE 11.1

1. (a) $1000   (b) $1000   (c) $914.86   (d) $1067.10   (e) $871.65   (f) $1122.89
   (g) $935.82
2. (a) premium   (b) $100   (c) $1091.28   (d) $91.28
3. $1182.57   4. $830.49   5. $1317.82
6. (a) $829.73   (b) $701.23   (c) $1229.39   (d) 1543.62

EXERCISE 11.2

1. (a) 9%   (b) 11%   (c) 9.5%   (d) 8.5%   (e) 10%   (f) 10.5%
2. 12.3%   3. 8.9%   4. 7.6%   5. 8.16%   6. 8.2%   7. 11.2%
8. 10.59%   9. 9%   10. 12%

EXERCISE 11.3

1. (a) $620   (b) $340   (c) $500   (d) $2880
2. (a) $2.00   (b) $3.00   (c) $2.50   (d) $1.56
3. (a) $0.50   (b) $0.25   (c) $0.20   (d) $0.67
4. (a) $2750   (b) $2975   (c) $475   (d) $450   (e) $1350
5. (a) $1800   (b) $27 000   (c) $21 750   (d) $34 000   (e) $18 500   (f) $136 250
   (g) $8900   (h) $26 100   (i) $8625
6. (a) $4150   (b) $4768.75   (c) $2550
7. (a) $3225   (b) $7275   (c) $5100
8. (a) $1750   (b) $3100   (c) −$750

EXERCISE 11.5

1. (a) $2000   (b) $1820   (c) $7200   (d) $5100   (e) $2100
2. (a) $1800   (b) $6250   (c) $5700   (d) $3450   (e) $1325   (f) $975
   (g) $7800
3. (a) $196.00   (b) $202.50   (c) $32.75   (d) $65.25   (e) $90.00
4. (a) $137.50   (b) $106.88   (c) $78.44   (d) $750.00
5. (a) $1802.16   (b) $6032.75   (c) $10 352.81   (d) $1917.19
6. (a) $1764.00   (b) $4569.25   (c) $5659.50   (d) $4851.00   (e) $16 047.50
7. (a) $192.56   (b) $604.75   (c) −$430.20   (d) −$565.25   (e) $988.91
8. (a) $14 348.25   (b) $5 666.29   (c) $7013.16   (d) $7815.63   (e) $8512.35
   (f) $12 940.63

EXERCISE 11.6

1. (a) $200; 11.11%   (b) $1460; 10.07%   (c) $300; 8.33%   (d) −$448; 4.15%
   (e) $1530; 11.33%
2. (a) 10.60   (b) 6.84   (c) 6.45   (d) 13.86   (e) 9.31
3. 4.19%   4. 8.19   5. 8.05%   6. 3.15

EXERCISE 11.7

2. $1273   5. $3750   9. $1659

EXERCISE 11.8

1. (a) $67 000　(b) $621.76　(c) $573.82　(d) $47.94　(e) $66 952.06
   (f) $573.41
   (g)

Payment Number	Total Payment($)	Interest Portion($)	Principal Portion($)	Outstanding Principal($)
1	621.76	573.82	47.94	66 952.06
2	621.76	573.41	48.35	66 903.71
3	621.76	573.00	48.76	66 854.95
4	621.76	572.58	49.18	66 805.77
5	621.76	572.16	49.60	66 756.17
6	621.76	571.73	50.03	66 706.14

2. (a) $64 000　(b) $73 000
   (c) No — only principal portion　(d) Yes
3. (a) $240 000
   (b) $2311.20
   (c) $2151.21
   (d) $159.99
   (e) $239 840.01
   (f) $2149.78
   (g)

Payment Number	Total Payment($)	Interest Portion($)	Principal Portion($)	Outstanding Principal($)
1	2311.20	2151.21	159.99	239 840.01
2	2311.20	2149.78	161.42	239 678.59
3	2311.20	2148.33	162.87	239 515.72
4	2311.20	2146.87	164.33	239 351.39
5	2311.20	2145.40	165.80	239 185.59
6	2311.20	2143.91	167.29	239 018.30

11.10 REVIEW EXERCISE

1. (a) 10.12%　(b) 11.52%　(c) 8%　(d) 9%　(e) 9.51%　(f) 11%
2. $37 500
3. (a) $1 102 500　(b) $735
4. (a) $550　(b) $210　(c) $740
5. (a) $9945　(b) $12 954　(c) $4590　(d) $1147.50　(e) $16 243.50
6. (a) $884.25　(b) $3045.75　(c) $786　(d) $28 492.50　(e) $2136.94
7. $13 731
8. (a) $4500　(b) $7040　(c) $4080
9. (a) $800　(b) 4.7%
10. (a) 13.6%　(b) 11.37%
11. (a) $840.15　(b) $945.74　(c) $740.20
12. $850.97
13. $16 855.71
14. $1835.54
15. (a) $1600; 14.29%　(b) $2625; 17.5%　(c) $12 000; 16.56%
16. $1515
17. (a) $3.31 per share　(b) $33 150　(c) $6691.87
18. (a) $774.00　(b) $731.91　(c) $42.09
   (d)

Payment Number	Total Payment($)	Interest Portion($)	Principal Portion($)	Outstanding Principal($)
1	774.00	731.91	42.09	74 957.91
2	774.00	731.50	42.50	74 915.41
3	774.00	731.08	42.92	74 872.49
4	774.00	730.67	43.33	74 829.16
5	774.00	730.24	43.76	74 785.40
6	774.00	729.82	44.18	74 741.22

19. (a) $200 000.00   (b) $2204.00   (c) $93.79   (d) $199 906.21   (e) $2109.23
    (g)

Payment Number	Total Payment($)	Interest Portion($)	Principal Portion($)	Outstanding Principal($)
1	2204.00	2110.20	93.79	199 906.21
2	2204.00	2109.23	94.77	199 811.44
3	2204.00	2108.23	95.77	199 715.67
4	2204.00	2107.21	96.79	199 618.88
5	2204.00	2106.19	97.81	199 521.07
6	2204.00	2105.16	98.84	199 422.23

20. $645.16   21. $2302.63   22. $21 720

11.11 CHAPTER 11 TEST

1. (a) 10.51%   (b) 9%
2. $1170.27
3. 9%
4. $597
5. (a) $3834.38   (b) $2849.25   (c) $850   (d) −$985.13
6. (a) 164, 181   (b) $1035   (c) $920
7. $476.69

REVIEW AND PREVIEW TO CHAPTER 12

CALCULATOR MATH

1. (a) −106.96   (b) 10.84   (c) 16.25   (d) 16 411.6   (e) 46.87   (f) −6.00
   (g) 459.49   (h) 2475.07   (i) 42.67   (j) −0.22
2. (a) 0.29   (b) 12.16   (c) 322.25   (d) 63.19
3. (a) 10.21 cm²   (b) 22.72 cm²   (c) 22.91 cm²
4. (a) 16 015.75 cm²   (b) 4   (c) 511 347 437.5 km²
   (d) 150 336 146.7 km²
5. (a) 446 092.22 cm³   (b) 8   (c) 1 087 295 101 000 km³
6. 41 912.72 km
7. Number of moves (M)—7, 15, 31, 63, 127, 255, 511, 1023

EXERCISE 12.1

1. (a) 6   (b) 4   (c) $\frac{2}{3}$ (d) $\frac{1}{3}$

2. (a) $\frac{2}{5}$ (b) $\frac{3}{5}$

3. The outcomes are not equally likely.

4. (a) {6, 8, 9, 3, 5, 1, 2}   (b) {9, 3, 5, 1}   (c) $\frac{4}{7}$

   (d) $\frac{3}{7}$

5. (a) {S, T, A, T, I, S, T, I, C, S}   (b) $\frac{3}{10}$

   (c) $\frac{7}{10}$

6. (a) $\frac{1}{3}$ (b) $\frac{1}{2}$ (c) $\frac{1}{6}$ (d) $\frac{2}{3}$ (e) $\frac{1}{2}$ (f) $\frac{5}{6}$

7. (a) {HHH, HHT, HTH, THH, HTT, THT, TTH, TTT}
   (b) (i) $\frac{1}{8}$ (ii) $\frac{3}{8}$ (iii) $\frac{1}{2}$

8. $\frac{1}{9}$

9. (a) $\frac{1}{2}$ (b) $\frac{1}{4}$ (c) $\frac{3}{4}$ (d) $\frac{1}{13}$ (e) $\frac{1}{52}$ (f) $\frac{1}{26}$
   (g) $\frac{12}{13}$ (h) $\frac{3}{13}$

10. (a) $\frac{1}{2}$ (b) $\frac{1}{2}$ (c) $\frac{1}{5}$ (d) $\frac{1}{4}$ (e) $\frac{1}{10}$ (f) $\frac{3}{10}$

11. (a) $\frac{1}{2500}$ (b) $\frac{1}{25\,000}$ (c) $\frac{1}{12\,500}$

EXERCISE 12.2

1. (a) 0.983   (b) 0.779

2. (a) 0.981   (b) 0.955   (c) 0.886   (d) 0.744

3. (a) (i) 0.914   (ii) 0.603   (iii) 0.067

    (b) (i) 0.955   (ii) 0.777   (iii) 0.170

EXERCISE 12.3

1. (a), (d), (e)

2. (a) $\frac{1}{4}$ (b) $\frac{5}{12}$ (c) $\frac{7}{12}$ (d) $\frac{5}{8}$ (e) $\frac{3}{4}$

3. (a) $\frac{2}{13}$ (b) $\frac{15}{52}$

4. (a) $\frac{2}{9}$ (b) $\frac{1}{12}$

5. (a) $\frac{5}{18}$ (b) $\frac{7}{12}$ (c) $\frac{5}{12}$

6. (a) 0   (b) $\frac{1}{216}$ (c) $\frac{1}{72}$ (d) $\frac{1}{36}$ (e) $\frac{5}{108}$

7. (a) $\frac{3}{8}$ (b) $\frac{1}{4}$ (c) $\frac{11}{16}$

EXERCISE 12.4

1. $\frac{1}{8}$ 2. (b) $\frac{1}{4}$ 3. (a) $\frac{1}{36}$ (b) $\frac{1}{18}$

4. $\frac{1}{36}$ 5. (a) $\frac{25}{676}$ (b) $\frac{441}{676}$

6. $\frac{1}{1296}$

7. (a) $\frac{36}{169}$ (b) $\frac{9}{169}$ (c) $\frac{18}{169}$ (d) $\frac{36}{169}$ (e) $\frac{49}{169}$

8. (a) $\frac{5}{26}$ (b) $\frac{1}{26}$ (c) $\frac{3}{26}$ (d) $\frac{3}{13}$ (e) $\frac{7}{26}$

9. (a) $\frac{1}{16}$ (b) $\frac{1}{16}$ (c) $\frac{1}{8}$ (d) $\frac{1}{169}$ (e) $\frac{2}{169}$ (f) $\frac{1}{2704}$

    (g) $\frac{1}{52}$

10. (a) $\frac{3}{51}$ (b) $\frac{13}{204}$ (c) $\frac{13}{102}$ (d) $\frac{1}{221}$ (e) $\frac{2}{221}$ (f) 0

    (g) 0

11. (a) $\frac{1}{1250}$ (b) $\frac{3}{6\,248\,750}$ (c) $\frac{1}{5\,205\,208\,750}$

12. (a) $\frac{1}{36}$ (b) $\frac{1}{6}$ (c) $\frac{1}{2}$ (d) $\frac{1}{6}$

13. (a) $\frac{1}{24}$ (b) $\frac{1}{24}$ (c) $\frac{1}{6}$

14. (a) $\frac{1}{48}$ (b) $\frac{1}{156}$ (c) $\frac{1}{624}$

15. $\frac{1}{2}$

EXERCISE 12.5

1. Answers vary

2. (a) Answers vary

3. Answers vary

4. surveying only the subscribers

5. (a) population too large   (b) Answers vary   (c) Answers vary

EXERCISE 12.7

1. (a) mean   (b) mode   (c) mean   (d) mean   (e) mean

2. (a) Mean: 11.4, Median: 12, Mode: 12   (b) Mean: 12.3, Median: 12.5, Modes: 10, 13

    (c) Mean: 15, Median: 15.5   (d) Mean: 4.8, Median: 5, Mode: 2

5. (a) Mean: $42 916.67, Median: $35 000, Mode: $30 000

6. (a) Mean: 10.9, Median: 10.5, Mode: 9

    (b) 9

EXERCISE 12.8

1. (a) 21   (b) 43   (c) 53

2. (a) 2.8   (b) 4.4

3. (a) 16%   (b) 48%

4. (a) 2%   (b) 16%

5. (a) 48%   (b) 2%   (c) 480

6. (a) 84%　　　　(b) 2%　　　　(c) 12　　　　(d) 408
7. (a) Mean: 5, $\sigma = 1.4$　　(b) Mean: 7, $\sigma = 0.58$　　(c) —　　(d) —　　(e) —

12.9 REVIEW EXERCISE

1. (a) $\frac{1}{4}$　　　　(b) $\frac{1}{2}$　　　　(c) 0　　　　(d) 1

2. (a) $\frac{3}{10}$　　　　(b) $\frac{3}{8}$　　　　(c) $\frac{1}{8}$　　　　(d) $\frac{1}{5}$　　　　(e) $\frac{27}{40}$　　　　(f) $\frac{4}{5}$

　　(g) $\frac{1}{2}$　　　　(h) 0　　　　(i) $\frac{7}{10}$

3. (a) 123, 132, 213, 231, 312, 321
　　(b) (i) 1　　(ii) $\frac{2}{3}$　　(iii) $\frac{1}{2}$　　(iv) 0

4. (a) $\frac{1}{2}$　　　　(b) $\frac{5}{6}$　　　　(c) $\frac{7}{8}$　　　　(d) $\frac{12}{13}$　　　　(e) $\frac{5}{9}$　　　　(f) $\frac{5}{26}$

5. $\frac{3}{11}$

6. (a) $\frac{1}{5}$　　　　(b) $\frac{1}{125}$　　　　(c) $\frac{64}{125}$

7. $\frac{1}{1024}$

8. (a) $\frac{1}{676}$　　　　(b) $\frac{1}{17\,576}$　　　　(c) $\frac{1}{456\,976}$

9. (a) $\frac{1}{650}$　　　　(b) $\frac{1}{15\,600}$　　　　(c) $\frac{1}{358\,800}$

10. (a) $\frac{1}{8}$　　　　(b) $\frac{7}{8}$　　　　(c) $\frac{1}{2}$　　　　(d) $\frac{1}{8}$

11. $\frac{1}{6}$

12. (a) $\frac{1}{8}$　　　　(b) $\frac{7}{8}$　　　　(c) $\frac{1}{8}$

13. (a) $\frac{1}{9}$　　　　(b) $\frac{1}{6}$　　　　(c) 0

14. (a) Mean: 10.3, Median: 10, Mode: 5　　　　(b) Mean: 16.2, Median: 15, Mode: 12
　　(c) Mean: 11.5, Median: 10.5, Modes: 9, 10
15. (a) Range: 6, σ : 2　　　　(b) Range: 15, σ: 4.6
18. (a) 98%　　　(b) 84%　　　(c) 378　　　(d) 216
19. (a) Mean: 50, σ: 1.9

12.10 CHAPTER 12 TEST

1. (a) $\frac{1}{3}$　　　　(b) $\frac{2}{3}$

2. (a) $\frac{5}{36}$　　　　(b) $\frac{31}{36}$　　　　(c) $\frac{10}{36}$

3. (a) $\frac{1}{25}$　　　　(b) $\frac{1}{45}$

4. Mean: 7, Median: 8, Mode: 8
5. Range: 6, σ: 2.2
6. (a) 68%　　　(b) 16%

REVIEW AND PREVIEW TO CHAPTER 13

SIMILAR TRIANGLES
1. (a) 2.70 cm　　　(b) 2.14 cm　　　(c) 8.10 cm　　　(d) 12.6 cm　　　(e) 7.04 cm
2. (a) x = 2.92 cm, y = 8.17 cm　　　(b) x = 5.14 cm, y = 4.06 cm　　　(c) x = 7.2 cm, y = 11.7 cm

THE PYTHAGOREAN THEOREM
1. (a) 9.43 m　　　(b) 10.9 m　　　(c) 15.0 cm　　　(d) 6.71 m　　　(e) 24.0 cm　　　(f) 70.7 m
　　(g) 6.37 m　　　(h) 23.1 cm　　　(i) 1.98 m

EQUATIONS
1. (a) 2.2　　　(b) 5.09　　　(c) 8.87　　　(d) 15.9　　　(e) 10.5　　　(f) 0.556
　　(g) 9.71　　　(h) 5.22　　　(i) 1.70

EXERCISE 13.1

1. (a) $\sin \theta = \dfrac{5}{\sqrt{41}}$, $\cos \theta = \dfrac{4}{\sqrt{41}}$, $\tan \theta = \frac{5}{4}$, $\csc \theta = \dfrac{\sqrt{41}}{5}$, $\sec \theta = \dfrac{\sqrt{41}}{4}$, $\cot \theta = \frac{4}{5}$

　　(b) $\sin \theta = \dfrac{2}{\sqrt{13}}$, $\cos \theta = \dfrac{3}{\sqrt{13}}$, $\tan \theta = \frac{2}{3}$, $\csc \theta = \dfrac{\sqrt{13}}{2}$, $\sec \theta = \dfrac{\sqrt{13}}{3}$, $\cot \theta = \frac{3}{2}$

(c) $\sin \theta = \frac{4}{7}$, $\cos \theta = \frac{\sqrt{33}}{7}$, $\tan \theta = \frac{4}{\sqrt{33}}$, $\csc \theta = \frac{7}{4}$, $\sec \theta = \frac{7}{\sqrt{33}}$, $\cot \theta = \frac{\sqrt{33}}{4}$

(d) $\sin \theta = \frac{7}{\sqrt{74}}$, $\cos \theta = \frac{5}{\sqrt{74}}$, $\tan \theta = \frac{7}{5}$, $\csc \theta = \frac{\sqrt{74}}{7}$, $\sec \theta = \frac{\sqrt{74}}{5}$, $\cot \theta = \frac{5}{7}$

(e) $\sin \theta = \frac{1}{\sqrt{2}}$, $\cos \theta = \frac{1}{\sqrt{2}}$, $\tan \theta = 1$, $\csc \theta = \sqrt{2}$, $\sec \theta = \sqrt{2}$, $\cot \theta = 1$

(f) $\sin \theta = \frac{\sqrt{3}}{2}$, $\cos \theta = \frac{1}{2}$, $\tan \theta = \sqrt{3}$, $\csc \theta = \frac{2}{\sqrt{3}}$, $\sec \theta = 2$, $\cot \theta = \frac{1}{\sqrt{3}}$

(g) $\sin \theta = \frac{3}{5}$, $\cos \theta = \frac{4}{5}$, $\tan \theta = \frac{3}{4}$, $\csc \theta = \frac{5}{3}$, $\sec \theta = \frac{5}{4}$, $\cot \theta = \frac{4}{3}$

(h) $\sin \theta = \frac{8}{17}$, $\cos \theta = \frac{15}{17}$, $\tan \theta = \frac{8}{15}$, $\csc \theta = \frac{17}{8}$, $\sec \theta = \frac{17}{15}$, $\cot \theta = \frac{15}{8}$

2. (a) 10, $\sin \theta = \frac{3}{5}$, $\cos \theta = \frac{4}{5}$, $\tan \theta = \frac{3}{4}$, $\csc \theta = \frac{5}{3}$, $\sec \theta = \frac{5}{4}$, $\cot \theta = \frac{4}{3}$

(b) 13, $\sin \theta = \frac{12}{13}$, $\cos \theta = \frac{5}{13}$, $\tan \theta = \frac{12}{5}$, $\csc \theta = \frac{13}{12}$, $\sec \theta = \frac{13}{5}$, $\cot \theta = \frac{5}{12}$

(c) 15, $\sin \theta = \frac{15}{17}$, $\cos \theta = \frac{8}{17}$, $\tan \theta = \frac{15}{8}$, $\csc \theta = \frac{17}{15}$, $\sec \theta = \frac{17}{8}$, $\cot \theta = \frac{8}{15}$

(d) 15, $\sin \theta = \frac{3}{5}$, $\cos \theta = \frac{4}{5}$, $\tan \theta = \frac{3}{4}$, $\csc \theta = \frac{5}{3}$, $\sec \theta = \frac{5}{4}$, $\cot \theta = \frac{4}{3}$

3. (a) 5, $\sin A = \frac{4}{5}$, $\cos A = \frac{3}{5}$, $\tan A = \frac{4}{3}$, $\csc A = \frac{5}{4}$, $\sec A = \frac{5}{3}$, $\cot A = \frac{3}{4}$

$\sin B = \frac{3}{5}$, $\cos B = \frac{4}{5}$, $\tan B = \frac{3}{4}$, $\csc B = \frac{5}{3}$, $\sec B = \frac{5}{4}$, $\cot B = \frac{4}{3}$

(b) 12, $\sin A = \frac{12}{13}$, $\cos A = \frac{5}{13}$, $\tan A = \frac{12}{5}$, $\csc A = \frac{13}{12}$, $\sec A = \frac{13}{5}$, $\cot A = \frac{5}{12}$

$\sin B = \frac{5}{13}$, $\cos B = \frac{12}{13}$, $\tan B = \frac{5}{12}$, $\csc B = \frac{13}{5}$, $\sec B = \frac{13}{12}$, $\cot B = \frac{12}{5}$

(c) 8, $\sin B = \frac{8}{17}$, $\cos B = \frac{15}{17}$, $\tan B = \frac{8}{15}$, $\csc B = \frac{17}{8}$, $\sec B = \frac{17}{15}$, $\cot B = \frac{15}{8}$

$\sin C = \frac{15}{17}$, $\cos C = \frac{8}{17}$, $\tan C = \frac{15}{8}$, $\csc C = \frac{17}{15}$, $\sec C = \frac{17}{8}$, $\cot C = \frac{8}{15}$

(d) 8, $\sin P = \frac{4}{5}$, $\cos P = \frac{3}{5}$, $\tan P = \frac{4}{3}$, $\csc P = \frac{5}{4}$, $\sec P = \frac{5}{3}$, $\cot P = \frac{3}{4}$

$\sin R = \frac{3}{5}$, $\cos R = \frac{4}{5}$, $\tan R = \frac{3}{4}$, $\csc R = \frac{5}{3}$, $\sec R = \frac{5}{4}$, $\cot R = \frac{4}{3}$

4. $\cos \theta = \frac{\sqrt{3}}{2}$, $\tan \theta = \frac{1}{\sqrt{3}}$, $\csc \theta = 2$, $\sec \theta = \frac{2}{\sqrt{3}}$, $\cot \theta = \sqrt{3}$

5. $\sin \theta = \frac{12}{13}$, $\cos \theta = \frac{5}{13}$, $\tan \theta = \frac{12}{5}$, $\csc \theta = \frac{13}{5}$, $\cot \theta = \frac{5}{12}$

6. $\sin \theta = \frac{2}{\sqrt{5}}$, $\cos \theta = \frac{1}{\sqrt{5}}$, $\csc \theta = \frac{\sqrt{5}}{2}$, $\sec \theta = \sqrt{5}$, $\cot \theta = \frac{1}{2}$

7. $\sin \theta = \frac{1}{2}$, $\tan \theta = \frac{1}{\sqrt{3}}$, $\csc \theta = 2$, $\sec \theta = \frac{2}{\sqrt{3}}$, $\cot \theta = \sqrt{3}$

8. (a) $\sin B = \frac{b}{a}$, $\cos B = \frac{c}{a}$, $\tan B = \frac{b}{c}$, $\csc B = \frac{a}{b}$, $\sec B = \frac{a}{c}$, $\cot B = \frac{c}{b}$

$\sin C = \frac{c}{a}$, $\cos C = \frac{b}{a}$, $\tan C = \frac{c}{b}$, $\csc C = \frac{a}{c}$, $\sec C = \frac{a}{b}$, $\cot C = \frac{b}{c}$

(b) (i) $\cos B = \sin C$ (ii) $\sin B = \cos C$ (iii) $\tan C = \cot B$

9. (a) $\sin A = \frac{5}{\sqrt{26}}$, $\cos A = \frac{1}{\sqrt{26}}$, $\tan A = 5$, $\csc A = \frac{\sqrt{26}}{5}$, $\sec A = \sqrt{26}$, $\cot A = \frac{1}{5}$

$\sin B = \frac{1}{\sqrt{26}}$, $\cos B = \frac{5}{\sqrt{26}}$, $\tan B = \frac{1}{5}$, $\csc B = \sqrt{26}$, $\sec B = \frac{\sqrt{26}}{5}$, $\cot B = 5$

EXERCISE 13.2

1. (a) $\frac{\sqrt{6}}{8}$ (b) 1   (c) $\frac{\sqrt{3}}{2}$ (d) 1   (e) $\frac{\sqrt{3}}{2}$ (f) 10

2. (a) $4\frac{1}{4}$ (b) $3\frac{1}{2}$ (c) $\frac{3}{4}$ (d) 3   (e) 3   (f) $4 - \frac{1}{\sqrt{2}}$

(g) 1   (h) $1\frac{1}{6}$ (i) $4\frac{1}{4}$

EXERCISE 13.3

1. (a) 0.4226   (b) 3.732   (c) 1.000   (d) 2.130   (e) 2.790   (f) 0.7431

(g) 1.012   (h) 0.5878   (i) 1.133   (j) 0.2588   (k) 3.078   (l) 0.9511

(m) 0.7536   (n) 1.236   (0) 0.6018   (p) 1.600

2. (a) 65°   (b) 60°   (c) 66°   (d) 25°   (e) 39°   (f) 16°
    (g) 65°   (h) 79°   (i) 35°   (j) 35°   (k) 25°   (l) 50°
    (m) 68°   (n) 1°   (o) 39°   (p) 25°

EXERCISE 13.4

1. (a) 39°   (b) 53°   (c) 60°   (d) 35°   (e) 45°   (f) 57°
    (g) 65°   (h) 28°

2. (a) 142 m   (b) 13.4 m   (c) 35.5 cm   (d) 89.9 cm   (e) 51.8 m   (f) 23.5 cm
    (g) 11.2 cm   (h) 161 m

3. (a) $\angle A = 39°$, $\angle C = 51°$, $b = 32$ m   (b) $\angle A = 45°$, $a = 80.6$ m, $c = 80.6$ m
    (c) $\angle A = 25°$, $a = 6.34$ m, $b = 13.6$ m   (d) $\angle B = 52°$, $\angle C = 38°$, $c = 11.7$ cm
    (e) $\angle A = 50°$, $\angle C = 40°$, $b = 5.39$ m   (f) $\angle A = 40°$, $\angle B = 50°$, $c = 3.91$ m
    (g) $\angle A = 27°$, $a = 3.72$ cm, $c = 7.31$ cm   (h) $\angle B = 50°$, $a = 8.87$ cm, $b = 6.79$ cm

4. (a) $\angle B = 47°$, $\angle C = 43°$, $c = 10.2$ m   (b) $\angle F = 55°$, $d = 70.0$ cm, $e = 122$ cm
    (c) $\angle G = 10°$, $g = 21$ cm, $h = 118$ cm   (d) $\angle J = 56°$, $\angle K = 34°$, $l = 33.2$ m

5. (a) $\angle B = 35°$, $\angle C = 55°$, $c = 3.07$ m   (b) $\angle A = 33°$, $\angle C = 57°$, $b = 39.2$ cm
    (c) $\angle C = 52°$, $a = 5.52$ m, $b = 3.40$ m   (d) $\angle B = 23°$, $a = 9.85$ cm, $b = 4.18$ cm

6. (a) 165 m   (b) 90.2 m   (c) 17.99 m   (d) 6.62 m

EXERCISE 13.5

1. 166 m   2. 78°   3. 369 m   4. 187 m   5. 89 m   6. 57.1 m
7. 39°   8. (a) 29 m   (b) 37 m   9. 196 m   10. 85 m   11. 7.5 m
12. 380 m   13. (a) 124 m   (b) 13.8 m   14. 150 km   15. 2412 m   16. 111 m
17. 21 m   18. 285 m   19. 35 m

13.6 REVIEW EXERCISE

1. (a) 10, $\sin \theta = \frac{4}{5}$, $\cos \theta = \frac{3}{5}$, $\tan \theta = \frac{4}{3}$, $\csc \theta = \frac{5}{4}$, $\sec \theta = \frac{5}{3}$, $\cot \theta = \frac{3}{4}$

    (b) 15, $\sin \theta = \frac{15}{17}$, $\cos \theta = \frac{8}{17}$, $\tan \theta = \frac{15}{8}$, $\csc \theta = \frac{17}{15}$, $\sec \theta = \frac{17}{8}$, $\cot \theta = \frac{8}{15}$

    (c) 2, $\sin \theta = \frac{5}{\sqrt{29}}$, $\cos \theta = \frac{2}{\sqrt{29}}$, $\tan \theta = \frac{5}{2}$, $\csc \theta = \frac{\sqrt{29}}{5}$, $\sec \theta = \frac{\sqrt{29}}{2}$, $\cot \theta = \frac{2}{5}$

2. (a) 3, $\sin A = \frac{4}{5}$, $\cos A = \frac{3}{5}$, $\tan A = \frac{4}{3}$, $\csc A = \frac{5}{4}$, $\sec A = \frac{5}{3}$, $\cot A = \frac{3}{4}$
        $\sin B = \frac{3}{5}$, $\cos B = \frac{4}{5}$, $\tan B = \frac{3}{4}$, $\csc B = \frac{5}{3}$, $\sec B = \frac{5}{4}$, $\cot B = \frac{4}{3}$

    (b) 24, $\sin S = \frac{5}{13}$, $\cos S = \frac{12}{13}$, $\tan S = \frac{5}{12}$, $\csc S = \frac{13}{5}$, $\sec S = \frac{13}{12}$, $\cot S = \frac{12}{5}$
        $\sin R = \frac{12}{13}$, $\cos R = \frac{5}{13}$, $\tan R = \frac{12}{5}$, $\csc R = \frac{13}{12}$, $\sec R = \frac{13}{5}$, $\cot R = \frac{5}{12}$

    (c) 34, $\sin R = \frac{8}{17}$, $\cos R = \frac{15}{17}$, $\tan R = \frac{8}{15}$, $\csc R = \frac{17}{8}$, $\sec R = \frac{17}{15}$, $\cot R = \frac{15}{8}$
        $\sin P = \frac{15}{17}$, $\cos P = \frac{8}{17}$, $\tan P = \frac{15}{8}$, $\csc P = \frac{17}{15}$, $\sec P = \frac{17}{8}$, $\cot P = \frac{8}{15}$

3. $\sin \theta = \frac{12}{13}$, $\cos \theta = \frac{5}{13}$, $\tan \theta = \frac{12}{5}$, $\sec \theta = \frac{13}{5}$, $\cot \theta = \frac{5}{12}$

4. $\sin \theta = \frac{3}{\sqrt{10}}$, $\cos \theta = \frac{1}{\sqrt{10}}$, $\csc \theta = \frac{\sqrt{10}}{3}$, $\sec \theta = \sqrt{10}$, $\cot \theta = \frac{1}{3}$

5. (i) 1   (ii) 1   (iii) $\frac{\sqrt{2} + \sqrt{6}}{2}$ (iv) $\frac{2}{3}$ (v) $\frac{1}{4}$

6. (a) 0.5878   (b) 0.3090   (c) 1.2349   (d) 5.2408   (e) 1.3902   (f) 0.1584

7. (a) 61°   (b) 37°   (c) 59°   (d) 10°   (e) 76°   (f) 44°

8. (a) $\angle A = 52°$, $b = 94$ m, $c = 152$ m   (b) $\angle B = 38°$, $\angle C = 52°$, $b = 71.6$ m
    (c) $\angle C = 23°$, $a = 25.6$ cm, $b = 23.6$ cm   (d) $\angle B = 45°$, $\angle C = 45°$, $b = 1000$ m
    (e) $\angle A = 20°$, $a = 45$ cm, $c = 133$ cm   (f) $\angle C = 52°$, $a = 905$ m, $b = 1158$ m

9. 81 m
10. 385 m
11. 6040 m
12. 173.4 m
13. 1520 m
14. 425 m

13.7 CHAPTER 13 TEST

1. $\sin A = \frac{12}{13}$, $\cos A = \frac{5}{13}$, $\tan A = \frac{12}{5}$, $\csc A = \frac{13}{12}$, $\sec A = \frac{13}{5}$, $\cot A = \frac{5}{12}$

$\sin C = \frac{5}{13}$, $\cos C = \frac{12}{13}$, $\tan C = \frac{5}{12}$, $\csc C = \frac{13}{5}$, $\sec C = \frac{13}{12}$, $\cot C = \frac{12}{5}$

2. (a) $\frac{3}{4}$ (b) 8
3. $\angle R = 56°$, $t = 8.39$ cm, $r = 12.44$ cm
4. $\angle A = 56°$, $\angle C = 34°$, $c = 9.95$ m
5. 128 m
6. 34 m

13.8 CUMULATIVE REVIEW FOR CHAPTERS 9 TO 13

1. (a) -4, -1, 4, 11, 20   (b) 5, 8, 11, 14, 17
2. (a) 1, 4, 9, 16, 25   (b) 2, 4, 8, 16, 32
3. (a) 3, 15, 75, 375, 1875   (b) -2, 0, 2, 4, 6   (c) 1, 2, 0, -2, -4
    (d) 3, 3, 6, 9, 15
4. (a) 5, 11, 17; $t_n = 6n - 1$ (b) -3, 1, 5; $t_n = 4n - 7$
    (c) -6, -8, -10; $t_n = -2n - 4$ (d) $x + y$, $2x$, $3x - y$; $t_n = nx + (2 - n)y$
5. (a) 1, 2, 4; $t_n = 2^{n-1}$ (b) 2, -6, 18; $t_n = 2(-3)^{n-1}$ (c) -3, -9, -27; $t_n = -3^n$
    (d) 1, $\frac{1}{x}$, $\frac{1}{x^2}$; $t_n = x^{1-n}$
6. (a) $a = \frac{11}{2}$, $d = \frac{3}{2}$, $t_n = \frac{1}{2}(3n + 8)$ (b) $a = -36$, $d = \frac{16}{3}$, $t_n = \frac{1}{3}(n + 134)$
7. (a) $a = 2$, $r = 2$, $t_n = 2^n$ (b) $a = -486$, $r = -\frac{1}{3}$, $t_n = 2(-3)^{6-n}$
8. (a) 15   (b) $\frac{163}{60}$ (c) 55
9. (a) 15 050   (b) 166 833   (c) 500 500
10. (a) 258   (b) 63.75
11. (a) 2, -11.2, -24.4, -37.6 (b) 2, -4, 8, -16
12. (a) $6605.33   (b) $17 169.23   (c) $15 337.86
13. (a) $2468.11   (b) $887.45   (c) $1578.82
14. 9.5 a
15. (a) $2697.35   (b) $7958.56
16. (a) $2903.81   (b) $4653.67
17. (a) $87.31   (b) $2948.35   (c) $1743.69
18. (a) $i = 0.1$ (b) $i = 0.05$
19. (a) $577 500.00   (b) $330.00
20. $9486.00
21. $20 586.50
22. $4279.04
23. $2647.50, profit
24. 52.65, 53.5, 29, 24.04
25. (a) $\frac{8}{15}$ (b) $\frac{1}{3}$ (c) $\frac{13}{15}$ (d) 0   (e) $\frac{7}{15}$ (f) $\frac{7}{15}$
26. 0.25
27. (a) $\angle A = 55°$, $BC = 8.57$, $AB = 10.5$ (b) $\angle R = 25°$, $PQ = 42.3$, $QR = 90.6$
    (c) $\angle T = 37°$, $\angle U = 53°$, $TU = 10$ (d) $\angle X = 45°$, $XZ = YZ = 109.6$
28. $\sin \theta = 0.866$, $\tan \theta = 1.732$, $\csc \theta = 1.15$, $\sec \theta = 2$, $\cot \theta = 0.577$
29. (a) $\frac{1}{\sqrt{2}}$, $\frac{1}{\sqrt{2}}$, 1, $\sqrt{2}$, $\sqrt{2}$, 1   (b) $\frac{1}{2}$, $\frac{\sqrt{3}}{2}$, $\frac{1}{\sqrt{3}}$, 2, $\frac{2}{\sqrt{3}}$, $\sqrt{3}$ (c) $\frac{\sqrt{3}}{2}$, $\frac{1}{2}$, $\sqrt{3}$, $\frac{2}{\sqrt{3}}$, 2
30. (a) $\frac{1}{2}(\sqrt{3} + 1)$ (b) 1   (c) 1
31. 299 m
32. 41 m
33. 64 m

Table I SQUARES AND SQUARE ROOTS

n	n²	√n	√10n	n	n²	√n	√10n
1.0	1.00	1.000	3.162	5.5	30.25	2.345	7.416
1.1	1.21	1.049	3.317	5.6	31.36	2.366	7.483
1.2	1.44	1.095	3.464	5.7	32.49	2.387	7.550
1.3	1.69	1.140	3.606	5.8	33.64	2.408	7.616
1.4	1.96	1.183	3.742	5.9	34.81	2.429	7.681
1.5	2.25	1.225	3.873	6.0	36.00	2.449	7.746
1.6	2.56	1.265	4.000	6.1	37.21	2.470	7.810
1.7	2.89	1.304	4.123	6.2	38.44	2.490	7.874
1.8	3.24	1.342	4.243	6.3	39.69	2.510	7.937
1.9	3.61	1.378	4.359	6.4	40.96	2.530	8.000
2.0	4.00	1.414	4.472	6.5	42.25	2.550	8.062
2.1	4.41	1.449	4.583	6.6	43.56	2.569	8.124
2.2	4.84	1.483	4.690	6.7	44.89	2.588	8.185
2.3	5.29	1.517	4.796	6.8	46.24	2.608	8.246
2.4	5.76	1.549	4.899	6.9	47.61	2.627	8.307
2.5	6.25	1.581	5.000	7.0	49.00	2.646	8.367
2.6	6.76	1.612	5.099	7.1	50.41	2.665	8.426
2.7	7.29	1.643	5.196	7.2	51.84	2.683	8.485
2.8	7.84	1.673	5.292	7.3	53.29	2.702	8.544
2.9	8.41	1.703	5.385	7.4	54.76	2.720	8.602
3.0	9.00	1.732	5.477	7.5	56.25	2.739	8.660
3.1	9.61	1.761	5.568	7.6	57.76	2.757	8.718
3.2	10.24	1.789	5.657	7.7	59.29	2.775	8.775
3.3	10.89	1.817	5.745	7.8	60.84	2.793	8.832
3.4	11.56	1.844	5.831	7.9	62.41	2.811	8.888
3.5	12.25	1.871	5.916	8.0	64.00	2.828	8.944
3.6	12.96	1.897	6.000	8.1	65.61	2.846	9.000
3.7	13.69	1.924	6.083	8.2	67.24	2.864	9.055
3.8	14.44	1.949	6.164	8.3	68.89	2.881	9.110
3.9	15.21	1.975	6.245	8.4	70.56	2.898	9.165
4.0	16.00	2.000	6.325	8.5	72.25	2.915	9.220
4.1	16.81	2.025	6.403	8.6	73.96	2.933	9.274
4.2	17.64	2.049	6.481	8.7	75.69	2.950	9.327
4.3	18.49	2.074	6.557	8.8	77.44	2.966	9.381
4.4	19.36	2.098	6.633	8.9	79.21	2.983	9.434
4.5	20.25	2.121	6.708	9.0	81.00	3.000	9.487
4.6	21.16	2.145	6.782	9.1	82.81	3.017	9.539
4.7	22.09	2.168	6.856	9.2	84.64	3.033	9.592
4.8	23.04	2.191	6.928	9.3	86.49	3.050	9.644
4.9	24.01	2.214	7.000	9.4	88.36	3.066	9.695
5.0	25.00	2.236	7.071	9.5	90.25	3.082	9.747
5.1	26.01	2.258	7.141	9.6	92.16	3.098	9.798
5.2	27.04	2.280	7.211	9.7	94.09	3.114	9.849
5.3	28.09	2.302	7.280	9.8	96.04	3.130	9.899
5.4	29.16	2.324	7.348	9.9	98.01	3.146	9.950
5.5	30.25	2.354	7.416	10	100.00	3.162	10.000

Table II TRIGONOMETRIC RATIOS

0°	sin θ	cos θ	tan θ	cot θ	sec θ	cosec θ
0	0.0000	1.0000	0.0000	—	1.0000	—
1	0.0175	0.9999	0.0175	57.290	1.0001	57.299
2	0.0349	0.9994	0.0349	28.636	1.0006	28.654
3	0.0523	0.9986	0.0524	19.081	1.0014	19.107
4	0.0698	0.9976	0.0699	14.301	1.0024	14.335
5	0.0872	0.9962	0.0875	11.430	1.0038	11.474
6	0.1045	0.9945	0.1051	9.5144	1.0055	9.5668
7	0.1219	0.9926	0.1228	8.1443	1.0075	8.2055
8	0.1392	0.9903	0.1405	7.1154	1.0098	7.1853
9	0.1564	0.9877	0.1584	6.3137	1.0125	6.3924
10	0.1737	0.9848	0.1763	5.6713	1.0154	5.7588
11	0.1908	0.9816	0.1944	5.1445	1.0187	5.2408
12	0.2079	0.9782	0.2126	4.7046	1.0223	4.8097
13	0.2250	0.9744	0.2309	4.3315	1.0263	4.4454
14	0.2419	0.9703	0.2493	4.0108	1.0306	4.1336
15	0.2588	0.9659	0.2680	3.7320	1.0353	3.8637
16	0.2756	0.9613	0.2867	3.4874	1.0403	3.6279
17	0.2924	0.9563	0.3057	3.2708	1.0457	3.4203
18	0.3090	0.9511	0.3249	3.0777	1.0515	3.2361
19	0.3256	0.9455	0.3443	2.9042	1.0576	3.0715
20	0.3420	0.9397	0.3640	2.7475	1.0642	2.9238
21	0.3584	0.9336	0.3839	2.6051	1.0711	2.7904
22	0.3746	0.9272	0.4040	2.4751	1.0785	2.6695
23	0.3907	0.9025	0.4245	2.3558	1.0864	2.5593
24	0.4067	0.9136	0.4452	2.2460	1.0946	2.4586
25	0.4226	0.9063	0.4663	2.1445	1.1034	2.3662
26	0.4384	0.8988	0.4877	2.0503	1.1126	2.2812
27	0.4540	0.8910	0.5095	1.9626	1.1223	2.2027
28	0.4695	0.8830	0.5317	1.8807	1.1326	2.1300
29	0.4848	0.8746	0.5543	1.8040	1.1433	2.0627
30	0.5000	0.8660	0.5774	1.7320	1.1547	2.0000
31	0.5150	0.8572	0.6009	1.6643	1.1666	1.9416
32	0.5299	0.8481	0.6249	1.6003	1.1792	1.8871
33	0.5446	0.8387	0.6494	1.5399	1.1924	1.8361
34	0.5592	0.8290	0.6745	1.4826	1.2062	1.7883
35	0.5736	0.8192	0.7002	1.4281	1.2208	1.7434
36	0.5878	0.8090	0.7265	1.3764	1.2361	1.7013
37	0.6018	0.7986	0.7536	1.3270	1.2521	1.6616
38	0.6157	0.7880	0.7813	1.2799	1.2690	1.6243
39	0.6293	0.7772	0.8098	1.2349	1.2867	1.5890
40	0.6428	0.7660	0.8391	1.1917	1.3054	1.5557
41	0.6561	0.7547	0.8693	1.1504	1.3250	1.5242
42	0.6691	0.7431	0.9004	1.1106	1.3456	1.4945
43	0.6820	0.7314	0.9325	1.0724	1.3673	1.4663
44	0.6947	0.7193	0.9657	1.0355	1.3902	1.4395
45	0.7071	0.7071	1.0000	1.0000	1.4142	1.4142

0°	sin θ	cos θ	tan θ	cot θ	sec θ	cosec θ
46	0.7193	0.6947	1.0355	0.9657	1.4395	1.3902
47	0.7314	0.6820	1.0724	0.9325	1.4663	1.3673
48	0.7431	0.6691	1.1106	0.9004	1.4945	1.3456
49	0.7547	0.6561	1.1504	0.8693	1.5242	1.3250
50	0.7660	0.6428	1.1917	0.8391	1.5557	1.3054
51	0.7772	0.6293	1.2349	0.8098	1.5890	1.2867
52	0.7880	0.6157	1.2799	0.7813	1.6243	1.2690
53	0.7986	0.6018	1.3270	0.7536	1.6616	1.2521
54	0.8090	0.5878	1.3764	0.7265	1.7013	1.2361
55	0.8192	0.5736	1.4281	0.7002	1.7434	1.2208
56	0.8290	0.5592	1.4826	0.6745	1.7883	1.2062
57	0.8387	0.5446	1.5399	0.6494	1.8361	1.1924
58	0.8481	0.5299	1.6003	0.6249	1.8871	1.1792
59	0.8572	0.5150	1.6643	0.6009	1.9416	1.1666
60	0.8660	0.5000	1.7320	0.5774	2.0000	1.1547
61	0.8746	0.4848	1.8040	0.5543	2.0627	1.1433
62	0.8830	0.4695	1.8807	0.5317	2.1300	1.1326
63	0.8910	0.4540	1.9626	0.5095	2.2027	1.1223
64	0.8988	0.4384	2.0503	0.4877	2.2812	1.1126
65	0.9063	0.4226	2.1445	0.4663	2.3662	1.1034
66	0.9136	0.4067	2.2460	0.4452	2.4586	1.0946
67	0.9205	0.3907	2.3558	0.4245	2.5593	1.0864
68	0.9272	0.3746	2.4751	0.4040	2.6695	1.0785
69	0.9336	0.3584	2.6051	0.3839	2.7904	1.0711
70	0.9397	0.3420	2.7475	0.3640	2.9238	1.0642
71	0.9455	0.3256	2.9042	0.3443	3.0715	1.0576
72	0.9511	0.3090	3.0777	0.3249	3.2361	1.0515
73	0.9563	0.2924	3.2708	0.3057	3.4203	1.0457
74	0.9613	0.2756	3.4874	0.2867	3.6279	1.0403
75	0.9659	0.2588	3.7320	0.2680	3.8637	1.0353
76	0.9703	0.2419	4.0108	0.2493	4.1336	1.0306
77	0.9744	0.2250	4.3315	0.2309	4.4454	1.0263
78	0.9782	0.2079	4.7046	0.2126	4.8097	1.0223
79	0.9816	0.1908	5.1445	0.1944	5.2408	1.0187
80	0.9848	0.1737	5.6713	0.1763	5.7588	1.0154
81	0.9877	0.1564	6.3137	0.1584	6.3924	1.0125
82	0.9903	0.1392	7.1154	0.1405	7.1853	1.0098
83	0.9926	0.1219	8.1443	0.1228	8.2005	1.0075
84	0.9945	0.1045	9.5144	0.1051	9.5668	1.0055
85	0.9962	0.0872	11.430	0.0875	11.474	1.0038
86	0.9976	0.0698	14.301	0.0699	14.335	1.0024
87	0.9986	9.0523	19.081	0.0524	19.107	1.0014
88	0.9994	0.0349	28.636	0.0349	28.654	1.0006
89	0.9999	0.0175	57.290	0.0175	57.299	1.0001
90	1.0000	0.0000	—	0.0000	—	1.0000

Table III

$(1 + i)^n$

i / n	$\frac{1}{2}\%$	1%	$1\frac{1}{2}\%$	2%	$2\frac{1}{2}\%$	3%	$3\frac{1}{2}\%$	i / n
1	1.005 00	1.010 00	1.015 00	1.020 00	1.025 00	1.030 00	1.035 00	1
2	1.010 03	1.020 10	1.030 23	1.040 40	1.050 63	1.060 90	1.071 23	2
3	1.015 08	1.030 30	1.045 68	1.061 21	1.076 89	1.092 73	1.108 72	3
4	1.020 15	1.040 60	1.061 36	1.082 43	1.103 81	1.125 51	1.147 52	4
5	1.025 25	1.051 01	1.077 28	1.104 08	1.131 41	1.159 27	1.187 69	5
6	1.030 38	1.061 52	1.093 44	1.126 16	1.159 69	1.194 05	1.229 26	6
7	1.035 53	1.072 14	1.109 84	1.148 69	1.188 69	1.229 87	1.272 28	7
8	1.040 71	1.082 86	1.126 49	1.171 66	1.218 40	1.266 77	1.316 81	8
9	1.045 91	1.093 69	1.143 39	1.195 09	1.248 86	1.304 77	1.362 90	9
10	1.051 14	1.104 62	1.160 54	1.218 99	1.280 08	1.343 92	1.410 60	10
11	1.056 40	1.115 67	1.179 95	1.243 37	1.312 09	1.384 23	1.459 97	11
12	1.061 68	1.126 83	1.195 62	1.268 24	1.344 89	1.425 76	1.511 07	12
13	1.066 99	1.138 09	1.213 55	1.293 61	1.378 51	1.468 53	1.563 96	13
14	1.072 32	1.149 47	1.231 76	1.319 48	1.412 97	1.512 59	1.618 69	14
15	1.077 68	1.160 97	1.250 23	1.345 87	1.448 30	1.557 97	1.675 35	15
16	1.083 07	1.172 58	1.268 99	1.372 79	1.484 51	1.604 71	1.733 99	16
17	1.088 49	1.184 30	1.288 02	1.400 24	1.521 62	1.652 85	1.794 68	17
18	1.093 93	1.196 15	1.307 34	1.428 25	1.559 66	1.702 43	1.857 49	18
19	1.099 40	1.208 11	1.326 95	1.456 81	1.598 65	1.753 51	1.922 50	19
20	1.104 90	1.220 19	1.346 86	1.485 95	1.638 62	1.806 11	1.989 79	20
21	1.110 42	1.232 39	1.367 06	1.515 67	1.679 58	1.860 29	2.059 43	21
22	1.115 97	1.244 72	1.387 56	1.545 98	1.721 57	1.916 10	2.131 51	22
23	1.121 55	1.257 16	1.408 38	1.576 90	1.764 61	1.973 59	2.206 11	23
24	1.127 16	1.269 73	1.429 50	1.608 44	1.808 73	2.032 79	2.283 33	24
25	1.132 80	1.282 43	1.450 95	1.640 61	1.853 94	2.093 78	2.363 24	25
26	1.138 46	1.295 26	1.472 71	1.673 42	1.900 29	2.156 59	2.445 96	26
27	1.144 15	1.308 21	1.494 80	1.706 89	1.947 80	2.221 29	2.531 57	27
28	1.149 87	1.321 29	1.517 22	1.741 02	1.997 50	2.287 93	2.620 17	28
29	1.155 62	1.334 50	1.539 98	1.775 84	2.046 41	2.356 57	2.711 88	29
30	1.161 40	1.347 85	1.563 08	1.811 36	2.097 57	2.427 26	2.806 79	30
31	1.167 21	1.361 33	1.586 53	1.847 59	2.150 01	2.500 08	2.905 03	31
32	1.173 04	1.374 94	1.610 32	1.884 54	2.203 76	2.575 08	3.006 71	32
33	1.178 91	1.388 69	1.634 48	1.922 23	2.258 85	2.652 34	3.111 94	33
34	1.184 80	1.402 58	1.659 00	1.906 68	2.315 32	2.731 91	3.220 86	34
35	1.190 73	1.416 60	1.683 88	1.999 89	2.373 21	2.813 86	3.333 59	35
36	1.196 68	1.430 77	1.709 14	2.039 89	2.432 54	2.898 28	3.450 27	36
37	1.202 66	1.445 08	1.734 78	2.080 69	2.493 35	2.985 23	3.571 03	37
38	1.208 68	1.459 53	1.760 80	2.122 30	2.555 68	3.074 78	3.696 01	38
39	1.214 72	1.474 12	1.787 21	2.164 74	2.619 57	3.167 03	3.825 37	39
40	1.220 79	1.488 86	1.814 02	2.208 04	2.685 06	3.264 04	3.959 26	40

i / n	4%	4½%	5%	5½%	6%	7%	8%	i / n
1	1.040 00	1.045 00	1.050 00	1.055 00	1.060 00	1.070 00	1.080 00	1
2	1.081 60	1.092 03	1.102 50	1.113 03	1.123 60	1.144 90	1.166 40	2
3	1.124 86	1.141 17	1.157 63	1.174 24	1.191 02	1.225 04	1.259 71	3
4	1.169 86	1.192 52	1.215 51	1.238 82	1.262 48	1.310 80	1.360 49	4
5	1.216 65	1.246 18	1.276 28	1.306 96	1.338 23	1.402 55	1.469 33	5
6	1.265 32	1.302 26	1.340 10	1.378 84	1.418 52	1.500 73	1.586 87	6
7	1.315 93	1.360 86	1.407 10	1.454 68	1.503 63	1.605 78	1.713 82	7
8	1.368 57	1.422 10	1.477 46	1.534 69	1.593 85	1.718 19	1.850 93	8
9	1.423 31	1.486 10	1.551 33	1.619 09	1.689 48	1.838 46	1.999 00	9
10	1.480 24	1.552 97	1.628 89	1.708 14	1.790 85	1.967 15	2.158 93	10
11	1.539 45	1.622 85	1.710 34	1.802 09	1.898 30	2.104 85	2.331 64	11
12	1.601 03	1.695 88	1.795 86	1.901 21	2.012 20	2.252 19	2.518 17	12
13	1.665 07	1.772 20	1.885 65	2.005 77	2.132 93	2.409 85	2.719 62	13
14	1.731 68	1.851 94	1.979 93	2.116 09	2.260 90	2.578 53	2.937 19	14
15	1.800 94	1.935 28	2.078 93	2.232 48	2.396 56	2.759 03	3.172 17	15
16	1.872 98	2.022 37	2.182 87	2.355 26	2.540 35	2.952 16	3.425 94	16
17	1.947 90	2.113 38	2.292 02	2.484 80	2.692 77	3.158 81	3.700 02	17
18	2.025 82	2.208 48	2.406 62	2.621 47	2.854 34	3.379 93	3.996 02	18
19	2.106 85	2.307 86	2.526 95	2.765 65	3.025 60	3.616 53	4.315 70	19
20	2.191 12	2.411 71	2.653 30	2.917 76	3.207 14	3.869 68	4.660 96	20
21	2.278 77	2.520 24	2.785 96	3.078 23	3.399 56	4.140 56	5.033 83	21
22	2.369 92	2.633 65	2.925 26	3.247 54	3.603 54	4.430 40	5.436 54	22
23	2.464 72	2.752 17	3.071 52	3.426 15	3.819 75	4.740 53	5.871 46	23
24	2.563 30	2.876 01	3.225 10	3.614 59	4.048 93	5.072 37	6.341 18	24
25	2.665 84	3.005 43	3.386 35	3.813 39	4.291 87	5.427 43	6.848 48	25
26	2.772 47	3.140 68	3.555 67	4.023 13	4.549 38	5.807 35	7.396 35	26
27	2.883 37	3.282 01	3.733 46	4.244 40	4.822 35	6.213 87	7.988 06	27
28	2.998 70	3.429 70	3.920 13	4.477 84	5.111 69	6.648 84	8.627 11	28
29	3.118 65	3.584 04	4.116 14	4.724 12	5.418 39	7.114 26	9.317 27	29
30	3.243 40	3.745 32	4.321 94	4.983 95	5.743 49	7.612 26	10.062 66	30
31	3.373 13	3.913 86	4.538 04	5.258 07	6.088 10	8.145 11	10.867 67	31
32	3.508 06	4.089 98	4.764 94	5.547 26	6.453 39	8.715 27	11.737 08	32
33	3.648 38	4.274 03	5.003 19	5.852 36	6.840 59	9.325 34	12.676 05	33
34	3.794 32	4.446 36	5.253 35	6.174 24	7.251 03	9.978 11	13.690 13	34
35	3.946 09	4.667 35	5.516 02	6.513 83	7.686 09	10.676 58	14.785 34	35
36	4.130 93	4.877 38	5.791 82	6.872 09	8.147 25	11.423 94	15.968 17	36
37	4.268 09	5.096 86	6.081 41	7.250 05	8.636 09	12.223 62	17.245 63	37
38	4.438 81	5.326 22	6.385 48	7.648 80	9.154 25	13.079 27	18.625 28	38
39	4.616 37	5.565 90	6.704 75	8.069 49	9.703 51	13.994 82	20.115 30	39
40	4.801 02	5.816 36	7.039 99	8.513 31	10.285 72	14.974 46	21.724 52	40

Table IV

$$\frac{1}{(1 + i)^n}$$

i / n	$\frac{1}{2}$%	1%	$1\frac{1}{2}$%	2%	$2\frac{1}{2}$%	3%	$3\frac{1}{2}$%	i / n
1	0.995 02	0.990 10	0.985 22	0.980 39	0.975 61	0.970 87	0.966 18	1
2	0.990 07	0.980 30	0.970 66	0.961 17	0.951 81	0.942 60	0.933 51	2
3	0.985 15	0.970 59	0.956 32	0.942 32	0.928 60	0.915 14	0.901 94	3
4	0.980 25	0.960 98	0.942 18	0.923 85	0.905 95	0.888 49	0.871 44	4
5	0.975 37	0.951 47	0.928 26	0.905 73	0.883 85	0.862 61	0.841 97	5
6	0.970 52	0.942 05	0.914 54	0.887 97	0.862 30	0.837 48	0.813 50	6
7	0.965 69	0.932 72	0.901 03	0.870 56	0.841 27	0.813 09	0.785 99	7
8	0.960 89	0.923 48	0.887 71	0.853 49	0.820 75	0.789 41	0.759 41	8
9	0.956 10	0.914 34	0.874 59	0.836 76	0.800 73	0.766 42	0.733 73	9
10	0.951 35	0.905 29	0.861 67	0.820 35	0.781 20	0.744 09	0.708 92	10
11	0.946 61	0.896 32	0.848 93	0.804 26	0.762 14	0.722 42	0.684 95	11
12	0.941 91	0.887 45	0.836 39	0.788 49	0.743 56	0.701 38	0.661 78	12
13	0.937 22	0.878 66	0.824 03	0.773 03	0.725 42	0.680 95	0.639 40	13
14	0.932 56	0.869 96	0.811 85	0.757 88	0.707 73	0.661 12	0.617 78	14
15	0.927 92	0.861 35	0.799 85	0.743 01	0.690 47	0.641 86	0.596 89	15
16	0.923 30	0.852 82	0.788 03	0.728 45	0.673 62	0.623 17	0.576 71	16
17	0.918 71	0.844 38	0.776 39	0.714 16	0.657 20	0.605 02	0.557 20	17
18	0.914 14	0.836 02	0.764 91	0.700 16	0.641 17	0.587 39	0.538 36	18
19	0.909 59	0.827 74	0.753 61	0.686 43	0.625 53	0.570 29	0.520 16	19
20	0.905 06	0.819 54	0.742 47	0.672 97	0.610 27	0.553 68	0.502 57	20
21	0.900 56	0.811 43	0.731 50	0.659 78	0.595 39	0.527 55	0.485 57	21
22	0.896 08	0.803 40	0.720 69	0.646 84	0.580 86	0.521 89	0.469 15	22
23	0.891 62	0.795 44	0.710 04	0.634 16	0.566 70	0.506 69	0.453 29	23
24	0.887 19	0.787 57	0.699 54	0.621 72	0.552 88	0.491 93	0.437 96	24
25	0.882 77	0.779 77	0.689 21	0.609 53	0.539 39	0.477 61	0.423 15	25
26	0.878 38	0.772 05	0.679 02	0.597 58	0.526 23	0.463 69	0.408 84	26
27	0.874 01	0.764 40	0.668 99	0.585 86	0.513 40	0.450 19	0.395 01	27
28	0.869 66	0.756 84	0.659 10	0.574 37	0.500 88	0.437 08	0.381 65	28
29	0.865 33	0.749 34	0.649 36	0.563 11	0.488 66	0.424 35	0.368 75	29
30	0.861 03	0.741 92	0.639 76	0.552 07	0.476 74	0.411 99	0.356 28	30
31	0.856 75	0.734 58	0.630 31	0.541 25	0.465 11	0.399 99	0.344 23	31
32	0.852 48	0.727 30	0.620 99	0.530 63	0.453 77	0.388 34	0.332 59	32
33	0.848 24	0.720 10	0.611 82	0.520 23	0.442 70	0.377 03	0.321 34	33
34	0.844 02	0.712 97	0.602 77	0.510 03	0.431 91	0.366 04	0.310 48	34
35	0.839 82	0.705 91	0.593 87	0.500 03	0.421 37	0.355 38	0.299 98	35
36	0.835 64	0.698 92	0.585 09	0.490 22	0.411 09	0.345 03	0.289 83	36
37	0.831 49	0.692 00	0.576 44	0.480 61	0.401 07	0.334 98	0.280 03	37
38	0.827 35	0.685 15	0.567 92	0.471 19	0.391 28	0.325 23	0.270 56	38
39	0.823 23	0.678 37	0.559 53	0.461 95	0.381 74	0.315 75	0.261 41	39
40	0.819 14	0.671 65	0.551 26	0.452 89	0.372 43	0.306 56	0.252 57	40

i / n	4%	4½%	5%	5½%	6%	7%	8%	i / n
1	0.961 54	0.956 94	0.952 38	0.947 87	0.943 40	0.934 58	0.925 93	1
2	0.924 56	0.915 73	0.907 03	0.898 45	0.890 00	0.873 44	0.857 34	2
3	0.889 00	0.876 30	0.863 84	0.851 61	0.839 62	0.816 30	0.793 83	3
4	0.854 80	0.838 56	0.822 70	0.807 22	0.792 09	0.762 90	0.735 03	4
5	0.821 93	0.802 45	0.783 53	0.765 13	0.747 26	0.712 99	0.680 58	5
6	0.790 31	0.767 90	0.746 22	0.725 25	0.704 96	0.666 34	0.630 17	6
7	0.759 92	0.734 83	0.710 68	0.687 44	0.665 06	0.622 75	0.583 49	7
8	0.730 69	0.703 19	0.676 84	0.651 60	0.627 41	0.582 01	0.540 27	8
9	0.702 59	0.672 90	0.644 61	0.617 63	0.591 90	0.543 93	0.500 25	9
10	0.675 56	0.643 93	0.613 91	0.585 43	0.558 39	0.508 35	0.463 19	10
11	0.649 58	0.616 20	0.584 68	0.554 91	0.526 79	0.475 09	0.428 88	11
12	0.624 60	0.589 66	0.556 84	0.525 98	0.496 97	0.444 01	0.397 11	12
13	0.600 57	0.564 27	0.530 32	0.498 56	0.468 84	0.414 96	0.367 70	13
14	0.577 48	0.539 97	0.505 07	0.472 57	0.442 30	0.387 82	0.340 46	14
15	0.555 26	0.516 72	0.481 02	0.447 93	0.417 27	0.362 45	0.315 24	15
16	0.533 91	0.494 47	0.458 11	0.424 58	0.393 65	0.338 73	0.291 89	16
17	0.513 37	0.473 18	0.436 30	0.402 45	0.371 36	0.316 57	0.270 27	17
18	0.493 63	0.452 80	0.415 52	0.381 47	0.350 34	0.295 86	0.250 25	18
19	0.474 64	0.433 30	0.395 73	0.361 58	0.330 51	0.276 51	0.231 71	19
20	0.456 39	0.414 64	0.376 89	0.342 73	0.311 80	0.258 42	0.214 55	20
21	0.438 83	0.396 79	0.358 94	0.324 86	0.294 16	0.241 51	0.198 66	21
22	0.421 96	0.379 70	0.341 85	0.307 93	0.277 51	0.225 71	0.183 94	22
23	0.405 73	0.363 35	0.325 57	0.291 87	0.261 80	0.210 95	0.170 32	23
24	0.390 12	0.347 70	0.310 07	0.276 66	0.246 98	0.197 15	0.157 70	24
25	0.375 12	0.332 73	0.295 30	0.262 23	0.233 00	0.184 25	0.146 02	25
26	0.360 69	0.318 40	0.281 24	0.248 56	0.219 81	0.172 20	0.135 20	26
27	0.346 82	0.304 69	0.267 85	0.235 60	0.207 37	0.160 93	0.125 19	27
28	0.333 48	0.291 57	0.255 09	0.223 32	0.195 63	0.150 40	0.115 91	28
29	0.320 65	0.279 02	0.242 95	0.211 68	0.184 56	0.140 56	0.107 33	29
30	0.308 32	0.267 00	0.231 38	0.200 64	0.174 11	0.131 37	0.099 38	30
31	0.296 46	0.255 50	0.220 36	0.190 18	0.164 25	0.122 77	0.092 02	31
32	0.285 06	0.244 50	0.209 87	0.180 27	0.154 96	0.114 74	0.085 20	32
33	0.274 09	0.233 97	0.199 87	0.170 87	0.146 19	0.107 23	0.078 89	33
34	0.263 55	0.223 90	0.190 35	0.161 96	0.137 91	0.100 22	0.073 05	34
35	0.253 42	0.214 25	0.181 29	0.153 52	0.130 11	0.093 66	0.067 63	35
36	0.243 67	0.205 03	0.172 66	0.145 52	0.122 74	0.087 54	0.062 62	36
37	0.234 30	0.196 20	0.164 44	0.137 93	0.115 79	0.081 81	0.057 99	37
38	0.225 29	0.187 75	0.156 61	0.130 74	0.109 24	0.076 46	0.053 69	38
39	0.216 62	0.179 67	0.149 15	0.123 92	0.103 06	0.071 46	0.049 71	39
40	0.208 29	0.171 93	0.142 05	0.117 46	0.097 22	0.066 78	0.046 03	40

Table V

AMOUNT OF AN ANNUITY OF 1 $S_{\overline{n}|}i$

n	$\frac{1}{2}$%	1%	$1\frac{1}{2}$%	2%	$2\frac{1}{2}$%	3%
1	1.000 000	1.000 000	1.000 000	1.000 000	1.000 000	1.000 000
2	2.005 000	2.010 000	2.015 000	2.020 000	2.025 000	2.030 000
3	3.015 025	3.030 100	3.045 225	3.060 400	3.075 625	3.090 900
4	4.030 100	4.060 401	4.090 903	4.121 608	4.152 516	4.183 627
5	5.050 251	5.101 005	5.152 267	5.204 040	5.256 329	5.309 136
6	6.075 502	6.152 015	6.229 551	6.308 121	6.387 737	6.468 410
7	7.105 879	7.213 535	7.322 994	7.434 283	7.547 430	7.662 462
8	8.141 409	8.285 671	8.432 839	8.582 969	8.736 116	8.892 336
9	9.182 116	9.368 527	9.559 332	9.754 628	9.954 519	10.159 106
10	10.228 026	10.462 213	10.702 722	10.949 721	11.203 382	11.463 879
11	11.279 167	11.566 835	11.863 262	12.168 715	12.483 466	12.807 796
12	12.335 562	12.682 503	13.041 211	13.412 090	13.795 553	14.192 030
13	13.397 240	13.809 328	14.236 830	14.680 332	15.140 442	15.617 790
14	14.464 226	14.947 421	15.450 382	15.973 938	16.518 953	17.086 324
15	15.536 548	16.096 896	16.682 138	17.293 417	17.931 927	18.598 914
16	16.614 230	17.257 864	17.932 370	18.639 285	19.380 225	20.156 881
17	17.697 301	18.430 443	19.201 355	20.012 071	20.864 730	21.761 588
18	18.785 788	19.614 748	20.489 376	21.412 312	22.386 349	23.414 435
19	19.879 717	20.810 895	21.796 716	22.840 559	23.946 007	25.116 868
20	20.979 115	22.019 004	23.123 667	24.297 370	25.544 658	26.870 374
21	22.084 011	23.239 194	24.470 522	25.783 317	27.183 274	28.676 486
22	23.194 431	24.471 586	25.837 580	27.298 984	28.862 856	30.536 780
23	24.310 403	25.716 302	27.225 144	28.844 963	30.584 427	32.452 884
24	25.431 955	26.973 465	28.633 521	30.421 862	32.349 038	34.426 470
25	26.559 115	28.243 200	30.063 024	32.030 300	34.157 764	36.459 264
26	27.691 911	29.525 632	31.513 969	33.670 906	36.011 708	38.553 042
27	28.830 370	30.820 888	32.986 679	35.344 324	37.912 001	40.709 634
28	29.974 522	32.129 097	34.481 479	37.051 210	39.859 801	42.930 923
29	31.124 395	33.450 388	35.998 701	38.792 235	41.856 296	45.218 850
30	32.280 017	34.784 892	37.538 681	40.568 079	43.902 703	47.575 416
31	33.441 417	36.132 740	39.101 762	42.379 441	46.000 271	50.002 678
32	34.608 624	37.494 068	40.688 288	44.227 030	48.150 278	52.502 759
33	35.781 667	38.869 009	42.298 612	46.111 570	50.354 034	55.077 841
34	36.960 575	40.257 699	43.933 092	48.033 802	52.612 885	57.730 177
35	38.145 378	41.660 276	45.592 088	49.994 478	54.928 207	60.462 082
36	39.336 105	43.076 878	47.275 969	51.994 367	57.301 413	63.275 944
37	40.532 785	44.507 647	48.985 109	54.034 255	59.733 948	66.174 223
38	41.735 449	45.952 724	50.719 885	56.114 940	62.227 297	69.159 449
39	42.944 127	47.412 251	52.480 684	58.237 238	64.782 979	72.234 233
40	44.158 847	48.886 373	54.267 894	60.401 983	67.402 554	75.401 260

n	$3\frac{1}{2}\%$	4%	$4\frac{1}{2}\%$	5%	$5\frac{1}{2}\%$	6%
1	1.000 000	1.000 000	1.000 000	1.000 000	1.000 000	1.000 000
2	2.035 000	2.040 000	2.045 000	2.050 000	2.055 000	2.060 000
3	3.106 225	3.121 600	3.137 025	3.152 500	3.168 025	3.183 600
4	4.214 943	4.246 464	4.278 191	4.301 125	4.342 266	4.374 616
5	5.362 466	5.416 323	5.470 710	5.525 631	5.581 091	5.637 093
6	6.550 152	6.632 975	6.716 892	6.801 913	6.888 051	6.975 319
7	7.779 408	7.898 294	8.019 152	8.142 008	8.266 894	8.393 838
8	9.051 687	9.214 226	9.380 014	9.549 109	9.721 573	9.897 468
9	10.368 496	10.582 795	10.802 114	11.026 564	11.256 260	11.491 316
10	11.731 393	12.006 107	12.288 209	12.577 893	12.875 354	13.180 795
11	13.141 992	13.486 351	13.841 179	14.206 787	14.583 498	14.971 643
12	14.601 962	15.025 805	15.464 032	15.917 127	16.385 591	16.869 941
13	16.113 030	16.626 838	17.159 913	17.712 983	18.286 798	18.882 138
14	17.676 986	18.291 911	18.932 109	19.598 632	20.292 572	21.015 066
15	19.295 681	20.023 588	20.784 054	21.578 564	22.408 664	23.275 970
16	20.971 030	21.824 531	22.719 337	23.657 492	24.641 140	25.672 528
17	22.705 016	23.697 512	24.741 707	25.840 366	26.996 403	28.212 880
18	24.499 691	25.645 413	26.855 084	28.132 385	29.481 205	30.905 653
19	26.357 181	27.671 229	29.063 562	30.539 004	32.102 671	33.759 992
20	28.279 682	29.778 079	31.371 423	33.065 954	34.868 318	36.785 591
21	30.269 471	31.969 202	33.783 137	35.719 252	37.786 076	39.992 727
22	32.328 902	34.247 970	36.303 378	38.505 214	40.864 310	43.392 290
23	34.460 414	36.617 889	38.937 030	41.430 475	44.111 847	46.995 828
24	36.666 528	39.082 604	41.689 196	44.501 999	47.537 998	50.815 577
25	38.949 857	41.645 908	44.565 210	47.727 099	51.152 588	54.864 512
26	41.313 102	44.311 745	47.570 645	51.113 454	54.965 981	59.156 383
27	43.759 060	47.084 214	50.711 324	54.669 126	58.989 109	63.705 766
28	46.290 627	49.967 583	53.993 333	58.402 583	63.233 510	68.528 112
29	48.910 799	52.966 286	57.423 033	62.322 712	67.711 354	73.639 798
30	51.622 677	56.084 938	61.007 070	66.438 848	72.435 478	79.058 186
31	54.429 471	59.328 335	64.752 388	70.760 790	77.419 429	84.801 677
32	57.334 502	62.701 469	68.666 245	75.298 829	82.677 498	90.889 778
33	60.341 210	66.209 527	72.756 226	80.063 771	88.224 760	97.343 165
34	63.453 152	69.857 909	77.030 256	85.066 959	94.077 122	104.183 755
35	66.674 013	73.652 225	81.496 618	90.320 307	100.251 364	111.434 780
36	70.007 603	77.598 314	86.163 966	95.836 323	106.765 189	119.120 867
37	73.457 869	81.702 246	91.041 344	101.628 139	113.637 274	127.268 119
38	77.028 895	85.970 336	96.138 205	107.709 546	120.887 324	135.904 206
39	80.724 906	90.409 150	101.464 424	114.095 023	128.536 127	145.058 458
40	84.550 278	95.025 516	107.030 323	120.799 774	136.605 614	154.761 966

Table VI

PRESENT VALUE OF AN ANNUITY OF 1 $a_{\overline{n}|}i$

n	$\frac{1}{2}\%$	1%	$1\frac{1}{2}\%$	2%	$2\frac{1}{2}\%$	3%
1	0.995 025	0.990 099	0.985 222	0.980 392	0.975 610	0.970 874
2	1.985 099	1.970 395	1.955 883	1.941 561	1.927 424	1.913 470
3	2.970 248	2.940 985	2.912 200	2.883 883	2.856 024	2.828 611
4	3.950 496	3.901 966	3.854 385	3.807 729	3.761 974	3.717 098
5	4.925 866	4.853 431	4.782 645	4.713 460	4.645 829	4.579 707
6	5.896 384	5.795 476	5.697 187	5.601 431	5.508 125	5.417 191
7	6.862 074	6.728 195	6.598 214	6.471 991	6.349 391	6.230 283
8	7.822 959	7.651 678	7.485 925	7.325 481	7.170 137	7.019 692
9	8.779 064	8.566 018	8.360 517	8.162 237	7.970 866	7.786 109
10	9.730 412	9.471 305	9.222 185	8.982 585	8.752 064	8.530 203
11	10.677 027	10.367 628	10.071 118	9.786 848	9.514 209	9.252 624
12	11.618 932	11.255 077	10.907 505	10.575 341	10.257 765	9.954 004
13	12.556 151	12.133 740	11.731 532	11.348 374	10.983 185	10.634 955
14	13.488 708	13.003 703	12.543 382	12.106 249	11.690 912	11.296 073
15	14.416 625	13.865 053	13.343 233	12.849 264	12.381 378	11.937 935
16	15.339 925	14.717 874	14.131 264	13.577 709	13.055 003	12.561 102
17	16.258 632	15.562 251	14.907 649	14.291 872	13.712 198	13.166 118
18	17.172 768	16.398 269	15.672 561	14.992 031	14.353 364	13.753 513
19	18.082 356	17.226 009	16.426 168	15.678 462	14.978 891	14.323 799
20	18.987 419	18.045 553	17.168 639	16.351 433	15.589 162	14.877 475
21	19.887 979	18.856 983	17.900 137	17.011 209	16.184 549	15.415 024
22	20.784 059	19.660 379	18.620 824	17.658 048	16.765 413	15.936 917
23	21.675 681	20.455 821	19.330 861	18.292 204	17.332 110	16.443 608
24	22.562 866	21.243 387	20.030 405	18.913 926	17.884 986	16.935 542
25	23.445 638	22.023 156	20.719 611	19.523 456	18.424 376	17.413 148
26	24.324 018	22.795 204	21.398 632	20.121 036	18.950 611	17.876 842
27	25.198 028	23.559 608	22.067 617	20.706 898	19.464 011	18.327 031
28	26.067 689	24.316 443	22.726 717	21.281 272	19.964 889	18.764 108
29	26.933 024	25.065 785	23.376 076	21.844 385	20.453 550	19.188 455
30	27.794 054	25.807 708	24.015 838	22.396 456	20.930 293	19.600 441
31	28.650 800	26.542 285	24.646 146	22.937 702	21.395 407	20.000 428
32	29.503 284	27.269 589	25.267 139	23.468 335	21.849 178	20.388 766
33	30.351 526	27.989 693	25.878 954	23.988 563	22.291 881	20.765 792
34	31.195 548	28.702 666	26.481 728	24.498 592	22.723 786	21.131 837
35	32.035 371	29.408 580	27.075 595	24.998 619	23.145 157	21.487 220
36	32.871 016	30.107 505	27.660 684	25.488 842	23.556 251	21.832 253
37	33.702 504	30.799 510	28.237 127	25.969 453	23.957 318	22.167 235
38	34.529 854	31.484 663	28.805 052	26.440 641	24.348 603	22.492 462
39	35.353 089	32.163 033	29.364 583	26.902 589	24.730 344	22.808 215
40	36.172 228	32.834 686	29.915 845	27.355 479	25.102 775	23.114 772

n	$3\frac{1}{2}\%$	4%	$4\frac{1}{2}\%$	5%	$5\frac{1}{2}\%$	6%
1	0.966 184	0.961 538	0.956 938	0.952 381	0.947 867	0.943 396
2	1.899 694	1.886 095	1.872 668	1.859 410	1.846 320	1.833 393
3	2.801 637	2.775 091	2.748 964	2.723 248	2.697 933	2.673 012
4	3.673 079	3.629 895	3.587 526	3.545 951	3.505 150	3.465 106
5	4.515 052	4.451 822	4.389 977	4.329 477	4.270 284	4.212 364
6	5.328 553	5.242 137	5.157 872	5.075 692	4.995 530	4.917 324
7	6.114 544	6.002 055	5.892 701	5.786 373	5.682 967	5.582 381
8	6.873 955	6.732 745	6.595 886	6.463 213	6.334 566	6.209 794
9	7.607 687	7.435 332	7.268 791	7.107 822	6.952 195	6.801 692
10	8.316 605	8.110 896	7.912 718	7.721 735	7.537 626	7.360 087
11	9.001 551	8.760 477	8.528 917	8.306 414	8.092 536	7.886 875
12	9.663 334	9.385 074	9.118 581	8.863 252	8.618 518	8.383 844
13	10.302 738	9.985 648	9.682 852	9.393 573	9.117 079	8.852 683
14	10.920 520	10.563 123	10.222 825	9.898 641	9.589 648	9.294 984
15	11.517 411	11.118 387	10.739 546	10.379 658	10.037 581	9.712 249
16	12.094 117	11.652 296	11.234 015	10.837 770	10.462 162	10.105 895
17	12.651 321	12.165 669	11.707 191	11.274 066	10.864 609	10.477 260
18	13.189 682	12.659 297	12.159 992	11.689 587	11.246 074	10.827 603
19	13.709 837	13.133 939	12.593 294	12.085 321	11.607 654	11.158 116
20	14.212 403	13.590 326	13.007 936	12.462 210	11.950 382	11.469 921
21	14.697 974	14.029 160	13.404 724	12.821 153	12.275 244	11.764 077
22	15.167 125	14.451 115	13.784 425	13.163 003	12.583 170	12.041 582
23	15.620 410	14.856 842	14.147 775	13.488 574	12.875 042	12.303 379
24	16.058 368	15.246 963	14.495 478	13.798 642	13.151 699	12.550 358
25	16.481 515	15.622 080	14.828 209	14.093 945	13.413 933	12.783 356
26	16.890 352	15.982 769	15.146 611	14.375 185	13.662 495	13.003 166
27	17.285 365	16.329 586	15.451 303	14.643 034	13.898 100	13.210 534
28	17.667 019	16.663 063	15.742 874	14.898 127	14.121 422	13.406 164
29	18.035 767	16.983 715	16.021 889	15.141 074	14.333 101	13.590 721
30	18.392 045	17.292 033	16.288 889	15.372 451	14.533 745	13.764 831
31	18.736 276	17.588 494	16.544 391	15.592 811	14.723 929	13.929 086
32	19.068 865	17.873 552	16.788 891	15.802 677	14.904 198	14.084 043
33	19.390 208	18.147 646	17.022 862	16.002 549	15.075 069	14.230 230
34	19.700 684	18.411 198	17.246 758	16.192 904	15.237 033	14.368 141
35	20.000 661	18.664 613	17.461 012	16.374 194	15.390 552	14.498 246
36	20.290 494	18.908 282	17.666 041	16.546 852	15.536 068	14.620 987
37	20.570 525	19.142 579	17.862 240	16.711 287	15.673 999	14.736 780
38	20.841 087	19.367 864	18.049 990	16.867 893	15.804 738	14.846 019
39	21.102 500	19.584 485	18.229 656	17.017 041	15.928 662	14.949 075
40	21.355 072	19.792 774	18.401 584	17.159 086	16.046 125	15.046 297

INDEX